THE DIAMOND HUNTERS

&

A FALCON FLIES

The novels of Wilbur Smith

THE COURTNEYS

Birds of Prey
When the Lion Feeds
The Sound of Thunder
A Sparrow Falls

THE COURTNEYS OF AFRICA

The Burning Shore
Power of the Sword
Rage
A Time to Die
Golden Fox

THE BALLANTYNE NOVELS

Men of Men
The Angels Weep
The Leopard Hunts in Darkness

also

The Dark of the Sun
Shout at the Devil
Gold Mine
The Diamond Hunters
The Sunbird
Eagle in the Sky
The Eye of the Tiger
Cry Wolf
Hungry as the Sea
Wild Justice
Elephant Song
River God
The Seventh Scroll
Warlock

Wilbur Smith was born in Central Africa in 1933. He was educated at Michaelhouse and Rhodes University.

He became a full-time writer in 1964 after the successful publication of *When the Lion Feeds*, and has since written twenty-six novels, all meticulously researched on his numerous expeditions worldwide. His books are now translated into twenty-six languages.

He owns a farm and game reserve and has an abiding concern for the peoples and wildlife of his native continent, an interest strongly reflected in his novels.

WILBUR SMITH

THE DIAMOND HUNTERS
&
A FALCON FLIES

PAN BOOKS

The Diamond Hunters first published 1971 by Heinemann
First published by Pan Books 1973
A Falcon Flies first published 1980 by Heinemann
First published by Pan Books 1981

This edition published 2004 by Pan Books
an imprint of Pan Macmillan Ltd
Pan Macmillan, 20 New Wharf Road, London N1 9RR
Basingstoke and Oxford
Associated companies throughout the world

ISBN 0 330 43259 1

1 3 5 7 9 8 6 4 2

A CIP catalogue record for this book is available from
the British Library.

Typeset by SetSystems Ltd, Saffron Walden, Essex
Printed and bound in Great Britain by
Mackays of Chatham plc, Chatham, Kent

THE DIAMOND HUNTERS

For DEE

His flight had been delayed for three hours at Nairobi, and despite four large whiskies he slept only fitfully until the intercontinental Boeing touched down at Heathrow. Johnny Lance felt as though someone had thrown a handful of grit in each eye, and his mood was ugly as he came through the indignity of Customs and Immigration into the main hall of the international terminus.

The Van Der Byl Diamond Company's London agent was there to meet him.

'Pleasant trip, Johnny?'

'Like the one to Hell,' Johnny grunted.

'Good practice for you.' The agent grinned. The two of them had seen some riotous times together.

Reluctantly Johnny grinned back at him.

'You got me a room and a car?'

'Dorchester – and Jag.' The agent handed over the car keys. 'And I've got two first-class seats reserved on tomorrow's nine o'clock flight back to Cape Town. Tickets at the hotel reception desk.'

'Good boy.' Johnny dropped the keys into the pocket of his cashmere overcoat and they started for the exit. 'Now where is Tracey van der Byl?'

The agent shrugged. 'Since I wrote to you she has dropped out of sight. I don't know where you can start looking.'

'Great, just great!' said Johnny bitterly as they came out into the car park. 'I'll start with Benedict.'

'Does the Old Man know about Tracey?'

Johnny shook his head. 'He's a sick man. I didn't tell him.'

'Here's your car.' The agent stopped by the pearl-grey Jaguar. 'Any chance of a drink together?'

'Not this trip, sorry.' Johnny slipped in behind the wheel. 'Next time.'

'I'll hold you to that,' said the agent and walked away.

It was almost dark by the time Johnny crossed the Hammersmith flyover in the moist smoky grey of the evening, and he lost himself twice in the maze of Belgravia before he found the narrow mews behind Belgrave Square and parked the Jaguar.

The exterior of the flat had been lavishly redecorated since his last visit, and Johnny's mouth twisted. He might not be so hot at earning the stuff – but our boy Benedict certainly was a dab hand at spending it.

There were lights burning and Johnny hit the door knocker half a dozen lusty cracks. It echoed hollowly about the mews, and in the silence that followed Johnny heard the whisper of voices from behind the curtains, and a shadow passed quickly across the window.

Johnny waited three minutes in the cold, then he stepped back into the middle of the mews.

'Benedict van der Byl,' he bellowed. 'I'll give you a count of ten to get this door open. Then I'll kick the bloody thing down.'

He drew breath, and bellowed again.

'This is Johnny Lance – and you know I mean it.'

The door opened almost immediately. Johnny pushed his way through it, not glancing at the man who held it, and started for the lounge.

'Dammit, Lance. You can't go in there.' Benedict van der Byl started after him.

'Why not?' Johnny glanced back at him. 'It is a Company flat – and I'm the General Manager.'

Before Benedict could reply, Johnny was through the door.

One of the girls picked up her clothing from the floor

and ran naked into the bedroom passage. The other pulled a full-length caftan over her head and glared at Johnny sulkily. Her hair was in wild disorder, fluffed out into a grotesque halo of stiff curls.

'Nice party,' said Johnny. He glanced at the movie projector on the side table, and then at the screen across the room. 'Films and all.'

'Are you the Fuzz?' demanded the girl.

'You've got an infernal cheek, Lance.' Benedict van der Byl was beside him, tying the belt of his silk dressing-gown.

'Is he Fuzz?' the girl demanded again.

'No,' Benedict assured her. 'He works for my father.' With the statement he seemed to gather self-assurance, drawing himself up to his full height and smoothing his long dark hair with one hand. His voice regained its polish and lazy inflection. 'Actually, he is Daddy's messenger boy.'

Johnny turned to him, but he addressed the girl without looking at her.

'Beat it, girlie. Follow your friend.'

She hesitated.

'Beat it!' Johnny's voice crackled like a bush fire, and she went.

The two men stood facing each other. They were the same age, in their early thirties – both tall, both dark-haired – but different in every other way.

Johnny was big in the shoulder and lean across the hips and belly, his skin polished and browned by the desert sun. The line of his heavy jawbone stood out clearly, and his eyes seemed still to seek far horizons. His voice clipped and twanged with the accents of the other land.

'Where is Tracey?' he asked.

Benedict lifted one eyebrow in a pantomine of arrogant surprise. His skin was pale olive, unstained by sunlight for it was months since he had last visited Africa. His lips were very red, as though they had been painted, the classical lines of his features were blurred by flesh. There were soft little pouches under his eyes, and a plumpness beneath the silk dressing-gown that suggested he ate and drank often and exercised infrequently.

'My dear chap, what on earth makes you think I know where my sister is? I haven't seen her for weeks.'

Johnny turned away and crossed to one of the paintings on the far wall. The room was hung with good original South African artists – Alexis Preller, Irma Stern and Tretchikof – an unusual mixture of techniques and styles, but someone had convinced the Old Man they were sound investments.

Johnny turned back to face Benedict van der Byl. He studied him as he had the paintings, comparing him with the clean young athlete he had been a few years before. A clear mental image in his mind pictured Benedict moving with leopard grace across the green field of play under the packed grandstands, turning smoothly beneath the high floating arc of the ball to gather it neatly, head high, and break back infield to open the line for the return kick.

'You're getting fat, Laddy Buck,' he said softly, and Benedict's anger stained his cheeks dull red.

'Get out of here,' he snapped.

'In a minute – tell me about Tracey first.'

'I've told you – I don't know where she is. Whoring it up around Chelsea, I expect.'

Johnny felt his own anger surge fiercely, but his voice remained level.

'Where is she getting the money, Benedict?'

'I don't know – the Old Man—'

Johnny cut him short. 'The Old Man is keeping her on an allowance of ten pounds a week. From what I hear she's throwing more than that around.'

'Christ, Johnny,' Benedict's tone became conciliatory, 'I don't know. It's not my business. Perhaps Kenny Hartford is—'

Again Johnny interrupted impatiently. 'Kenny Hartford is giving her nothing. That was part of the divorce agreement when they split up. Now I want to know who is subsidizing her trip to oblivion. How about it, big brother?'

'Me?' Benedict was indignant. 'You know there is no love wasted between us.'

'Must I spell it out?' Johnny asked. 'All right, then. The Old Man is dying – without losing his horror of all weakness and sin. If Tracey turns into a drug-soaked little tramp – then there's a good chance that our boy Benedict will come back into full favour. It would be a good gamble on your part to lay out a few thousand now, to send Tracey to Hell. Cut her off completely from her father – and all those nice fat millions.'

'Who said anything about drugs?' Benedict blustered.

'I did.' Johnny stepped up to him. 'You and I have a little unfinished business. It would give me intense pleasure to take you to pieces and see what makes you work.'

He held Benedict's eyes for long seconds, then Benedict looked down and fiddled with the cord of his dressing-gown.

'Where is she, Benedict?'

'I don't know, damn you!'

Johnny moved softly across to the movie projector and picked up a reel of film from the table beside it. He peeled off a few feet of celluloid from the reel and held it up to the light.

'Pretty!' he said, but the line of his mouth tightened with disgust.

'Put that down,' snapped Benedict.

'You know what the Old Man thinks about this sort of thing, don't you, Benedict?'

Suddenly Benedict went pale.

'He wouldn't believe you.'

'Yes, he would.' Johnny tossed the reel on the table and turned back to Benedict. 'He believes me because I've never lied to him.'

5

Benedict hesitated, wiped his lips nervously with the back of his hand.

'I haven't seen her for two weeks. She was renting a place in Chelsea. Stark Street. No. 23. She came to see me.'

'What for?'

'I lent her a couple of pounds,' Benedict muttered sulkily.

'A couple of pounds?' Johnny asked.

'All right, a couple of hundred. After all, she is my sister.'

'Damn decent of you,' Johnny lauded him. 'Write down the address.'

Benedict crossed to the leather-topped writing desk and scribbled on a card. He came back and handed the card to Johnny.

'You like to think you're big and dangerous, Lance.' His voice was pitched low but it shook with fury. 'Well, I'm dangerous too – in a different sort of way. The Old Man can't live for ever, Lance. When he's gone I'm coming after you.'

'You frighten the hell out of me.' Johnny grinned at him, and went down to the car.

The traffic was solid in Sloane Square as Johnny eased the Jaguar slowly down towards Chelsea. There was plenty of time to think; to remember how close they had been – the three of them. He and Tracey and Benedict.

Running together as wild young things with the endless beaches and mountains and sun-washed plains of Namaqualand as their playground. That was before the Old Man made the big strike on the Slang River, before there was money for shoes. When Tracey wore dresses made from flour sacks sewn together, and they rode to school each day, all three of them bare-back on a single pony like a row of bedraggled little brown sparrows on a fence.

He remembered how the long sun-drenched weeks while the Old Man was away were spent in laughter and secret games. How they climbed the *kopje* behind the mud-walled shack each evening and looked towards the north across

6

the limitless land, flesh-coloured and purple in the sunset, searching for the wisp of dust in the distance that would mean the Old Man was coming home.

Then the almost painful excitement when the dusty, rackety Ford truck with its mudguards tied on with wire was suddenly there in the yard, and the Old Man was climbing down from the cab, a sweat-stained hat on the back of his head and the dust thick in the stubble of his beard, swinging Tracey squealing above his head. Then turning to Benedict, and lastly to Johnny. Always in that order – Tracey, Benedict, Johnny.

Johnny had never wondered why sometimes he was not first. It was always that way. Tracey, Benedict, Johnny. The same way as he had never wondered why his name was Lance and not van der Byl. Then it had come to an end suddenly, the whole brightly sunlit dream of childhood was gone and lost.

'Johnny, I'm not your real father. Your father and mother died when you were very young.' And Johnny had stared at the Old Man in disbelief.

'Do you understand, Johnny?'

'Yes, Pa.'

Tracey's hand groped for his beneath the table top like a little warm animal. He jerked his own hand away from it.

'I think you'd better not call me that any more, Johnny.' He could remember the exact tone of the Old Man's voice, neutral, matter of fact, as it splintered the fragile crystal of his childhood to fragments. The loneliness had begun.

Johnny accelerated the Jaguar forward and swung into the King's Road. He was surprised that the memory hurt so intensely – time should have mellowed and softened it.

His life from then on had become a ceaseless contest to win the Old Man's approval – he dare not hope for his love.

Soon there were other changes, for a week later the old Ford had come roaring unexpectedly out of the desert in the night, and the barking of the dogs and the Old Man's

7

shouted laughter had brought them, sleepy-eyed, tumbling from their bunks.

The Old Man had lit the Petromax lamp and sat them on the kitchen chairs about the scrubbed deal table. Then with the air of a conjuror he had lain something that looked like a big lump of broken glass on the table.

The three sleepy children had stared at it solemnly, not understanding. The harsh glare of the Petromax was captured within the crystal, captured, repeated, magnified and thrown back at them in fire and blue lightning.

'Twelve carats – ' gloated the Old Man, 'blue-white and perfect, and there is a cartload more where that came from.'

After that there were new clothes and motor cars, the move to Cape Town, the new school and the big house on Wynberg Hill – but always the contest. The contest that did not earn the Old Man's approval as it was designed to do, but earned instead Benedict van der Byl's jealousy and hatred. Without his drive and purpose, Benedict could not hope to match Johnny's achievements in the classroom and on the sports field. He fell far behind the pace that Johnny set – and hated him for it.

The Old Man did not notice for he was seldom with them now. They lived alone in the big house with the thin silent woman who was their housekeeper, and the Old Man came infrequently and for short periods. Always he seemed tired and distracted. Sometimes he brought presents for them from London and Amsterdam and Kimberley, but the presents meant very little to them. They would have liked it better had it stayed the way it was in the desert.

In the void left by the Old Man the hostility and rivalry between Johnny and Benedict flourished to such proportions that Tracey was forced to choose between them. She chose Johnny.

In their loneliness they clung to each other.

The grave little girl and the big gangling boy built

8

together a castle against the loneliness. It was a bright secure place where the sadness could not reach them – and Benedict was excluded from it.

Johnny swung the Jaguar out of the line of traffic into Old Church Street, down towards the river in Chelsea. He drove automatically and the memories came crowding back.

He tried to recapture and hold the castle of warmth and love that he and Tracey had built so long ago, but instantly his mind leapt to the night on which it had collapsed.

One night in the old house on Wynberg Hill Johnny had come awake to the sound of distant weeping. He had gone barefooted in his pyjamas, following that heart-rending whisper of grief. He was afraid, fourteen years old and afraid in the dark house.

Tracey was weeping into her pillow and he had stooped over her.

'Tracey. What is it? Why are you crying?'

She had jumped up, kneeling on the bed, and flung both arms about his neck.

'Oh, Johnny. I had a dream, a terrible dream. Hold me, please. Don't go away, don't leave me.' Her whisper was still thick and muffled with tears. He had gone into her bed and held her until at last she slept.

Every night after that he had gone to her room. It was innocent and completely childlike, the twelve-year-old girl and the boy who was her brother, in fact if not in name. They held each other in the bed, and whispered and laughed secretly until sleep carried them both away.

Then suddenly the castle was blasted by the bright electric glare of the overhead light. The Old Man was standing in the door of the bedroom, and Benedict was behind him in his pyjamas dancing with excitement and chanting triumphantly.

'I told you, Pa! I told you so!'

The Old Man was shaking with rage, the bush of grey

hair standing erect like the mane of a wounded lion. He had dragged Johnny from the bed, and struck away Tracey's clinging hands.

'You little whore,' he bellowed, holding the terrified boy easily with one hand and leaning forward to strike his daughter in the face with his open hand. Leaving her sobbing, face down on the bed, he dragged Johnny down the passages to the study on the ground floor. He threw him into the room with a violence that sent him staggering against the desk.

The Old Man had gone to the rack and taken out a light Malacca cane. He came to Johnny and, taking a handful of his hair, threw him face down over the desk.

The Old Man had beaten him before, but never like this. Mad with rage the Old Man's blows had been unaimed, some fell across Johnny's back.

Yet in the agony it was deadly important to the boy that he should not cry out. He bit through his lip so the taste of blood was salt and copper in his mouth. He must not hear me cry! And he choked back the moans feeling his pyjama trousers hanging heavy and sodden with blood.

His silence served only as a goad to the Old Man's fury. Flinging the cane aside, he pulled the boy upright and attacked him with his hands. Slamming Johnny's head from side to side with full, open-handed blows that burst in Johnny's skull with blinding flashes of light.

Still Johnny kept on his feet, clinging to the edge of the desk. His lips broken and swollen and his face bloated and darkening with bruises, until at last the Old Man was driven far beyond the borders of sanity. He bunched his fist and drove it into Johnny's face – and with a wonderful sense of relief Johnny felt the pain go out in a warm flood of darkness.

Johnny heard voices first. A strange voice:

'– As though he's been savaged by a wild beast. I'll have to inform the police.'

Then a voice he recognized. It took him a little time to place it. He tried to open his eyes but they seemed locked tight, his face felt enormous, swollen and hot. He forced the fat lids of his eyes back and recognized Michael Shapiro, the Old Man's secretary. He was talking quietly to the other man.

There was the smell of antiseptic and the doctor's bag lay open on the table beside the bed.

'Listen, Doctor. I know it looks bad – but hadn't you better talk to the boy before you stir up the police?'

They both looked towards the bed.

'He's conscious.' The doctor came to him quickly. 'What happened to you, Johnny? Tell us what happened. Whoever did this to you will be punished – I promise you.'

The words were wrong. Nobody must ever punish the Old Man.

Johnny tried to speak but his lips were stiff and swollen. He tried again.

'I fell,' he said. 'I fell. Nobody! Nobody! I fell down.'

When the doctor had gone Mike Shapiro came and stood over him. His Jewish eyes were dark with pity, and something else – anger perhaps, or admiration. 'I'm taking you to my house, Johnny. You will be all right now.'

He stayed two weeks under the care of Michael Shapiro's wife, Helen. The scabs came away, the bruises faded to a dirty yellow, but his nose stayed crooked with a lump at the bridge. He studied his new nose in the mirror, and liked it. It made him look like a boxer, he thought, or a pirate, but it was many months before the tenderness passed and he could finger it freely.

'Listen, Johnny, you are going to a new school. A fine boarding-school in Grahamstown.' Michael Shapiro tried to sound enthusiastic. Grahamstown was five hundred miles

11

away. 'In the holidays you'll be going to work in Namaqualand – learning all about diamonds and how to mine them. You'll enjoy that, won't you?'

Johnny had thought about it for a minute, watching Michael's face and reading in it his shame.

'I won't be going home again then?' By home he meant the house on Wynberg Hill. Michael shook his head.

'When will I see—' Johnny hesitated as he tried to find the right words, ' – when will I see them again?'

'I don't know, Johnny,' Michael answered him honestly.

As Michael had promised it was a fine school.

On his first Sunday after the church service, he had followed the other boys back to their classroom for the session of compulsory letter-writing. The others had immediately begun dashing off hasty scribbles to their parents. Johnny sat miserably until the master in charge stopped at his desk.

'Aren't you going to write home, Lance?' he asked kindly. 'I'm sure they'll all want to hear how you are.'

Johnny picked up his pen obediently, and puzzled over the blank writing-pad.

He wrote at last:

Dear Sir,
    I hope you will be pleased to hear that I am now at school. The food is good, but the beds are very hard.
    We go to church every day and play rugby football.
                                        Yours faithfully,
                                                Johnny.

From then until he left school and went up to University three years later, he wrote every week to the Old Man. Every letter began with the same salutation and went on, 'I hope you will be pleased to hear—' There was never a reply to any of these letters.

Once each term he received a typewritten letter from

Michael Shapiro setting out the arrangements that had been made for the school holidays. Usually these involved a train journey hundreds of miles across the Karroo to some remote village in the vast dry wasteland, where a light aircraft belonging to Van Der Byl Diamonds was waiting to fly him still deeper into the desert to one of the Company's concession areas. Again, as Michael Shapiro had promised, he learned about diamonds and how to mine them.

When the time to move on to University arrived, it was completely natural that he chose to take a degree in Geology.

During all that time he was an outcast from the van der Byl family. He had seen none of them – not the Old Man, nor Tracey, nor even Benedict.

Then, in one long eventful afternoon, he saw all three of them. It was his final year at University. His degree was a certainty. He had headed the lists at every examination from his first year onwards. He had been elected the senior student of Stellenbosch University, but now there was a further honour almost within his grasp.

In ten days' time the National Selectors would announce the rugby team to meet the New Zealand All Black touring team – and Johnny's place at flank forward was as certain as his degree in Geology.

The sporting press had nicknamed Johnny 'Jag Hond' after that ferocious predator of the African wilds, the Cape hunting dog; an animal of incredible stamina and determination that savages its prey on the run. The nickname had stuck fast, and Johnny was a favourite of the crowds.

In the line-up of the team from Cape Town University was another crowd-pleaser whose place in the National Side to meet the All Blacks seemed equally assured. From his position at full back Benedict van der Byl dominated the field of play with a grace and artistry that were almost god-like. He had grown tall and wide-shouldered, with long powerful legs and dark brooding good looks.

13

Johnny led the visiting team out on to the smooth green velvet field, and while he jogged and flexed his back and shoulders he looked up at the packed stands seeking assurance that the high priests of rugby football were all there. He saw Doctor Danie Craven sitting with the other selectors in their privileged position below the Press enclosure. While in front of the Doctor, leaning back to exchange a few words with him, sat the Prime Minister.

This meeting between the two universities was one of the high spots of the rugby season, and the *aficionados* travelled thousands of miles to watch it.

The Prime Minister smiled and nodded, then leaned forward to touch the shoulder of the big white-headed figure that sat in the row below him.

Johnny felt an electric tingle run up his spine as the white head lifted and looked directly at him. It was the first time he had seen the Old Man in the seven years since that terrible night.

Johnny lifted an arm in salute, and the Old Man stared at him for long seconds before he turned away to speak with the Prime Minister.

Now the drum majorettes came out in ranks on to the field. White-booted, dressed in Cape Town University colours with short swinging skirts and tall hats, they high-stepped and paraded, lovely young girls flushed with excitement and exertion.

The roar of the crowd drummed with the blood in Johnny's ears, for Tracey van der Byl was leading the first rank. He knew her instantly, despite the passage of years in which she had grown to young womanhood. Her legs and arms were sun-bronzed and her dark hair hung glossily to her shoulders. She cavorted and kicked and stamped shouting the traditional cheers, jiggling her firm young bottom with innocent abandon while the crowds screamed and writhed, beginning already to work themselves into an hysterical frenzy. Johnny watched Tracey. He stood completely still in the thunderous uproar. She was the most beautiful woman he had ever seen.

Then the show was over, the drum majorettes retreating back through the stadium entrance, and the home team trotted on to the field.

The presence of the Old Man and Tracey added intensity to the glare of hatred that Johnny turned on the tall white-clad figure that fell back to take control of the Cape Town back field.

Benedict van der Byl reached his position and turned. From inside his calf-length sock he took a comb and ran it through his dark hair. The crowd bellowed and whistled, loving this little theatrical gesture. Benedict returned the comb to his sock and posed with one hand on his hip, his chin lifted arrogantly as he surveyed the opposition.

Suddenly he intercepted Johnny's glare, and the pose altered as he dropped his eyes and shuffled his feet a little.

The whistle fluted, and play began. It was everything the crowds had hoped for, a match that would long be remembered and gloated over. Massive Panzer offensives by the forwards, long probing raids by the backs with the oval ball flickering from hand to hand, until a bone-jarring tackle smashed the carrier to the turf. Hard and fast and clean play swung from side to side, a hundred times the crowd came up on its feet as one, eyes and mouths wide in screaming unbearable tension, to sink back with a groan as the ball was held by a desperate defence within inches of the try line.

No score and three minutes of play left, Cape Town attacking from a set scrummage, driving through a gap in the defence and then putting the ball in the air with a long raking pass, taken cleanly by the Cape Town wing without a break in his stride. His feet twinkled across the green turf, and again the crowd came up with a gasp.

Johnny hit him low, just above the knee, with his shoulder. The two of them rolled together out of the field of play lifting a puff of white lime from the line and the crowd groaned and sank back.

While they waited to receive the throw, Johnny whispered hoarse orders. His gold and maroon jersey was soaked with sweat, and blood from a grazed hip stained his white shorts.

'Get it back fast. Don't run with it. Give it to Dawie. Kick high and deep, Dawie.'

Johnny leapt high to the flight of the thrown ball, with a bunched fist he punched it back accurately into Dawie's hands, at the same moment twisting his body to block the attackers.

Dawie fell back two paces and kicked. The power of the kick swung his right foot above his head, and the impetus flung him forward to put his forwards 'on side'.

The ball climbed slowly, flying like a dart with no wobble or roll in the air, reaching the zenith of its trajectory high over mid-field, then floating back to earth.

Twenty thousand heads followed its flight, a hush had fallen over the field – and in the unnatural silence Benedict van der Byl was drifting back deep into his own territory, anticipating the drop of the ball with deceptively unhurried strides, yet timing it with the precision of the gifted athlete.

The ball slotted neatly into his arms, and he began moving lazily infield to open his angle for the return kick. Still a tense throbbing mesmeric hush hung over the field, Benedict van der Byl was at the focus of attention.

'Jag Hond!' A single voice in the crowd alerted them, and twenty thousand heads swung downfield.

'Jag Hond!' A roar now. Johnny was well clear of the pack, arms pumping and legs churning as he bore down on Benedict. It was a futile effort, he could not hope to intercept a player of Benedict's calibre from such long range, yet Johnny was burning the last of his physical reserves in that charge. His face was a sweat-shining mask

16

of determination, and clods of torn grass flew from under his savagely driving boots.

Then something happened which was unaccountable, almost past belief. Benedict van der Byl glanced round and saw Johnny. He broke his stride, two clumsy shuffling paces, and tried to pivot away deeper into his own ground. All the assurance had gone from his body, all the skill and grace. He tripped and stumbled, almost fell and the ball popped out of his hands, bouncing awkwardly.

Benedict scrambled after it, groping blindly, looking back over his shoulder. Now on his face was an expression of naked terror. Johnny was very close. Grunting at each stride like a gut-shot lion, massive shoulders already bunching for the strike, his lips drawn back into a murderous parody of a grin.

Benedict van der Byl dropped to his knees and covered his head with both arms, cringing down on to the green turf.

Johnny swept past him without a check, stooping easily in his run to gather the bouncing ball.

When Benedict uncovered his head and, still kneeling, looked up, Johnny stood ten yards away between the goal posts watching him. Then, deliberately, Johnny placed the ball between his feet to complete the formality of the touchdown.

Now, as if by agreement, both Johnny and Benedict looked towards the main grandstand. They saw the Old Man rise from his seat and make his way slowly through the ecstatic crowds towards the exit.

The day after the match, Johnny went back into the desert.

He was down in the bottom of a fifteen-foot prospect trench that had been dug across the grain of the country rock. It was oppressively hot in the confines of the trench and Johnny was stripped to a skimpy pair of khaki shorts, his sun-browned muscles oily with sweat, but he worked steadily at his sampling. He was establishing the contours and profile of an ancient marine terrace that the ages had buried beneath the sand. It was here on the bedrock that he expected to find the thin layer of diamond-bearing gravel.

He heard the Jeep pull up at ground level above him, and the crunch of footsteps. Johnny straightened up and held his aching back muscles.

The Old Man stood at the edge of the trench and looked down at him. He held a folded newspaper in his hand. This was the first time Johnny had seen him at close range in all the years, and he was shocked at the change. The mass of bushy hair was so white, and his features were folded and creased like those of a mastiff, leaving the big hooked nose standing like a hillock from his face. But there was no wasting or deterioration in his body, and his eyes were still that chilling enigmatic blue.

He dropped the newspaper into the trench and Johnny caught it, still staring up at the Old Man.

'Read it!' said the Old Man. The paper was folded to the sports page, and the headline was thick and bold.

JAG HOND IN. VAN DER BYL OUT.

The shock was as delicious as the plunge into a mountain stream. He was in – he would carry the gold and green, and wear the leaping Springbok on his blazer pocket.

He looked up, proud and happy, standing bare-headed in the sun waiting for the Old Man to speak.

'Make up your mind,' said the Old Man softly. 'Do you want to play ball – or work for Van Der Byl Diamonds? You can't do both.' And he walked back to the Jeep and drove away.

Johnny cabled his withdrawal from the team to the Doctor personally. The storm of outraged protest and abuse in the national press, and the hundreds of viperous letters Johnny received accusing him of cowardice and treachery and worse made him thankful for the sanctuary of the desert.

Neither Johnny nor Benedict had ever played the game again. Thinking about it, even at this remove of time, Johnny felt the sting of disappointment. He had wanted that green and gold badge of honour so very deeply. Brusquely he pulled the Jaguar off the road and scanned the street map of London and found Stark Street tucked away off the King's Road. He drove on remembering how it had been after the Old Man had taken it from him. The agony of mind had been scarcely endurable.

His companions in the desert were Ovambo tribesmen from the north, and a few of those taciturn white men that the desert produces, as hardy and uncompromising as her vegetation or her mountain ranges.

The deserts of the Namib and the Kalahari are amongst the loneliest places on earth, and the desert nights are long. Not even the day's unremitting physical labour could tire Johnny sufficiently to drug his dreams of a lovely girl in a short white skirt and high boots – or an old white-headed man with a face like a granite cliff.

Out of those long days and longer nights came solid achievements to stand like milestones marking the road of his career. He brought in a new diamond field, small but rich, in country which no one else had believed would yield diamonds. He pegged a uranium lode which Van Der Byl

Diamonds sold for two and a half millions, and there were other fruits from his efforts as valuable if not as spectacular.

At twenty-five, Johnny Lance's name was whispered in the closed and forbidding halls of the diamond industry as one of the bright young comers.

There were approaches – a junior partnership in a firm of consulting geologists, field manager for one of the struggling little companies working marginal ground in the Murderers' Karroo. Johnny turned them down. They were good offers, but he stayed on with the Old Man.

Then the big Company noticed him. A century ago the first payable pipe of 'blue ground' in Southern Africa was discovered on a hard scrabble farm owned by a Boer named De Beer. Old De Beer sold his farm for £6,000, never dreaming that a treasure worth £300,000,000 lay beneath the bleak dry earth. The strike was named De Beers New Rush, and a horde of miners, small businessmen, drifters, chancers, rogues and scoundrels moved in to purchase and work minute claims, each the size of a large room.

From this pretty company of fortune's soldiers two men rose high above the others, until between them they owned most of the claims in De Beers New Rush. When these two, Cecil John Rhodes and Barney Barnato, at last combined their resources, a formidable financial enterprise was born. From such humble beginnings the Company has grown to awesome respectability and dignity. Its wealth is fabled, its influence immeasurable, its income is astronomical. It controls the diamond supply to the world. It controls also mineral concessions over areas of Central and Southern Africa which total hundreds of thousands of square miles, and its reserves of unmined precious and base minerals cannot be calculated. Small diamond companies are allowed to co-exist with the giant until they reach a certain size – then suddenly they become part of it, gobbled up as a tiger shark might swallow any of its pilot fish who become too large and daring. The big Company can afford to buy the best prospects, equipment – and men. It reached out one of its myriad tentacles to draw in Johnny Lance. The price they set on him was twice his present salary, and three times his future prospects.

Johnny turned it down flat. Perhaps the Old Man did not notice, perhaps it was mere coincidence that a week later Johnny was promoted Field Manager of Beach Operation. The nickname that went with the job was 'King Canute'.

Van Der Byl Diamonds had thirty-seven miles of beach concession. The tiny ribbon of shoreline, one hundred and twenty feet above high-water mark, and one hundred and twenty feet below low-water mark. Inland the concession belonged to the big Company. It had purchased the land, a dozen vast ranches, simply to obtain the mineral rights. The sea concessions, territorial up to waters twelve miles off shore, belonged to them also. Granted to them by Government charter twenty years before. But Van Der Byl Diamonds had the Admiralty strip – and it was 'King Canute's' job to work it.

The sea-mist came smoking in like ground pearl dust off the cold waters of the Benguela current. From out of the mist bank the high unhurried swells marched in towards the bright yellow sands and the tall wave-cut cliffs of Namaqualand.

The swells peaked up sharply as they felt the land. Their crests trembled and turned luminous green, began to dissolve in plumes of wind-blown spray, arched over and slid down upon themselves in the roar and rumble of white water.

Johnny stood on the driver's seat of the open Land-Rover. He wore a sheepskin jacket against the chill of the dawn mist, but his head was bare and his dark hair fluttered nervously against his forehead in the wind.

His heavy jaw was thrust forward, and his hands in the pockets of the sheepskin jacket were balled into fists. He

21

scowled aggressively as he measured the height and push of the surf. With his crooked nose he looked like a boxer waiting for the gong.

Suddenly with an awkward angry movement he jerked his left hand from his pocket and looked down at the dial of his wrist watch. Two hours and three minutes to low tide. He pushed his fist back into his pocket, and swivelled quickly to look at his bulldozers.

There were eleven of them, big bright yellow D.8 Caterpillars, lined up along the high-water mark. The operators sat goggled and tense in their high stern seats. They were all watching him anxiously.

Beyond them, standing well back, were the earthloaders. They were ungainly, pregnant-looking machines with swollen bellies, and heavily lugged tyres that stood taller than a man. When the time came they would rush in at thirty miles an hour, drop a steel blade beneath their bellies and scrape up a fifteen-ton load of sand or gravel, race back inland and drop their load, turn and rush back for another gargantuan bite out of the earth.

Johnny was steeling himself, judging the exact moment in which to hurl a quarter of a million pounds' worth of machinery into the Atlantic Ocean, in the hope of recovering a handful of bright pebbles.

The moment came, and Johnny spent half a minute of precious time in scrutinizing his preparations before committing himself to action.

Then 'GO!' he shouted into his loudhailer and windmilled his right arm in the unmistakable command to advance.

'Go!' he shouted again, but his voice was lost. Even the sound of the wild surf was lost in the bull bellow of the diesels. Lowering their massive steel blades, a chorus line of steel monsters, they crawled forward.

Now the golden sand curled before the scooped blades, like butter from the knife. It built up before the monstrous machines, becoming a pile and then a high wall. Thrusting, pulling back, butting, worrying, the bulldozers swept the wall of sand forward. The arms of the operators pumping the handles of the controls like mad barmen drawing a thousand pints of beer, the diesels roaring and muttering and roaring again.

The wall of sand met the first low push of sea water up the beach and smothered it. In seeming astonishment and uncertainty the sea pulled back, swirling and creaming before the advancing dyke of sand.

The bulldozers were performing a complicated but smoothly practised ballet now. Weaving and crossing, blades lifting and falling, backing and advancing, all under the supervision of the master choreographer, Johnny Lance.

The Land-Rover darted back and forward along the edge of the huge pit that was forming, with Johnny roaring orders and instructions through the electric loudhailer.

Gradually a sickle-shaped dyke of sand was thrown out into the sea, while behind it the bulldozer blades cut down, six, ten, fifteen feet through the loose yellow sand.

Then suddenly they hit the oyster line, that thin layer of fossilized oyster shell that so often covers the diamond gravels of South West Africa.

Johnny saw the change in the character of his pit, saw the shell curling from the blades of the bulldozers.

With half a dozen orders and hand signals he had his 'dozers flatten a ramp at each end of his pit, to give the earthloaders access. Then he ordered them away to hold the dyke against the sea.

He glanced at his watch. 'One hour thirteen minutes,' he muttered. 'We're running tight!'

Quickly he checked his pit. Two hundred yards long, fifteen feet deep, the overburden of sand stripped away, the oyster line showing clean and white in the sun, the bulldozers clear of the pit bottom – fighting back the sea.

'Right,' he grunted. 'Let's see what we've got.'

He turned to face the two earthmovers waiting expect-antly above the high-water mark.

'Go in and get it!' he shouted, and gave the windmill arm signal.

Nose to tail the earthloaders roared forward, swinging wide at the head of the pit, then swooping down the ramp and dashing along the bottom. They scooped up a load of shell and gravel without checking their speed and went bellowing up the far ramp, swinging again to race up and deposit their load below the cliff, but above the high-water mark.

Round they went, and round again, chasing their tails, while the bulldozers held back the sea which was now becoming angry – sending its cohorts to skirmish along the dyke, seeking a weak place to attack.

Johnny glanced at his watch again.

'Three minutes to low water,' he spoke aloud, and grinned. 'We're going to make it – I think!'

He lit a cigarette, relaxing a little now. He dropped into the driving seat and swung the Land-Rover up the beach, parking it beyond the mountain of gravel that the earth-loaders were building.

He climbed out and took up a handful of the gravel.

'Lovely!' he whispered. 'Oh sweet! Sweet!'

It was right. All the signs were good. In the single handful he identified a small garnet, and a larger lump of agate.

He scooped another handful.

'Jasper,' he gloated. 'And banded ironstone!'

All these stones were the team-mates of the diamond, you found them together.

The shape was right also, the stones polished round and shiny as marbles, not flattened like coins which would mean they had washed in only one direction. Round stones meant a wave action zone – a diamond trap!

'We've hit a jewel box – I'll take Lysol on that!'

24

From thirty-seven miles of beach Johnny had picked a two-hundred-yard stretch, and hit it right on the nose. A choice not by luck, but by careful study of the configuration of the coastline, aerial photographs of the wave patterns and bottom contours of the sea, an analysis of the beach sands, and finally by that indefinable 'feel' for ground that a good diamond man has.

Johnny Lance was mightily delighted with himself as he climbed back into the Land-Rover. The earthloaders had scraped the gravel down to bedrock. Their job was finished, and they pulled out of the pit and stood with panting exhausts beside the enormous pile of gravel they had recovered.

'Bottom boys!' roared Johnny, and the patient army of Ovambo tribesmen who had been squatting above the beach came swarming down into the pit. Their job was to sweep and clean the pit bottom, for a high proportion of the diamonds would have worked their way down through the gravel into the crevices and irregularities of the bedrock.

The sea changed its mood, furious at the brutal rape of its beaches it came hissing and tearing at the sand dyke. The tide was making now, and the bulldozers had to redouble their efforts to keep it out.

In the pit the Ovambos worked in a frenzy of activity, sparing only an occasional apprehensive glance for the wall of sand that held the Atlantic at bay.

Now Johnny was tensing up again. If he pulled them out early he would be leaving diamonds down there, if he left them in too late he might drown machinery – and men.

He cut it fine, just a fraction too fine. He pulled the bottom boys out with the sea beginning to break over the dyke, and to seep through under it.

Then he began to pull out his bulldozers, ten of them out – one still coming infinitely slowly, waddling across the wide bottom of the empty pit.

The sea broke through, it broke simultaneously in two places and came boiling into the pit in a waist-high wave.

The bulldozer operator saw it, hesitated one second, then his spirit failed him and he jumped down from his machine; abandoning it to the sea he sprinted ahead of the wave, making for the steep nearest side of the pit.

'Bastard!' swore Johnny as he watched the operator scramble to safety. 'He could have made it.' But his anger was against himself also. His decision to withdraw had been too long delayed, that was £20,000 worth of machinery he had sacrificed to the sea.

He slammed the Land-Rover into gear, and put her to the pit. She went off the edge like a ski jump, falling fifteen feet before she hit the bottom, but her fall was cushioned by the slope of sand and she sprang forward bravely to meet the rush of sea water.

It broke over the bonnet, slewing the vehicle viciously, but Johnny fought her head round and kept her going towards the stranded bulldozer.

The engine of the Land-Rover had been sealed and water-proofed against just such an emergency, and now she ploughed forward throwing a sheet of water to each side. But her forward rush faltered as the green water poured over her.

Now suddenly the entire sand dyke collapsed under the white surf and the Atlantic took control. The tall wave of green water that raced across the pit hit the Land-Rover, upending her, throwing Johnny into the jubilant frothing water, while the Land-Rover rolled over on her back, pointing all four wheels to the sky in surrender.

Johnny went under but came up immediately. Half swimming, half wading, battered by the boisterous sea he struggled on towards the yellow island of steel.

The sea struck him down, and he went under again. Found his feet for a moment, then had them cut from under him once more.

Then suddenly he had reached the bulldozer and was dragging himself up over the tracks to the driver's seat. He was coughing and vomiting sea water, as he reached the controls.

The bulldozer sat immovable, held down by her own twenty-six tons of dead weight on to the hard bedrock of the pit. Although the sea burst over her, and swirled through her tracks, it could not move her.

Through eyes blurred and swimming with salt water and his own tears, Johnny briefly checked the gauges on the instrument panels. She had oil pressure and engine revs, and high above his head the exhaust pipe chugged blue smoke.

Johnny coughed again. Vomit and sea water shot up his throat in a scalding jet, but he pushed the throttle wide and threw in both clutch levers.

Ponderously the great machine ground forward, almost contemptuously shouldering the sea aside, her tracks solidly gripping the bedrock.

Johnny looked about him quickly. The sand ramps at each end of the pit were washed away. The sides were sheer now, and behind him the sea was rushing unimpeded into the pit.

A wave broke over his head, and Johnny shook the water from his hair like a spaniel and looked around with mounting desperation for an avenue of escape.

With a shock of surprise he saw the Old Man. He had thought him to be four hundred miles away in Cape Town, but here he was on the edge of the pit. The white hair shone like a beacon.

Instinctively Johnny swung the bulldozer in his direction, crawling through the turbulent waters towards him.

The Old Man was directing two of the other bulldozers, reversing them as close as he dared to the lip of the bank of sand, while from the service truck parked below the cliff a line of Ovambos came staggering down the beach with the heavy tractor tow chain over their shoulders. They shuffled bow-legged under the tremendous weight of the chain, sinking ankle deep into the sand with each step.

The Old Man roared at them, urging them on, but the words were lost in the thunder of diesel engines and the ranting of the wind and the sea. Now he turned back to Johnny.

'Get her in close,' the Old Man yelled through cupped hands. 'I'll bring the end of the chain down to you!'

Johnny waved an acknowledgement, then grabbed at the controls as the force of the next wave pushed even the giant tractor off its line, and Johnny felt the diesel falter for the first time – the water had found its way in through the seals at last.

Then he was under the high bank of yellow sand that towered twenty feet above him and he scrambled forward over the engine bonnet to meet the Old Man.

The Old Man was poised on the lip of the pit with the end of the chain draped in a loop over both shoulders. He was stooped beneath its weight, and when he stepped forward the sand crumpled away beneath him and he came sliding and slipping down the steep incline, buried waist deep, the great chain snaking after him.

Judging the rush of the sea Johnny jumped down to help him. Together, battered by the sea, they dragged the chain to the bulldozer.

'Fix it on to the blade arm,' grunted the Old Man, and they got a double turn of chain around the thick steel arm.

'Shackle!' Johnny snapped at him, and while the Old Man untied the length of rope which secured the steel shackle around his waist, Johnny looked up at the cliff of sand that hung over them.

'Christ!' he said softly, the sea was attacking it – and now it was soft and trembling above them, ready to collapse and smother them.

The Old Man passed him the huge shackle, and Johnny

began with numbed hands to secure the end of the chain. He must pass the thick case-hardened pin through two links and then screw it closed. It was a Herculean task under these conditions, with the surf bursting over his head, the drag of the sea on the chain, and the cliff of sand threatening to fall on them at any moment. From twenty feet above them Johnny's foreman was watching anxiously, ready to pass the word to the two waiting bulldozers to throw their combined weights on the chain.

The thread of the pin caught, half a dozen turns would secure it, he would have finished the job by the time the word was passed to the 'dozer operators.

'Okay,' he nodded and gasped at the Old Man. 'Pull!'

The Old Man lifted his head and bellowed up the bank, 'Pull!'

The foreman acknowledged with a wave.

'Okay.' And his head disappeared behind the bank as he ran back to the bulldozers, and at that moment the surf swung the chain. A movement of a few inches, but enough to catch Johnny's left index finger between two of the links.

The Old Man saw his face, saw him struggling to free himself.

'What is it?'

Then the water sucked back for a moment, and he saw what had happened. He waded forward to help – but from above them came the throaty roar of the diesels and the chain began running away, snaking and twisting up the bank like a python.

The Old Man reached Johnny and caught him about the shoulders to steady him. They braced themselves in horror, staring at the captive hand.

The chain jerked taut, severing the finger cleanly in a bright burst of scarlet, and Johnny reeled back into the Old Man's arms. The great yellow bulk of the bulldozer was dragged relentlessly down on top of them, threatening to crush them both, but using the next break and push of the

29

sea the Old Man dragged Johnny clear – and they were carried sideways along the bank, tumbled helplessly by the strength of the water out of the bulldozer's path.

Johnny clutched his injured hand to his chest, but it hosed a bright stream of blood that discoloured the water about them. His head went under and salt water shot down his throat into his lungs. He felt himself drowning, the strength oozing out of him.

He surfaced again, and through bleary eyes saw the glistening wet bulldozer half-way up the sand bank. He felt the Old Man's arms about his chest and he went under again relaxing as the darkness closed over his eyes and brain.

When the darkness cleared from his eyes, he was lying on the dry sand of the beach and the first thing he saw was the Old Man's face above him, furrowed and pouched, his silver white hair plastered across his forehead.

'Did we get her out?' Johnny asked thickly.

'Ja,' the Old Man answered. 'We got her out.' And he stood up, walked to the Jeep, and drove away, leaving the foreman to tend to Johnny.

Johnny grinned at the memory, and lifting his left hand off the driving-wheel of the Jaguar he licked the shiny stump of his index finger.

'It was worth a finger,' he murmured aloud, and still searching for road signs he drove on slowly.

He smiled again comfortably, shaking his head with amusement as he remembered his hurt and disappointment when the Old Man had walked away and left him lying on the beach. He had not expected the Old Man to fall on his shoulders sobbing his gratitude and begging forgiveness for all the years of misery and loneliness – but he had expected something more than that.

After a two-hundred-mile round Jeep-journey through the desert night to the nearest hospital where they had trimmed and bound the stump, Johnny was back at the workings the next day in time to watch the first run of gravel from the beach.

In his absence, the gravel had been screened to sieve off all the over-size rock and stone, then it had been puddled through a tank of silicon mud to float off all the material with a specific gravity less than 2.5, then finally what was left had been run through a ball mill – a long steel cylinder containing steel balls the size of baseballs. The cylinder revolved continually and the steel balls crushed to powder all substance softer than 4 on Mohs hardness scale.

Now there was a residue, a thousandth part of the gravel they had won from the sea. In this remainder would be the diamonds – if diamonds there were.

When Johnny arrived back at the shed of galvanized iron and wood on the cliff above the beach that housed his separation plant, he was still half groggy from the anaesthetic and lack of sleep.

His hand throbbed with the persistence of a lighthouse, his eyes were reddened and a thick black stubble covered his jaws.

He went to stand beside the grease table that filled half the shed. He was swaying a little on his feet, as he looked around at the preparations. The massive bin at the head of the table was filled with the concentrated diamond gravels, the plates greased down, and his crew was standing ready.

'Let's go!' Johnny nodded at his foreman, who immediately threw in the lever that set the table shaking like an old man with palsy.

The table was a series of steel plates, each slightly inclined and thickly coated with dirty yellow grease. From the bin at the head of the shuddering table a mixture of gravel and water began to dribble, its consistency and rate of flow carefully regulated by the foreman. It spread over the greased table like spilled treacle, dropping from one plate to the next, and finally into the waste bin at the end of the table.

A diamond is unwettable, immerse it in water, scrub it, – but it comes out dry. A coat of grease on a steel plate is also unwettable, so wet gravel and sea shell will slide over it and keep moving across the agitating, sloping table.

But a diamond when it hits grease sticks like a half-sucked toffee to a woollen blanket.

In the excitement and anxiety of the moment Johnny felt his weariness recede, even the pain in his stump was muted by it. His eyes and whole attention were fastened on that glistening yellow sheet of grease.

The little stuff under a carat in weight, or the industrial black diamond and boart would not be visible on the table; the agitation was too rapid – blurring with speed, and the flow of loose material would disguise them.

So complete was his absorption that it was some seconds before he was aware of a presence beside him. He glanced up quickly.

The Old Man was there, standing with the wide stance and tension-charged attitude that was his own special way.

Johnny was acutely conscious of the Old Man's bulk beside him – and he felt the first flicker of alarm. What if this was a barren run? He needed diamonds now – as he had never needed anything in his life. He scanned the blurring plates of yellow grease, seeking the purchase price that could buy back the Old Man's esteem. The speckled gravel flowed imperturbably across the plates, and Johnny felt a flutter of panic.

Then from across the table the foreman let out a whoop, and pointed.

'Thar she blows!'

Johnny's eyes darted to the head of the table. There beneath the outlet from the bin, half buried in the thick grease by its own weight, anchored solidly while the worthless gravel washed past it, was a diamond.

A big fat five-carat thing, that glowed sulky and yellow, like a wild animal resenting its captivity.

Johnny sighed softly and darted a sideways glance at the Old Man. The Old Man was watching the table without expression, and though he must have been conscious of Johnny's scrutiny, he did not look up. Johnny's eyes were dragged irresistibly back to the table.

By some freakish chance, the next diamond fell from the bin directly on to the one already anchored in the grease.

When diamond strikes diamond it bounces like a golf ball off a tarmac road.

The second diamond, a white beauty the size of a peach pip, clicked loudly as it struck the other then spun head high in the air.

Both Johnny and the foreman laughed involuntarily with delight at the beauty of that twinkling drop of solid sunlight.

Johnny reached across the table with his good hand, and snatched it out of the air. He rubbed it between his fingers – revelling in the soapy feel of it, then turned and offered it to the Old Man.

The Old Man looked at the diamond, nodded in acknowledgement. Then he pulled back the cuff of his coat and checked his wrist watch.

'It's late. I must get back to Cape Town.'

'Won't you stay for the rest of the run, sir?' Johnny realized his tone was too eager. 'We could have a drink together afterwards.' He remembered after he had spoken, that the Old Man abhorred alcohol.

'No.' The Old Man shook his head. 'I have to get back by this evening.' Now he looked steadily into Johnny's eyes. 'You see, Tracey is getting married tomorrow afternoon and I must be there.'

Then he smiled, watching Johnny's face, but nobody could ever guess the meaning of a smile on the Old Man's lips – for it never showed in his eyes.

'Didn't you know?' he asked, still smiling. 'I thought you had received an invitation.' And he went out of the shed

33

to where his Jeep stood in the bright sunshine waiting to take him out to the aircraft landing-strip among the sand dunes.

The pain in his injured hand, and the Old Man's words denied Johnny the sleep he so desperately needed, but it was two o'clock in the morning before he threw back his blankets and lit the lamp beside his camp bed.

'He said I had been invited – and, by God, I'll be there.'

He drove through the night, and the next morning. The first two hundred miles were on desert tracks of sand and stone, then he reached the metalled highway in the dawn and turned south across the great plains and over the mountains. It was noon before he saw the squat blue silhouette of Table Mountain on the skyline dwarfing the city that huddled beneath it.

He checked in at the Vineyard Hotel, and hurried to his room to bath and shave and change into a suit.

The grounds of the old house were crowded with expensive automobiles, and the overflow was parked along both sides of the street outside, but he found a space for the dusty Land-Rover. He walked up through the white gates and across the green lawns.

There was a band playing in the house, and a hubbub of voices and laughter drifted out through the windows of the ballroom.

He went in through the side door. The passages were thronged with guests, and he made his way amongst them seeking a familiar face in the groups of loud-voiced gesticulating men and giggling women. At last he found one.

'Michael.' And Michael Shapiro looked round, recognizing him and letting the conflicting emotions of pleasure, surprise and alarm show clearly on his face.

'Johnny. It's good to see you.'

'Is the ceremony over?'

'Yes, and the speeches also – thank God.' He took Johnny's arm and led him aside. 'Let me get you a glass of

34

champagne.' Michael hailed a waiter and put a crystal glass into Johnny's hand.

'Here's to the bride,' Johnny murmured and drank.

'Does the Old Man know you are here?' Michael came out with the question that was burning his mouth, and when Johnny shook his head, Michael's expression became thoughtful.

'What's he like, Michael, Tracey's husband?'

'Kenny Hartford?' Michael considered the question. 'He's all right, I suppose. Nice-looking boy, plenty of money.'

'What's he do for a crust of bread?'

'His daddy left him the whole loaf – but to fill in the time he does fashion photography.' And Johnny pulled down the corners of his mouth.

Michael frowned. 'He's all right, Johnny. The Old Man picked him carefully.'

'The Old Man?' Johnny's jaw thrust out.

'Of course, you know him – he wouldn't leave an important decision like that to anybody else.'

Johnny finished his champagne in silence, and Michael watched his face anxiously.

'Where is she? Have they left yet?'

'No.' Michael shook his head. 'They're still in the ballroom.'

'I think I'll go and wish luck to the bride.'

'Johnny.' Michael caught hold of his elbow. 'Don't do anything stupid – will you?'

Johnny stood at the head of the marble staircase that led down into the ballroom. The floor was crowded with dancing couples and the music was loud and merry. The bridal party sat at a raised table across the floor.

Benedict van der Byl saw Johnny first. His face flushed and he leaned quickly to whisper to the Old Man, then began to rise from his seat. The Old Man placed a restraining hand on Benedict's shoulder, and smiled across the room at Johnny.

Johnny went down the stairs and made his way through the dancers. Tracey had not seen him. She was talking to the silky-faced young man who sat beside her. He had wavy blond hair.

'Hello, Tracey.' She looked up at Johnny and caught her breath. She was more beautiful than he remembered.

'Hello, Johnny.' Her voice was almost a whisper.

'May I dance with you?' She was pale now, and her eyes went to the Old Man, not to her new husband. The gleaming white bush of hair nodded slightly, and Tracey stood up.

They made one circuit of the dance floor before the band stopped playing. Johnny had planned a hundred different things to say to her, but he was dumb until the music ended and the opportunity was passing. Hurriedly now in the few seconds that were left Johnny told her: 'I hope you will be happy, Tracey. But if you ever need help – ever – I will come, I promise you that.'

'Thank you.' Her voice was husky, and for a moment she looked like the little girl who had cried in the night. Then he took her back to her husband.

The promise had been made five years ago, and now he had come to London to honour it.

Number 23 Stark Street was a neat double-storeyed cottage with a narrow front. He parked outside it. It was dark now and lights burned on both floors. He sat in the parked Jaguar, suddenly reluctant to go further. Somehow he knew that Tracey was here, and he knew it would not be pretty. For a moment he recaptured the image of her as a lovely young woman in a wedding dress of white satin, then he climbed out of the Jaguar and went up the steps to the front door. He reached for the bell before he noticed with surprise that the door was ajar. He pushed it open and walked into a small sitting-room furnished with feminine taste.

The room had been hastily ransacked, one of the curtains was spread on the floor and on it were piled books and

36

ornaments. Pictures had been taken down from the walls and stacked ready for removal.

Johnny picked up one of the books, and opened the cover. On the fly leaf was a handwritten name. 'Tracey van der Byl.' He dropped it back on the pile as he heard footsteps on the stairs from the floor above.

A man came down the stairs. He was dressed in soiled green velvet trousers, sheepskin boots, and shabby frock coat of military cut frogged with tarnished gold braid. He was carrying an armful of women's dresses.

He saw Johnny and stopped nervously, his pink lips opened in vacant surprise but his eyes were beady and bright under the thatch of lank blond hair.

'Hello,' Johnny smiled pleasantly. 'Are you moving out?' And he drifted quietly closer to the man on the stairs and stood looking up at him.

Suddenly from the floor above a low wail echoed down the stairs. It was an eerie sound, without passion or pain, as though steam were escaping from a jet, only just recognizable as human. Johnny went rigid at the sound, and the man on the stairs glanced nervously over his shoulder.

'What have you done to her?' Johnny asked softly, without menace.

'No. Nothing! She's on a trip. A bad trip.' The man's denial was frantic. 'It's her first time on acid.'

'So you're cleaning the place out, are you?' Johnny asked mildly.

'She owes me plenty. She can't pay. She promised – and she can't pay.'

'Oh,' said Johnny. 'That's different. I thought you were hitting the place.' He reached into his overcoat and brought out his wallet, riffling the wad of banknotes. 'I'm a friend of hers. How much does she owe you?'

'Fifty nicker.' The man's eyes sparkled when he saw the wallet. 'I gave her credit.'

Johnny counted off ten fivers, and held them out. The

man dropped the bundle of clothing over the banisters and came eagerly down the last few stairs.

'Did you sell her the stuff – the acid?' Johnny asked, and the man stopped a pace from him, his expression stiffening with suspicion.

'Oh, for God's sake.' Johnny grinned. 'We are not children – I know the score.' He offered the banknotes. 'Did you get the stuff for her?'

The man grinned back at him weakly, and nodded as he reached for the money. Johnny's free hand snapped closed on the thin wrist and he swung him off his feet, forcing his wrist up between his shoulder-blades.

Johnny stuffed the money into his pocket, and marched him up the stairs.

'Let's go and have a look, shall we?'

There was a mattress on the iron bedstead covered with a grey army blanket. Tracey sat cross-legged on the blanket. She wore only a thin cotton slip and her hair hung lank and lustreless to her waist. Her arms crossed over her chest were thin and white as sticks of chalk. Her face also was pale, the skin translucent in the light of the electric bulb. She was rocking gently back and forth and wailing softly, her breath steaming in the icy cold room.

It was her eyes that shocked Johnny the most. They seemed to have expanded to an enormous size, and beneath each was a dark bruised-looking smear. The pupils of the eyes were distended and glittery with the same adamant sheen as uncut diamonds.

The big glittery green eyes fastened on Johnny and the man in the doorway, and the wail rose abruptly to a shriek. The shriek died away, and she bowed forward and buried her face in her hands, covering her eyes.

'Tracey,' said Johnny softly. 'Oh God, Tracey!'

'She'll be all right,' the man whimpered and twisted in his grip. 'It's the first time – she'll be all right.'

'Come.' Johnny dragged him out of the room, and pushed

the door closed with his foot. He held him against the wall, and his face was set and pale, his eyes merciless – but he spoke quietly, patiently as though he was explaining to a child.

'I'm going to hurt you now. I'm going to hurt you very badly. Just as badly as I can without killing you. Not because I enjoy it, but because that girl is a very special person to me. In the future when you think about giving poison to another girl – I want you to remember what I did to you tonight.'

Johnny held him with his left hand against the wall and he used his right hand, punching up under the ribs at an angle to tear the stomach muscles. With three or four blows he was too high, and he felt ribs crack and snap under his fist.

When he stepped back the man sagged slowly face forward, and Johnny caught him cleanly in the mouth snapping his teeth off at the gums, splitting his lips open like the petals of a rose. The man had made a lot of noise. Johnny looked into Tracey's room to make sure she had not been disturbed, but she was still bowed forward, rocking rhythmically on her haunches.

He found the bathroom and dampened his handkerchief to wipe the blood off his hands and the front of his overcoat. He came out into the passage again and stooped over the unconscious body to check the pulse. It was strong and regular, and he felt a lift of relief as he dragged the man's face out of the puddle of his own vomit and blood to prevent him drowning.

He went through to Tracey and, despite her frantic struggles, wrapped her in the greasy army blanket and carried her down to the Jaguar.

She quietened down and lay like a sleeping child in the back seat while he tucked the blanket round her, then he went back into the house and phoned 999, giving the address and hanging up immediately.

He left Tracey in the car outside the Dorchester, while he went in to speak to the reception clerk. Within minutes Tracey was in a wheelchair on her way up to the two-bedroom suite on the second floor. The doctor was there fifteen minutes later.

After the doctor had gone Johnny bathed, and carrying a tumbler of Chivas Regal in one hand he went into Tracey's room and stood by her bed. Whatever the doctor had given her had put her out cleanly. She lay gaunt and pale – yet with a strangely fragile beauty that seemed enhanced by the bruised discoloration of her eye sockets.

He stooped to brush the hair from her cheek, and her breath was light and warm on his hand. He felt such an infinity of tenderness for her then as he had never known for any other person, he was amazed by the strength of it.

He stooped over her and gently brushed her lips with his own. Her lips were dry and flaky white, and their touch was harsh as sandpaper.

Johnny straightened up and went to the armchair across the room. He sank into it wearily, and sipped the whisky, feeling its warmth spread from his belly and untie the knots in his muscles. He watched the pale ruined face on the pillows.

'We are in a hell of a mess, you and I,' he spoke aloud, and felt anger again. For long minutes it was undirected, but slowly it gelled and found an object to focus on.

For the first time in his life he was angry with the Old Man.

'He has brought you to this,' he said to the girl on the bed. 'And me—'

The reaction was swift, his loyalty was a thing grown part of his existence. Always he had trained himself to believe that the Old Man's machinations were just and wise – even if at times the justice and wisdom were hidden from him. Mortal man does not doubt the omnipotence of his gods.

Sickened by his own treachery, he began to examine the Old Man's motives and actions under the bright light of reason.

Why had the Old Man sent Michael Shapiro to fetch him out of the desert?

'He wants you in Cape Town, Johnny. Benedict didn't measure up. The Old Man has given him the London Office, it's a form of exile. He's picked you to take over the Company,' Michael explained. 'Tracey is out of the way. She and her husband are in London also. I guess the Old Man thinks it's safe to have you back in Cape Town now.'

Michael watched Johnny's undisguised joy and went on slowly.

'I'm speaking out of turn, perhaps. Mr van der Byl is a strange man. He's not like other people. I know how you feel about him, I've watched it all, you know – but listen, Johnny, you can go anywhere on your own now. There are a lot of other companies that want you—' But he had seen the expression on Johnny's face, and stopped ' – okay, Johnny. Forget I ever said it. I only spoke because I like you.'

Thinking on it now, there had been substance in Michael's warning. Certainly he was General Manager of Van Der Byl Diamonds, but he was no nearer to the Old Man than he had ever been. He lived under the mountain but the mountain was remote and he had not been able to scale the lowest slopes.

He had found the city as lonely as the desert, and he was ripe for the first attractive woman who set her snares for him.

Ruby Grange was tall and slim with hair the colour they call 'Second Cape' in a diamond, like sunlight through a crystal glass of champagne.

He wondered now at his own naïvety. That he should be so easily misled, and should have rushed so headlong into her web. After the wedding she had revealed herself,

41

exposing the deeply calculating greed, the driving hunger for flattery and material possessions which was her main-spring, and her complete absorption with herself – Johnny had not been able to believe it. For months he fought off the growing certainty until it could be denied no longer, and he looked with chilled dismay on the shallow selfish little creature he had married.

He had withdrawn from her and flung all his energies into the Company.

This, then, was his life and he saw that it was an empty thing, hollowed out by the Old Man's hand.

For the first time his mind skirted the idea that it was a carefully calculated and sadistic revenge for the innocent action of a half-grown boy.

As though it were an escape from thoughts too dreadful to be borne, he fell asleep in the chair and the glass fell from his hand.

Jacobus Isaac van der Byl sat in a leather chair before the X-ray viewer. Fear had blasted the granite of his features, leaving them cracked and sagging, recognizable but subtly alerted below the gleaming white mane.

Fear was in his eyes also, moving below the surface like slimy water creatures in the pale blue pools. With the fear chilling and numbing his limbs he watched the cloudy and swirling images on the screen.

The specialist was talking softly, impersonally, as though he were lecturing one of his classes.

' – enveloping the thymus here, and extending beyond the trachea.'

The point of his gold pencil followed the ghostly outline on the screen. The Old Man swallowed with an effort. It seemed to be swelling in his throat as he listened, and his voice was hoarse and blurred to his own ears.

'You will operate?' he asked, and the specialist paused in his explanation. He glanced at the surgeon across the desk. The exchange was as guilty as that of conspirators.

The Old Man swivelled his chair and faced the surgeon.

'Well?' he demanded harshly.

'No.' The surgeon shook his head apologetically. 'It's too late. If only you had—'

'How long?' The Old Man overrode his explanation.

'Six months, not more.'

'You are certain?'

'Yes.'

The Old Man's chin sank on to his chest and he closed his eyes. There was complete silence in the room, they watched him with professional pity and interest as he reached his own personal acceptance of death.

At last the Old Man opened his eyes and stood up slowly. He tried to smile but his lips would not hold the shape.

'Thank you, gentlemen,' he croaked in this new rough voice. 'Will you excuse me, please. There are many things to arrange now.'

He went down to where the Rolls waited at the entrance. He walked slowly, shuffling his feet and the chauffeur came to him quickly, but the Old Man shrugged away his helping hands and climbed into the back seat of the car.

Michael Shapiro was waiting for him in the study of the big house. He saw the change in him immediately and jumped up from his chair. The Old Man stood in the doorway, his body seemed to have shrunk.

'Six months,' he said. 'They give me six months.' He said it as though he had expected to buy off death, and they had tricked him. He closed his eyes again, and when he opened them there was a glint of cunning in them, even his face had a pinched foxy look to it.

'Where is he? Is he back yet?'

'Yes, the Boeing got in at nine this morning. He's at the

43

office now.' Michael was shocked, it was the first time he had seen the Old Man without the mask.

'And the girl?' He had not called her 'daughter' since the divorce.

'Johnny has her in a private nursing home.'

'Worthless slut,' said the Old Man softly, and Michael stilled the protest before it reached his lips. 'Get your pad. I want you to take something down.' The Old Man chuckled hoarsely. 'We'll see!' he said, making it sound like a threat. 'We'll see!'

Johnny's doctor was waiting at Cape Town airport.

'Take her, Robin. Dry her out, and fatten her up. She's up to her gills with drugs and she probably hasn't eaten for a month.'

Tracey showed her first spark of spirit.

'Where do you think—'

'Into a nursing home.' Johnny anticipated her questions. 'For as long as is necessary.'

'I'm not—'

'Oh, yes, you bloody well are.' He took her arm, and Robin grabbed the other. They walked her, still protesting weakly to the car park.

'Thanks, Robin Old Soldier, give her the full workout.'

'I'll send her back to you like new,' Robin promised and drove away. Johnny took a few moments to look at the massive square silhouette of the mountain – his own private home-coming ceremony. Then he fetched the Mercedes from the airport garage, and hesitated between home or the office, decided he was not up to an interrogation from Ruby and chose the office. He kept a clean shirt and shaving tackle in his private bathroom there.

They descended on him like a tribe of man-eating Amazons as he came in through the glass doors into the lusciously furnished and carpeted reception area of Van Der Byl Diamonds head office.

The two pretty little receptionists began yipping joyously in chorus.

'Oh, Mr Lance, I have a whole sheaf of messages—'

'Oh, Mr Lance, your wife—'

Trying not to run he made it to within ten feet of his own door, when the Old Man's secretary popped out of ambush from behind her frosted-glass panel.

'Mr Lance, where on earth have you been? Mr van der Byl has been asking—'

Which alerted Lettie Pienaar, his own secretary.

'Mr Lance, thank goodness you're back.'

Johnny stopped and held up his hands in an attitude of surrender.

'One at a time, ladies. There is enough to go round – don't panic.'

Which broke the reception team into a quivering jelly of giggles, and sent the Old Man's watchdog back behind her panel sniffing disgustedly.

'Which is the most important, Lettie?' he asked as he went to his desk and flipped through his mail, shrugged out of his coat and began stripping tie and shirt as he headed for his bathroom.

They shouted at each other through the open door of the bathroom, as Johnny shaved quickly and showered, Lettie bringing him up to date on every aspect of Company and domestic business.

'Mrs Lance has phoned regularly. She called me a liar when I told her you were at Cartridge Bay.' Lettie was silent a moment, then as Johnny came out of the bathroom she asked, 'By the way, where have you been?'

'Don't you start that.' Johnny stood over the desk, and began flipping through the accumulated papers. 'Get my wife on the phone, please – no, hold it. Tell her I'll be home at seven.'

Lettie saw she had lost his attention, and she stood and went out. Johnny settled down behind his desk.

Van Der Byl Diamonds was a sick company. Despite Johnny's protests the Old Man had been drawing off its reserves and feeding them into his other ventures – the property-developing company, the clothing factory, Van Der Byl fisheries, the big irrigation scheme on the Orange River – and now the cupboard was almost bare.

The beach concessions were reaching the end of a short but glorious life. They were starting to work break-even ground. The Old Man had sold the Huib Hoch concession to the big Company for a quick profit – but the profit had been just as quickly transferred out of Johnny's control.

There was only one really fat goose left in his pen, and it wasn't laying eggs yet.

Eighteen months earlier Johnny had purchased two offshore diamond grounds from a company which had died in attempting to work them. It had been strangled by its own inefficiency.

Taking diamonds from the sea is about eight times more expensive than working them from a dry opencast. One must dredge the gravel from the wild and unpredictable waters of the Skeleton Coast, load it into dumb barges, tow the barges to a safe base, off-load it and then begin the recovery process – or rather, that was the method the defunct company had attempted.

Johnny had dreamed up, and then ordered a vessel which was completely self-contained. It could lie out at sea, suck up the gravel and process it, spilling the waste gravel back into the sea as rapidly as it was sucked aboard. It was fitted with a sophisticated recovery plant that was completely computerized and contained within the ocean-going hull. It needed only a small crew, and it could work in all weather conditions short of a full tornado.

The *Kingfisher* was lying at Portsmouth dockyards rapidly

nearing completion. Her trials were scheduled for early August.

Financing the building of this vessel had been a nightmare for Johnny. The Old Man had been unhelpful, when he wasn't being downright obstructive. He never discussed the venture without that little smile twitching at his lips. He had restricted Van Der Byl Diamonds' monetary involvement in the project so severely that Johnny had been forced to raise two millions outside the company.

He had found the money, and the Old Man had smiled again.

Kingfisher should have been lying on the grounds three months ago, sucking up diamonds. The whole financial structure of the scheme was based on her completion on schedule, but *Kingfisher* was running six months behind and now the foundations were shivering.

Sitting at his desk Johnny was working out how to shore up the whole edifice and keep it from collapsing before he could get *Kingfisher* working. The creditors were rumbling and creaking, and Johnny had only his own enthusiasm and reputation left to keep them quiet.

Now he must ask them to defer their interest payments for another three months. He picked up the telephone.

'Get me Mr Larsen at Credit Finance,' he said, steeling himself as he did so, jutting out his jaw and thrusting one bunched fist into his jacket pocket.

At five o'clock he stood up from his desk and went to the cabinet. He poured three fingers of whisky and went back to lower himself wearily into his swivel chair. He felt no elation at having won another reprieve, he was too tired.

The unlisted telephone on his desk rang and he picked it up.

'Lance,' he said.

'How was London?' he recognized the voice instantly, feeling no surprise that the Old Man knew about his

47

journey. The Old Man knew everything. Before he could answer the hoarse croak came again, 'Come up to the house – now!' And the receiver clicked dead.

Johnny looked at the whisky in his hand regretfully and set it down untouched. The Old Man would smell it and smile.

C loud was blowing over the mountain, and the setting sun turned it to the colour of tangerine and peaches. The Old Man stood at the window and watched the cloud cascade down into the valley, dispersing as it fell.

He turned from the window as Johnny entered the study and instantly Johnny was aware that something momentous had taken place in his absence.

He glanced quickly at Michael Shapiro for a cue, but Michael's grey-streaked head was bowed over the papers he held on his lap.

'Good evening.' Johnny addressed the Old Man.

'Sit there.' The Old Man indicated the Spanish leather chair opposite his desk.

'Read it,' the Old Man ordered Michael, and Michael cleared his throat and patted the papers into a neat square before he began.

The Old Man sat with his eyes on Johnny's face. It was a candid, intimate scrutiny, but Johnny felt no discomfort under it. It was almost as though the Old Man's eyes were caressing him.

Mike Shapiro read intelligently, bringing out the meaning of the involved and convoluted legal phrases. The document was the Old Man's Last Will and Testament, and it took twenty minutes for Mike to complete the reading of it. When he had finished there was silence in the room, and the Old Man broke it at last.

'Do you understand?' he asked. There was a gentleness about him that there had never been before. He seemed to have shrunk, the flesh withering on his bones and leaving them dry and light – like the sun-dried bones of a long dead seabird.

'Yes, I understand.' Johnny nodded.

'Explain it to us simply, not in your lawyer's gobbledy-gook, just to be certain,' the Old Man insisted, and Mike began to speak.

'Mr van der Byl's private estate, with the exception of his shares in Van Der Byl Diamond Co. Ltd, after taxes and expenses, is placed in Trust for his two children, Tracey—'

The Old Man interrupted impatiently, swatting Mike's words out of the air as though they were flies.

'Not that. The Company. Tell him about the shares in the Company.'

'Mr van der Byl's shares in the Company are to be divided equally between you and the two van der Byl children, Tracey—'

Again the Old Man interrupted.

'He knows their names, dammit.'

It was the first time ever that either of them had heard him swear. Mike grinned ruefully at Johnny, as though asking for his sympathy, but Johnny was intent on the Old Man, studying his face, feeling the deep satisfying thrill swelling within his chest.

A third share in Van Der Byl Diamonds was no great fortune – nobody knew that better than Johnny.

However, by placing Johnny's name on the list with Tracey and Benedict – he had made him his own. This was what he had worked for all these years. The declaration was public, an acknowledgement to the world.

Johnny Lance had a father at last. He wanted to reach out and touch the Old Man. His chest felt swollen, tight with emotion. Behind his eyelids was a slow soft burning. Johnny blinked.

'This is—' His voice was ragged, and he coughed. 'I just don't know how to tell you—'

The Old Man interrupted him impatiently, silencing him with an imperious gesture, and he croaked at Mike.

'Now read him the codicil to the Will. No, don't read it. Explain it to him.'

Michael's expression changed; he looked down at his papers as he spoke, as though reluctant to meet Johnny's eyes. He cleared his throat unnecessarily and shifted in his seat.

'"By the codicil to the Will, dated the same date, and duly signed by Mr van der Byl the bequest of shares in Van Der Byl Diamond Co. Ltd to JOHN RIGBY LANCE, is made conditional on the issue by the said JOHN RIGBY LANCE of a personal guarantee for the debts of the company, including the present loan account and amounts outstanding to tributary companies for royalties and options."'

'Christ,' said Johnny, stiffening in his chair and turning to stare incredulously at the Old Man. The tightness in his chest was gone. 'What are you trying to do to me?'

The Old Man dismissed Mike Shapiro quietly, without even looking at him. 'I'll call you when I want you.' And when he had gone he repeated Johnny's question.

'What am I trying to do to you?' he asked. 'I am trying to make you responsible for debts totalling about two and a half million Rand.'

'No creditor would come to me for half a million, I would be hard pressed to raise ten thousand on my personal account.' Johnny shook his head irritably, the whole thing was nonsensical.

'There is one creditor who could come to you, and subject you to the full process of law. Not to receive payment in cash – but in personal satisfaction. He would smash you – and delight in doing so.'

Johnny's eyes narrowed disbelievingly. 'Benedict?'

The Old Man nodded. 'For once Benedict will hold the top cards. He won't be able to dislodge you from the management of the company, because Tracey will support you as she always has done – but he will be able to watch every move you make from his seat on the Board of Directors. He will be able to hound you, bring you and the Company down without suffering financial loss himself. And when you fall – you know better than to expect mercy from him. You will be devoured by the ogre you have created.'

'Created?' Johnny's voice was shocked. 'What do you mean?'

'You turned him into what he is now. You broke his heart, made him weak and useless—'

'You are crazy.' Johnny came to his feet. 'I have never done anything to Benedict. It was he who—'

But the Old Man's husky croak brushed aside Johnny's protests. 'He tried to run with you – but could not. He gave up, became small and vicious. Oh, I know about the way he is – how you made him.'

'Please, listen to me. I did not—'

But the Old Man went on remorselessly. 'Tracey also, you have ruined her life. You enslaved her, in your sin—'

'That night!' Johnny shouted at him. 'You never let me explain. We never—'

Now the Old Man's voice was a whiplash.

'Silence!' And Johnny could not defy him, the habit was too deeply engrained. The Old Man was trembling, his eyes glittering with passion. 'Both my children! You have plagued me and my family. My son is a weak-willed drifter, trying to hide his hurt in a hunt for pleasure. I have given him the instruments to destroy you, and when he does so perhaps he will become a man.'

The Old Man's voice was strained now, rusty and pain-racked. He swallowed with an effort, his throat convulsing – but there was no softening of the glitter in his eyes.

'My daughter also, tortured by her lust. A lust which you awakened – she also seeks an escape from her guilty passion. Your destruction will be her release.'

'You're wrong,' Johnny cried out, half in protest, half in entreaty. 'Please, let me explain—'

'This is how it will work. I have made you vulnerable, linking you to a crippled and foundering enterprise. This time we will be rid of you.' He stopped to pant quickly, like a running dog. His breathing was strangled, harsh-sounding. 'Benedict will cut you down, and Tracey will have to watch you go. She cannot help you, her inheritance is carefully tied up, she has no control of the capital. Your only hope is the *Kingfisher*. The *Kingfisher* will turn into a vampire and suck your life blood! You asked why I was systematically transferring the assets of Van Der Byl Diamonds to my other companies? Well, now you know the answer.'

Johnny's lips moved. He was very pale. His voice came out small and whispery.

'I could refuse to sign the guarantee.'

The Old Man smiled bleakly, a drawing back of the lips that was without warmth or humour.

'You will sign it.' His voice was wheezy. 'Your pride and conceit will not let you do otherwise. You see, I know you. I've studied you all these years. But if you refuse to sign the guarantee, I will still have smashed you. Your shares will go to Benedict. You will be out. Out! Gone! We will be finished with you at last.' Then his voice dropped, 'But you will sign. I know it.'

Involuntarily Johnny lifted his hands towards the Old Man, a gesture of supplication. 'In all this time. When I stayed with you, when I—' His voice went husky and dried up. 'Did you never feel anything for me – anything at all?'

The Old Man sat up in his chair. He seemed to regain his bulk and he began to smile. He spoke quietly now, he did not have to shout.

'Get out of my nest, Cuckoo. Get out and fly!' he said.

Slowly Johnny's expression changed, the line of his jaw hardened, thrusting out aggressively. His shoulders went back. He pushed his hands into his pockets, balling his fists into bony hammers.

He nodded once in understanding.

'I see.' He nodded again, and then he started to grin. It was an unconvincing grin, that twisted his mouth out of shape and left his eyes dark and haunted.

'All right, you mean old bastard, I'll show you.'

He turned and walked from the room without looking back.

The Old Man's expression lit in deep satisfaction. He chuckled, then his breath caught. He began to cough, and the pain ripped his throat with a violence that left him clinging weakly to the edge of his desk.

He felt the crab of death move within his flesh, sinking its claws more deeply into his throat and lungs – and he was afraid.

He called out in his pain and fear, but there was nobody in the old house to hear him.

*K*ingfisher was launched in August and ran her trials in the North Sea. Benedict was aboard, by the Old Man's express command. With a vessel of such complexity, and of such revolutionary design, it would have been a miracle had she functioned perfectly. August that year was not the month for miracles. At the end of the trials Johnny had compiled a list of twenty-three modifications that were necessary.

'How long?' he asked the representative of the shipyard.

'A month.' The reply was hesitant.

'You mean two,' said Benedict and laughed out loud. Johnny looked at him thoughtfully, he guessed that the Old Man had spoken to him.

'I'll tell you something, Johnny.' Benedict was still laughing. 'I'm glad this cow isn't my dream of paradise.'

Johnny froze. Those words were the Old Man's, repeated parrot fashion. It was all the confirmation that he needed.

Johnny flew back to Cape Town to find his creditors on the verge of mutiny. They wanted to sell out, and take the loss.

Johnny spent two whole precious days on Larsen's wine farm at Stellenbosch calming his fears. When Fifi Larsen, twenty years younger than her husband, squeezed Johnny's thigh under the lunch table he knew it would be all right – for another two months.

During the next hectic, strength-sapping week of argument and negotiation, Johnny made time to see Tracey.

She had been out of the nursing-home for a month now, staying with friends on a small farm near Somerset West.

When Johnny climbed out of the Mercedes, and Tracey came down from the stoop to greet him, he had his first real lift of pleasure in a long time.

'God,' he said. 'You look great.'

She was dressed in a cotton summer dress with open sandals on her feet. Her friends were away for the day, so they walked through the orchards. He studied her openly, noticing how her cheeks and arms had filled out and the colour had come back into them. Her hair was bright and springing with lights in the sun, but there were still the dark smears under her eyes, and she smiled only once when he picked a sprig of peach blossoms for her. She seemed to be afraid of him, and unsure of herself.

At last he faced her and placed his hands on her shoulders. 'All right. What's eating you?'

It came out in a quick staccato rush of words.

'I want to thank you for coming to find me. I want to explain why I was – like that. In that state. I don't want you to believe – well, bad things about me.'

'Tracey, you don't have to explain to me.'

'I want to. I must.' And she told him, not looking at his face, twisting and tearing the blossoms in her hands.

'You see, I didn't understand, I thought all men were like that. Not wanting, I mean not doing it—' She broke off, and started again. 'He was kind, you understand. And there were lots of parties and friends around all the time, every night. Then he wanted to go to London – for his career. There was not enough scope here. Even then I didn't know. Well, I knew he had lots of men friends and that some of them were different – but . . . Then I went to his studio and found them, and they laughed, Kenny and the boy twined together like snakes. "But you must have known," he said. Something just snapped in my head, I felt spoiled, dirty and horrible and I wanted to die. There was nobody to go to and I didn't want anybody – I just wanted to die.' She stopped and stood waiting for him to speak.

'Do you still want to die?' he asked gently, and she looked up startled and shook her shining hair.

'I don't want you to die either.' And suddenly they were both laughing. After that it was good between them and they talked with all the strangeness gone until it was almost dark.

'I must go,' Johnny said.

'Your wife?' she asked, the laughter fading.

'Yes. My wife.'

It was dark when Johnny went in through the front door of the new split-level ranch type in Bishopscourt which was his house but not his home; the telephone was ringing. He picked up the receiver.

'Johnny?'

'Hello Michael.' He recognized the voice.

'Johnny, get up here to the old house right away.'

Michael Shapiro's voice was strained.

'Is it the Old Man?' Johnny asked anxiously.

'No talk – just come, quickly!'

T he curtains were drawn, and a log fire roared on the stone hearth. But the Old Man was cold. The coldness was deep inside him where the flames could not warm it. His hands shook as he picked sheets of paper from the open document box, glanced at them and then dropped them into the fire. They exploded into orange flame, then curled and blackened to ash. At last the box was empty but for a thick wad of multi-coloured envelopes bound together with a ribbon. He loosened the knot, picked out the first envelope, and slipped from it a single sheet of writing-paper.

'Dear Sir,
I hope you will be pleased to hear that I am now at school. The food is good but the beds are very hard—'

He dropped both envelope and letter on the fire and selected another. One at a time he read and then burned them.

'—that I have been selected to play for the first fifteen—'

Sometimes he smiled, once he chuckled.

'—I was top in all subjects except history and religious teaching. I hope to do better next—'

When there was one envelope left he held it a long time in his blue-veined bony hands. Then with an impatient flick of his wrist he threw it on to the fire and reached up to the mantelpiece to pull himself to his feet. As he stood he looked into the gilt-framed mirror above the fireplace.

He stared at his reflection, mildly surprised by the change that the last few weeks had wrought in his appearance. His eyes had lost the sparkle of life, fading to a pale dirty brownish blue – the colour of putrefaction. They bulged from the sockets, in the glassy startled stare which is peculiar to the later stages of terminal cancer.

The watery feeling of limb, and the coldness were not the result of the pain-killing drugs, he knew. Nor was the shuffling feet-dragging gait with which he crossed the thick Bokhara carpet to the stinkwood desk.

He looked down at the oblong leather case with its brass-bound corners, and he coughed, a single flesh-tearing bark. He caught at the desk to steady himself, waiting for the pain to pass before he sprang the catch on the case and laid the lid back.

His hands were quite steady as he took the barrel and butt section of the Purdy Royal twelve-bore shotgun from the case and fitted them together.

He died the way he had lived – alone.

'God, how I hate black.' Ruby Lance stood in the centre of the bedroom floor, staring at the clothing laid out on the double bed. 'It makes me look so washed out.'

She swung her head from side to side, setting the champagne-coloured cascade of her hair swinging. She turned and moved lazily across the room to the tall mirrors. She smiled at herself, a languid slanting of the eyes, and then she spoke over the shoulder of her own image.

'You say that Benedict van der Byl has arrived from England?'

'Yes,' Johnny nodded. He sat slumped in the chair beside his dressing-room door, pressing his fingers into his eyes.

Ruby came up on her toes, pulling in her stomach and pushing forward her small hard breasts.

'Who else will be there?' she asked, cupping her hands under her breasts and squeezing out the nipples between thumb and forefinger, inspecting them critically. Johnny took his hand from his eyes.

'Did you hear me?' Ruby's voice took on a sharp admonishing note. 'I'm not talking to myself, you know.'

She turned away from the mirror to face him. Standing long and slim and golden as a leopard, even her eyes had the yellow intentness of a leopard's stare. She gave the impression that at any moment she would draw her lips back in a snarl.

'It's a funeral,' he said quietly. 'Not a cocktail party.'

'Well, you can't expect me to die of sorrow. I couldn't stand him.' She crossed to the bed and picked up the pair of peach-coloured panties and rubbed the glossy material against her cheek. Then she stepped into them with two long-legged strides.

'At least I can wear something pretty under the weeds.' She snapped the elastic against her sun-gilt belly, and the almost colourless blonde curls were flattened beneath the sheer silk.

Johnny stood up slowly, and went into his dressing-room. Scornfully she called after him. 'Oh for God's sake, Johnny Lance, stop dragging that long face around as though it's the end of the world. Nobody owes that old devil a thing – he collected all his debts long before they fell due.'

They were a few minutes early, and they stood together beneath the pine-trees outside the entrance to the chapel.

When the pearl-grey Rolls drew up at the gate and brother and sister stepped down and came up the paved path, Ruby could not contain her interest.

'Is that Benedict van der Byl?'

Johnny nodded.

'He's very good-looking.'

But Johnny was looking at Tracey. The change in her appearance since he had last seen her was startling. She walked like a desert girl again, straight and proud. She came directly to Johnny and stopped in front of him. She removed her dark glasses, and he could see she had been weeping, for her eyes were slightly puffy. She wore no make-up, and with the dark scarf framing her face she looked like a nun. The marks that sorrow had left gave her face maturity.

'I did not think this day would ever come,' she said softly.

'No,' Johnny agreed. 'It was as though he would live for ever.'

Tracey moved a step closer to him, she reached out as if to touch Johnny's arm but her fingers stopped within inches of his sleeve. Johnny understood the gesture, it was a sharing of sorrow, an understanding of mutual loss, and an unstated offer of comfort.

'I don't think we have met.' Ruby used her sugar and arsenic tone. 'It is Miss van der Byl, isn't it?'

Tracey turned her head and her expression went flat and neutral. She replaced the dark glasses, masking her eyes.

'Mrs Hartford,' she said. 'How do you do.'

Mike Shapiro stood beside Johnny in the pew. He spoke without moving his lips, just loud enough for Johnny to catch the words.

'Benedict knows the conditions of the Will. You can expect his first move immediately.'

'Thanks, Mike.'

Johnny kept his eyes on the massive black coffin. The candlelight glinted and sparkled on the elaborate silver handles.

As yet he could find no interest for the conflict that lay

ahead. That would come. Now he was too deeply involved in the passing of an era, his life had reached another point of major departure. He knew it would change, had already changed.

He looked across the aisle suddenly, his gaze drawn intuitively.

Benedict van der Byl was watching him, and at that moment the priest asked for the pallbearers.

They went to stand beside the coffin, Benedict and Johnny on opposite sides of the polished black casket among the massed display of arum lilies. They watched each other warily. It seemed to Johnny that the whole scene was significant. The two of them standing over the Old Man's corpse, facing each other, with Tracey looking on anxiously.

Johnny glanced back into the body of the church, looking for Tracey. Instead he found Ruby. She was watching them both, and Johnny knew suddenly that the board had changed more than he realized. A new piece had been added to the game.

He felt Mike Shapiro nudge him, and he stooped forward and grasped the silver handle. Between them they carried the Old Man out into the sunshine.

The handle had cut into his palm with the weight of the coffin. He went on massaging it, even after the coffin had gone down into the pit. The crude mounds of fresh earth were covered with blankets of bright green artificial grass. The mourners began to drift away, but Johnny went on standing there bare-headed. Until Ruby came to touch his arm.

'Come on.' Her voice pitched low, but stinging. 'You're making a fool of yourself.'

Benedict and Tracey were waiting under the pine-trees by the churchyard gate, shaking hands and talking quietly to the departing mourners.

'You are Ruby, of course.' Benedict took her hand, smiling a little, urbane and charming. 'The flattering reports I've had of you hardly do you justice.' And Ruby glowed, seeming like a butterfly to spread her wings to the sun.

'Johnny.' Benedict turned to him, and Johnny was taken off balance by the friendly warmth of his smile and the grip of his hand. 'Michael Shapiro tells me that you have accepted my father's legacy and the conditions attached to it – you have signed the guarantee. It's wonderful news. I don't know what we would have done without you in Van Der Byl Diamonds. You are the only one that can pull the Company though this difficult period. I want you to know I am behind you all the way, Johnny. I intend becoming much more involved with the Company now, giving you all the help you need.'

'I knew I could depend on you, Benedict.' Johnny accepted the challenge as smoothly as it was thrown down. 'I think everything is going to turn out all right.'

'We have a meeting on Monday, then I must return to London on Thursday – but I hope you can have dinner with me before then – you and your lovely wife, of course.'

'Thank you.' Ruby seeing the refusal on Johnny's lips, interrupted quickly. 'We'd enjoy that.'

'Y ou were going to refuse, weren't you?' She sat with her legs curled up sideways under her, watching him from the passenger seat of the Mercedes with the slanting eyes of a Persian cat.

'You're damn right.' Johnny nodded grimly.

'Why?'

'Benedict van der Byl is poison.'

'You say so.'

'Yes, I say so.'

'Could be you're jealous of him.' Ruby lit one of her gold-tipped cigarettes, putting the smoke through her lips.

'Good God!' Johnny gave one harsh snort of laughter, then they were silent awhile, both staring ahead.

'I think he's pretty dreamy.'

'You can have him.' Johnny's tone was disinterested, but her retort was shrill.

'I could too – if I wanted to. Anyway you and that Tracey creature mooning—'

'Cut it out, Ruby.'

'Oh my, I've said the wrong thing. The precious Mrs Hartford—'

'Cut it out, I said.' Johnny's tone was sharp.

'Little Miss Fancy Pants – God! She almost had them down for you in the bloody graveyard—'

'Shut up, damn you.'

'Don't you swear at me.' And she lashed out at him flat-handed, leaning forward across the seat to strike him in the mouth. His lower lip smeared against his teeth, and the taste of blood seeped into his mouth. He took the handkerchief from his breast pocket and held it to his mouth, steering the Mercedes with one hand.

Ruby sat curled in her corner of the seat, puffing quickly at the cigarette. Neither of them spoke again until he drew up in front of the double garage. Then Ruby slipped out of the Mercedes and ran across the lawns to the front door. She slammed it behind her, with a force that rattled the full-length glass panel.

Johnny parked the Mercedes, closed the garage door and followed her slowly into the house. She had kicked off her shoes on the wall-to-wall carpet in the lounge, and run through on to the patio beside the swimming pool. She stood barefooted staring down into the clear water, hugging herself about the shoulders.

'Ruby.' He came up behind her, forcing the anger out of

his voice with an effort, trying to keep it conciliatory. 'Listen to me—'

She spun around to face him, eyes blazing like a cornered leopard.

'Don't try and gentle talk me, you bastard. What do you think I am – your damned servant. When did I last get to do anything I wanted?'

With Ruby he had long ago realized that placation was the short cut to peace, so he was roused by the implication.

'I've never stopped you from—'

'Good! That's just fine! Then you won't stop me going away.'

'What do you mean?' He was caught between shock and a sneaking sense of hope. 'Are you talking about divorce—'

'Divorce? Are you out of your little mind! I know all about the big bagful of goodies the Old Man left you in his will. Well, little Ruby is getting her pinkies into that bag – starting right now.'

'What do you want exactly?' His voice was cold and flat.

'A new wardrobe, and a quick whip around all those nice places you go to all the time – London, Paris and the rest. That will do for a start.'

He thought a moment, assessing how far he could stretch his overdraft; since his marriage his bank statement had seldom been typed in black. It was worth it, he decided. He could afford no distraction over the next few months. He could move faster and think quicker without having Ruby Lance sitting between his shoulder-blades – much better she should go.

'All right,' he nodded. 'If that's what you want.'

Her eyes narrowed slightly and her mouth pinched in as she studied his face.

'That was too easy,' she said. 'You want to get rid of me? Don't get any ideas, Johnny boy, you put one finger – or anything else – out of line and I'll chop it off.'

'There is a Mrs Hartford to see you, sir.' Lettie Pienaar's voice whispered through on the intercom, then just audibly she added, 'Lucky you!'

Johnny grinned. 'You're fired for insolence – but send her in before you go.'

He stood up as Tracey came in, and went around his desk to meet her. She wore a no-nonsense grey suit, with her hair scraped back from her face. She should have looked like a school marm – but she didn't.

'You've got your times mixed up, Tracey. The Directors' meeting is at two this afternoon.'

'That's a sweet greeting.' She sat down in an egg-shaped swivel chair, crossing long legs which Johnny dragged his eyes off with an effort. 'I've come looking for a job.'

'A job?' He stared at her blankly.

'Yes, a job. You know – work? Employment?'

'What on earth for?'

'Well, now that you've dragged me back from the bright lights with all the finesse of a caveman – you don't expect me to sit around until I drop dead of boredom. Besides, your tame doctor feels that good healthy employment is essential to the completion of my – ah – cure.'

'I see.' He sank back into his own chair. 'Well – what can you do?'

'Mr Lance.' Tracey widened her eyes suggestively, but made her voice prim. ' – Really!'

'All right,' Johnny chuckled. 'What are your qualifications?'

'You may or may not know that I have a law degree from the University of Cape Town.'

'I didn't.'

'Also, it occurred to me that during the next few months you might need someone around whom you can trust.' She was serious now, and Johnny's smile faded also. 'Like the old days.' She added quietly. They were silent for a few seconds.

'It just so happens that we are looking for a personal assistant in our legal department,' Johnny murmured, and then softly, 'Thanks, Tracey.'

The Board Room of Van Der Byl Diamonds was furnished in soft forest colours, browns and greens. A long luxurious room that reflected the opulence of the days when the Company had been glutted with capital. But now the air was charged with a tension that crackled in the air like static electricity.

The subject of debate was the diamond recovery vessel, *Kingfisher*. The Company's last hope. Her only substantial asset, and Johnny's personal cross.

'This vessel should have been in operation nine months ago. All the estimates were based on that assumption – yet, she is still lying awaiting completion on the slips at Portsmouth.' Benedict was speaking with unconcealed relish. 'In consequence, the interest charges that are accruing put us in a position—'

'The shipyard was out on strike for a total of four months during construction – in addition they were working to rule for—' Johnny's jaw was thrust out, he was ready to fight.

'Ah! I don't think we are particularly interested in the unpredictability of the British workman – the contract should have gone to the Japanese company. Their tender was lower—'

'It would have,' grated Johnny, 'if your father had not insisted—'

'Please, let us not attempt to lay the blame at the door of a dead man.' Benedict's tone was sanctimonious. 'Let us rather try and rectify a grievous situation. When will *Kingfisher* be ready for sea?'

'On the thirteenth of September.'

'It had better be.' Benedict dropped his eyes to his notes.

'Now, this man whom you have engaged to captain the vessel – Sergio Caporetti – let us hear a little about him, please.'

'Fifteen years' experience on offshore oil-drilling vessels in the Red Sea. Three years as Captain of Atlantis Diamonds' offshore dredger operating off the West Coast. He's one of the best, no doubt about it.'

'All right.' Almost reluctantly Benedict accepted this, and consulted his notes. 'Now, we have two sea-concession areas. No. 1 area off Cartridge Bay; No. 2 some twenty miles north of that. Judging by your prospecting results you will elect to work No. 1 area first.'

Johnny nodded, waiting for the next attack to develop. Benedict sat back in his chair.

'Atlantis Diamonds Ltd went broke working our No. 1 area – what makes you think you can succeed where they failed?'

'We've been over this before,' Johnny snapped.

'I wasn't there, remember? Humour me, please. Go over it again.'

Quickly Johnny explained that Atlantis Diamonds' costs had been inflated by their method of operation. Their dredgers were not self-propelled but had to be towed by tugs. The gravel they recovered had been stored, taken into Cartridge Bay in bulk, transhipped ashore to be processed at a land-based plant. *Kingfisher* was a self-propelled and self-contained vessel. She would suck up the gravel, process it through the most sophisticated system of cyclone and X-ray equipment and dump the waste overboard.

'Our costs will be one quarter those of Atlantis Diamonds,' he finished.

'And our loan account is a mere two millions,' Benedict murmured dryly. Then he looked towards Mike Shapiro at the bottom of the table. 'Mr Secretary, please note the following motion – "That this Company proceed to sell the vessel *Kingfisher* presently building at Portsmouth. That it then sells all diamond concessions at the most advantageous terms negotiable, and goes into voluntary liquidation forthwith." Have you got that?'

It was a direct frontal attack. Clearly if the motion succeeded the Company was worthless. They could not recover the price of the *Kingfisher* on a forced sale. There would be a shortfall – and Johnny had signed the guarantee. It was a straight test that Benedict was making. A setting of the lines of battle. Tracey held the balance between Johnny and Benedict. He was forcing her to declare herself.

Benedict watched her while the motion was put to the vote. He leaned forward in the padded leather chair, a slightly amused smile on the full red lips. Beautifully groomed and tailored, with the grace that wealth and position give to a man and which cannot be counterfeited. But the clean athletic lines of his body were fractionally blurred by indulgence, and there was a little too much flesh along the line of his jaw that gave him the petulant look of a spoilt child.

Tracey voted with Johnny Lance, not hesitating a second before lifting her hand. Returning Benedict's smile levelly, and watching her brother's smile alter subtly – become wolfish, for Benedict did not like to lose.

'Very well, my darling sister. Now we know how we stand at least.' He turned easily to Johnny. 'I presume you wish me to continue with my duties in London.'

For years now Benedict had handled the London sales of the Company's stones. It was an unexacting task, which the Old Man had judged within his capabilities.

'Thank you, Benedict,' Johnny nodded. 'Now, I have a proposal to put to the meeting – "That the Directors of this Company, as a gesture of solidarity, agree to waive their rights to Directors' fees until such time as the Company's financial position is on a more sound footing."'

It was a puny counter-attack, but the best he could mount at the moment.

Take-off was in the first light from Youngsfield, and Johnny swung the twin-engined Beechcraft on to a northerly heading, leaving the blue massif of Table Mountain on the left hand.

Tracey wore an anorak over her rose-coloured shirt, and the bottoms of her denim pants were tucked into soft leather boots, her dark hair caught at the nape of her neck with a leather thong.

She sat very still, looking ahead through the aircraft's windshield at the dawn-touched contours of the land ahead. At the bleak lilac and purple mountains and the great lion-coloured plains spreading down to meet the mists that hung over the cold Atlantic.

In her stillness Johnny sensed her excitement and found it infectious.

The sun exploded over the horizon, washing golden and bright over the plains, and tipping the mountains with flame.

'Namaqualand.' Johnny pointed ahead.

She laughed with excitement, like a child at Christmas, turning in the seat to face him.

'Do you remember—' She began, then stopped in confusion.

'Yes,' said Johnny. 'I remember.'

They landed before noon on a rough airstrip bulldozed out of the wilderness. There was a Land-Rover waiting to take them down to the beach to inspect the progress of the workings.

There was little remaining along the thirty-seven-mile Admiralty strip worth working. It was a clean-up and shut-down operation.

When the reigning 'King Canute' handed over the parcel of diamonds that made up the month's recovery, he was apologetic.

'You took out all the plums, Johnny. It's not like the old days.'

Johnny prodded the pathetic pile of small, low-grade stones with his forefinger.

'No, it's not,' he agreed. 'But every little bit helps.'

They climbed back into the Beechcraft and flew on northwards.

Now they passed over areas where the desert had been scratched and torn over wide areas.

The tractors had left centipede tracks in the soft earth.

'Ours?' asked Tracey.

'I wish they were. We'd have no worries then. No, all this belongs to the big Company.'

Johnny checked his watch, automatically comparing the progress of the flight to his estimates. Then he lifted the microphone from the R/T set.

'Alexandra Bay Control. This is Zulu Sugar Peter Tango Baker.'

He knew that they had him on the radar plot, and were watching him – not because they were worried about his safety, but because he was now over the Proclaimed Diamond Area of South West Africa – that vast jealously guarded tract of nothingness.

The radio crackled back at him instantly, demanding his permit number, his flight plan, querying his intentions and his destination.

Having convinced Control of his innocence, and received their permission to continue his flight, he switched off his R/T set, and grimaced at Tracey.

He felt ruffled by this small brush with Olympus. He knew that most of it was professional jealousy. He smarted under the knowledge that he was working ground that the big Company would despise as not sufficiently lucrative to bother about.

Sometimes Johnny dreamed about discovering a fault in a land title, or an error in a survey that had been casually performed seventy years previously before the value of this parched denuded earth had been realized. He imagined himself being able to claim the mineral rights to a few square miles plumb in the middle of the big Company's richest field. He shivered voluptuously at the thought, and Tracey looked at him enquiringly.

He shook his head, then his line of thought took him on to a further destination.

He banked the aircraft, crossing the coastline with its creamy lines of surf running in on the freezing white sands of the beach.

'What?' She was expectant, receptive to the new tone in his voice.

'Thunderbolt and Suicide,' he said, and she made a small grimace of incomprehension.

'There.' He pointed ahead, and through the light smoke of the sea mist she saw them show bare – white and shiny, like a pair of albino whales.

'Islands?' she asked. 'What's so special about them?'

'Their shape,' he answered. 'See how they lie like the mouth of a funnel, with a small opening between them.'

She nodded. The two islands were almost identical twins, two narrow wedges of smooth granite, each about three miles long, lying in a chevron pattern to each other – but not quite meeting at the peak. The mighty Atlantic swells bore up from the south and ran into the mouth of the funnel. Finding themselves trapped in this granite corral, the swells reared up wildly and hurled themselves on the cliffs in massive bomb-bursts of spray before streaming out in white foam through the narrow opening between the two islands.

'I can see how Thunderbolt gets its name.' Tracey eyed the wild booming surf with awe. 'But how about Suicide?'

'The old guano collectors must have called it that, after they tried landing on it.'

'Guano,' Tracey nodded. 'That accounts for the colour.'

Johnny put the Beechcraft into a shallow dive, hurtling in low over the green water. Ahead of them the seabirds rose in alarm, streaming in a long black smear into the sky, the cormorants and gannets whose excreta through the ages had painted the rocks that glaring white.

As they flashed through the gap below the level of the cliffs, Tracey exclaimed, 'There's some sort of tower there – look! In the back of the island.'

'Yes,' Johnny agreed. 'It's an old wooden gantry they used for loading the guano into the longboats.'

He pulled the Beechcraft up in a climbing turn, gaining height to look down on the two islands.

'Do you see where the surf comes through the gap? Now look beneath the surface, can you see the reefs under the water?'

They lay like long dark shadows through the green water, at right angles to the drift of white foam.

'Well, you are looking at the most beautifully designed natural diamond trap in the world.'

'Explain,' Tracey invited.

'Down there,' he pointed south, 'are the big rivers. Some of them dried up a million years ago, but not before they had spat the diamonds they carried into the sea. The tide and the wind has been working them up towards the north for all these ages. Throwing some of them back on the beach but carrying others up this way.'

He levelled the Beechcraft out and resumed their interrupted flight northwards.

'Then suddenly they run up against Thunderbolt and Suicide. They are concentrated and squeezed through the gap, then they are confronted by a series of sharp reefs across their path. They cannot cross them – they just settle down in the gullies and wait for someone to come and suck them out.'

He sighed like a man crossed in love.

'My God, Tracey. The smell of those diamonds reeks in my nostrils. I can almost see the shine of them through a hundred and sixty feet of water.'

He shook himself as though waking from a dream.

'I've been in the game all my life, Tracey. I've got the "feel", the same as a water-diviner has. I tell you with absolute certainty there are millions of carats of diamonds lying in the crotch of Thunderbolt and Suicide.'

'What's the snag?' Tracey asked.

'The concession was granted twenty years ago to the big Company.'

'By whom?'

'The Government of South West Africa.'

'Why aren't they mining it?'

'They will – sometime in the next twenty years. They aren't in any hurry.'

They lapsed into silence, staring ahead, though once Johnny clucked his tongue irritably and shook his head – still thinking about Thunderbolt and Suicide.

To distract him Tracey asked, 'Where do they come from in the first place – the diamonds?'

'Volcanic pipes,' Johnny answered. 'There are more than a hundred known pipes in Southern Africa. Not all yield stones, but then some do. New Rush – Finsch – Dutoitspan – Bulfontein – Premier – Mwadui. Great oval-shaped treasure chests, filled with the legendary "Blue Ground" – the mother lode of the diamond.'

'There are no pipes here – surely?' Tracey turned towards him in his seat.

'No,' Johnny agreed. 'We are after the alluvial stones. Some of those ancient pipes exploded with the force of a hydrogen bomb, spraying diamonds over hundreds of square miles. Others were sub-marine pipes that discharged their treasure into the restless sea. Others of the more passive volcanic pipes were simply eroded away by wind and water and the diamonds were exposed.'

'Then they were washed down to the sea?' she guessed.

Johnny nodded. 'That's right. Over millions of years they

were moved infinitely slowly by landslides, floods, rivers and rainwater. Where all the other pebbles and stones were abraded and worn away to nothingness – the diamonds, four hundred times harder than any other natural substance on earth, were unmarked. So at last they reached the sea and mingled with the others from the sub-marine pipes, to be laid down by wave action on the beaches, or finally to come up against a place like Thunderbolt and Suicide.'

Tracey opened her mouth to ask another question, but Johnny interrupted.

'Here we are. There is Cartridge Bay.' And he pushed the nose of the aircraft down slightly. It was more a lagoon than a bay. Separated from the sea by a narrow sandspit, it spread away into the treeless waste, an enormous extent of quiet shallow water in tranquil contrast to the unchecked surf that burst on the sandspit. There was a deep water entrance through the sandspit, and a channel meandered across the lagoon to where a cluster of lonely whitewashed buildings sprang up on the edge of the desert.

Johnny banked steeply towards the buildings, and below them flocks of black and white pelicans and pink flamingoes rose in panic from the shallows.

Johnny landed and taxied across to the waiting Land-Rover with the white lightning insignia of Van Der Byl Diamonds painted on its side.

Lugging the cool box that contained their lunch, Johnny led Tracey to the vehicle and introduced her to his foreman. Then they climbed in and went bumping down to the buildings on the lagoon. Johnny received from his foreman a report on progress of the work. The buildings had been abandoned by the defunct Atlantis Diamond Company. Johnny was renovating them to serve as a base for the *Kingfisher*; a rest and recreation centre for the crew, a radio centre, a refuelling depot and a workshop to handle running maintenance and repairs. In addition he was putting a jetty out into the lagoon for the converted seventy-foot pilchard trawler that would be *Kingfisher*'s service boat – acting as tender and ferry.

They ran an extensive inspection of the base. Johnny was pleased with the interest Tracey showed, and he enjoyed answering her questions for his own enthusiasm was high. It was nearly two o'clock before they had finished.

'How are the watchtowers coming?' Johnny asked.

'All up, ready and waiting.'

And suddenly Johnny had a two-edged inspiration.

'Might as well go and have a look.' He made it casual.

'Okay, I'll fetch the Land-Rover,' the foreman agreed.

'I know the way.' Johnny put him off. 'You go and get your lunch.'

'It's no trouble—' the foreman began, caught Johnny's frown, cut himself short, then glanced at Tracey. 'Yeah! Sure! Fine! Okay – here are the keys.' He handed Johnny the Land-Rover keys, and disappeared into his own quarters.

Johnny checked the grub box, and they climbed into the open Land-Rover.

'Where are we going?' Tracey asked.

'Inspect the watchtowers along the sandspit.'

'Watchtowers?'

'We've put up a line of fifty-foot wooden towers along the beach. From them we will take continual bearings on *Kingfisher* when she is working offshore. By radio we will be able at any time to give her the exact position over the bottom to within a few feet, as a check to the computer.'

'My, you are clever.' Tracey fluttered her eyelashes at him in mock admiration.

'Silly wench,' said Johnny, and let out the clutch. He swung down past the radio shack on to the hard wet sand at the edge of the lagoon; accelerating he hit second then third and they went away around the curve of the lagoon, headed towards the great yellow wind-carved dunes that lined the coast.

Tracey stood up on her seat, clutching the edge of the windscreen, and the wind snatched at her hair. She pulled the retaining thong from it, and shook it out into a shiny black flag that snapped and snaked behind her.

'Look! Look!' she cried as the flocks of startled flamingoes lurched into flight, streaming white and pink and black over the glossy silver water.

Johnny laughed with her, and swung the Land-Rover towards the dunes.

'Hold on!' he shouted, and she clung to the windscreen, shrieking in delicious terror as they flew up the steep side of a dune, spinning a cloud of sand from the rear wheels and then dropped over the crest in a stomach-churning swoop. They crossed the sandspit and hit the beach, racing along it, playing tag with the waves that shot up the sand.

Five miles up the beach Johnny parked above the high-water mark and they ate cold chicken and drank a bottle of chilled white wine sitting side by side in the sand, leaning against the seat cushions from the Land-Rover. Then they went down to the edge of the sea to wash the chicken grease from their fingers.

'Yipes! It's cold.' Tracey scooped a double handful of sea water. Then she looked at Johnny and her expression became devilish.

He backed away, but not quickly enough. The icy water hit him in the chest, and he gasped.

'War!' It was their childhood cry.

Tracey whirled and went off long-legged along the beach, with Johnny pounding after her. She sensed him gaining on her, and shouted.

'It was a mistake! I didn't mean it! I'm sorry!'

At the last moment as he reached out to grip her shoulder, she jinked and ran knee-deep into the sea. Turning at bay to face him, she kicked a spray of water at him, shouting defiance and laughter.

'All right, come on then!'

Braving the flying spray, he reached her and picked her up kicking and struggling and waded out waist deep.

'No, no – please. Johnny. I give in – I'll do anything.'

At that moment a freak wave, bigger and stronger than the others, knocked Johnny's legs out from under him. They went under, and were rolled up the beach, to stagger out, completely soaked, clinging together, helpless with laughter.

They stood beside the Land-Rover trying to wring the water out of their clothing.

'Oh, you beast!' sobbed Tracey through her laughter. Her hair was a sodden mass, and drops of sea water clung in her eyelashes like dew.

Johnny took her in his arms and kissed her, and they stopped laughing.

She went loose against his chest, her eyes tightly closed and her lips, salty with sea water, opened against his.

The radio telephone in the Land-Rover beside them began to bleat fretfully, flashing its little red warning light.

They drew apart slowly, reluctantly, and stared at each other with dazed, bemused eyes.

Johnny reached the Land-Rover, unhooked the microphone and lifted it to his lips.

'Yes?' his voice cracked. He cleared his throat and repeated. 'Yes?'

The foreman's voice was distorted and scratchy through the speaker.

'Mr Lance, I'm sorry to have—' he was clearly about to finish ' – interrupted you.' But he stopped abruptly, and began again. 'It's just that I think you should know we've had a gale warning. Northerly gale building up quickly. If you want to get back to Cape Town you had better get airborne before it hits us – otherwise you could be shut in for days.'

'Thanks. We'll be back right away.' He hung up, and

Tracey smiled shakily. Her voice was also husky and unnatural-sounding.

'And a damn good thing too!'

Tracey's hair was still damp, and the borrowed polo-neck jersey swamped her. The grey trousers were also borrowed, rolled up to show her bare feet.

She sat very quietly and thoughtfully in the passenger seat of the Beechcraft. Far below them a small fishing vessel lay with a white cloud of seabirds hovering over it, and she watched it with exaggerated attention. There was a heavy feeling of restraint between them now, they could no longer meet each other's eyes.

'Pilchard trawler.' Johnny noticed her gaze.

'Yes,' said Tracey, and they were silent again.

'Nothing happened.' Johnny spoke again gruffly.

'No,' she agreed. 'Nothing happened.' Then shyly she reached out and took his hand. Lightly she rubbed the stump of his missing finger.

'Still friends?' she asked.

'Still friends.' He grinned at her with relief, and they flew on towards Cape Town.

Hugo Kramer watched the aircraft through his binoculars, balancing easily against the roll and pitch of the bridge.

'Police patrol?' asked the man at the helm beside him.

'No,' Hugo replied without lowering the glasses. 'Red and white twin Beechcraft. Registration ZS – PTB. Private aircraft, probably one of the diamond companies.'

He lowered the glasses, and crossed to the wing of the bridge. 'Anyway, we are well outside territorial waters.'

The drone of the aircraft engine faded away, and Hugo transferred his attention to the frantic activity on the deck below him.

The trawler, *Wild Goose*, lay heeled over under the weight of fish that filled her purse seine-net; at least a hundred tons of seething silver pilchards bulging the net out alongside the trawler into a round bag fifty feet across. While above it a shrieking canopy of seabirds swirled and wheeled and dived, frantic with greed.

Three of the crew on a scoop-net which hung from the overhead derrick were dipping the fish out of the net, swinging a ton of fish at each scoop over the side, and dropping them like a silver cloudburst into the trawler's hold. The donkey engine on the winch clattered harshly in time to their movements.

Hugo watched with satisfaction. He had a good crew, and although the fishing was only a cover for the *Wild Goose* – yet Hugo took pride in his teutonic thoroughness which dictated that the cover should be as solid as possible. In any case, all profits from fishing were for his personal account. It was part of the agreement with the Ring.

He packed the binoculars carefully into their leather case, and hung them behind the chartroom door. Then he clambered swiftly down the steel ladder to deck level, moving with catlike grace despite the heavy rubber hip boots he wore.

'I'll take her here for a while,' he told the man at the winch controls. He spoke in Afrikaans, but his accent was shaded with the German of South West Africa.

Wide-shouldered under the blue fisherman's jersey, he worked with smooth economic movements. His hands on the winch control were rough and reddened by wind and sun, for his skin was too fair to weather. The skin of his face was also red, and half-boiled, peeling so there were pinky raw places on his cheeks and black scabs on his lips.

The hair that hung out under his cap was white as

bleached sisal, and his eyelashes were thick and colourless, giving him a mild near-sighted look. His eyes were the palest of cornflower blue, yet without being weak and watery as those of most albinos; they were slitted now, as he judged the roll and dip of the boat – engaging the clutch to meet the movement, or pulling on the drum brake.

'Skipper!' A shout from the bridge above him.

'*Ja.*' Hugo did not allow his attention to waver as he replied, 'What is it?'

'Gale warning! There is a northerly buster building up.'

And Hugo grinned, pulled on the brake and shut the throttle.

'All right, clean up. Cut the purse rope, let the fish go free.'

He turned and swarmed up the ladder to the bridge, and went to his chart table.

'It will take us three hours to get in position,' he muttered aloud, leaning over the chart, then he barged out on to the wing of the bridge again to chase up his crew.

They had cut the purse rope on the net, allowing the net to fall open like a woman's skirt, and the fish were pouring out, a dark spreading stain through the gap. Two men had the pressure hose on, washing loose fish from the deck into the sea, others were slamming the hatch-covers closed.

Within forty minutes *Wild Goose* was running south under full throttle to take up her waiting station.

The diamond coast of South West Africa lies in the belt of the Trades. The prevailing wind is the south-easter, but periodically the wind system is completely reversed and a gale comes out of the north, off the land.

It is a Scirocco-type wind like the 'Khamsin' of the Libyan desert, or the 'Simoom' of Tripoli.

It was the same searing dry wind out of the desert, filling the sky with brooding dust and sand clouds, smothering everything beneath a hellish pall like the smoke from a great battlefield.

The dust clouds were part of the design, the Ring had taken account of them when they planned the system – for the north wind lifted into the sky such a quantity of mica dust that the radar screens of the diamond security police were cluttered and confused, throwing up phantom echoes and making it impossible to pick up the presence of a small airborne object.

Turn Back Point was three miles inland, and sixty miles north of the Orange River. The name was given by the first travellers, and expressed their views on continuing a journey northwards. Those old travellers had not known that they stood in the centre of an elevated marine terrace, an ancient beach now lifted above the level of the sea, and that it was the richest prospect of an area so diamond-rich that it was to be ring-fenced, patrolled by Jeep and dog and aircraft, guarded by gun and radar, a laager so secure that a man leaving it would have to submit to X-ray, and take nothing out with him but the clothes he wore.

At Turn Back Point was one of the four big separation plants where all the gravel from the big Company's workings from miles around was processed. The settlement was comparatively large, with plant, workshops and stores, and accommodation for five hundred men and their families. Yet not all the Company's efforts to make it attractive and liveable could alter the fact that Turn Back Point was a hell-hole in a savage and forbidding desert.

Now with the north wind blowing, what had been unpleasant before was almost unbearable. The buildings were tightly sealed, even the joints around the windows and doors were plugged with cloth or paper – and yet the red dust seeped in to powder the furniture, the desks, the bed linen, even the interior of the refrigerators, with a thin gritty film. It settled in the hair, was sugary between the teeth, clogged the nostrils – and with it came that searing heat that seemed to dry the moisture from the eyeballs.

Outside the dust was a red glittering fog which reduced visibility to a dozen yards. Men who were forced out into that choking dry soup wore dust goggles to protect their eyes, and the mica dust covered their clothing with a shiny coating that glittered even in that dun light.

Beyond the settlement a man moved now through the fog, carrying a small cylindrical object. He leaned forward into the wind, moving slowly away into the desert. He reached a shallow depression and went down into it. Setting his burden on the sand, he rested a moment. Then he knelt over the cylinder. He appeared monstrous under his leather jacket and cap, his face covered by goggles and a scarf.

The fibre-glass cylinder was painted with yellow fluorescent paint. At one end was a transparent plastic bubble which housed an electric globe, at the other end was a folded envelope of rubberized nylon material attached to the cylinder by a stainless steel coupling, and linked to the coupling was a small steel bottle of hydrogen gas. The whole assembly was eighteen inches long, and three inches in diameter. It weighed a few ounces more than fifteen pounds.

Within the cylinder were two separate compartments. The larger contained a highly sophisticated piece of transistorized electronic equipment which would transmit a homing signal, light or extinguish its lamp on long-distance radio signal command, and also at command it would control the inflow of hydrogen gas into the nylon balloon through the connecting coupling.

The smaller compartment held simply a sealed plastic container into which were packed twenty-seven diamonds. The smallest of these stones weighed fourteen carats, the largest a formidable fifty-six carats. Each of these stones had been selected by experts for colour, brilliance, and perfection. These were all first-water diamonds, and once they were cut they would fetch in the open market between seven hundred thousand and a million pounds – depending on the skill of the cutting.

There were four members of the Ring at Turn Back Point. Two of them were long service and trusted diamond sorters employed behind the guarded walls of the processing plant. They worked together, to check each other, for the Company operated a system of employee double check – which was completely useless when there was collusion. These men selected the finest stones and got them out of the plant.

The third member of the Ring was a diesel mechanic in the Company workshops. It was his job to receive and assemble the equipment which arrived concealed in a marked drum of tractor grease. He also packed the stones into the cylinder and passed it on to the man who was now kneeling out in the desert, preparing to launch the cylinder into the swirling dust fog.

His final check completed, the man stood up and went to the lip of the depression and peered out into the dust storm. At last he seemed satisfied, and hurried back to the yellow cylinder. With an incisive twist of the bevelled release ring he opened the valve on the bottle of hydrogen gas. There was a snakelike hiss, and the nylon balloon began to inflate. The folds of material crackled as they filled. The balloon lifted, eager to be gone, but the man restrained it with difficulty until the balloon was smooth and tight. He let go, and the balloon with its dangling cylinder leapt into the sky, and almost instantly was gone into the dust clouds.

The man stood with his face lifted to the dark furnace-red sky. His goggles glinted blindly, but his attitude was one of triumph, and when he turned away he walked lightly with the step of a man freshly released from danger.

'One more package,' he promised himself. 'Just one more,

and I'll pull out. Buy that farm on the Olifants River, do a bit of fishing, take a shooting trip every year—'

He was still dreaming when he reached the parked Land-Rover and climbed into the driver's seat. He started the engine, switched on the headlights, and drove slowly down the track towards the settlement.

The sign on the rear of the departing Land-Rover was in white paint so that it showed clearly through the haze of red dust.

'SECURITY PATROL,' it read.

*W*ild *Goose* lay on station, with her diesels throbbing softly, ticking over to hold her head into the wind. Even twenty miles out at sea the wind was searing hot and the occasional splatter of spray on Hugo's face was refreshing.

He stood in the corner of the bridge where he could watch the sea and the helmsman, but he was anxious. *Wild Goose* had been lying on station for fifteen hours now, during ten of which the norther had howled dismally through her rigging.

He was always anxious at the beginning of a pick-up. There was so much that could have gone wrong, anything from a police sweep to a tiny electrical fault in the equipment.

'What time is it, Hansie?' he shouted and the helmsman glanced up at the chronometer above his head.

'Three minutes after six, skipper.'

'Dark in half an hour,' Hugo grunted disgustedly, slitted those pale-lashed eyes into the wind once more, then shrugged and ambled back into the bridge house.

He stopped at the console beside his chart table. Even to an experienced eye the machine was an ordinary 'Fish-Finder', an adaptation of the old wartime anti-submarine device, the ASDIC, to the more prosaic business of plotting the depth and position of the pilchard shoals beneath the surface.

However, this model had undergone a costly and specialized conversion. The Ring had flown an expert out from Japan to do the work.

Now that the set hummed softly, its control panel lit soft green by the internal light, but the sound was neutral, and the circular glass screen was blank.

'You want some coffee, Hansie?' Hugo asked the old coloured man at the wheel. His crew were handpicked, loyal and trusted. They had to be – one loud mouth could blow a multi-million pound business.

'*Ja dankie*, skipper.' The old man creased his weather-battered face in appreciation, and Hugo shouted down the companionway to the galley.

'Cooky, how's it for a pot of coffee?'

But the reply was lost, for at that moment the console came to life dramatically. A row of lights blinked on above the control panel, the muted hum changed to a rapid beep-beep signal, and the screen glowed ghostly green.

'She's up!' Hugo shouted his relief, and ran to the set. His first mate rushed through from his cabin behind the bridge, tucking his shirt into unbuttoned trousers, his face puckered with sleep.

'About bloody time,' he blurted, groggily.

'Take over from Hansie,' Hugo told him, and he settled into the padded seat in front of the ASDIC set.

'Right, bring her round two points to port and open her up.'

The *Wild Goose* swung her head into the sea, and her motion changed from easy swoop and glide to a crabbing butting lunge, and the spray burst over the glass of the bridge.

Sitting before the console Hugo was tracking the flight of the balloon and keeping *Wild Goose* on an interception course.

Driven by the forty-knot norther the balloon crossed the coastline, climbing swiftly to three thousand feet. Hugo manipulated the knob on the console which sent the balloon a command to release gas and maintain altitude. Her response was recorded immediately on the screen.

'Good,' Hugo whispered. 'Good girl.' Then louder. 'Bring her round a bit, Oscar – the balloon is drifting to the south.'

For twenty minutes more they butted through the swells.

'Okay,' Hugo broke the silence. 'I'm going to ditch her.' He twisted the knob clockwise slowly, expelling all the gas from the nylon balloon.

'Ja. That's it. She's down.' He looked out of the window above the set. The dust-laden clouds had brought the night on prematurely. It was dark outside, with a low black ceiling through which no star showed.

Hugo turned his attention back to the set.

'That's it, Oscar. You're right on course. Hold her there.'

Then he glanced across at old Hansie and another younger crewman. They were sitting patiently on the bench against the far bulkhead. Both of them were clad in full oilskins, shiny yellow plastic from head to ankle, with gumboots below that.

'Okay, Hansie,' Hugo nodded. 'You can get up in the bows. We are only a mile or so away now.'

They climbed down on to the wave-swept deck, and Hugo watched them scuttling forward between each green burst of water and crouching in the bows. Both of them ducking each time another swell poured over the top of them, their yellow plastic suits showing clearly in the murky deck lights.

'I'm going to switch her on now,' Hugo warned the helm. 'We should have her on visual.'

'Right.' Oscar peered ahead, and Hugo flicked a switch on the panel, commanding the balloon to turn on her guide light.

Almost immediately there was a shout from Oscar.

'There she is. Dead ahead!'

Hugo jumped up and ran forward. It took a few seconds for his eyes to adjust, then he made out the tiny red firefly of light ahead of them in the vast blackness of sea and sky. It showed for a second then was gone in the trough of the next wave.

'I'll take her.' Hugo replaced Oscar at the wheel. 'You get on the spotlight.'

The beam of the spotlight was a solid white shaft through the darkness. The fluorescent yellow paint of the cylinder glowed in the circle of the spot.

Hugo lay *Wild Goose* upwind of the cylinder, and then allowed her to drift down gently on it. Hansie and his assistant were ready with the twenty-foot boat-hook.

Delicately Hugo manoeuvred down over the bobbing yellow cylinder, and grunted with satisfaction as the boat-hook slipped through the recovery ring and the cylinder was hauled in over the bows.

He watched while the two dripping oilskin-clad figures clambered back up the ladder into the wheelhouse, and laid the cylinder on the chart table.

'Good! Good!' Hugo slapped their backs heartily. 'Now, go and get dry – both of you!' They climbed down the companionway, and Hugo handed the wheel over to Oscar.

'Home!' he instructed him. 'As quick as you like.' And he carried the cylinder through into his cabin.

Sitting at the fold-down table in his cabin, Hugo unscrewed the lower section of the cylinder and took out the plastic container. He opened it and spilled the contents out on the table top.

He whistled softly, and picked up the biggest stone. Although he was no expert he knew instinctively that it was a brilliant of exceptional quality. Even the roughness of its exterior could not mask the fire in its depths.

To him it was worthless, there was nowhere he could market a stone like that. There was no temptation to take it out of the Ring – all it would mean for him was fifteen years at hard labour.

The Ring was based on this mutual reliance, no one part of it could function without all the others – yet each part was self-contained and watertight. Only one man knew all its parts, and nobody knew who that one man was.

From the drawer beside him Hugo took out his tools and set them on the table. He lit the spirit stove and set the pot of paraffin wax in the gimbal above it to heat.

Then he poured the diamonds into a shiny metal can. It was the type of ordinary commercial can used for packing and preserving foodstuffs.

Balancing against the ship's motion, he lifted the pot from the stove, and poured the steaming liquid wax over the diamonds, filling the can to rim level.

The wax cooled and solidified quickly, turning opaque and white. The stones were now incorporated in a cake of wax that would prevent them rattling, and would give the sealed can authentic weight.

Hugo lit a cigarette and crossed the cabin to look out into the wheelhouse. The helmsman winked at him and Hugo smiled.

He went back to the table, the can was cool enough to handle. He placed the circular lid over it, and moved to the portable Jenny bolted to a chest of drawers. Carefully he clinched the lid into place, his eyes squinting at the smoke from the cigarette that dangled from his lips.

Satisfied at last he set the sealed can on the table, while he went to where his jacket hung on the door. From the inside pocket he pulled out a manilla envelope, then from the envelope he drew a printed, colour-screened label. He came back to the bench and meticulously pasted the label around the can. On the label was a highly glamorized artist's conception of a leaping pilchard, making it look like a Scottish salmon.

'Pilchards in Tomato Sauce.' Hugo read the label aloud,

as he leaned back to admire his work. 'A product of South West Africa.' He smiled with satisfaction and began packing his equipment away.

'How much?' The foreman of the fish pump called across the narrowing gap between *Wild Goose* and the jetty.

'About fifty tons,' Hugo shouted back. 'Then the norther chased us home.'

'*Ja*. None of the boats stayed out.' The foreman watched his gang secure the mooring ropes, and swing the hose of the vacuum pump over *Wild Goose*'s hold to begin pumping out her pilchards.

'Take over, Oscar.' Hugo picked up his jacket and cap. 'I'll be back tomorrow.' He jumped down on to the jetty and strode down towards the canning factory with its awesome stink of pilchard oil. His jacket was slung over his shoulder, one finger hooked through the tag.

He went down an alley between the boiler rooms and the fish-drying plant, across a wide yard where the bags of fish meal were piled to the height of a double-storey building. He turned in through the double doors of the cavernous warehouse filled to roof height with cardboard cartons, each stencilled with the words:

1 gross cans. Pilchards in tomato sauce.
Consign to: Vee Dee Bee Agencies Ltd.
              32, Bermondsey Street,
              London, S.E.1.

He went into the cubicle that served the warehouse storeman as an office.

'Hello, Hugo. Good trip?' The storeman was Hugo's brother-in-law.

'Fifty ton.' Hugo hung his coat casually on the hook behind the door. 'I've got to take a leak,' he said, and went to the latrine across the floor of the warehouse.

He came back, and drank a cup of tea with his brother-in-law. Then he stood and said, 'Jeannie will be waiting.'

'Give her my love.'

'She don't need yours. She's going to get plenty of mine!' Hugo winked, and took his coat from the hook. It was lighter now, the can was gone from the pocket.

He went through the main gates of the harbour, exchanging a casual greeting with the customs officer, and went to the battered early model convertible in the car park.

He kissed the girl at the driving-wheel, threw his coat on the back seat and climbed in over the door.

'You drive,' he told her, grinning. 'I want both hands free.'

She squeaked and pulled his hand out of her skirts.

'Can't you wait till we get home?'

'I've been at sea for five days and I'm hungry as hell.'

'You're a caution, you are.' She laughed at him and started the car.

This was Sergio Caporetti, the man Johnny had chosen to captain *Kingfisher*. He was a round man, the same shape as a snowman. He filled the doorway of Johnny's office, and his great belly bulging into the room ahead of him. His face was round also, like a baby's – but the beautiful dark Italian eyes fringed with thick lashes like a girl's.

'Come in, Sergio,' Johnny greeted him. 'Nice to see you.'

The Italian crossed the room deceptively quickly, and Johnny's hand was completely engulfed by the enormous hairy paw.

'So, at last we are ready,' Sergio grunted. 'Three months

I sit on bum – do nothing. Look at me – ' He slapped his belly with a sound like a pistol shot. ' – fat! No good.'

'Well, not quite ready.' Johnny qualified the statement. He was flying Sergio and his crew over to England well ahead of time. He wanted the big Italian to have plenty of opportunity to study and get to know the revolutionary new equipment with which *Kingfisher* was fitted. Then when the vessel was ready for sea, Sergio would sail her out to Africa.

'Sit down, Sergio. Let's go over the crew list—'

When Sergio left an hour later, Johnny went as far as the lift with him.

'If you have any problems phone me, Sergio.'

'*Si*.' Sergio shook hands. 'Don't worry – Caporetti is in charge. All is well.'

On his way back Johnny stopped at the reception desk.

'Is Mrs Hartford in today?' he asked one of the little receptionists, and both of them replied in chorus like Tweedledum and Tweedledee.

'No, Mr Lance.'

'Has she phoned to say where she is?'

'No, Mr Lance.'

Tracey had disappeared. Five days now there had been no sign of her, her new office was deserted and unused. Johnny was worried and angry. He was worried that she had gone on another binge, and he was angry because he missed her.

He was scowling ferociously as he went back into his office.

'Goodness me.' Lettie Pienaar stood beside his desk with a batch of mail in her hand. 'We do look happy. Here's something to cheer you.'

She handed him a postcard with a colour picture of the Eiffel Tower. It was the first word from Ruby since she had left. Johnny read it quickly.

'Paris – ' he said, ' – is fun, it seems.' He tossed the card on to the desk and plunged back into the day's work.

He worked late, stopped at a steakhouse to eat, then drove back to the silent house in Bishopscourt.

The crunch of tyres on the gravel drive and headlights flashing across the bedroom wall woke him. He sat up in bed as the front-doorbell began a series of urgent peals and he switched on the bedside light. Two o'clock – Christ!

He pulled a dressing-gown over his nudity and tottered down the passage, switching on lights as he went. The doorbell kept ringing.

He turned the front door key. The door flew open and Tracey came in like a strong wind, clutching a briefcase to her chest.

'Where the hell have you been?' Johnny was suddenly fully awake, angry and relieved.

'Johnny! Johnny!' She was dancing with excitement, incoherent, her cheeks flaming and eyes shining. 'I've got them – at least, it, both of them.'

'Where have you been?' Johnny was not to be so easily sidetracked, and with an obvious effort Tracey brought her excitement under control, but she was still smiling and gave the impression of humming like an electric motor.

'Come.' She took his hand and dragged him into the lounge. 'Get yourself a large whisky and sit down,' she ordered, imperious as a queen.

'I don't want a whisky, and I don't—'

'You'll need one,' she interrupted, and went to the open liquor cabinet, poured a massive whisky into a crystal glass, squirted soda into it, and brought it back to Johnny.

'Tracey, what the hell is going on?'

'Please, Johnny. It's so wonderful, don't spoil it for me. Just sit there, please!'

Johnny sank reluctantly into the chair, and Tracey

slipped the catch on her briefcase and drew out a sheaf of documents. She stood in the centre of the floor, and took up the pose of a Victorian actress.

'This – ' she explained, ' – is a translation from the original German of a proclamation by Governor in Council dated 3rd May 1899 and issued at Windhoek. I will leave out the preamble and go straight to the meat.'

She cleared her throat and began reading:

'In consideration of the sum of 10,000 marks which is hereby paid and received, the rights to mine, win, recover, collect or carry away all metals, whether base or precious, stones whether base, semi-precious or precious, minerals, guano, vegetation and other substances organic or inorganic for a period of Nine Hundred and Ninety-Nine years is granted to Messrs Farben, Hendryck and Mosenthal S.A., Guano Merchants of 14 Bergenstrasse, Windhoek, in respect of a circular area ten kilometres in radius whose centre shall be a point situated at the highest elevation of the island lying on latitude 23° 15′ South and longitude 15° 12′ East.'

Tracey paused and looked at Johnny. He was frozen, stony-faced, staring at her with all his attention. She went on quickly, gabbling it out.

'All the old German mineral concessions and rights were ratified by the Union Parliament when the Union of South Africa took over the mandate after the Great War.'

He nodded, unable to speak. Tracey's smile kept breaking out.

'That concession still has all the force of law behind it. The grant of any subsequent rights is invalid, and although the original grant was mainly for the recovery of guano – yet it covers precious stones also.'

Again Johnny nodded, and Tracey put the document at the bottom of the sheaf of papers in her hand.

'The concession Company, Farben, Hendryck and

Mosenthal S.A., is still in existence. The Company's only remaining asset, apart from any long-forgotten concessions, is an old building at 14 Bergenstrasse, Windhoek.'

Tracey seemed to change the subject suddenly.

'You asked me where I have been, Johnny. Well I've been to Windhoek, and over most of the worst roads in South West Africa.

'The Farben, Hendryck and Mosenthal Company is owned by the brothers Hendryck, a couple of Karakul fur farmers. They are a pair of horrible old men, and when I saw them slitting the throats of those poor little Persian lambs just to prevent the fur uncurling, well—' Tracey paused, and gulped. 'Well, I didn't explain to them about the concession. I just offered to buy the Company, and they asked twenty thousand and I said "sign", and they signed and I left them chuckling with glee. They thought they'd been terribly clever. There! It's all yours!'

Tracey handed the Agreement to Johnny and while he read it she went on.

'I made the Agreement in the name of Van Der Byl Diamonds, I signed it as a Director – I hope you don't mind.'

'Christ!' Johnny took a long deep swallow of whisky, then set the glass down and stood up.

'Mind?' he repeated. 'You bring me the concession to Thunderbolt and Suicide – and ask me if I mind.'

He reached for her and eagerly she went to him.

'Tracey, you're wonderful.' They hugged each other ecstatically, and Johnny swung her off her feet. Without either of them planning it they were suddenly lying, still in each other's arms, on the couch. Then they were kissing, and the laughter dwindled into small murmurs and incoherent sounds.

Tracey pulled away from him at last, and slipped off the couch. She stood in the centre of the floor. Her breathing was ragged. Her hair was a dark tangle.

'Whoa! That's enough.'

'Tracey.' He started up from the couch, wild for her, but she held him at the full stretch of her arms, her hands flat on his chest, backing away in front of him.

'No, Johnny, no!' She shook her head urgently. 'Listen to me.'

He stopped. The wild look faded from his eyes.

'Look, Johnny, God knows I'm no saint, but – well, I don't want us to – well, not on a couch in some other woman's house. That's not how I want it to be.'

B enedict eased the big honey-coloured Bentley out of the traffic stream that clogged Bermondsey Street and turned into the gates of the warehouse. He parked beside the loading bay and climbed out.

As he pulled off his gloves, he glanced along the bank. Mountains of goods were stacked ready for distribution. Cases of Cape wine and spirits, canned fruit in brown cartons, canned fish, forty-gallon drums of fish oil, bundles of raw hides stiff as boards, and cases of indefinable goods – all of it the produce of Southern Africa.

Vee Dee Bee Agencies had grown out of all recognition in the ten years since Benedict had launched it.

Benedict climbed the steps to the bank three at a time, and strode down between the towering stacks of goods that reached up into the murk of the high ceiling. He walked with the assurance of a man on his own ground, the skirts of his overcoat swirling about his knees, broad-shouldered and tall. The storemen and porters greeted him deferentially as he passed, and when he entered the main office there was a stirring and whispering among the ranks of typists as though a wind had blown through a forest.

The Managing Director hurried out of his office to greet Benedict and usher him in.

'How are you, Mr van der Byl? There's tea coming now.' And he stood behind Benedict to take his coat.

The meeting lasted half an hour, Benedict reading through the weekly sales and cost reports, querying an item here, remarking on a figure with pleasure or displeasure as it deserved. Many people watching him work would have been startled. This was not the indolent playboy they thought they knew, this was a hard-eyed businessman coldly and unemotionally milking the maximum profit from his enterprise.

There would have been others who wondered where Benedict had found the capital to finance a business of this magnitude, especially if they had known that he owned the premises, and that Vee Dee Bee Agencies was by no means his only stake in the world of business. He had not received money from his father – the Old Man had not believed Benedict capable of successfully negotiating the purchase of a pound of butter.

The meeting ended, and Benedict stood up and shrugged on his overcoat, while his Managing Director went to the grey steel safe in the corner, tumbled the combination and swung the heavy door open.

'The shipment arrived yesterday,' he explained as he reached into the safe and brought out the can. 'On the SS *Loch Elsinore* from Walvis Bay.'

He handed the can to Benedict, who examined it briefly, smiling a little at the painting of the leaping fish and the lettering 'Pilchards in Tomato Sauce'.

'Thank you.' He slipped the can into his briefcase, and the Managing Director walked with him back to the Bentley.

Benedict left the Bentley in a garage in Broadwick Street, and walked through the jostle of Soho until he reached the grimy brick building behind the square. He pressed the bell opposite the card that read Aaron Cohen, Manufacturing Jeweller, and when the door opened he climbed the stairs to the top and fourth floor. Again he rang, and after a while an eye peeped at him through the peephole – but the door opened almost immediately.

'Hello, Mr van der Byl. Come in! Come in!' The young doorman welcomed him in and locked the door behind him. 'Papa is expecting you!' he went on as they both looked up at the eye of the closed-circuit TV camera above the wrought-iron grille that barred the passage.

Whoever was viewing the screen was satisfied, for there was an electrical buzz and the grille swung open. The doorman led Benedict down the passage.

'You know the way. Papa is in his office.'

Benedict was in a shabby little reception room, with a threadbare carpet and a pair of chairs that looked like Ministry of Works rejects. He turned to the right-hand door and went through it into a long room that clearly occupied most of the top floor of the building.

Along one side of the room ran a narrow bench, to which were bolted twenty small lathes. Each machine ran off a belt from a central drive below the bench. The man tending the machines wore a white dust jacket, and he grinned at Benedict. 'Hello, Mr van der Byl, Papa is expecting you.' But Benedict delayed a moment to watch the operation of the saws. In the jaws of each lathe was set a diamond, and spinning against the diamond was a circular blade of phosphor bronze. As Benedict watched, the man turned back to the task of spreading a fine paste of olive oil and diamond dust on to the cutting edge of each blade – for it was not the bronze that cut. Only a diamond will cut a diamond.

'Some nice stones, Larry,' Benedict remarked, and Larry Cohen nodded.

'All of them between four and five carats.'

Benedict leaned close and examined one of the diamonds. The line of the cut was marked with Indian ink on the stone. Benedict knew what heart-searching and discussion, what examination and drawing upon the rich storehouses of experience had preceded the positioning of that ink line. It might take two days to saw through each diamond, so Benedict left the bench and moved on.

In a row down the other side of the room sat the other Cohen brothers. Eight of them. Old Aaron was a great breeder of boys. They ranged in age from nearly forty to nineteen years and there were a couple who were still in school and hadn't yet come into the business.

'How do you like this one, Mr van der Byl?' Michael Cohen looked up as Benedict approached. Michael was shaping a fine diamond, cutting it into a round using a lesser stone as a blade. A small tray beneath the lathe caught the dust from the two stones. This dust would be used later for sawing and polishing.

'A beauty,' said Benedict. These men were of the brotherhood, working with diamonds all their lives and loving them as other men loved women or horses and fine paintings.

He moved on down the room, greeting each of the brothers, stopping to watch for a minute the loving care with which the elder boys, master craftsmen each of them, were cutting the precisely angled facets that make up the perfect round brilliant. The fifty-eight facets – table, stars, pavilions and the others which endow a cut stone with its mystic 'life' and 'fire'.

Leaving them crouched over their wheels, so similar to those of a potter, he went through the door at the end of the room.

'Benedict, my friend.' Aaron Cohen came from his desk

to embrace him. He was a tall thin man in his late sixties with a thick silver-grey mane of hair, round-shouldered from years of crouching over a diamond wheel. 'I did not know you were in London, they told me you were in Cape Town. Yesterday was Ruby's birthday. If we had known—'

Benedict took the envelope from his pocket and shook twenty-seven diamonds on to the blotter of the desk.

'What do you think of those, Papa?'

'*Shu! Shu!*' Papa patted his own cheeks with delight, and he reached instinctively for the biggest stone.

'I should live to see such a stone!' He screwed a jeweller's *loupe* into his eye, turning to catch the natural light from the high windows, and scrutinized the diamond through the eyepiece.

'Ah, yes. There is a feather*, but small. V.V.S.I.† But we will cut through it. Yes, we will take two gems from this stone. Two perfect diamonds of ten or twelve carats each, and perhaps five smaller ones.' More than half a diamond's bulk is lost in the cutting.

'Yes! Yes! From this stone we will sell a hundred thousand pounds' worth of polished diamonds!'

Aaron crossed to the door. 'Boys! Come see! I will show you a prince among diamonds.' And his sons crowded into the office. Michael took it first and gave his opinion. 'A good stone, yes. But not of the same water as the stone we had in the last batch. You remember that octahedron crystal—'

'What you talking!' his father interrupted. 'You wouldn't know a diamond from a piece of gorgonzola cheese already!'

'He is right, Papa.' Larry joined in the discussion. 'The other stone was better.'

'So the Big Lover argues with his father! Little shiksas with skirts up around their tochis you know all about.

---

* Semi-transparent veinlike flaw.

† Very very slight imperfection.

Dancing the Watsui and the Cha-Cha. Yes! But diamonds you know from nothing.' This declaration precipitated a full-scale, family argument in which each of the brothers joined with gusto.

'Shuddup! Shuddup! Back to work all of you! Out! Out!' Aaron broke up the meeting, driving his sons from the office and slamming the door behind them.

'Shu!' He looked to heaven. 'What a business! Now we can weigh the stones.'

When they had weighed and tallied the stones, and Aaron had locked them into the safe, Benedict told him:

'I am thinking of breaking up the Ring.'

Aaron froze and looked across the desk at Benedict. Between them there was always the pretence that their relationship was legitimate. They never spoke about the Ring, or where the unregistered stones came from, or how the finished gems were sent to Switzerland.

'Why?' Aaron asked carefully.

'I am a rich man now. With my father's money, and what I have made from the Ring and invested. A very rich man. I no longer need to take the risks.'

'Such problems I wish I had. But perhaps you are wise – I would not think to argue with you.'

'There will be one or two more packages. Then it will be finished.'

Aaron nodded. 'I understand,' he said. 'Like all good things it must end.'

It was a little after noon when Benedict parked the Bentley outside the mews flat off Belgrave Square. He went to shower immediately he was home. In all the years he had lived here he had never grown accustomed to London's grime-laden atmosphere, and he bathed or showered at least three times a day.

He sang in the shower, and then enveloped in a huge bath towel he left a string of damp footprints through to the lounge where he mixed a Martini, and screwed up his eyes at the first stinging taste of the drink.

The phone rang.

'Van der Byl,' he said into the mouthpiece, and then his expression changed as he listened. Quickly he put down the glass and used both hands to hold the telephone receiver.

'What on earth are you doing here?' His tone of astonishment was not faked.

'What a wonderful surprise. When can I see you? How about right now – for lunch? That's great! No, nothing I can't put off – this is a pretty special occasion, you know. Where are you staying? The Lancaster. Fine. Look, give me forty-five minutes, and I'll meet you in the Looking-Glass Room on the top floor. Yes, ten past one. God, what a delightful – I've said that already. See you in three-quarters of an hour.'

He replaced the receiver, swallowed the remains of his Martini and headed for his bedroom suite. This would make a good day into a truly remarkable one, he thought, as he quickly selected a silk shirt. He looked at his reflection in the mirror and grinned.

'The ball has really started bouncing your way, Benedict,' he whispered.

She was not at the bar, nor in the Looking-Glass Room. Benedict crossed to the tall picture windows for a glimpse of one of London's finest views across Hyde Park and the Serpentine. It was a smoky blue day, and the pale sun added bronze to the autumn shades of gold and red in the park.

He turned from the windows, and she was crossing the room towards him. His stomach swooped with delight for she also was pale gold with the coppery sheen of sun on her long legs and bare arms. The grace of her carriage was as he remembered it, the precise lifting and laying down of narrow feet on the thick pile of the carpet.

He stood quite still, letting her come to him. Heads turned all across the room, for she was a splendid golden creature. Benedict knew suddenly and clearly that he wanted this woman for himself.

'Hello, Benedict,' she said, and he stepped forward to take her hand in both of his.

'Ruby Lance!' He squeezed her long fingers gently. 'It's so good to see you again.'

The use of her surname was the clue to the strength of his reaction. She belonged to the one man in the world that Benedict most envied and hated. For this reason she was infinitely desirable.

'Let us celebrate with a little drink. I think the occasion deserves at the very least a champagne cocktail.'

She sat with those long slim legs neatly crossed, leaning back in her chair, holding the stemmed glass with tapering fingers. Her hair hung straight to her shoulders, like some rare silken tapestry in white gold, and her eyes watched him with a catlike candour, a calm feline intentness that seemed to look into his soul.

'I should not have bothered you,' she said. 'But I know so few people here.'

'How long can you stay?' He brushed aside the disclaimer. 'I must cancel my other arrangements.'

'A week.' She made it sound like an offer that was subject to negotiation.

'Oh, no!' His voice was mock distressed. 'You can do better than that – we won't be able to do half what I had planned in so short a time. You can stay longer, surely?'

'Perhaps,' she agreed, and lifted the glass an inch. 'It's good to see you.'

'And you.' Benedict agreed with emphasis. They sipped the sparkling wine watching each other's eyes.

W here others must wait weeks and months Benedict went in immediately as though it were his right. A smile and a murmured word, and theatre tickets were his or the doors of fashionable restaurants opened magically.

That first night he took her to the National Theatre, then for dinner at Le Coeur de France where a very famous movie actor stopped at their table.

'Hello, Benedict. We are all going on to the yacht later for a bit of a party. Join us, won't you?' And those legendary eyes turned to Ruby. 'And bring your beautiful friend with you.'

They ate breakfast under the awning on the after-deck of the yacht, eggs and bacon and Veuve Clicquot champagne, and watched the hubbub of dawn traffic on the wondrously smelly old Thames. Ruby was the only girl in the party without a fur to cheat the chill of the river dawn. Benedict made a mental note of the fact.

On the way home she sat with those long legs curled under her in the seat of the Bentley, still sleek and golden despite the night's exertions, but with the lightest touch of blue beneath her eyes.

'I can't remember having enjoyed an evening so much, Benedict.' She patted a tiny pink-lipped yawn. 'You're a wonderful companion.'

'Tonight again?' he asked.

'Yes, please,' she murmured.

S he sensed an urgency in him, when she came down into the lobby of the Lancaster that evening. He came quickly to meet her as she stepped out of the lift, and the quiet assurance with which he kissed her cheek and took her arm surprised her.

They were silent as he snaked the Bentley through the

evening traffic. Ruby realized that at the tips of her long tapered fingers, within touching distance, was a fortune such as she had never before allowed herself to dream about. She was deadly afraid. A wrong move, even a wrong word might drive that fortune beyond her grasp for ever. She would never have a chance like this again, and she was afraid to move, almost paralysed with fear. The decision she knew she would have to make very soon would be fateful. Must she encounter his advance with withdrawal, or must she meet it as frankly as it was made.

She was so deeply involved with her thoughts that when the Bentley came to a standstill she looked up with surprise. They were parked in a mews outside an expensive-looking flat.

Benedict came round and opened her door, and led her without protest into the flat.

She looked about her curiously, recognizing some of the art works on display in the entrance hall. Benedict took her through into the long lounge and settled her solicitously into the tapestry-covered chair which dominated the room like a throne, and suddenly her fear was gone. She felt queenlike in her control. She knew with certainty that this would all be hers.

Benedict stood in the centre of the room, almost a petitioner in his attitude, and he began to speak. She listened quietly, her expression showing no hint of the triumphant surge of her spirits, and when he stopped to wait for her reply she did not hesitate.

'Yes,' she said.

'I will be with you when you tell him,' Benedict promised.

'It won't be necessary,' Ruby assured him. 'I can handle Johnny Lance.'

'No.' Swiftly Benedict crossed to her chair and took her hands, drawing her to her feet. 'I must be there with you. Promise me that.'

Then it became suddenly evident to Ruby that the strength of her position was unassailable. Benedict needed her not for any physical reasons – but merely because she belonged to Johnny Lance.

Looking steadily into Benedict's eyes she determined to test her intuition.

'He does not have to know about you,' she said. 'I could arrange a divorce with him.'

'He *must* know about me. That's what I want, don't you see?'

'Yes, I see.' She was secure.

'It is agreed?' He could barely conceal his anxiety.

'Yes,' she nodded. 'It's agreed.' And they smiled at each other – each completely satisfied.

'Come.' He led her almost reverently into his bedroom, and Ruby stopped in the doorway with a little cry of delight.

The double bed was a mountain of glowing fur in a score of shades ranging from soft pinkish cream, through beige and oyster, pale smoky blue to midnight and jet glossy black.

'Choose one!' he ordered her. 'To seal our bargain.'

She moved like a sleepwalker towards the bed, but as she reached the centre of the Khedive carpet Benedict called out softly.

'Wait.'

She stopped obediently and he came up behind her. She felt his hands on the back of her neck, and she lowered her chin, shaking her hair forward so that he could unhook the clasp and draw down the fastening of her dress.

She stepped out of the dress as it dropped around her ankles, then waited passively as he carefully removed her brassiere.

'Now,' he said. 'Try them on.'

In her stockings and high-heeled shoes she went to the bed, subtly emphasizing the lilt of her movements, and took up the first fur.

Benedict was sprawled in the wing-back chair across the room as she glanced back at him. His face was gloating and flushed, so that his features seemed swollen and coarsened as he stared at her. She understood now that this was a form of ritual in which they were engaged. Like a victorious Roman general, Benedict was conducting his own personal triumph, reviewing the spoils and the plunder. It had no basis in sexual or physical desire, but was rather a service of worship to Benedict himself. She was the priestess of this rite.

Yet, knowing this, Ruby felt no resentment. Rather, she found herself excited by the cold perversity of the pageant. As she paraded and postured, turned and swirled and flared the skirts of a wild mink, she was very conscious of his eyes upon her body. She knew it was perfect, and his scrutiny stirred her physically for the first time in her life. She felt her blood quicken and pound, felt her heart flutter within its cage of ribs like a captive bird, and her loins tighten like a clenching first. For her also the ritual was narcissistic – satisfying her own deep emotional need.

As she discarded each of the coats she dropped it in the centre of the floor, until there was a knee-high pile of precious fur.

At last she faced him, hugging the soft creamy cloud tightly about her body. Then she opened her arms, and the coat also, standing on tip-toe to tighten and highlight the hard muscle in her legs and flanks.

'This one,' she whispered, and he came out of the chair, picked her up in his arms and laid her, still wrapped in mink, on the great pile of furs.

Ruby woke in the double bed to a feeling of excitement and enormous well-being such as she had not experienced since she was a schoolgirl on the first morning of a holiday.

The morning was far advanced, pale sunshine in a square shaft poured through the open window like a stage effect.

Benedict in a yellow silk dressing-gown stood beside the bed watching her with an unfathomable expression which changed immediately he realized that she was awake.

'My man has collected your luggage from the Lancaster. Your toilet things are in the bathroom, your clothing in the dressing-room.'

He sat down carefully on the edge of the bed, and leaned forward to kiss her forehead and then each cheek.

'We will breakfast when you are ready.' He sat back and watched her eyes; clearly he was waiting for her to say something important. Immediately she was on guard, wary of making a mistake, seeking a clue in his expression.

'Last night,' he asked. 'Was it as good for you, as it was for me?'

Understanding washed over her in a warm wave. He wanted assurance, a comparison between himself and Johnny Lance.

'I have never in my life – ' She placed the emphasis carefully. ' – experienced anything like it.'

He nodded, relieved, pleased – and stood up.

'After breakfast we will go to town.'

This morning, Edmund, Benedict's man, chauffeured the Bentley. When they alighted at the north end of Bond Street and walked arm in arm along the pavement, Edmund tailed them at a dignified crawl, steadfastly ignoring the abuse of other drivers.

The morning was cool enough for Ruby to wear her new cream mink, and the looks of admiration and envy she drew from the other strollers delighted Benedict. He wanted to impress her, he wanted to flaunt his wealth.

'The wife of a diamond man must have diamonds.' He spoke on impulse as they came up to an expensive-looking jewellers. Ruby squeezed his arm and turned to look into the window.

'Good Lord,' Benedict laughed. 'Not here!' And Ruby looked at him with surprise.

Mockingly Benedict began reading the signs in the window.

'"Paradise Jewellers. A large selection of blue-white gems. Certificate of flawlessness with every diamond. Perfect flawless stones at bargain prices as advertised on T.V. and in the national Press. Small deposit secures your ring now. A diamond is forever – show her you really care."'

'But they are such a well-known firm. They have branches all over the world – even in South Africa!' Ruby protested, and bridled a little as Benedict smiled patronizingly.

'Let me explain about diamonds. They are bought for two reasons by two different types of people. Firstly by rich men as investments that will not erode and can only increase in value. These men buy notable stones on the advice of experts, the best of the gem diamond production goes to them. So when Richard Burton buys Liz a £300,000 diamond he is not being extravagant – on the contrary he is being ultra-conservative and thrifty with his money.'

'That's the kind of meanness I like,' Ruby laughed, and Benedict smiled at her honesty.

'You may find me as thrifty,' he promised.

'Go on,' she said, 'tell me more about diamonds.'

'Well, there is another type who buys diamonds. Usually just one in his life, luckily for him – and he very seldom tries to resell it again or he would get a nasty shock. This type is Joe Everybody – who wants to get married. He usually goes to somebody like Paradise Jewellers.' Benedict poked a derisive finger at the sign in the window. 'Because he has seen it on telly and he can get a ring on the instalment plan. In many cases the deposit covers the dealer for the cost of the stone – the rest goes on advertising, finance charges and, of course, profit.'

'How do you know Paradise Jewellers are that type?' Ruby's attention was wide-eyed and girlish.

'You recognize them firstly by the big advertising splurge, then by their language.' Again he studied the notices in the window. '"A large selection of blue-white gems" – of every thousand stones of jewellery quality produced only one is fine enough in colour to be termed blue-white. It is unlikely they have a large selection. "Gem" is a special term reserved for a diamond which is in every way superb. "Flawless stones at bargain prices" – the lack of flaws in a diamond is only one of many factors governing its value. As for bargain prices – there ain't no such animal. Prices are maintained at the lowest level by fierce competition among expert and canny dealers, and there are no "sales" or special prices for anyone.'

'But where should a person buy a diamond?' Ruby was impressed and dismayed despite herself.

'Not here.' Benedict chuckled. 'Come, I will show you.' And before she could protest he had taken her arm and swept her into the shop, to be greeted with enthusiasm by the manager who must have noticed Ruby's mink and the attendant Bentley, which was causing a small traffic jam outside the shop.

'Good morning, madam and sir. May I be of service to you?'

'Yes,' Benedict turned to Ruby, as they settled down in the manager's office with a tray of diamond rings in front of them. 'You cannot examine a diamond properly in its setting.'

He selected the biggest diamond, took from his pocket a

gold-plated penknife fitted with a special tool and prised open the claws of the setting to a chorus of horrified squeals from the staff.

'I will make good any damage,' he snapped, and they subsided as Benedict took the loose stone and laid it on the velvet-covered tray.

'Firstly, size. This stone is about one carat.' He looked for confirmation to the manager who nodded. 'Let us say the value of this stone is £500. Ten similar stones will be worth £5,000, right? However, a ten-carat stone may be worth as much as £75,000. So the price per carat rises sharply as the total weight of the stone increases. If I were investing I would not touch a stone under three carats.'

The staff were listening now with as much attention as Ruby.

'Next, colour,' said Benedict, and glanced at the manager. 'A sheet of clean white paper, please.'

The manager scratched in his drawer and laid a sheet of paper in front of Benedict who placed the stone upon it, bottom upwards.

'We compare the colour it "draws" from white paper in good natural light.' He looked up at the manager. 'Switch off the fluorescent lights, and open those curtains, please.'

The manager obeyed with alacrity.

'This is a matter of experience. The colour is judged by a standard. We forget about the fancy rare colours like blue and red and green, and take our top standard as blue-white. A stone so white as to appear slightly blue, after which the distinctions drop to "fine white", and "white". Then stones which "draw" a yellowish tinge which we call "Cape" – in different shades, then finally stones which "draw" a brown colour – which will reduce the value of a stone by up to eighty per cent.'

Benedict fished in his fob pocket and pulled out a guinea case which he opened.

'Every expert carries a special diamond which he uses as

a gauge for colour by which to judge all other stones. This is mine.'

The staff exchanged apprehensive looks as Benedict placed a small diamond beside the other. He studied them a moment then replaced his gauge in the case.

'Second Silver Cape, I'd say,' he grunted, and the staff looked suitably abashed. 'Now we consider the stone's perfection.' He looked at the manager. 'Please lend me your *loupe*.'

'*Loupe?*' The manager was mystified.

'Yes, your jeweller's glass.'

'I—' the manager was deeply embarrassed.

'You sell diamonds – yet you do not own a *loupe*.' Benedict shook his head in disapproval. 'No matter, I have my own.'

Benedict took the glass from his inner pocket and placed it in his eye.

'Imperfections can be almost negligible – a "natural" at the girdle, or a bubble or pinpoint of carbon in the stone, on the other hand they can be gross "cracks", "clouds", "ice", or "feathers" which will ruin the value of the stone. But this one is flawless – so when the certificate of flawlessness is issued there will be no misrepresentation.' Benedict closed the glass and tucked it back into his pocket. 'However, in order to produce a flawless stone, the cut has been squeezed.'

He held up the stone between thumb and forefinger.

'The cut or "make" of a stone is the fourth and final decider of its value. The "make" should conform closely to the "ideal". This stone has been cut to exclude a flaw, and in consequence it is badly proportioned – heavy and out of round. I would prefer to see a graceful stone which includes a slight imperfection rather than a grotesque little cripple like this.'

He put the diamond down on the desk.

'The asking price by Paradise Jewellers for this stone is

£500 – which would be fair and correct for a gem. However, the colour is poor and although it is flawless it is of ungainly make. Its true value would be about – ah, let's see – £185 approximately.'

There was another chorus of protest from the assembled staff, led by the manager.

'I assure you, sir, that all our stones have been most carefully appraised.'

'How long have you been with Paradise Jewellers?' Benedict demanded brusquely. 'Four months, isn't it?'

The manager gaped at him.

'Before that you were a salesman in the showroom of a large firm of embalmers and undertakers.'

'I, well – I mean.' The manager fluttered his hands weakly. 'How did you know that?'

'I like to know about all my employees.'

'Employees?' The manager looked stunned.

'That's correct. My name is Benedict van der Byl. I own Paradise Jewellers.'

Ruby clapped her hands and cooed her applause.

'What a bundle of surprises you are!' she exclaimed.

Benedict smiled in acknowledgement and inclined his head.

'Now,' he said, as he rose and helped Ruby to her feet. 'We will go and buy some real diamonds.'

Aaron Cohen sold them two fine white twin marquise-cut brilliants, and Ruby chose the mounting for a pair of white-gold earrings from a leather-bound catalogue.

Benedict gave Aaron his cheque for twenty thousand pounds, then turned to Ruby.

'Now,' he said. 'We'll have lunch at the Celeste Grill-room. The food is bloody awful – but the decor is stupendous. We had best phone and reserve a table – it isn't really necessary but they get terribly hurt if you don't.'

As they settled back in the lush upholstery of the Bentley, Benedict instructed the chauffeur.

'Go past Trafalgar Square, Edmund. I want to pick up the newspapers from South Africa House.'

Edmund double-parked outside the Ambassador's entrance, and the doorman recognized the car and hurried inside to fetch the bundle of newspapers.

As they pulled away around the square towards Haymarket, Benedict selected a copy of the *Cape Argus*.

'Let's see what's happened at home.' He glanced at the front page, and stiffened perceptibly.

'What is it?' Ruby leaned towards him anxiously, but he ignored her. His eyes were darting across the page like the shuttle of a loom. She saw the colour fade from his face, leaving it white and intent. He finished reading, and pushed the paper towards her. She spread the page.

'VAN DER BYL DIAMONDS WIN VALUABLE CONCESSION.
APPEAL COURT SUPPORTS KAISER'S MINERAL GRANT.
LANCE GETS THUNDERBOLT AND SUICIDE.
    Bloemfontein, Thursday.

'In an urgent application by the Central Diamond Mines Ltd, to prevent Van Der Byl Diamonds Ltd prospecting and mining a concession area off the South West African Coast, Mr Justice Tromp today dismissed the application with costs stating in his judgment: "The original concession granted by German Imperial Decree in 1899, and subsequently ratified by Act of the Union Parliament in Act 24 of the 1920 must hold good in law, and will take precedent over any subsequent grant or concession purporting to have been made by the Minister of Mines to any other party."

'The area in dispute covers 100 square kilometres surrounding two small islands lying some fifteen miles south of Cartridge Bay and five miles offshore. The islands are known as Thunderbolt Island and Suicide Island, and at the turn of the century were the site of considerable exploitation by a German guano company. Mr John Rigby Lance, the General Manager of Van Der Byl Diamond Co. Ltd, acquired the rights to the concession when he took over the inoperative guano company.

'In Cape Town today Mr Lance stated: "It's the opportunity I have waited for all my life. All indications are that Thunderbolt and Suicide will prove to be one of the richest marine diamond fields in the world."

'Van Der Byl Diamond Co. had a diamond-dredging vessel nearing completion in the United Kingdom, and Mr Lance stated that he hoped to begin recovery operations off Thunderbolt and Suicide Islands before the end of the year.'

R uby lowered the paper and looked at Benedict. What she saw was an intense physical shock.

Benedict had crumpled down in the seat. Gone was all the assurance and *savoir faire*. His face was deathly pale, but now his lips trembled and with disgust she saw that his eyes were swimming with tears. He hunched forward over his hands, shaking his head gently and hopelessly.

'The bastard,' he whispered, and his voice was soggy and muffled. 'He does it every time. I thought that I had him at last but – Oh God, I hate him.'

He looked at her, his face soft with self-pity. 'He does it every time. Often I've thought I had him, but he just—'

She was mystified by his reaction.

'Aren't you pleased? Van Der Byl Diamonds will make millions—'

'No! No!' he cut in savagely, and then the years of hatred and frustration and humiliation began pouring out. Ruby listened quietly, slowly beginning to understand it all, marvelling at the accumulation of pain and hatred that he exposed for her. He remembered conversations from twenty years ago. Small childhood episodes, innocent remarks that had festered and rankled for a decade.

'You don't want him to succeed, is that it?' she asked.

'I want to crush him, break him, humiliate him.'

For ten seconds Ruby was silent.

'Well, what are we going to do about it?' she asked flatly.

'Nothing, I suppose.' Benedict's tone irritated her. 'He always comes out on top, you just can't—'

'Nonsense,' Ruby snapped. She was angry now. 'Let's go over it carefully, and see how we can stop him. He is only human, and you have shown me enough to prove you are a brilliant and successful businessman.'

Benedict's expression changed, becoming trusting and animated. He turned to her almost eagerly: he blinked his eyes. 'Do you really believe that?'

The bunk was too narrow, Sergio Caporetti decided, much too narrow. He would have one of the carpenters alter it today.

He lay on his back, wedged in firmly, with the blanket-covered mound of his belly blocking his view southwards. He lay and assessed his physical condition. It was surprisingly good. There was but a small blurred pain behind his eyes and the taste of stale cigars and rank wine in the back of his throat was bearable. The leaden feeling in his lower limbs alarmed him until he realized that he was still wearing his heavy fisherman's boots. He remembered one of the girls complaining about that.

He hoisted himself on one elbow, and looked at the girls. One on each side of him, jamming him solidly into the bunk with magnificent hillocks of pink flesh. Big strong girls both of them, he had chosen them with care, neither of them an ounce under twelve stone. Sergio sighed happily — it had been a wonderful weekend. The girls were snoring, in such harmony that it might have been a rehearsed stage act. He listened to them with mild admiration for a few minutes, then crawled over the outside girl and stood in the centre of the cabin, clad only in his heavy boots. He yawned extravagantly, scratching the thick black wiry curls that covered his chest and belly, and cocked an eye at the bulkhead clock. Four o'clock on a Monday morning, but it had been a truly memorable weekend.

The table was hidden under a forest of empty wine bottles, and dirty plates. There was a congealed mass of cold spaghetti bolognaise in a dish and he picked it up. As he clumped out of the cabin on to *Kingfisher*'s bridge he was scooping up spaghetti with his fingers and cramming it into his mouth.

He stood at the rail of the bridge, a naked hairy figure in tall black boots clutching a dish of spaghetti to his chest, and looked around the dockyard.

Kingfisher was in stocks undergoing the modifications that Johnny Lance had ordered. She was standing high above the level she would attain when she was launched. Although she was a vessel of a mere 3,000-ton displacement, she appeared black and monstrous in the floodlights that illuminated the ship-builders' yard. It was obvious from her unusual silhouette that she was designed for a special purpose. Her superstructure was situated well aft like that of an oil tanker, while her foredeck was crowded by the huge gantry which would control the dredge, and by the massive storage tanks for the compressed air.

At this hour of the morning the shipyard was deserted, and wisps and tendrils of sea mist drifted about *Kingfisher*'s bulk.

Standing fifty feet above the dry dock, still wolfing cold spaghetti, Sergio urinated over the rail – deriving a simple honest pleasure from the long arching stream and the tinkle of liquid striking the concrete far below.

He clumped back into his cabin, and looked down fondly

on his two sleeping Valkyries while he finished the last of the spaghetti. Then he wiped his fingers carefully on his chest hair and called to them gently.

'Come, my kittens, my little doves, the time for play she has passed – the time for work she commences.'

With Latin gallantry he bundled them into a taxi at the dockyard gates, pressing on to each of them a lusty kiss, a banknote, a bottle of Chianti, protestations of deep affection, and the promise of another party next Friday night.

He picked his way back through the dockyard jungle of machinery and buildings, lighting a long black cigar and inhaling smoke pleasurably until he came in sight of *Kingfisher* and halted with surprise and annoyance. There was a big honey-coloured Bentley parked near the gangway that led up to *Kingfisher*'s deck. He resented visits from the Company bosses, especially this one, and especially at this ridiculous hour on a Monday morning.

The hose spiralled down into greenness, and they followed it down holding hands. Tracey was still a little nervous. This was not like the Mediterranean, a warm blue friendly embrace of waters to welcome the diver – it was the wild Atlantic, coldly menacing, green and untamed. It frightened her, and Johnny's hand gave her comfort.

Their Draeger demand valves repeated their breathing in a singing metallic wheeze, and icy leaks and rivulets kept finding their way into the cuffs and neck of Tracey's rubber suit.

Sixty feet below the surface Johnny paused, and peered into the glass window of her mask. He grinned at her, his mouth distorted by the bulky mouthpiece, and she gave him a thumbs-up sign. They both looked upwards. The surface was silvered like an imperfect mirror, and the black cigar shape of the boat was lapped in strange light. The hose and anchor chain pierced the silver ceiling and hung down into the shady green depths.

Johnny pointed downwards, and she nodded. They put their heads down, pointed their flippers to the surface, and still hand in hand they paddled steadily towards the sea bed.

Tracey was aware of a crackling hissing sound now, and from out of the greeny blackness below them scudded clouds of silver bubbles twisting and writhing towards the surface. She strained her eyes downwards, following the line of the hose, and slowly out of the murk materialized the black rubber-clad forms of the two men working at the end of the hose; they appeared weird and mystical like black priests performing a satanical mass.

She and Johnny reached the sea bed and hung just above it, a little way off from the two men on the hose. Johnny indicated the depth gauge that he wore like a wrist watch. It showed a depth of 120 feet. Then he turned and by a hand signal showed her the direction of the reefs.

They were in a valley between these long peaked underwater ridges of black rocks, the same reefs that Tracey had seen from the air. There was a distinct pull of water as the current drifted at right angles to the direction of the reefs.

Johnny squeezed her hand, and then pulled her down. They lay on their bellies on the floor of the sea, and Johnny scooped a handful of the white sand, washed it quickly so that the smaller particles were carried away in a cloud on the current, then he showed her the coarse gravel which remained. Again he grinned, and she returned his smile.

Still leading her by the hand, he swam slowly towards the two men working on the hose, and stopped to watch them.

Attached to the end of the hose was a rigid steel pipe two inches in diameter, and twenty feet long – although

117

now only half of its length was visible above the sand bottom. The two divers were forcing it down through sand and gravel to reach bedrock. The hose itself was attached to a compressor on the deck of the boat which was generating a vacuum in the hose and sucking up the sand and gravel as the steel pipe was forced downwards.

They were prospecting the Thunderbolt and Suicide field. Taking these two-inch samples at 500-foot intervals to ascertain the depth of water, the thickness of the overburden, and the content of the gravel beds. They were also mapping and plotting the reefs, so that by the time *Kingfisher* arrived they would have a fairly clear picture of the topography and aspect of the field. They would know where to begin dredging, and roughly what to expect when they did.

So far the results had endorsed Johnny's most optimistic expectations. There was a good thick catchment of gravel in all the gullies between the reefs. As he had expected, the heavier gravels had been laid down in the gullies closest to the gap between Thunderbolt and Suicide, and the smaller and lighter gravels had been carried further. In some of the gullies the gravel beds were fifteen feet deep, and the types of stone present were all highly promising. He had isolated garnet, jasper, ironstone, beryl chips and titanium dust.

However, the conclusive and definite proof had also come up through that two-inch hose out of the depths. They had already pulled the first diamonds from the Thunderbolt and Suicide fields. When you considered the odds against finding a stone in a two-inch sample at 500-foot centres and that payable gravel contained one part diamond in fifty million, it was exciting and encouraging that they had already recovered four diamonds. Small stones, to be sure, not one of them more than half a carat, but diamonds for all that, and some of them of excellent quality.

One of the men on the hose turned and gave Johnny a flat-handed cut-out sign. The pipe was on bedrock. Johnny nodded and jerked a thumb upwards, and drawing Tracey with him, started for the surface.

They climbed the ladder over the survey boat's counter, moving clumsily under the weight of the air bottles strapped to their backs, but there were willing hands to help them aboard and strip off the heavy equipment, and unzip the clinging rubber suits.

Tracey accepted a towel gratefully from one of the crew, and while she tilted her head to dry her sodden mane of hair, she looked across half a mile of green sea to the two white whale-backed islands with their attendant clouds of seabirds. The wave bursts on the cliffs sounded like distant artillery, or far thunder.

'God, this is a wild and exciting place.' Her voice bubbled with excitement as she scrubbed at her hair. 'It makes one come alive.'

Johnny understood her feelings, it was the forbidding restless sea and the harsh land that promised danger and adventure. He was about to reply, but the two hose men came aboard at that moment, the taller of them spitting out his mouthpiece and letting it fall to his chest.

'We'll move up to the next point, if it's okay by you, Mr Lance?' The man pulled off his mask and hood, exposing white-blond hair and a sun-broiled face.

'Fine, Hugo,' Johnny agreed, and watched approvingly as Hugo Kramer gave the orders to get the anchor and the hose up before taking *Wild Goose* seawards to her next prospecting point. Johnny had been reluctant to charter *Wild Goose* as the prospecting vessel and as the service boat for *Kingfisher*. He did not know Hugo Kramer, and Benedict van der Byl's insistence on the man had made him suspicious.

However, it was natural that they should use a skipper from the van der Byl fleet and Johnny was now prepared to admit he had been wrong. Kramer was an intelligent and willing worker, resourceful and trustworthy, a fine seaman who handled *Wild Goose* with all the skill it would need to bring her alongside *Kingfisher* in a heavy sea. His unfortunate physical appearance Johnny hardly noticed any more, although the original shock of that pink face, white hair and those blind-looking eyes had been considerable.

Tracey was not so charitable. The man made her uneasy. There was a wild-animal ferocity about him, a barely controlled violence. The way he looked at her sometimes made her skin prickle. He did it now; turning back from issuing his orders he ran his eyes over her body. In the black silk costume her good round breasts showed at their best, and Hugo Kramer looked at them with those white-fringed bland eyes. Instinctively she covered them with the towel, and it seemed as though his lips twitched with amusement as he turned to Johnny.

'They tell me this dredger of yours is something special, Mr Lance?'

'She is, Hugo. Not like the other half-baked barges and bastardized conversions that have been tried by other companies. She's the first diamond recovery vessel designed expressly for the job.'

'What's different about her?'

'Nearly everything. Her hose is operated off a gantry on the foredeck, it goes out through a well pierced through her hull.'

'What kind of hose?'

'Eighteen-inch armoured woven steel with rubber liner. We can get it down to a hundred fathoms, and it has a compensating section in it to stop it plunging with the wave action of the hull.'

'Eighteen inches is pretty big. How will you build up vacuum?'

'That's the point, Hugo. We don't suck – we blow! We evacuate water from the hose by purging it with compressed air, the inrush of water into the opening of the hose sucks in the gravel.'

'Hey, that's neat. So the deeper you work the more effective it will be!'

'Right.'

'What about the actual recovery? Are you going to have the usual screening, ball mill, and grease table arrangement?'

'That's what killed the other companies – trying to separate by the old methods. No. We've got a cyclone to start with.'

'Cyclone?'

'You know a cream separator?'

'Yeah.'

'Same principle. Just spin the gravel in a circular tank and float off everything with a specific gravity of less than 2.5. Take what is left, dry it, spread it on a conveyor belt and run it under an X-ray machine which pinpoints every single diamond. As you know, diamonds fluoresce under X-rays and they show up crisply. The X-ray machine reports the diamond to the central computer. . . ' Johnny's voice and whole attitude was charged with enthusiasm which was impossible for his listeners to resist. Tracey was carried along with him, watching his eyes and his mouth as he talked, smiling when he smiled, her lips following his faithfully.

'This is the cyclone room,' Benedict van der Byl explained as, with a hand on her elbow, he helped Ruby Lance down off the last rung of the ladder. 'I explained to you how it worked.'

'Yes.' Ruby nodded, and looked around the room with interest. The roughly riveted and grey painted plating of *Kingfisher*'s hull formed a square metal box, in the centre of which stood the cyclone. It was also painted battleship grey, a ten-foot-high cone-shaped circular tower.

'The gravel is blown in through here.' Benedict indicated the eighteen-inch pipe which entered the cyclone room through the forward bulkhead, then connected to the bottom of the cyclone. 'Up it goes.' Benedict flung his hand upwards. 'And round it goes.' He made a stirring motion. 'The heavy stuff is thrown off and led away through that.' A smaller pipe emerged from the shoulder of the cyclone and disappeared through the farther bulkhead. 'While the lighter stuff shoots out through the top and is sprayed overboard again.'

'I understand. Now, where is the weak spot?' Ruby asked.

'Come.' Benedict led her across the room, picking their way among the litter left by the workmen who still swarmed through *Kingfisher*. They reached a steel door in the bulkhead.

'Watch your head.' They ducked into a long passageway with doors at both ends. On their right hand was an enclosed tunnel that ran the length of the room.

'This is the conveyor room,' Benedict explained. 'The concentrated gravels fall through a hot air draught from an electric furnace to dry them. They are gathered on a conveyor belt, concealed in that tunnel, and carried through into the X-ray room.'

'This is where you will fit it?' Ruby asked.

'Yes. In the conveyor tunnel. It will mean moving that inspection hatch back twelve feet to give us the space.'

Ruby nodded. 'The man who will do the work – can you trust him?'

'Yes. He has worked for me before.' Benedict did not add that the same man had designed the electronic equipment for the balloons used by the Ring, and had flown out from Japan to convert the ASDIC equipment on *Wild Goose*.

'All right.' Ruby seemed satisfied. More and more she was becoming the driving force in the alliance, bolstering

Benedict's resolution when he showed timidity or when he tried to evade the actions which must, in time, lead to a confrontation with Johnny Lance.

'Let's see the X-ray room.'

It was a tiny cupboard-like compartment. The floor, roof and all four walls were clad with thick sheet lead. Suspended from the roof was the X-ray machine, and under it a circular table the surface of which was covered with a honeycomb-patterned stainless steel sheet.

'The concentrated gravel spills on to the table, and the table revolves under the X-ray machine which fluoresces each diamond and the computer picks it up and reports its size and exact position on the table. The computer then commands one of those – ' Benedict pointed to a forest of hard plastic tubes, each attached to a metal arm, ' – to swing out over the table exactly above the diamond and suck it up. The computer selects the correct diameter of tube for the size of the diamond – and, after the tube has obeyed the computer, the table passes under a second X-ray machine which confirms that the diamond has been collected. If, by chance, the tube fails to suck up the stone, then the computer automatically sends the table on another circuit. If, however, the diamond is safely gathered then the waste material is scraped from the table and it swings round to pick up more gravel from the cyclone room – and repeat the whole process. The system is 100 per cent effective. Every single diamond is recovered by it. Even stones as small as sugar grains.'

'Where is the computer?' Ruby asked.

'There.' Benedict pointed through the small leaded glass window which overlooked the X-ray table. Beyond it was another small compartment. Ruby flattened her nose against the glass, and peered in. The computer occupied most of the room, a huge glossy enamelled cabinet not unlike a refrigerator despite the switches and dials. Benedict peered in beside her.

'The computer runs the entire operation. It controls the flow of compressed air into the dredger pipe, it regulates the cyclone, runs the X-ray machine and the table, it weighs and counts the diamonds recovered before depositing them in a safe, and it even navigates the *Kingfisher* and reports to the bridge her exact position over the sea bed, it checks the lubrication and temperature of the engines and power plant and on request will make a complete and immediate report of the whole or any part of the operation.'

Ruby was still peering into the computer room.

'What happens to the diamonds once they have been picked off the revolving table?' she asked.

'They are sucked through an electronic scale which weighs each stone, then they are carried through into the computer room and deposited in that safe.' Benedict pointed out the steel door set in the bulkhead. 'The safe has a time and combination lock. So the system works without a diamond being touched by human hand.'

'Let's go and talk to the Italian peasant,' suggested Ruby, and as she turned from the window Benedict slipped his arm about her shoulders and hugged her possessively.

'Not now,' snapped Ruby irritably, shrugging off his arm, and she led the way out of the X-ray compartment, passing the locked door of the computer control room opposite the door to the conveyor room. She was impressed with the ingenuity of the system – but the fact that it had been constructed by Johnny Lance made her angry.

Her loyalties had changed completely, going to the highest bidder.

S ergio Caporetti felt a small twinge of pity when he looked at Ruby Lance. So thin, and with a backside like a boy. She would be little comfort to a man on a cold night. Sergio worked the cheroot from one corner of his mouth to other, anointing the stub with saliva in the process. Also she was cold-blooded, he decided. Sergio had a very sensitive intuition when it came to judging the temperature of a woman's passion. Cold like a snake, he decided, his pity giving way to revulsion. He repressed a small shudder as he watched her settle on to the day couch in his cabin, and cross her long golden legs precisely. Just like a snake, she would eat a man as though he were a little hopping frog. Sergio had admiration for Johnny Lance, but – he decided – not even he would be safe with a woman like this.

'You like my ship?' he asked, an attempt at friendliness. 'She is *very* fine ship.'

Sergio actually used a more forceful adjective than *very*, one that suggested *Kingfisher* was capable of procreation, and Ruby's lips curled with disgust. She ignored the question and lit a cigarette, swinging one leg impatiently, and turned her head to stare through the porthole.

Sergio was hurt by the rebuff, but he had no time to brood on it for Benedict van der Byl came to stand in the centre of the cabin with his hands clasped lightly behind his back.

'Mr Caporetti—' he asked quietly. 'How much do you like money?'

Sergio grinned, and pushed the grubby maritime cap to the back of his head. 'I like it pretty good, I like it better than mother – and I love my mother like my life,' he said.

'Would you like to become a rich man?' Benedict asked, and Sergio sighed wistfully.

'*Si!*' he nodded. 'But it is the impossible thing. There is too much *vino*, too much lovely girls, and the cards they are cruel like – ' Sergio paused to find a suitable simile and glanced at Ruby, ' – like a thin woman. No. Money she does not stay long, she comes and she goes.'

'What would you do for £25,000?' Benedict asked.

'For twenty-five thousand – ' Sergio's eyes were dark liquid and lovely as those of a dying gazelle or a woman in love, ' – there is nothing I will not do.'

*K*ingfisher sailed for Africa on the 4th of October. As the representative of the owners, Benedict van der Byl drove down from London to bid her *bon voyage*, and he spent an hour behind locked doors with Sergio Caporetti before the departure of the vessel.

Kingfisher made good time southwards on her first leg of the voyage, but the unscheduled delay of ten days at the island of Las Palmas infuriated Johnny. His urgently cabled enquiries from Cape Town elicited the reply that there were teething troubles in *Kingfisher*'s engine room which were being attended to in the Las Palmas dockyards. The voyage would be resumed as soon as the repairs had been effected.

The Japanese gentleman who welcomed *Kingfisher* to Las Palmas was named Kaminikoto. This was too much for Sergio's tongue, so he called him 'Kammy'.

Sergio's crew was sent ashore with the excuse that the work on *Kingfisher* was dangerous. They were installed in the best tourist hotel and liberally supplied with intoxicating liquor. Sergio did not see them for the next ten days that he and Kammy were busy on the modifications to *Kingfisher*'s computer, and recovery equipment.

During those ten days Sergio and Kammy discovered that despite physical appearances they were brothers.

Kammy had mysterious packing-cases brought on board and they worked like furies from dawn until after dark each day. Then they relaxed.

Kammy was half Sergio's size with a face like a mischievous monkey. At all times he wore a Homburg hat. On the one occasion that Sergio saw him in his bath without his headgear he discovered that Kammy was as bald as St Peter's dome.

Kammy's abundant tastes in women were identical to Sergio's. This made the hiring of partners an easy matter, for what suited the one suited the other. Sergio took south with him fond memories of the little Japanese clad only in his Homburg hat, uttering bird-like cries of encouragement and excitement, while perched like a jockey on top of a percheron mare.

When at last Sergio shepherded his debauched crew back aboard *Kingfisher* the only obvious sign of their labours was that the inspection hatch on the conveyor tunnel had been moved back twelve feet.

'It is my best work,' Kammy told Sergio. Already he was sad at the prospect of parting. They were brothers. 'I signed my name. You will remember me when you see it.'

'You good guy, Kammy. The best!' Sergio embraced him, lifting him off his feet and kissing him heartily on each cheek while Kammy clutched desperately at his Homburg.

They left him standing on the wharf, a forlorn and solitary figure, while *Kingfisher* butted out into the Atlantic and swung away southwards.

Ruefully Johnny Lance glanced over at the mountain of empty champagne bottles beyond the barbecue pits. The bill for this little party would be in the thousands, but it was not an extravagance. The guest list included all Van Der Byl Diamond Company's major creditors and their wives. Johnny Lance was showing them all what they were getting for their money. To appear prosperous was almost as reassuring to a creditor as being prosperous. He was going to stuff them full of food and champagne, show them over the *Kingfisher* and fly them back home, hoping sincerely that they would be sufficiently impressed to stop badgering him for a while – and let him get on with the business of taking the Company out into the clear.

Tracey caught his eye. Her humorous roll of the eyes was a plea for sympathy, for she was surrounded by a pack of middle-aged bankers and financiers whom champagne had made susceptible to her charms. Johnny winked at her in reply, then glanced around guiltily to find Ruby, and was relieved that she was in deep conversation with Benedict van der Byl in a far corner of the marquee.

He made his way out of the crowd to the edge of the dune, and lit a cigarette while he looked back across Cartridge Bay.

The chartered Dakotas that had flown the guests and caterers up from Cape Town were standing on the airstrip beyond the buildings.

The marquee was situated on the crest of a sand dune overlooking the narrow entrance to the bay. The dune had been bulldozed to accommodate the tent, the laden tables, and the barbecue pits around which white-clad servants were busy, and the spitted carcasses of three sheep and a young ox were already browning crisply and emitting a cloud of fragrant steam.

Tracey watched Johnny standing out on the edge of the dune. He looked tired, she thought. The strain of the last few months had worn him down. Looking back on it now she realized that every few days had thrust a crisis upon him. The terrible worry of the court case that had won them Thunderbolt and Suicide had barely ended before Johnny had faced the delays in the construction of *Kingfisher*, the bullying of creditors, the sniping of Benedict and a hundred other worries and frustrations.

He was like a prize-fighter coming out to the bell of the last round, she thought tenderly, as she studied the profile of his face now staring out to sea. His stance was still aggressive, the big jaw pushed out and the hand with the missing finger that held the cigarette balled into a fist, but there were blue shadows under his eyes and lines of tension at the corners of his mouth.

Suddenly, there was an alertness in Johnny's attitude, he shaded his eyes with a hand before turning back towards the marquee.

'All right, everyone!' he called, stilling the babble of their voices. 'Here she comes.'

Immediately the uproar was redoubled and the whole party trooped out into the sunlight, their excitement and the shrillness of their voices enhanced by the Pommery they had been walloping back since mid-morning.

'Look! There she is!'

'Where? Where?'

'I can't see her.'

'Just to the left of that cloud on the horizon.'

'Oh yes! Look! Look!'

Tracey took a second glass of champagne from one of the waiters, and carried it across to Johnny.

'Thanks.' He smiled at her with the ease that now existed between them.

'It's taken her long enough to get here.' Tracey picked out the faraway speck on the green ocean that was *Kingfisher*. 'When will she begin working?'

'Tomorrow.'

'How long will it be before we know – well, if it has come off?'

'A week.' Johnny turned to her. 'A week to be certain, but we'll know in a day or two how it's shaping up.'

They were silent then, staring out at the gradually approaching speck. The crowd lost interest quickly and drifted back to the liquor, and the fragrant steaming platters of golden-brown meat that were coming from the barbecue pits.

Tracey broke the long friendly silence at last. She spoke hesitantly, as though reluctant to bring up a painful subject.

'How long has Ruby been back now – ten days?'

'About that,' Johnny agreed, glancing at her quickly. 'I haven't seen much of her,' he admitted. 'But she seems to be a lot more relaxed – and at least she's kept off my back.'

'She and Benedict seem to have become very pally.' Tracey glanced across to where the other couple were now included in a boisterous circle of revellers.

'She bumped into him in London,' Johnny agreed, sounding offhand. 'She tells me they had lunch a couple of times.'

She waited for him to continue, to express some suspicion or reservation, but he seemed to have no further interest in the subject; instead he began running over the day's further arrangements with her.

'I'm relying on you to take charge of the wives when we go aboard. Keep an eye on Mrs Larsen particularly – she's up to her gills in bubbly.'

For the next two hours that it took *Kingfisher* to make her approach and enter the channel of Cartridge Bay, Johnny hardly took his eyes from her unusual silhouette. She was not a pretty vessel but the white lightning insignia of Van Der Byl Diamonds on her funnel gave her a special beauty in his eyes. As she passed below them and entered the bay, Larsen proposed a toast to her successful career, then they all descended the dune and climbed into the waiting Land-Rovers and drove round the bay to meet her. By the time they arrived *Kingfisher* had made fast alongside the jetty, and Captain Sergio Caporetti was waiting to welcome them aboard.

He stood at the head of the gangway, and sensible to the importance of the moment he was decked in his finest and best; a double-breasted suit with a cream and lilac pinstripe set off the tomato-red silk tie, but his two-tone black and white crocodile skin shoes drew attention to his large feet and his gait was that of an emperor penguin. A liberal application of a hair pomade with a penetrating smell of violets had flattened his black hair into a shiny slick, bisected by the ruler-straight line of white scalp which was his parting. However, the aroma of the pomade was at odds with the particularly stinky cheroot of a brand which Sergio reserved for weddings, funerals and other special occasions.

His beautiful gazelle eyes became passionate and dark as they lit on Fifi Larsen. Mrs Larsen's tight-fitting slacks moved as though they were full of live rabbits and her pink sweater was straining its seams. Her eyes were sparkling with champagne and she giggled without apparent reason, flushing under Sergio's scrutiny.

The tour of *Kingfisher* began, Sergio Caporetti taking up an escort position directly behind Mrs Larsen. They had hardly descended the first ladder when Mrs Larsen let out a small squeak and shot about eighteen inches into the air, before coming back to earth with all her plentiful womanhood aquiver.

'My dear Fifi, whatever is wrong?' Her husband was all solicitude, while behind her Sergio Caporetti wore an expression of cherubic innocence. Johnny felt dizzy with alarm, for he had seen Sergio's great hairy paw settling comfortably on to those majestic buttocks. Mrs Fifi Larsen had been thoroughly goosed.

In relieved disbelief Johnny heard her reply, which was preceded by another giggle.

'I seem to have twisted my ankle. Perhaps there is somewhere I could sit down.'

Johnny looked around frantically for Tracey to get Mrs Larsen out of Sergio's clutching range, but before he could signal her, Fifi was limping away on Sergio Caporetti's arm, bravely declining all offers of help.

'Please don't let me spoil your fun. I'll just sit in the Captain's cabin for a few minutes.'

Quickly Johnny moved up beside the silver-haired Larsen

and resolved to stay close beside him. Even if he could not prevent Fifi visiting Sergio's quarters, he was going to make good and sure that the husband didn't join the party.

'This is the explosives locker.' Johnny took Larsen's arm and led him away. 'We keep a store of plastic explosives for underwater blasting—'

Larsen's concern at his wife's injury dissolved and he became immersed in the tour of *Kingfisher*. Johnny followed the line of production for him from the moment the gravel was sucked in through the dredge.

As they left the cyclone room Johnny preceded him, holding the steel door open for Larsen.

'From the cyclone the concentrates pass through here into—' He stopped with surprise as they entered the narrow compartment beyond the cyclone.

'What's wrong, Lance?' Larsen demanded.

'No. It's nothing,' Johnny assured him. After the surprise of finding that the inspection plate in the conveyor tunnel had been moved he realized that it was as well from a security angle. Probably the marine architects had ordered the modification. 'The concentrates are carried through into the next compartment to the X-ray room. This way, please.'

As Johnny led the way to the next door he resolved to check with the architects. Larsen asked a question and he replied and the conveyor tunnel was forgotten. They went through into the X-ray room.

'He noticed it.' Benedict puffed quickly and nervously at the cigarette cupped in his hand. 'He doesn't miss a thing. The bastard.'

'He noticed it, yes. But he accepted it.' Ruby was definite. 'I know him. I was watching him. He was disturbed for a second then he rationalized it. I could almost see his mind work. He accepted it.'

They stood together on the exposed angle of *Kingfisher*'s bridge. Suddenly Ruby laughed.

'Don't look so worried,' she warned him merrily. 'We are being watched by your sister again. She's down on the foredeck. Come.'

Still smiling she led him around the angle of the bridge house, and out of sight she was immediately deadly serious again.

'That sister of yours is getting suspicious. We must keep away from each other until you tell Johnny.'

Benedict nodded.

'When are you going to tell him?' she demanded.

'Soon.'

'How soon?' Ruby would not be able to rest until it was out in the open, until Benedict had committed himself, yet she must not push him too hard.

'As soon as *Kingfisher* runs the Company under. I will pick the moment that he is beaten financially, then I will tell him. I want it to be the *coup de grâce*.'

'When will it be, Benedict darling? I am so anxious to be with you – without all this subterfuge.'

Benedict opened his mouth to reply and froze like that, his expression changing slowly into that of a man who doubts the evidence of his own eyes. He was staring over Ruby's shoulder.

Ruby turned quickly. The curtain across the Captain's porthole behind her was open a chink. She looked in upon a spectacle of such whole-hearted rubicund magnitude that it should have occurred only to Olympus between Jupiter and Juno.

In the cabin Fifi Larsen was receiving treatment for her sprained ankle.

'Well, you've got your toy now. Let's hope for all our sakes you can do something with it,' Benedict smiled pleasantly as he came across to where Johnny stood with Larsen under the great gallows-shaped gantry on the foredeck of *Kingfisher* that would raise and lower the dredging head.

'Toy, Mr van der Byl?' Larsen's white eyebrows bristled. 'Surely you have no doubts? I mean, now that you've got this Thunderbolt and Suicide concession?'

'Oh, I wouldn't say doubts, Mr Larsen. Reservations perhaps, but not really doubts. Mr Lance has been the champion of this venture. His enthusiasm has carried it – in the face of all opposition. Even that of my late father.' Benedict turned to Larsen smoothly.

'Your father opposed the scheme? I didn't know that!' Larsen was perturbed.

'Not opposed it, Mr Larsen.' Benedict smiled reassuringly. 'Not really opposed it. But you will notice that he was prepared to risk your money – not his own. That will give you some idea of how he felt.'

There was a chilled silence, before Larsen turned to Johnny.

'Well, Lance, thank you for an interesting day. Very interesting. I'll be watching your progress with attention – close attention.' And he turned away and strode to where a subdued and demure Fifi was waiting with a group of the other wives.

'Thanks.' Johnny gave Benedict a bleak grin.

'Don't mention it.' Benedict smiled that charming boyish smile.

'At the end of this week I'll take that little speech of yours, roll it into a ball and ram it down your throat,' Johnny promised him softly, and Benedict's expression changed. His eyes slitted and his grin showed his teeth and tightened the soft line of his jaw.

'You're pretty slick with your mouth, Lance.'

They glared at each other, the antagonism so apparent, as elemental as a pair of rutting stags, that they were suddenly the centre of all attention. The guests stared curiously, aware of the drama but not understanding it.

Ruby started forward quickly to intervene, taking Johnny's elbow, her voice sugary.

'Oh, Benedict, do you mind if I talk to Johnny a moment? I have to know if he's returning to Cape Town with me this evening.'

She led him away, and the tension dissolved. The disappointed guests began drifting to the gangplank and filing down on to the jetty.

In the confusion of embarkation aboard the two Dakotas, Johnny managed to exchange a last word with Tracey.

'You'll stay here until you know?' she asked. He nodded.

'Good luck, Johnny. I'll pray for you,' she whispered, then followed Ruby Lance up into the fuselage of the Dakota.

Johnny watched the two big aircraft taxi down to the end of the strip, turn in succession and roar away into the purple and red sky of evening.

After they had gone it was very still, and the silence of the desert was complete. Johnny sat in the open Land-Rover and smoked a cigarette while the night came down around him.

He was uneasy, aware of a deep-down tickle of apprehension and foreboding which he could not pin down.

The last glow of sunset faded from the western sky, and the desert stars were bright and hard and close to the earth, silvering the dome of space with the splendour that the city-dweller would never guess at.

Still Johnny hunched in the seat of the open Land-Rover trying to explore the source of his uneasiness, but with so little success that at last he must attribute it to the strain and fatigue of the last few months, his involvement with Tracey, his steadily worsening relationship with Ruby – and the latest clash with Benedict.

He flicked the stub of his cigarette away, watching morosely the explosion of red sparks as it struck the earth, then he started the Land-Rover and drove slowly down towards the jetty.

Kingfisher's lights were smeared in paths of yellow and silver across the still waters of the bay. Every porthole was lit brightly, giving her the festive air of a cruise ship.

Johnny left the Land-Rover at the head of the jetty, and walked out to her. The muted throb of her engines cheered him a little, the knowledge that the vessel was preparing for the morrow.

On deck he paused beside the gigantic compressed air tanks, each the size of a steam locomotive, and checked the pressure gauges. The needles were moving perceptibly around the dials, and his mood lightened a little.

He went up the ladder to the bridge, and into the chart-room where Sergio and Hugo were drinking coffee.

'Not my fault, Mr Lance,' Sergio began defensively. 'I am a gentleman – I cannot refuse a lady.'

'You'll dig your own grave with that spade of yours one day,' Johnny warned him grimly, as he went to the chart table and hung over it.

'Now let's get cracking.' Johnny's sense of dread lifted completely as he looked down at the large-scale Admiralty chart. The twin humps of Thunderbolt and Suicide were clearly marked. 'Hugo, have you got the prospecting schedules?'

'There, on the table.' Hugo and Sergio came to stand on each side of Johnny while he opened the bound file of typewritten sheets.

'The soundings we made differ from the Admiralty chart. We'll put in our figures, before we plot the dredging pattern.'

The three of them settled over the chart with dividers

and parallels to begin marking in the path that *Kingfisher* would follow through the maze of reefs and gullies.

It was long after midnight before Johnny made his way wearily to the guest cabin below the bridge. He kicked off his shoes, lay on the bunk to rest a moment before undressing and fell into a deathlike sleep of exhaustion.

He was awakened by one of the crew with a mug of coffee and he pulled on a windbreaker before hurrying on to the bridge.

Kingfisher was just passing out through the channel of Cartridge Bay into the open sea, and Sergio grinned at him from where he stood beside the helm.

Dawn was only a lemon-coloured promise over the desert behind them, and the sea was black as washed anthracite, kicked into a chop by the small morning wind. They stood on the darkened bridge and sipped steaming coffee, cupping their hands around the enamel mugs.

Then they turned and ran south, parallel to the desert which was now touched with hot orange and violet. The seabirds were up, a flight of malgas turned to glowing darts of fire by the early sun as they winged swiftly across the bows.

With a dramatic suddenness the sun came up over the horizon, and highlighted the chalk-white cliffs of Thunderbolt and Suicide far ahead so that they shone like beacons on the cold green sea. The curtains of spray that burst on the islands flashed and faded as they shot into the sky and fell again.

Wild Goose was waiting for them lying under the lee of the islands, but she came out to meet them, staggering and plunging theatrically over the short uneasy sea that hooked around the islands or boiled through the gap between them.

The radio telephone began crackling and squawking as the sighting reports from Johnny's watchtowers on the shore started coming in, cross-referencing to give *Kingfisher* her position over the ground. There were short exchanges between Sergio and Hugo on *Wild Goose* as they came together, and the little trawler worked in close, ready to give assistance with the laying of the cables if she were needed.

But standing in the angle of the bridge, Sergio Caporetti had the situation under his control. The grubby marine cap pushed to the back of his head and a long black cheroot stuck in the side of his mouth, he stood balanced on the balls of his feet, his eyes darting from judging the set of the sea surf to the repeater of the computer which was feeding him his soundings and position – yet attentive to the R/T reports from shore and from *Wild Goose*.

Johnny was contented with his choice of man as he watched Sergio at work. *Kingfisher* crept slowly up into the lee of Suicide Island, half a mile from the pearly white cliffs, then she hung there a moment before Sergio punched one of the buttons on her control panel.

From forward there was the harsh metallic roar of an anchor cable running out, and as *Kingfisher* backed away leaving a yellow-painted buoy the size of a barrage balloon bobbing under the cliffs of Suicide so one of the massive deck winches began automatically paying out its six-inch steel cable.

Kingfisher backed and crept forward, drifted down on the current or butted up against it while she went through the laborious but delicate operation of laying her four anchors at each point of the compass. Chained above each anchor floated the huge yellow buoys, and from each buoy the steel cables led to the winches on *Kingfisher*'s deck. On instructions from the computer the winches on each quarter would pay out or reel in the cable to hold *Kingfisher* steady over the ground while she worked.

It was mid-afternoon before *Kingfisher* was ready, pinned down like an insect to a board, and the computer reported that she was directly over the gully that Johnny had selected as the starting point. She had twenty-five fathoms of water under her – and then the thick bank of gravel.

'All is ready.' Sergio turned to Johnny, who had stood by quietly all this time – not interfering in the task of positioning. 'You will begin the programme now?'

'Yes.' Johnny stirred himself.

'I would like to watch,' Sergio suggested, and Johnny nodded.

'All right, come.' Sergio handed over the bridge to the helmsman and they went down to the armoured door of the computer room.

Johnny opened the lock. There were only two keys to this compartment. Johnny had one and Benedict van der Byl had the other. He had insisted on having the duplicate, and Johnny had reluctantly agreed not knowing that the key would be used in Las Palmas.

The heavy steel door swung back, and Johnny stepped over the coaming and seated himself before the console of the computer. Covered in cellophane and suspended on a clip above the keyboard of the computer were the cards containing the various programme codes.

Johnny selected the sheet headed: PRIMARY OPERATION: DREDGING AND RECOVERY, and began feeding the code into the computer, punching it on the keyboard.

'Beta, stroke, oh, oh, seven, alpha.'

And within the enamelled console, a change of sound heralded the beginning of the new programme, the hum of her reels and the click of the selectors, while on her control panel the lights blinked and flashed.

Now the computer's screen began to answer the instruction, spelling out her response like a typewriter.

'New programme.'
'Primary Operation. Dredging and Recovery.'
'Phase One.'

'Initiate safety procedure:—

a) Report air pressure . . . 1

b) Report air pressure . . . 2'

Johnny leaned back in the padded stool and watched the exhaustive check that the computer now made of *Kingfisher*'s equipment, typing out the results on the screen.

'What she do now?' Sergio asked curiously, as though he had never spent ten days in this compartment assisting his Japanese friend. Johnny explained the procedure briefly.

'How come you know this so good?' Sergio enquired.

'I spent a month at the Computer Company's head office in America last year while they designed this machine.'

'You the only one in the Company who can work it?'

'Mr Benedict van der Byl has done the course as well as I.' Johnny told him, then he leaned forward again. 'Now she is set.'

The screen on the computer reported itself satisfied.

'Phase One Completed.

Initiate Phase Two.

Lowering and siting of dredge head.'

Johnny stood up. 'Okay, let's get upstairs.'

He locked the door of the computer room and followed Sergio up to the bridge.

Johnny went to stand beside the repeater screen on the bridge, which was relaying the computer's signals exactly as they were printed on the main screen in the compartment below. He could see out through the windows of the wheelhouse, and he watched the automatic response of the heavy equipment on the foredeck.

The gantry swung forward, and the steel arms picked the dredge head from its chocks, and lifted it with the armoured suction hose dangling behind it. Then the gantry swung back, and with a jerky mechanical movement lowered the head through the square opening in the deck. This well pierced the hull, and through it the hose began to snake – a monstrous black python sliding into its hole. The huge reels that held the hose revolved smoothly, as the dredge was lowered to the sea bed.

'Head on bottom.'

The computer screen reported, and the hose reels stopped abruptly.

'Phase Two Completed.
Initiate Phase Three.
Cyclone Revolutions 300.
Vent dredge pump.'

There was a rising high-pitched whine now, like the approach of a jet aircraft. The sound reached a peak, and steadied – and immediately another sound overlaid it. The dull roar of high-pressure air through water, a sound of such power and excitement that Johnny felt the hair on his forearms prickle erect. He stood still as a statue, his expression rapt and his lips set in a small secret smile. That sound was the culmination of two years of planning and endeavour, the sweet reward of the driving dedication that had made a dream into reality.

Suddenly he wished that Tracey was with him to share this moment, and then he knew instinctively that she had deliberately left him alone to savour his moment of triumph.

He grinned then, as he watched the thick black hose engorge and pulse with internal life, like a great artery – pumping, pumping, pumping.

In his imagination Johnny could see the rich porridgy mixture of sea water and mud and gravel that shot up the hose into the spinning cyclone, he could imagine the steel head on the sea bed below the hull surging rhythmically to stir the sand and pound loose any gravel that pressure had welded into a conglomerate.

From the waste pipe over *Kingfisher*'s stern poured a solid steam of dirty yellow water mixed with the sand and gravel that had been rejected and spun out of the cyclone. It stained the green sea with a cloudy fecal discharge, like the effluent from a sewage outlet.

For three days and two nights *Kingfisher*'s pumps roared, and she inched forward along the marine gully like a fussy housewife vacuuming every speck of dust from her floor. As the third evening spread its dark cloak over her, Johnny Lance sat on the padded seat in front of the computer console. He sat forward on the stool with his elbows on his knees and his face in his hands for a full hour. He sat like that in the attitude of despair.

When he lifted his head, his face was haggard and the lines of defeat were clearly cut into his features with the cold chisel of failure.

From the meagre recovery of small diamonds that *Kingfisher* had made in the last three days it was clear beyond reasonable doubt that despite all the indications the Thunderbolt and Suicide field would not support the running costs of the vessel, let alone cover the overheads, or the interest charges and capital repayments on the loan account.

Van Der Byl Diamond Company was finished – and Johnny Lance was financially ruined beyond any possibility of ever finding redemption.

It remained only for the jackals to assemble and squabble over the carcass.

Sergio Caporetti leaned over the railing of the bridge, blowing long streamers of blue cheroot smoke from mouth and nostrils to help foul a morning which was already thick and grey with sea-fret. The islands of Thunderbolt and Suicide were blanketed by the mist, but the surf broke against their hidden cliffs like distant artillery and the seabirds' voices were plaintive and small lost souls in the void.

Wild Goose came bustling up out of the mist, swinging in under *Kingfisher*'s side to hover there under power with two of her crew fending off.

Hugo Kramer stuck his white blond head out of the wheelhouse window, and shouted up at the deck.

'Okay, boss. Come on!'

Sergio watched the tall figure on *Kingfisher*'s deck rouse and look around like a man waking from sleep. Johnny Lance lifted his head and looked up at the bridge, and Sergio noticed that he was unshaven, a new beard darkened his jaw and emphasized its prominence. He looked as though he had not slept, and he hunched into the windbreaker with the collar turned up against the mist. He did not smile, but lifted one hand in farewell salute to Sergio – who noticed incongruously that the index finger was missing from the hand. Somehow, that pathetic little detail struck Sergio. He was sorry, truly sorry. But there is always a loser in every game, and twenty-five thousand pounds is a lot of money.

'Good luck, Johnny.'

'Thanks, Sergio.' Johnny went to the rail lugging his briefcase and swung over it; he dropped swiftly down the steel rungs set in *Kingfisher*'s hull and jumped the narrow gap of surging water to *Wild Goose*'s deck.

The trawler's engine bellowed, and she pulled away, rounding on to a course for Cartridge Bay. Johnny Lance stood on her open deck looking back at *Kingfisher*.

'He's a good guy.' Sergio shook his head with regret.

'He's a boss,' the helmsman grunted. 'No boss is any good.'

'Hey, you! I am also a boss,' Sergio challenged him.

'Like I said.' The helmsman suppressed a grin.

'I kiss your mother,' Sergio insulted him with dignity, then changed the subject. 'I go below now, take over.'

Sergio opened the door to the control room with the duplicate key. He closed the door behind him, seated himself at the console and took from his pocket a sheet of paper headed: KAMINIKOTO SECONDARY RECOVERY PROGRAMME.

Ten minutes later he came out of the control room and locked the door.

'Kammy, I love you,' he chuckled as he closed off the watertight doors that isolated this deck from the one above. He wound the locking bars into position to ensure that he was not interrupted by one of his own crew.

From the tool cupboard on the bulkhead he selected a pair of set spanners and went through into the conveyor room. It took him twenty minutes to unscrew the heavy, deeply threaded bolts that secured the hatch. It had been designed to resist easy entry – a deterrent to casual investigation, but at last Sergio could lift the steel plate off its seating.

He eyed the small square opening with distaste, and reflexively sucked in his pendulous belly. The hatch had not been designed to afford passage to a man of his dimensions.

He took off his cap and jacket and hung them on the cock of one of the pipes, then he ground out his cheroot under his heel and brushed back the hair from his forehead with both hands, checked that his flashlight was in his pocket, and committed himself to the hatch.

He wriggled and kicked, and grunted and built up a heavy sweat for five minutes before he had squeezed through into the conveyor belt tunnel. He squatted on his haunches, panting heavily and flashed his torch along the tunnel. Above his head the conveyor belt carrying the gravel ran smoothly, but the residual heat from the driers made it unbearably hot. He began to crawl rapidly to the end of the tunnel.

From the inside it was impossible to tell, without measuring, that the conveyor belt tunnel was shorter by twelve feet than the external length.

The end of the tunnel was false, and beyond it was a secret cubicle only just large enough to house Kaminikoto's equipment through which all the gravel passed on its journey to the X-ray room.

The Japanese genius for miniaturization was demonstrated by the equipment in this secret cubicle. It was an almost exact copy of the sorting equipment in the main X-ray compartment – except that it had been scaled down to one tenth of the size without affecting its efficiency; in addition, this miniature plant could discriminate in the diamonds it selected. It would not allow a stone over four carats to pass through, and it screened out fixed percentages of the smaller stones – allowing only a proportion of the smaller and less valuable diamonds to proceed through into the main X-ray room.

It was an amazing piece of electronic engineering, but Sergio was unimpressed as he lay on his side in the cramped hot tunnel and began laboriously to unscrew another smaller plate in the false bulkhead.

At last it was open, and he reached through the opening; after a few seconds of fiddling and groping and heavy breathing he brought out a stainless steel cup with a capacity of about two pints. There were clamps on the cup to hold it in position below the chute under Kaminikoto's machine.

The metal cup was heavy, and Sergio placed it carefully on the deck beside him before propping himself on an elbow and shining the flashlight into the cup and took something out of it, stared at it a moment then dropped it back.

'By the blood of all the martyrs!' he gasped with shock, and then immediately contrite for his blasphemy he crossed himself awkwardly with the hand holding the flashlight. Then again he shone the torch into the cup, and shook his head in disbelief. Quickly he pulled a canvas drawstring bag from his pocket, and lying on his side he carefully poured the contents of the cup into the bag, drew the string tight and stuffed it back into his pocket where it made a big hard bulge on his hip like a paper sack of rock-candy. He clamped the stainless steel cup back into position, screwed the coverplate over the opening, and backed away down the tunnel on hands and knees.

He very badly needed a cheroot.

Four hours later Hugo Kramer shinned up the ladder on to *Kingfisher*'s deck while his helmsman took the trawler down to leeward to wait for him.

Sergio shouted down from the bridge.

'Johnny he has gone?'

'*Ja!*' Hugo shouted back. 'He should almost be in Cape Town by now. That Beechcraft is a fast plane.'

'Good.'

'How did it go with you?' Hugo countered.

'Come up – I'll show you.'

Sergio led him into his cabin behind the bridge and locked the door carefully. Then he went to each of the portholes and drew the curtains across them, before crossing to his desk and switching on the reading lamp.

'Sit down.' Sergio indicated the chair opposite the desk. 'You want a drink, or something?'

'Come on,' Hugo grated impatiently. 'Stop mucking around, let's have a look.'

'Ah!' Sergio looked at him sadly. 'You Germans, you are always too much hurry. You cannot rest, enjoy life—'

'Cut the crap!' Hugo's pale eyes were on his face, and Sergio was suddenly aware that this man was dangerous, like a tiger-shark. Coldly dangerous, without malice or passion. Sergio was surprised he had not noticed it before. I must be careful with this one, he thought, and he unlocked the drawer of his desk and took out the canvas bag.

He loosened the drawstring and poured the diamonds on to the blotter. The smallest was the size of a match head – perhaps point one of a carat, and the poorest quality was black and granular-looking, ugly little industrial stones, for Kammy had been careful not to take out only the best and so distort the *Kingfisher*'s recovery as to arouse suspicion. There were hundreds of these tiny crystals and chips which would fetch a few pounds in the industrial market; but there were other stones in the full range of quality and shapes and sizes – as big as green peas, or as marbles, and a few bigger than that. Some of them were perfect octahedron crystals, others water-worn, chipped or amorphous in shape.

They formed a dully glittering pile in the centre of the blotter, in all perhaps five hundred diamonds, yet all of these were dwarfed by one single stone that lay in the centre of the pile, rising out of it the way Mount Everest rises from her foothills.

There are freak diamonds so large or unusual that they become legend. Diamonds who have their own names and whose histories are recorded and invested with romance. The great 'paragons' – stones of the first water whose cut and finished weight exceed one hundred metric carats.

Africa has produced many of them: the *Jonker Diamond*, a 726-carat rough cut to a brilliant of 125 carats that hangs about the throat of the Queen of Nepal; The *Jubilee Diamond*, a superb 245-carat cushion of unearthly fire fashioned out of 650-carat rough – then the biggest of them all, a monstrous rough stone of 3,106 carats, the *Cullinan* which yielded not one, but two paragons. The

Great Star of Africa at 530 carats and *The Cullinan II* at 317 carats. Both these stones grace the Crown Jewels of England.

Now on Sergio Caporetti's desk lay a rough stone which would add yet another paragon to the list.

'Have you weighed it?' asked Hugo, and Sergio nodded.

'How much?'

'Three hundred and twenty carats,' Sergio said softly.

'Jesus!' whispered Hugo, and Sergio crossed himself quickly to dissociate from the use of names.

Reverently Hugo Kramer leaned forward and picked up the big diamond. It filled the palm of his hand, the cleavage plane that formed its base was smooth and clean as an axe-stroke. There were bigger diamonds in history, but this diamond had a special feature which would set it in a niche of its own and endow it with peculiar value.

Its colour was the serene blue of a high summer sky.

This stone could pay half the total bill for *Kingfisher*'s construction – if it were ever used for that purpose.

Hugo replaced the blue diamond on the desk and lit a cigarette without taking his eyes from it.

'This field – it is bigger – richer, far richer than we had guessed.'

Sergio nodded.

'In three days we have taken diamonds that I hoped to see once in five years,' Hugo went on as he began picking out the larger diamonds from the pile and laying them in a line across the desktop in approximate order of size, while Sergio opened his desk drawer and took out a box of his special occasion cheroots.

'We must tell the boss,' Hugo decided. He began arranging the diamonds in a neat circle about the big blue, thinking deeply. 'He must know how rich it is before he talks to Lance. He must make arrangements. He will know what to do – he's a clever one.'

'What about these?' Sergio indicated the treasure on the desk. 'Are you going to take them off?'

Hugo hesitated. 'No,' he decided. 'We could never get rid of this big Blue through the usual channel, it is too big, too distinctive. We will keep it aboard. When the boss takes over the Company again – then we will just declare it all nice and legal. No trouble.' He stood up. 'Look after them. I must hurry if I am to get a message to Cape Town in time.'

'The Company bears my father's name, Mr Larsen. It's as simple as that.' Benedict's voice was husky with emotion, and he looked down at his hands. 'I have a duty to my father's memory.'

'My boy, well—' Larsen came to lay his hand on Benedict's arm. 'Well, I just don't know what to say. Honour is a rare and precious thing these days.' With his free hand he was groping almost frantically for the bell on the desk behind him. He must get this signed up solid before the youngster changed his mind.

'I tried to warn you, Mr Larsen. My father and I never had any faith in this marine recovery scheme. Lance pushed it through—'

'Yes, quite so,' Larsen agreed, and turned to his assistant who came into the office at the trot in response to the bell. 'Ah, Simon. The Van Der Byl Diamond loan. Will you have an agreement made out immediately – Mr van der Byl will take over the capital amounts, and the outstanding interest as well.' By rolling his eyes Larsen tried to convey to his assistant the deadly urgency of the situation. The young man understood and fifteen minutes later laid the Agreement on Larsen's desk. Larsen unscrewed the cap of his pen and handed it to Benedict.

Larsen and three of his young men ushered Benedict out through the glass doors of the bank and across the pavement to where the Rolls stood in the reserved parking bay in Adderley Street.

Benedict settled into the back seat, acknowledged the bank official's farewells, and tapped on the chauffeur's window. As they pulled away, Ruby Lance slipped her hand through his arm and squeezed it.

'Did you get it?' she asked.

Benedict grinned happily. 'I frightened five years' growth out of old Larsen. He almost broke his neck in the hurry to give it to me.'

'Now you've got it all.' Ruby snuggled a little closer to him on the soft leather upholstery. Benedict nodded, and checked his watch.

'The meeting is set for fifteen minutes' time. I'll go up the front way, but I want you to go up in the private lift from the basement garage, and wait in my office. We will be in the Board Room. I will ring you at the right moment.'

The Rolls picked its way slowly down around the Heerengracht and double-parked outside the building. The chauffeur came to open the door, but before he alighted Benedict smiled into Ruby's face.

'This will be one of the high moments of my life,' he said softly. 'This time I've got the bastard cold.'

'I'll be waiting for you,' Ruby said, and he climbed out of the Rolls. He waited until it had turned into the entrance ramp of the basement garage, then he crossed the pavement into the main lobby of the skyscraper. He strode to the elevator with long eager strides, and his mouth kept pulling into a small excited smile.

The Board Room was set high, and the picture windows looked up at the great squat mountain, whose sheer cliffs dropped directly to the wooded slopes up which the first buildings of the city straggled.

Johnny Lance stood at the head of the table. He had lost weight in the last few days, so that his shoulders appeared bony and gaunt under the white silk shirt. He had discarded his jacket, and pulled the knot of his tie down an inch. The bones of his cheeks and jaw made harsh angles that were accentuated and not softened by the deep shades of fatigue that darkened his eyes. His hands were thrust deep into his pockets, and he spoke without reference to the sheaf of paper that lay on the table before him.

'Our working costs are as close as dammit to a hundred pounds an hour; right Mike?' And Michael Shapiro nodded. 'Well, we worked the Suicide main gully for sixty-six hours, and we recovered a princely 200 carats of the lousiest pile of junk I've ever seen. If we get a thousand quid for the lot we'll be doing well. This for an expenditure of six and a half thousand.'

Johnny paused, and looked around the table. Michael Shapiro was doodling on his note pad with fierce concentration, Tracey van der Byl was pale, her eyes never left Johnny's face and her expression ached with pity and helpless compassion; Benedict van der Byl was looking out of the window at the mountain, he was slumped comfortably in his chair, smiling a little and listening politely.

'The Suicide main gully is one of the five most likely parts of the entire concession. It's no good, so the rest of the field may be useless. We have the two other concessions, the original fields, to try. However, it will take three or four days to get *Kingfisher* moved up the coast.' Johnny paused, and Benedict swivelled his chair, still with the small smile on his lips.

'The interest payments fall due on the 30th – three days'

time. Where are you going to find one hundred and fifty thousand Rand?'

'Yes,' Johnny nodded. 'I think I can persuade Larsen to extend for a few weeks; he will bloody well have to if he wants to protect his—'

'Hold on,' Benedict murmured. 'Larsen has got nothing to do with it.'

Johnny was silent, watching him warily. 'Explain,' he invited.

'I've taken over the loan from Larsen,' Benedict told him. 'I'm not interested in extending.'

'Larsen wouldn't have negotiated without warning me.' Johnny was stricken, his disclaimer was wrung from him in pain.

'Shapiro?' Benedict turned to Michael Shapiro for confirmation.

'Sorry, Johnny. It's true. I've seen the documents.'

'Thanks, Michael.' Johnny's voice was bitter with accusations. 'Thanks for letting me know.'

'He showed me a few minutes before the meeting, Johnny. I swear I didn't know.' Michael's expression was distressed.

'Right.' Benedict straightened up in his chair, his voice was brisk. 'Let's get down to first principles. You've ruined my father's Company, Lance – but, thank God, I may be able to retrieve the situation. Call it sentiment or what you like, but I want your shares and yours.' He turned to Tracey and nodded at her.

'No,' said Tracey sharply.

'Right.' Benedict smiled at her. 'Then I'm going to hammer Johnny Lance for his full obligations. That way I get the Company anyway, but I'll make damn sure he remains an unrehabilitated bankrupt for the rest of his life.'

Tracey lifted her hand to her throat, and turned her eyes to Johnny. Waiting for him to set a lead. There was a long stark silence, then Johnny Lance dropped his eyes.

'I've still got three days.' His voice was gruff and tired.

'Three days you have.' Benedict grinned coldly. 'And you're welcome to them.'

Johnny picked up his papers and put them under one arm; he took his jacket off the back of the chair and swung it over his shoulder.

'Wait,' ordered Benedict.

'What for?' Johnny's grin was twisted. 'You've had your fun.'

Benedict lifted the receiver of the telephone on the table and dialled swiftly.

'Come through, please, darling.' He spoke into the mouthpiece and smiled at Johnny as he hung up. Then as the door opened he went to meet Ruby Lance, and kissed her on the mouth. The two of them stood, arm in arm, and looked at Johnny.

'The Company is not the only thing I've taken from you,' Benedict said softly.

'I want a divorce.' Ruby looked steadily into Johnny's face. 'Benedict and I are going to get married.'

They were all watching Johnny, and they saw him flinch. He looked from one face to the other, then his mouth tightened and his forehead furrowed.

Tracey saw his anger mounting, and her eyes flicked to Benedict's face. He was leaning forward expectantly, his lips quivering with expectation, his eyes alight with triumph. Tracey wanted to scream out a warning to Johnny, stop him falling into the trap that Benedict had set so carefully.

Johnny took a pace forward, coming up on to the balls of his feet. He was about to make his defeat total and ineradicable. Then Benedict spoiled it for himself, he goaded once more.

'Game, set and match, Johnny Lance,' he crowed.

The effort of will that Johnny made to recover his reason was not shown on his face, he made the step forward seem natural and he continued towards the door.

'The house is in your name, of course, Ruby, so would you please send my things down to the Tulbagh Hotel,' he asked quietly.

He stopped in front of the couple and spoke to Ruby.

'You'll want to protect your reputation, of course, so I'll not sue for adultery. We'll call it desertion.'

'You're eating your guts out,' jeered Benedict. 'Lance can't keep his woman. Van der Byl took her away from him. Go on, sue for adultery – let the world hear it.'

'As you wish,' Johnny agreed.

And he walked out of the Board Room to the elevators.

Johnny flopped on to the bed fully dressed, and rubbed his closed eyes with his fingertips. He felt confused and off-balance, the edge of his mind which usually slashed quickly and incisively through a problem was dulled. This problem was so multiplied and tangled that he felt like a man in a thicket of African ebony trying to cut his way out with a blunt machete.

Without opening his eyes he groped for the telephone, and the girl on the hotel switchboard downstairs answered. He gave her a number in Kimberley.

'Person to person, Mr Ralph Ellison.'

'Fifteen minutes' delay, Mr Lance,' the girl told him.

'Okay,' Johnny replied. 'Ask room service to send me up a Chivas Regal and soda.' He suddenly needed liquor, something to dull the pain. 'Make that a double, honey – no, make it two doubles.'

He had drained both glasses by the time his Kimberley call came through.

'Ralph?' Johnny spoke into the receiver.

'Johnny, how nice of you to call.' There was an echo like distant laughter beneath those cool ambassadorial tones in Ralph Ellison's voice and Johnny knew instantly that the word was out. Damn it, he was slow – of course Benedict would have blocked him.

'Are you still interested in a deal on the Thunderbolt and Suicide Concession?' Johnny loosed a despairing long-range shot.

'Of course, you know we are always interested,' Ralph replied.

'The price is two million.' Johnny lost interest and lay back on the bed, closing his eyes again. He knew Ralph was having his revenge – you didn't take this boy to court and win without sowing yourself a minefield to retreat over.

'Two million,' Ralph murmured. 'Now that's a little high – for a field that's yielding 200 carats of small industrial diamonds per 10,000 loads, that's definitely on the high side. Of course we wouldn't want that battleship of yours either – we are not starting our own navy.' Ralph chuckled juicily. 'We could talk around fifty to a hundred grand – no more than that, Johnny.'

'Okay, Ralph.' Johnny spoke wearily. 'Thanks all the same. We'll have a drink together sometime.'

'Any time, Johnny,' Ralph agreed. 'Any time at all. You call me.'

Johnny dropped the receiver back on its hook and looked at the ceiling. He had heard that a gunshot wound was numb at first – he felt numb now. All that energy had seeped out of him, he had lost direction.

The telephone shrilled and he picked it up. The girl on the switchboard asked politely:

'Are you finished, Mr Lance?'

'Yes,' said Johnny. 'You might say that.'

'Is there anything else you require?' The girl sounded puzzled.

'Yes, honey, send up the hemlock.'

'I beg your pardon?'

'Two more big fat whiskies, please.'

He drank them in the bath, and while he dried himself

the doorbell tinkled. He wound the bath towel round his waist and went through to open the door.

Tracey stepped into the bedroom and closed the door behind her. They stood looking at each other for a long moment. Her eyes were big and dark, reflecting his agony faithfully.

'Johnny—' Her voice was husky, and she reached out and laid the palm of her hand on his cheek. His shoulders sagged, and he moved close to her, his forehead sinking on to her shoulder. He sighed, a ragged broken exhalation of breath.

'Come,' she said, and led him to the bed. Leaving him there she went to the windows and closed the curtains.

It was warm and safe in the half dark of the curtained room, and they held each other as they had done long years before. Clinging together, so their breathing mingled, and it was not necessary to speak.

When they became lovers it seemed that they had waited all their lives for that moment.

Afterwards he lay in her arms and he felt strength flowing back into him, drawing it from her. When he sat up in the bed the bemused, almost dazed expression had gone. His jaw was out and his eyes were bright.

'We've still got three days,' he said.

'Yes.' She sat up beside him. 'Go, Johnny. Go quickly, don't waste another moment.'

'I'll pull *Kingfisher* out of the main gully. I'll find those diamonds. They are there. I *know* they are there. I'll take her right down into the jaws of Thunderbolt and Suicide, I'll find those bloody diamonds – damned if I won't.' He swung his legs off the bed, reached for his clothes, glancing at his wrist watch as he did so. 'Four o'clock. I can get to Cartridge Bay a few minutes after dark. Will you call the communications office, ask them to radio Cartridge Bay and have a flare path set for me, and *Wild Goose* standing by to run me out?'

'I'll phone them from here. Then I'll take a bath – you go on. Don't waste time.' Tracey nodded eagerly, and Johnny let his eyes drop down over her body. He reached out and touched one big white breast almost diffidently.

'You are beautiful – I hate to go.'

'It will all be here, waiting for you, when you get back.'

'It wasn't the way I'd planned it. It wasn't good the way I'd dreamed about it.' Benedict paced angrily over the floor of the Old Man's study, swinging towards the windows and pausing to stare out at the mountain across the valley.

'You hurt him. You crushed him.' Ruby moved restlessly in her chair across the room from him, curling her long golden legs defensively under her, sitting like a cat in the big chair. She was worried, and it showed in the tiny crows' feet at the corners of her eyes and the way she held her lips pursed. She should have anticipated this reaction from him, she should have known that this moment of triumph could never match his anticipation of it – and that revenge must always be followed by sour distaste and a feeling of disappointment. She realized that her safest course was to leave him alone now, she should never have returned with him to the old house on Wynberg Hill. She stood up.

'Darling.' She crossed to him. 'I'm going home now. I want to pack his clothes and get rid of them. I want to wipe out every memory of him. From now on it's you and I – together.'

She stretched up to kiss him, but Benedict turned his face away.

'Oh! So you're going, are you?' His expression was petulant, his lips pouting spitefully.

'We're both tired, darling. Let us both rest a while – and I'll come back later this evening.'

157

'So now you're giving the orders, are you?' He laughed nastily.

'Darling—'

'And cut out all the darlings. We pulled a deal and it didn't work out. You were meant to be a club to break his skull – and do you know something? He didn't give a damn. I was watching him, he was pleased – yes! He was bloody well delighted to get rid of you.'

'Benedict—' She stepped back.

'Listen.' He stepped close to her, and pushed his face towards hers. 'If you're so bloody anxious to go, why don't you bloody well go – and keep going. If he doesn't want you – then I sure as hell don't want you either.'

'Benedict,' she whispered. The colour faded from her face, leaving it washed white as beach sand. She stared at him in horror, as her dreams began to fall into ruins around her. 'You don't mean that.'

'I don't? Is that so?' He threw back his head and laughed again. 'Listen, you got some nice diamonds and a mink coat. You got a big house in Bishopscourt – now, that's pretty good pay for a whore.'

'Benedict—' She gasped at the insult, but he wasn't listening.

'I proved I could have you, didn't I? I proved I could take you away from him – and that's what it was all about. Now, why don't you go on home like a good girl.'

'The machine. I know about that thing in *Kingfisher*.' It was a mistake. Until then she still had a chance. His face changed shape and the blood flooded into it. His voice when he spoke was unsteady, thick with rage for his lips seemed to have swollen.

'Try it,' he whispered back at her. 'Go on, try it. They'll give you fifteen years in a woman's prison, my beauty. And think about this also—' He showed her his hand, holding it like a blade before her eyes. 'I'll kill you. I swear before

God, I'll kill you with my bare hands. You know I'll do it —
you know enough about me now.'

She backed away from him, and he followed her still
holding his hand at her throat.

'You've been paid. Now get out.'

A few seconds longer she stood before him, and he was
not too far gone in rage to see the fear in her eyes mingle
with something else that made her eyes slit and drew her
lips back to show the little white teeth.

'All right,' she said. 'I'll go.' And she walked from the
room, stepping daintily, the long yellow hair swinging
against her shoulders.

R uby drove slowly for her vision was blurred with
her tears. Twice other drivers hooted at her but she
kept both hands clenched on the wheel and stared
ahead, following the De Waal Drive around the lower slopes
of the mountain. Before she reached the University she
swung off the road and drove up through the pine forests
until she reached the car park behind the Cecil Rhodes
Memorial. She left the car and walked down on to the wide
paved terrace below the Greek columns and stone steps
where the mounted statue eternally searched the horizon
with one hand lifted to shade his eyes.

She went to the parapet and looked across to the far blue
mountains of the Helderberg. She hugged herself about the
shoulders for the wind through her silk summer dress was as
cold as her misery.

Now the tears broke over her lids and slid down her
cheeks to fall unheeded on to the silken front of her dress.
They were tears of self-pity, but also the tears of an anger as
searing cold as dry ice.

'The swine,' she whispered through lips that trembled.

Near her two young students sitting on the parapet, kicking their legs over the drop beneath them and hugging each other in the abandon of first love, turned to glance at her.

The boy whispered to the girl, and she giggled in unthinking cruelty – but looked away as Ruby directed a long venomous glare at her. Then in embarrassment the couple scrambled off the parapet and moved away, leaving her alone.

Never for one moment did she consider standing aside. Benedict's threats meant nothing – her only concern was to take the action which would injure him most severely. The consequences to herself were not part of her calculations. She wanted only to select the swiftest and most terrible vehicle of retribution. As the dark clouds that fogged her reason slowly cleared, the means came to her readily.

Johnny was staying at the Tulbagh Hotel.

She turned and ran back to her car, the long yellow banner of her hair floating behind her like the pennant at the tip of a cavalry lance. She drove fast until she hit the downtown rush-hour crowds. The tears dried on her face as she crawled, fuming with impatience, along the slow river of traffic.

It was after five when she parked in a loading zone outside the entrance to the Tulbagh and ran into the lobby.

'What room is Mr Lance in?' she demanded of the girl at the desk.

'Mr Lance checked out an hour ago.' Curiously the girl examined Ruby's ruined make-up.

'Did he say where he was going?' Ruby snapped at her, feeling the sickening slide of disappointment within her.

'No, Madam.' The girl shook her head. 'But he was in a hurry.'

'Damn! Damn!' Ruby swore bitterly. She turned from the desk undecided where next to look. Perhaps Johnny had gone back to the office.

Across the lobby the elevator doors slid open and Tracey

Hartford stepped out. Even in her impatience Ruby recognized in the glow that seemed to emanate from her that Tracey was a girl freshly risen from the bed of the man she loved. There was not a vestige of doubt in Ruby's mind as to the identity of that man.

The shock of it paralysed her for a moment. Then she felt the urge to cross the floor and claw that smug smile from Tracey's face. She fought it down, and instead she stepped into her path as Tracey started for the glass outer doors.

'Where is Johnny?' she demanded, and Tracey came up short. Her little gasp of guilt confirmed Ruby's suspicion.

'Where is he, damn you!' Ruby's voice was pitched low but brittle with emotion.

'He's not here.' Tracey recovered herself, quickly masking her expression.

'Where has he gone? I must see him.'

'He's flown up to Cartridge Bay.'

'When did he leave? It's important – terribly important.'

'An hour ago. He'll be airborne already.'

'Can you get a message to him?' In her impatience Ruby caught Tracey's wrist, holding her in a grip that marked the skin.

'I can radio him—' Tracey pulled her hand free.

'No,' Ruby cut in quickly. She could not have her message shouted across the ether for all to hear. 'Can you follow him – charter plane?'

Tracey shook her head. 'They won't fly to an unscheduled airfield after dark.'

'Then you must follow him – by car. You must drive up there.'

'Why?' Tracey stared at her, puzzled by this strange insistence, noticing the dried tears and the wild look in Ruby's eyes. 'It's an eight-hour drive.'

'I'll tell you. Can we use Johnny's room?'

Tracey hesitated, remembering the unmade bed. Then

the hotel manager came into the lobby and Tracey turned to him with relief.

The Beechcraft bucked suddenly and dropped a wing, instinctively Johnny corrected the lunge with stick and rudder then glanced quickly at his instrument panel for an explanation. There was none to be found there, so he looked over the wing and for the first time noticed the dust on the great plains below him; it was moving low against the earth in long streamers like mist, and the setting sun turned it to mauve and old gold. With a prickle of alarm he scanned the horizon ahead, and saw it coming down from the north like a great moving range of blue mountains. Even as he watched it, it rolled across the low sun, turning it into a sullen red orb. The light in the cockpit changed to a weird glow as though the door to a furnace had been thrown open.

Again the Beechcraft crabbed awkwardly as another gust of high wind hit her, and at the same moment the radio crackled and came alive.

'Zulu Sugar Peter Tango Baker this is Alexandra Bay Control, come in please.'

The voice of the controller was almost unintelligible with storm static. Johnny reached for the transmit switch of the radio, then stopped his hand. He thought quickly. He could guess they were trying to reach him to cancel his flight approval. That was a big northern boiling down out of the desert. They would abort his flight, and divert him out of the path of the storm.

He checked his wrist watch. Twenty minutes' flying time to Cartridge Bay. No – he was flying full into the eye of the wind, say twenty-five or thirty minutes. Quickly he searched the coast on his port side and saw the long white lines of surf stretched ahead into the thickening purple gloom. The coast was still clear, it might stay that way for another thirty minutes.

'Zulu Sugar Peter Tango Baker this is Alexandra Bay. I say again – come in. Come in. Zulu Sugar Peter Tango Baker.' The agitation in the controller's voice came through over the static.

There was a fair chance of getting in to Cartridge Bay, racing the storm and winning. He could edge out to the westward and come in from the sea, pick up *Kingfisher's* riding lights as a beacon and sneak in under the leading edge of the dust clouds. If he missed he could turn and run with the wind for home. The radio was hissing and crackling angrily now, the controller's voice sometimes lost in the interference, sometimes coming through strongly.

' – Cancelled. I say again: your flight approval is cancelled. Do you read me, Zulu Sugar Peter Tango Baker. Come in, please. – Beaufort force seven. – visibility in the storm area – I say again, there is nil visibility in—'

The norther would roar for days now and with it would blow away his last chance of working the gap at Thunderbolt and Suicide.

Johnny switched off the radio, cut off contact with Control and immediately it was strangely quiet in the cockpit. He settled himself down into the bucket seat, and eased open the throttles, watching the needles creep up around the dials of the rev counters.

Now he was down to an altitude of three hundred feet and the Beechcraft was leaping about like a hooked marlin. He was flying her on instruments for outside the cockpit it was completely dark. He could not see his own wingtips, but above him the stars still showed. He was riding the vanguard of the storm, and the dust clouds were ahead, racing to meet him and blanket the flare path at Cartridge Bay.

Every few seconds he darted a quick glance ahead, hoping to pick up the lights, then his eyes flew back to the instrument panel.

'Now,' he thought grimly. 'It should be now. I should be over the grounds. Thirty seconds more and I'll know I've missed her.'

He looked up again, and there was *Kingfisher* dead ahead. All her lights were ablaze, a burning beacon of hope in the darkness. She appeared to be riding easily, for the wind had not yet had time to thrash the sea into a frenzy.

He flashed over her, seeming to graze her superstructure in passing, and now he was searching anxiously for the glow of the flare path on the land beyond.

It came up as a path of lesser darkness in the absolute blackness of the night. He steadied on course towards it, watching it change to a long double line of oil-burning flares that smoked and fluttered in the wind.

He flew her in fast, high above the stall and the shock of touchdown threatened to tear the undercarriage off her.

Then he was jolting and trundling down the earthen runway with the flares flashing past his wingtips.

'Lance, old man,' he murmured thankfully, 'that was a very shaky do!'

The wind hum against the body of the car, and the snarl of rubber on tarmac as the Mercedes snaked through the bends of the twisting mountain road were sounds to match the racing of Tracey's blood and the hammer of her heart.

She drove with an inspired abandon, watching the bends leap out at her out of the darkness, sensing the massive crags and cliffs that hung over the road and blotted out half the night sky.

The silver sheet of Clanwilliam Lake reflected the stars,

and then was left behind. Down from the mountains she went and over the Olifants River to make a brief fuel stop at Vanrynsdorp and scan the road map anxiously in the light of the gasoline pumps. She read with a sinking feeling the mileage figures printed along the little red ribbon of the road, and knew that for her each mile would be multiplied by her own urgency.

Then once more behind the wheel she faced the vast emptiness of Namaqualand – and sent the Mercedes flying across it.

' – There is some type of machine, I don't know how it works, but it filters out the diamonds. Benedict had it installed at Las Palmas—'

The headlights were puny little white shafts, and the road a long blue smear that went on endlessly. Tracey lit a cigarette with one hand, hearing Ruby's voice again in her ears.

' – There is one diamond amongst them. He called it "The Big Blue". Benedict says it's worth a million—' Tracey was not sure she believed it. It was the enormity of the treachery and deceit that she could not accept.

' – The Italian, the Captain, be careful of him. He works for Benedict. The other one also – Hugo – they are all in it. Warn Johnny.'

Benedict! Weak, spoiled Benedict, the playboy, the spendthrift. Could he have planned and carried this through?

A gust of wind hit the car from the side, taking her unawares, pushing the Mercedes off the tar on to the gravel. Tracey fought to hold the skid. Dust and gravel roared out in a cloud from under the wheels. Then she was back on the road, hurtling northwards.

'Warn Johnny! Warn Johnny!'

B enedict van der Byl sat in his father's chair, in his father's house, and he was alone. His loneliness ate deep into the fibres of his whole being. Before him on the stinkwood desk stood a crystal glass and a decanter. The brandy was no comfort, its warmth in his throat and belly seemed only to accentuate the icy cold of his loneliness. His fantasy showed him as a hollow man. He thought of himself as a husk, filled only with the cold of melancholia.

He looked about the room with its dark panelled woodwork and he smelled the musty dead smell. He wondered how many times his father had sat in this chair alone and lonely. Lonely and afraid as the cancer ate him alive.

He stood up and moved listlessly about the room, touching the furniture as if he were trying to communicate with the man who had lived and died here. He moved across and stood in front of the curtained windows. The rug was new. It replaced the other that they had been unable to clean.

'The Old Man had the right idea.' He spoke aloud, his voice sounding strange in his own ears.

Then on an impulse he crossed quickly to the cupboard that flanked the massive stone fireplace, and tried the door. It was locked.

Without passion he stood back and kicked in the panel. The wood splintered and he kicked again, smashing the door from its hinges.

The oblong leather case was on the top shelf, and he took it down and carried it to the desk. He sprang the catches and laid back the lid.

He lifted out the blue metalled double barrels of the Purdy Royal, and the gun oil was greasy on his hands.

'Jacobus Isaac van der Byl.' He read aloud the name in gold inlay set into the steel among the engraved pheasants and gundogs.

He smiled then.

'The old devil.' He shook his head smiling as though at some private joke, and began slowly to assemble the shotgun. He weighed it in his hands, feeling the sweet pure balance of the weapon.

'The old bastard made his own decisions.' And still smiling he carried the gun across to the new carpet. He placed it butt down between his feet with the barrels pointed at the ceiling and leaning slowly forward he opened his mouth and placed the muzzles between his lips, then reaching farther down he placed a thumb on each trigger and pushed them simultaneously.

Click! Click!

The firing pins fell on the empty chambers, and Benedict straightened up and wiped the taste of gun oil from his lips. He grinned again.

'That's the way he did it. Both barrels in the back of the throat. What a cure for tonsilitis!' he chuckled, and glanced across at the shattered cupboard door. The square packets of cartridges were on the second shelf.

He tucked the gun under his arm and went to the cupboard again, moving more purposefully now. He snatched down a packet of SSG and broke it open. Suddenly his hands were shaking and the fat red cartridges spilled on to the floor. He stooped and picked up two of them.

With mounting excitement and dread he broke open the shotgun and slipped the cartridges into the blank eyes of the breeches. They slid home against the seating with a solid double thunk, and he hurried back to the spot in front of the window.

His eyes were bright and his breathing quick as he pushed the safety catch on to 'Fire' and placed the butt between his feet once more.

He took the muzzles in his mouth again, in an obscene soul kiss and reached down for the triggers. They were cold and oily. He caressed them lightly, feeling the fine grooving in the curves of metal, thrilling to the touch and feel of them as he had never thrilled to the feel of a woman's body.

Then abruptly he stood up again. He was gasping for breath.

Unsteadily he carried the weapon back to the desk and laid it on the dark polished wood.

As he poured brandy into the crystal glass his eyes were fastened with perverse fascination on the beautiful glistening weapon.

The steam had fogged the mirrored walls of the bathroom, so her image was dewed and misty. Ruby Lance dried herself slowly with one of the thick fluffy towels. She was in no hurry; she wanted Tracey to have a start of at least four hours on her journey to Cartridge Bay. With a deep narcissistic pleasure she noticed in the mirrors how her whole body glowed with soft pink highlights from the hot waters of the bath.

Wrapping herself in the towel she went through into the dressing-room and picking up one of the silver-backed brushes began stroking it through her hair, moving across to the open wardrobe to select a dress for the occasion. It must be something special, perhaps the unworn full-length Louis Feraud of daffodil satin.

Still undecided she went back to seat herself at the dressing-table and began the complicated ritual of applying her make-up. She worked with meticulous care until at last she smiled at her reflection with satisfaction.

She dropped the towel, went back to the wardrobe, and stood slim and naked before it. Pouting slightly with concentration she decided against the Feraud. Then suddenly she smiled, and reached for Benedict's mink.

She wrapped herself in the pale cloud of fur, fluffing up

the collar to frame her face. It was perfect. Just the fur and a pair of golden slippers, pale gold, a perfect match for her hair.

Now suddenly she was eager to go. She ran from the house to where her car was parked in the driveway.

She switched off the headlights as she turned into the driveway that curved up to where the old house crouched on the top of Wynberg Hill. The whisper of the engine was unobtrusive and blended with the whimper of the night breeze in the chestnut trees that flanked the driveway.

She parked in the courtyard, and saw that Benedict's Rolls was still in the garage and a light burned in the window of the study, a yellow oblong behind the curtains. The front door was open. Her slippered feet made no sound along the gloomy passages, and when she tried the door to the study it swung open readily. She stepped into the room, and closed the door behind her. She stood with her back to the dark panelled wood. A single shaded lamp lit the room dimly.

Benedict sat behind the desk. The room was heavy with the smell of cigar smoke and brandy fumes. He had been drinking. His face was flushed, and the top button of his shirt was undone. On the desk in front of him lay a shotgun. Ruby was surprised at the presence of the weapon, it disconcerted her and the words she had prepared were forgotten.

Benedict looked up at her. His eyes were slightly unfocused and he blinked slowly. Then he grinned; it twisted his mouth and his voice when he spoke was slurred.

'So you've come back.'

Instantly her hatred returned in full flood. But she kept her face impassive. 'Yes,' she agreed. 'I've come back.'

'Come here.' He swivelled his chair to the side of the desk. Ruby did not move, she leaned back against the door.

'Come here.' Benedict's voice was stronger now, and suddenly Ruby smiled and obeyed.

She stood in front of him, huddled in the fur.

'Kneel down,' commanded Benedict, and she hesitated.

'Down!' his voice crackled. 'Down, damn you!'

Ruby sank to her knees in front of him, and he straightened up in the chair. She knelt in front of him in the attitude of submission, with her head forward so the golden hair hung like a curtain over her face.

'Say it,' he gloated. 'Ask me to forgive you.'

Slowly she lifted her face and looked up at him. She spoke softly.

'Tracey left for Cartridge Bay at five-thirty this evening.'

Benedict's expression changed.

'She has a start of four hours – she is half-way there already.'

He stared at her with his lips parting, soft and red and slack.

'She is going to Johnny,' Ruby went on. 'She knows about the thing in *Kingfisher*. She knows about the big blue diamond.'

He began to shake his head in disbelief.

'By dawn tomorrow Johnny will know also. So you see, my darling, you have lost again – haven't you? You can never beat him, can you, Benedict? Can you, my darling?'

Her voice was rising, ringing with triumph.

'You?' he croaked. 'You?'

And she laughed, nodding her head in agreement, unable to speak through her laughter.

Benedict lunged clumsily out of the chair, his hands going for her throat. She went over backwards with him on top of her. Her laughter died gurgling in her throat.

They rolled together on the floor. Benedict's hands locked on her neck, his voice rising in a scream of fury and despair. Her long legs kicking and thrashing, clawing at his face and hands, she fought him with the strength of a cornered animal.

They rolled back suddenly and Benedict's head struck

the solid leg of the desk with a crack that jarred his whole body. His grip on her throat loosened and she tore herself free with fresh breath hissing into her open mouth. She rolled away from him and in one fluid movement gained her feet, reeling back from him with the front of the mink torn open and her hair tangled across her face.

Benedict dragged himself up the desk on to his knees. He was still screaming, a high keening note without form or coherence, as Ruby spun away from him and stumbled to the door.

Blinded by her own hair, fighting for strangled breath, she fumbled for the door handle with her back turned to him.

Benedict reached up and lifted the shotgun off the desk. Still kneeling beside the desk he held the weapon across his hip. The recoil was a liquid pulsing jolt in his hands and the muzzle blast was thunderous in the confines of the room, the long yellow flame lighting the scene like a photographer's flash-bulb.

The heavy charge caught Ruby in the small of her back. At that range there was no spread of shot and it went through spine and pelvis in a solid shattering ball. It tore out through the front of her belly, spinning her sideways along the wall. She slid down into a sitting position, facing him with the mink flared open about her.

On his knees Benedict swung the gun to follow her fall and he fired the second barrel; again the brief thunder and flame of the muzzle blast flashed across the room.

At even closer range than the first charge it struck her full in her beautiful golden face.

Benedict stood in the garage with his forehead pressed against the cold metal of the Rolls-Royce. The shotgun was still in his hands, and his pockets were crammed with cartridges that he had picked up from the floor before leaving the study.

He was shivering violently, like a man in high fever.

'No!' he moaned to himself, repeating the single negative over and over again, leaning against the big car.

Abruptly he gagged, remembering the carnage he had created. Then he retched, still leaning against the Rolls, bringing up the brandy mingled with his horror.

It left him pale and weak, but steadier. Through the open window he threw the gun on to the back seat of the Rolls, and climbed shakily into the driver's seat.

He sat there bowed over the steering wheel, and now his instinct of self-preservation took hold of him.

It seemed to him there was but one avenue of escape still open to him. *Wild Goose* had the range to take him across an ocean – South America perhaps, and there was money in Switzerland.

He started the Rolls and reversed out of the garage, the spin of tyres against concrete burning blue smoke into the beams of the headlights.

The Mercedes crawled through the thick sand, the headlights probing ineffectually into the bright orange fog of dust that whipped endlessly over the track ahead. The hot gritty wind buffeted the car, rocking it on its suspension.

Tracey sat forward in the driver's seat peering ahead through eyes that felt raw and swollen with fatigue and mica dust.

From the main road to the coast this Jeep track was the only land access to Cartridge Bay. It was a hundred miles of tortuous trail, made up of deep sandy ruts and broken stone where it crossed one of the many rocky ridges.

The radiator of the Mercedes was boiling furiously, over-heating in the searing wind and the slogging low-gear grind through thick sand. In places Tracey followed the track only by driving through gaps in the stunted knee-high growth of desert bush. Every few minutes a tumbleweed, driven by the wind, would bowl across the track like a frightened furry animal.

At times she was sure she had missed a turning and was now grinding aimlessly out into the desert, then reassuringly the twin ruts would show up in the lights ahead of her. Once she did drive off the road, and immediately the Mercedes came to a gentle standstill with its rear wheels spinning helplessly in the soft sand. She had to climb out of the cab and, with her bare hands, scoop away the sand from behind the wheels and stuff bundles of tumbleweed into the depressions to give the wheels purchase. She almost wept with relief when the Mercedes pulled back sluggishly on to the trail again.

The slow dawn broke through the dust clouds and Tracey switched off the headlights and drove on until suddenly, and quite unexpectedly, she reached Cartridge Bay. The depot buildings loomed suddenly before her, and she left the Mercedes and ran to the living quarters. The foreman opened the door to her insistent hammering, and stared at her in astonishment before ushering her in. Tracey cut off his questions with her own.

'Where is *Wild Goose?*'

'She took Mr Lance out to *Kingfisher*, but she's back now lying at the jetty.'

'Hugo Kramer – the Captain?'

'He's aboard, holed up in his cabin.'

'Thanks.' Tracey left him, pushed the door open against the wind and ran out into the storm.

Wild Goose lay at her moorings, secured by heavy lines

to the bollards, but fidgeting and fretting at the push of the wind. There was a gangplank laid to her deck, and lights showed at her portholes. Tracey went aboard.

Hugo Kramer came to the doorway of his cabin in a suit of rumpled striped pyjamas. Tracey pushed past him.

'You took Lance out to *Kingfisher?*' she accused him, her voice sharp and anxious.

'Yes.'

'You idiot, didn't you realize there was something up? Good God, why otherwise would he fly in through this weather?'

Hugo stared at her, and instinctively she knew that what Ruby had told her was true.

'I don't know what you're talking about,' he blurted.

'You'll know all right when we are all sitting behind bars – we'll have fifteen long years to think about it. Lance has tumbled to it, you fool, I've got to stop him. Take me out to *Kingfisher.*'

He was confused – and afraid.

'I know nothing about—' Hugo started again.

'You're wasting time.' Brusquely Tracey brushed his protests aside. 'Take me out to *Kingfisher.*'

'Your brother – where is he? Why didn't he come?'

Tracey had anticipated the question. 'Lance beat him up – badly. He's in hospital. He sent me.'

Suddenly Hugo was convinced.

'*Gott!*' he swore. 'What are we going to do? This storm – I may be able to get you out there, but I won't be able to leave *Wild Goose.* My crew can't handle her in this sea. What can you do on your own?'

'Get me out there,' said Tracey. 'Get me aboard *Kingfisher* and you can come back. The Italian, Caporetti, he and I will take care of Lance. In this storm a man can be washed overboard very easily.'

'*Ja.*' Hugo's face lit with relief. 'That's it. The Italian!' And he a reached for his oilskins hanging on the bulkhead.

As he pulled them on over his pyjamas he looked at Tracey with new respect.

'You,' he said. 'I didn't know you were in it.'

'Did you think my brother and I would stand by and let a stranger take our birthright from us?'

Hugo grinned. 'You're a cool one, I'll say that for you. You had me fooled.' And he went out on to the bridge.

Johnny Lance and Sergio Caporetti stood shoulder to shoulder on *Kingfisher*'s bridge. The ship was taking the big green seas over her bows, solid walls of water, and the wind whipped spray that spattered the armoured glass windows of the bridge house.

Kingfisher had slipped her moorings the previous evening, leaving the big yellow buoys floating on their anchor cables and she was working free of her fetters. She was on computer navigation, holding her position over the ground against the swells and the wind by use of her engine and rudder.

'She is no good.' Sergio spoke morosely. 'We come too close to the rocks. I get sick in my heart looking at them.'

The dust clouds did not carry this far out to sea despite the vicious screeching of the wind. The visibility was a mile or more, quite enough to show the brooding twin hulks of Thunderbolt and Suicide. The storm-crazed swells burst against them, throwing white spray two hundred feet into the gloomy sky, then surging back to expose the gleaming white rock.

'Hold her,' growled Johnny. Twice during the night they had changed position, each time edging down closer on the gap between the two islands. *Kingfisher* was battling gamely to hold her ground against the insidious sucking current that added its pull to that of the swell and the wind.

Johnny was not attempting to work any one of the gullies extensively, he wanted only to sample as much of the field as possible in the time that was left to him. The storm would not stop him – for *Kingfisher* was constructed to work in worse weather than his. Her compensating hose section was keeping the dredge head on the bottom despite the lift and fall of her hull.

'Calm down, Sergio.' Johnny relented a little. 'The computer is foolproof.'

'The goddamned computer she no got eyes to see those rocks. Me, I got eyes – and it gives me a sick heart.'

Twice during the night Johnny had gone down into the control room and ordered the computer to report its recovery of diamonds. Each time the reply had been consistent – not a single stone over four carats, and a very precious few of any others.

'I'm going through to the plot. Watch her,' Johnny told Sergio, and staggering against the pitch and roll he went through the door behind the bridge.

He paused behind the repeater screen of the computer, and at a glance saw that *Kingfisher* was holding her primary operation and all departments were running normally. He passed the screen and leaned over the chart table.

The large-scale chart of the South West African coast between Luderitz and Walvis Bay was pinned down on the board. The *Wild Goose* soundings were pencilled in, and the pattern of *Kingfisher*'s sweeps were carefully plotted around the islands of Thunderbolt and Suicide.

Johnny picked up a pair of dividers and stared moodily at the chart. Suddenly a surge of anger rose in him against those two names. They had promised so much and delivered so little.

He stared at the names Thunderbolt and Suicide printed in italics among the maze of soundings, and his anger turned to blind red hatred.

With the points of the dividers he slashed at the chart, ripping the thick linen paper once, and twice, in a ragged cross-shaped tear.

This small act of violence dissipated his anger. He felt embarrassed, it had been a petty childish gesture. He tried to smooth the edges of the tear, and through the gap he felt another loose scrap of paper which someone had slipped under the chart. He probed a finger through the tear in the chart and wormed the scrap out. He glanced at the scribbled title and the lines of figures and numbers that followed.

The sheet was headed: KAMINIKOTO SECONDARY RECOVERY PROGRAMME.

He studied it, puzzled by the title but recognizing the numbers as a computer programme. The writing was in Sergio Caporetti's pointed continental style. The easiest way to resolve the mystery was to ask Sergio. Johnny started back for the bridge.

'Boss,' Sergio called anxiously, as Johnny stepped through the door. 'Look!'

He was pointing ahead into the eye of the wind. Johnny hurried to his side, the paper crumpled and forgotten in his hand.

'*Wild Goose*.' Sergio identified the small craft that was staggering and plunging towards them out of the gloom.

'What the hell is he doing here?' Johnny wondered aloud. *Wild Goose* was lost for long seconds behind the walls of green sea, then again she was lifted high into unnatural prominence, showing the red lead of her bottom as she rode the crests; water poured from her scuppers, before she shot down the steep slope of the next wave to bury her nose deep in frothing water. She came down swiftly on the wind, rounding to and beginning to edge in under *Kingfisher*'s counter.

'What the hell is he playing at?' Johnny protested, and then in disbelief he saw a slim figure dart from *Wild Goose*'s wheelhouse and run to the side nearest *Kingfisher*.

'It's Tracey,' shouted Johnny.

She reached the rail just as another swell burst over the

bows and smothered her. Johnny expected to see her washed away, but she was still there clinging to the rail.

Thrusting the page of paper into his pocket, Johnny went out through the wing of the bridge and swarmed down the steel ladder to the deck, jumping the last ten feet and running the instant he landed.

He reached the side and looked down on the drowned-kitten figure of Tracey.

'Go back,' he yelled. 'Go back. Don't try it.'

She shouted something that was lost in the next smother of spray, and when it cleared he saw her poising herself to jump the gap of surging water between the two vessels.

He flung himself over *Kingfisher*'s side and climbed swiftly down the steel rungs.

He was still ten feet above her as she gathered herself for the leap.

'Go back,' he shouted desperately.

She jumped, missed her hold and fell into the murderous stretch of water between the hulls. Her head bobbed below Johnny, and he was aware of the next swell bearing down on them. It would throw *Wild Goose* against the steel cliff of *Kingfisher*, crushing Tracey between them.

Johnny went down those last ten feet and hanging outwards by one arm he got his other arm around her, and with a heave that crackled in his muscles and joints he plucked her from the water just as the two vessels dashed together with a crunching impact that tore splinters from *Wild Goose*'s planking, and left a smear of alien paint on *Kingfisher*'s steel plating.

Wild Goose swung away, and with her diesels bellowing went bucking off into the wind.

With puddles of sea water forming around her feet from her sodden clothing, Tracey stood in the centre of *Kingfisher*'s guest cabin. Her dark hair was plastered down her face and neck, and she was shivering so violently from shock and the icy water that she could not talk. Her teeth chattered together, and her lips were blue with cold.

Desperately she was trying to form words, her eyes never leaving Johnny's face.

Quickly he stripped off her clothing and throwing one towel round her shoulders he began roughly to chafe warmth back into her with another.

'You little idiot,' he berated her. 'Are you stark staring bloody mad?'

'Johnny,' she gasped through her chattering teeth.

'Christ – that was so close,' he snarled at her as he knelt to rub her legs.

'Johnny, listen.'

'Shut up and dry your hair.'

Humbly she obeyed him, her shivers became controllable as he crossed to the locker and found a thick jersey which he pulled over her head. It hung almost to her knees.

'Now,' he said, taking her roughly by the shoulders. 'What the hell is this all about?'

And she told him in a rush of words that poured out like water from a broken dam. Then she burst into tears and stood there forlornly in the voluminous jersey with her damp hair dangling about her shoulders, sobbing as though her heart was breaking.

Johnny took her in his arms.

For a long minute Tracey revelled in his warmth and strength, but she was the first to pull away.

'Do something, Johnny,' she implored him, her voice still thick with tears. 'Stop them. You mustn't let them get away with it.'

He went back to the locker, and while he ransacked it

for clothing that might fit her, Johnny's mind was racing over the story she had told him.

He watched her pull on a pair of blue serge trousers and tie them at the waist with a length of cord. She folded back the cuffs and tucked them into thick woollen socks, before thrusting her feet into a pair of sea-boots that were only a few sizes too large for her.

'Where do we start?' she asked, and he remembered the sheet of paper. He fished it out of his pocket and flattened it on the table beside the bunk. Quickly he ran his eyes over the columns of figures. His first guess was right – it *was* a computer programme.

'Stay here,' he ordered Tracey.

'No.' Her response was immediate, and he grinned.

'Listen, I'm just going up on to the bridge to keep them busy there. I'll come back for you, I promise. You won't miss anything.'

'How is she, boss?' Sergio Caporetti's concern was genuine. Johnny realized that he must be worrying himself into a frenzy trying to guess the reason for Tracey's arrival.

'She is pretty shaken up,' Johnny answered.

'What she want – that was big chance she takes. Nearly fish food.'

'I don't know,' Johnny said. 'I want you to take over up here. Keep *Kingfisher* working. I'm going to get her to bed – I'll let you know what it's all about as soon as I find out.'

'Okay, boss.'

'Oh, and Sergio – keep an eye on those rocks. Don't let her drift down any closer.'

Johnny chose a powerful incentive to keep Sergio up on the bridge.

Johnny left him and went below, stopping only at the guest cabin.

'Come on.'

Tracey followed him, lurching unsteadily with *Kingfisher's* antics in the high sea.

Two decks down they reached the computer control room and Johnny unlocked the heavy steel door, then locked it again behind them.

Tracey wedged herself against the bulkhead and watched as Johnny seated himself at the console and clipped the rumpled sheet into the board.

Reading from the sheet he typed the first line of figures on the keyboard. Immediately the computer registered a protest.

'Operator error,' it typed back. Johnny ignored its denial and typed the second line. This time it was more emphatic.

'No procedure. Operator error.'

And Johnny typed the next line of figures. He guessed that whoever had stored this programme in the computer's memory would have placed a series of blocks to prevent accidental discovery. Again the denial flashed back at him.

'Operator error.'

And Johnny muttered, 'Thrice before the cock crows,' striking an incongruously biblical note in the tense atmosphere of the control room.

He typed the last line of figures and the denial faded from the screen. The console clicked like a monstrous crab, then suddenly it started to print again.

'KAMINIKOTO SECONDARY RECOVERY PROGRAMME. INSTALLED OCTOBER 1969. AT LAS PALMAS BY HIDEKI KAMINIKOTO. DOCTOR OF SCIENCE. TOKYO UNIVERSITY.'

The little Japanese had been unable to resist autographing his masterpiece. Tracey and Johnny crouched over the screen, staring at it with awful fascination as the computer began spelling out its report. It began with the number of hours worked, and the weight of gravel processed during that time. Next it reported the weight of concentrates recovered from the cyclone and finally, in a series of columns, it printed out the weights and sizes of all the diamonds won from the sea. The big Blue showed up in the place of honour, and wordlessly Tracey touched the figure 320 with a forefinger. Johnny nodded grimly.

The computer ended by giving the grand total of carats recovered, and Johnny spoke for the first time.

'It's true,' he said softly. 'It doesn't seem possible – but it is.'

The click and hum of the computer ceased, and the screen went blank.

Johnny straightened up in the chair.

'Where would they put it?' he asked himself, as he ran quickly over the line of recovery. He stood up from the chair and peered through the leaded glass peephole into the X-ray room. 'It must be this side of the cyclone, this side of the drier – ' He was speaking aloud. ' – Between the drier and the X-ray room.'

Then there bobbed to the surface of memory the modification in design which he had meant to query, but which he had forgotten.

'The inspection plate on the conveyor tunnel!' He punched his fist into his palm. 'They moved the inspection plate! That's it! It's in the conveyor tunnel.'

His hands were frantic with haste as he unlocked the steel door of the control room.

Sergio Caporetti paced his bridge like a captive bear, puffing so furiously on his cheroot that sparks flew from its tip. The wind howled hungrily around the wheelhouse, and the swells still marched in from the north.

Suddenly he reached a decision and turned to the helmsman.

'Watch those goddamn rocks – watch them good.'

The helmsman nodded and Sergio shambled through the chartroom to his own cabin. He locked the door behind him, and crossed to his desk. Fumbling with his keys he opened the bottom drawer of his desk and reaching under the pile of cheroot packets he brought out the canvas bag.

Weighing it thoughtfully in his hand, he looked about the cabin for a more secure hiding place. Through the canvas he could feel the nutty irregular shape of the stones.

'That Johnny, he a clever bastard,' he muttered. 'It better be good place.'

Then he reached a decision. 'Best place where I can watch them all a time.'

He opened his jacket and stuffed the bag into his inside pocket. He buttoned the jacket and patted the bulge over his heart.

'Fine!' he said. 'Good!' And stood up from the desk. He hurried back, unlocking the door into the chartroom, and headed for the bridge. He stopped in the middle of the chartroom, and his head swung towards the repeater screen of the computer. The buzzer was going like a rattlesnake, and the red bulb that warned of a new procedure was blinking softly.

Fearfully Sergio approached the screen and stooped over it. A single glance was enough, and he rushed from it to the chart table. He saw the cross-shaped tear in the chart.

'Mary Mother!'

He ripped back the thick crackling paper and searched under it. He stepped back from the table and hit himself across the chest.

'Fool!' he said. 'Idiot!' He spent ten seconds in self-castigation, then he looked about for a weapon. The locking handle of the cabin was a twelve-inch steel bar with a heavy head. He pulled out the pin and worked it loose. He slipped it into the waistband of his trousers.

'I'm going below,' he told the helmsman curtly, and clambered down the companionway. Swiftly he moved through the ship, balancing easily to her roll and pitch.

When he reached the lowest deck he became more stealthy, creeping silently forward. Now he carried the steel bar in his right hand. Every few paces he stopped to listen, but *Kingfisher*'s hull was groaning and popping as she worked in the swells.

He could hear no other sound. He crept up to the door of the control room and cautiously peered through the small armoured glass window. The control room was empty. He tried the handle, and found it locked.

Then he heard voices – from the open doorway of the conveyor room behind him. Quickly he crossed to it and flattened himself against the jamb.

Johnny's voice came muffled and indistinct:

'There's another hatch in here. Get me a half-inch spanner from the tool cupboard.'

'What's a half-inch spanner look like?'

'It's a big one. The size is stamped on it.'

Sergio glanced one-eyed around the door jamb. The cover was off the inspection hatch in the conveyor tunnel, and Tracey's head was thrust into the opening.

It was clear that Johnny Lance was in there, and that he had found the secret compartment.

Tracey drew her head out of the hatch, and Sergio ducked back and looked down the passageway. The tool cupboard was bolted to the bulkhead under the stairs from the deck above. He turned and darted around the corner of the passageway. Tracey came out of the conveyor room, and went to the cupboard. She opened the doors on the glittering array of tools, each clipped securely to its rack.

While she stood before the cupboard, completely absorbed in her search for a half-inch spanner, Sergio came from around the corner and crept up silently behind her.

He lifted the steel bar over his shoulder and came up on his toes, poised to strike.

Tracey was muttering softly to herself, head bowed slightly, handling the spanners – and Sergio knew the blow would crush her skull.

He closed his mind to the thought, and aimed carefully at the base of her skull. He started the blow, and then checked it. For a second that seemed to last for a long time he remained frozen. He couldn't do it.

With an exclamation of satisfaction Tracey found what she was searching for. As she turned away from the cupboard Sergio shrank back behind the angle of the bulkhead, and Tracey hurried back into the conveyor room.

'I've got it, Johnny,' she shouted into the hatch.

'Bring it to me. Hurry, Tracey. Sergio will be getting suspicious,' he shouted back, and Tracey hitched up her voluminous trousers and wriggled into the hatch.

On hands and knees she crawled up beside him. It was cramped and hot in the narrow tunnel. He took the spanner from her.

'Hold the flashlight.' She took it from him, holding the beam on the panel while he unscrewed the retaining bolts and lifted off the cover.

Lying on his side he peered into the opening.

'There's a container of sorts,' he grunted, and reached in. For a minute he struggled with the clamps, then slowly he lifted out the stainless steel cup.

At that moment *Kingfisher* reared and plunged to a freak wave and the cup slipped from Johnny's fingers, and from it spilled the diamonds. They cascaded over both of them, a glittering shower of stones of all sizes and colours. Some lodged in Tracey's damp hair, the rest rolled and bounced and scattered about them, catching the light from the torch and throwing it back in splinters of sunshine.

'Yipes!' gasped Tracey and laughed at Johnny's whoop of triumph.

Lying side by side they scrabbled and snatched at the treasure scattered around them.

'Look at this one,' exulted Tracey.

'And this.' They were crazy with excitement, hands filled with diamonds. They hugged each other and kissed ecstatically, laughing into each other's mouths.

Johnny sobered first, 'Come on. Let's get out of here.'

'What about the diamonds?'

'Leave them. There'll be plenty of time later.'

They crawled backwards down the tunnel, still laughing and exclaiming, and one after the other emerged into the conveyor room. While they straightened their clothing, and regained their breath, Tracey asked, 'What now?'

'First thing is to get young Sergio safely under lock and key, his crew also.' Johnny's face hardened. 'The bloody bastards,' he added angrily.

'Then?' Tracey asked.

'Then we'll pull up the hose, and sail *Kingfisher* back into Cartridge Bay. Then we'll call up the police on the radio. There's going to be an accounting with the whole gang of the bastards – your darling brother included.'

Johnny started for the door, asking as he crossed the deck:

'Why did you close the door, Tracey?'

'I didn't,' she replied as she hurried after him, and Johnny's expression changed. He ran to the heavy steel door and threw his weight on it. It did not move, and he swung round to face the door that led into the cyclone room.

It was closed also. He charged across the room and grabbed the handle, heaving at it with all his strength.

He stood back at last, and looked wildly about the long narrow cabin. There was no other opening, no hatch or porthole – nothing except the tiny square peephole in the centre of the steel door that led into the cyclone room beyond. The peephole was covered with three-inch armoured glass that was as strong as the steel that surrounded it. He looked through it.

The tall cyclone reached from floor to roof, dominating the room. Beyond it the steel pipe that carried the gravel from the sea bed pierced the roof from the deck above, but the cyclone room was deserted.

Johnny turned slowly back to Tracey and put an arm around her shoulders.

'We've got problems,' he said.

After closing and locking both the doors that led into the conveyor room, Sergio climbed quickly back to his bridge. The helmsman looked at him curiously.

'How's the lady?'

'Fine,' Sergio snapped at him. 'She's safe.' And then with unnecessary violence, 'Why you no mind your own business, hey? You think you Captain for this ship?'

Startled, the helmsman quickly transferred his attention back to the storm which still raged lustily about them. Sergio began to pace up and down the bridge, balancing easily and instinctively to her exaggerated motion. His smooth baby face was crumpled into a massive scowl, and he puffed on one of his cheroots. With all his soul Sergio Caporetti was lamenting his involvement in this business. He wished that he had never heard of *Kingfisher*. He would have traded his hopes of a life hereafter to be sitting on the seafront at Ostia, sipping *grappa* and watching the girls go by.

Impulsively he pulled open the storm doors at the angle of the bridge and went out on to the exposed wing. The wind buffeted him and set his soft hair dancing and flickering.

From inside his jacket he pulled the canvas bag.

'This is the trouble,' he muttered, looking at the bag in his hand. 'Bloody little stones.'

He threw back his arm like a baseball pitcher, set to hurl the bag out into that hissing green sea below him, but again he could not make the gesture. Swearing quietly to himself, he stuffed the stones back into his jacket, and went back into the wheelhouse.

'Call the radio operator,' he ordered, and the helmsman reached quickly for the voice tube.

The radio operator reached the bridge still owl-eyed with sleep and buttoning his clothing.

'Get on to *Wild Goose*,' Sergio told him.

'I won't be able to raise her in this,' the man protested, glancing out at the storm.

'Call her.' Sergio stepped towards him threateningly. 'Keep calling until you get her.'

*W*ild Goose staggered and wallowed through the turbulence at the entrance to Cartridge Bay, then fought her way into the sanctuary of the channel.

Hugo relaxed perceptibly. It had been a long hard run back from Thunderbolt and Suicide. Yet there was an uneasy feeling that still persisted. He hoped that the girl was able to handle Lance. He was a tough cookie that Lance, he wished that he had been able to go along with her and make sure of the business. Fifteen years was one hell of a long time – he would be almost fifty years old at the end of it.

Hugo followed the channel markers that appeared like milestones out of the dust clouds, until ahead he made out the loom of the jetty and the depot buildings.

There was a figure on the jetty, crouched beside the mountain of dieseline drums. With a prickle of alarm, Hugo strained his eyes in the bad visibility.

'Who the hell is it?' he puzzled aloud. The figure straightened and came forward to stand on the edge of the jetty. Bare-headed, dressed in rumpled dark business suit, the man carried a shotgun in one hand – and it was another few seconds before Hugo recognized him.

'Christ! It's the boss!' Hugo felt alarm flare in his stomach and chest, it tightened his breathing.

Benedict van der Byl jumped down on to the deck of *Wild Goose* at the moment she touched the jetty.

'What's happened?' Benedict demanded as he barged into the wheelhouse.

'I thought you were in hospital,' Hugo countered.

'Who told you that?'

'Your sister.'

'You've seen her? Where is she?'

'I took her out to *Kingfisher*. Like you said. She went out to deal with Lance.'

'Deal with Lance! She's with him, you idiot, she's not with us. She knows the whole deal. Everything!'

'She told me—' Hugo was appalled. But Benedict cut him short.

'The whole thing's blown up. We've got to clear out. Get your crew to load those drums of dieseline into the hold. How are your water tanks?'

'Full.'

'Food?'

'We are stocked up.'

'For how long?'

'Three weeks – at a push, four.'

'Thank God for that.' Benedict looked relieved. 'This storm will blow another three days – we'll have that much start. They'll never find us in this. By the time it clears we'll be well on our way.'

'Where to – Angola?'

'God, no! We have to get well clear. South America.'

'South America!'

'Yes – we can do it, carrying extra fuel.'

Hugo was silent a moment, becoming accustomed to the idea.

'We can do it,' Benedict repeated.

'Yes.' Hugo nodded. 'We can do it,' he agreed thoughtfully. For the first time he examined Benedict closely. He saw that he was in an emotional and physical mess, his bloodshot eyes were sunk into deep plum-coloured hollows, dark new beard covered his jowls, and there was a gaunt hunted look to him – like some fugitive animal.

He was filthy with dust, and there was a streak of something that could have been dried vomit down the front of his jacket.

'But what do we do when we get there?' For the first time since he had known Benedict he felt in control. This was the time to deal, to make bargains.

'We'll get ashore on some deserted spot, and then we split up and disappear.'

'What about money?' Hugo spoke carefully. He glanced down at the shotgun. Benedict's hands were fidgety and restless on the weapon.

'I've got money.'

'How much?' Hugo asked.

'Enough.' Benedict blinked cautiously.

'For me also?' Hugo prodded him, and Benedict nodded.

'How much for me?' Hugo went on.

'Ten thousand.'

'Pounds?'

'Pounds,' Benedict agreed.

'That's not enough.' Hugo shook his head. 'I'll need more than that.'

'Twenty.' Benedict increased his bid, but he knew he was playing from weakness into strength. Ruby was lying mutilated in his study, the net was probably being spread for him already.

'Fifty,' said Hugo decisively.

'I haven't that much.'

'Who are you kidding, Buster!' Hugo snorted. 'You've been stacking it away for years.'

Benedict let the barrels of the shotgun swing towards Hugo's belly suggestively.

'Go ahead,' Hugo grinned at him, screwing up his pale albino eyes. 'That'll leave you to paddle this canoe – you want to try it? You'd pile her up on the bar at the entrance – that's how far you'd get.'

Benedict swung the barrels aside.

'Fifty,' he agreed.

'Right!' Hugo spoke briskly. 'Let's get the hell out of here.'

*W*ild *Goose* was clear of the land, and of the towering blinding dust clouds. The following seas came sweeping up under her stern urging her on her westward flight, while the high-pitched shriek of the wind in her rigging cried to her to hasten.

'Why don't you get down below and grab some sleep?' Hugo said. He found Benedict's restless haunting presence in the crowded wheelhouse disconcerting.

Benedict ignored the suggestion. 'Switch on the radio,' he said.

'What for? You'll get nothing on the set.'

'We are out of the dust,' Benedict replied. 'We might pick up a police message.'

The image of Ruby was so clear in his mind. He wanted to know if they'd found her yet. He felt his gorge rising again. That head – oh God – that head! He crossed quickly to the radio set and switched it on.

'They won't be on to us yet,' said Hugo, but Benedict was manipulating the dials – searching the tortured radio waves. The static wailed and gibbered and shrieked like a maniac.

'Turn it off,' snapped Hugo, and at that moment a voice cut in on them.

' – *Wild Goose*,' said the voice from the loudspeaker quite clearly. Benedict crouched eagerly over the set, his hands busy on the dials, and Hugo came up beside him.

' – Come in, *Wild Goose*. This is *Kingfisher*. I repeat, come in *Wild Goose*—'

Benedict and Hugo looked at each other. 'Don't answer,' said Hugo, but he made no move to intervene as Benedict lifted the microphone off its hook.

'*Kingfisher*, this is *Wild Goose*.'

'Stand by, *Wild Goose*.' The answer came back immediately. 'Stand by for Captain Caporetti.'

'*Wild Goose* standing by.'

Hugo caught Benedict's shoulder and his voice was angrily uncertain.

'Leave it, don't be a fool.'

Benedict shrugged off the hand, and Sergio's voice boomed out of the speaker.

'This is Caporetti – who dat?'

'No names,' Benedict cautioned him. 'Where are your guests?'

'They safe – battened down nicely.'

'Safe? Are you certain? Both of them safe?'

'*Si*. I have them safe and sure.'

'Stand by.' Benedict crouched over the set, and his mind

192

was racing. Johnny Lance was in his power. This was the last chance he would ever have. Plans began to form, gelling quickly in his mind.

'The diamonds. Caporetti has the diamonds. That big Blue is worth a million on its own,' said Hugo. 'If Caporetti has taken care of the others – it would be worth the risk.'

'Yes.' Benedict turned to him, he had been puzzling how he could make Hugo turn back. He had forgotten the diamonds. 'It would be worth it,' he agreed.

'Just a quick pass alongside *Kingfisher* – pick up Caporetti with the diamonds and we'd be on our way.'

'I have to go aboard.' Benedict qualified the suggestion.

'Why?' Hugo asked.

'Wipe out the reel on the computer that carries the programme – it's got the Jap's name on it. They could trace him. I paid him on my Swiss bank. They'll find the account.'

Hugo hesitated. 'No killing – or anything like that. We've got enough trouble without that.'

'You think I'm crazy?' Benedict demanded.

'Okay, then,' Hugo agreed.

'*Kingfisher*,' Benedict spoke into the microphone. 'We are coming to you. I'll be coming on board to finalize matters.'

'Fine.' Through the static they could hear the relief in Sergio's voice. 'I'll be standing by.'

It took nearly two hours for *Wild Goose* to slug her way back to where *Kingfisher* lay beneath the ghostly white shapes of Thunderbolt and Suicide, and it was after midday before Hugo began manoeuvring *Wild Goose* into the big ship's lee.

'Don't waste time,' Hugo cautioned Benedict. 'The sooner we get on our way – the better for all of us.'

'I'll be about half an hour,' Benedict answered. 'You lay off and wait for us.'

'Are you taking that bloody shotgun?' Benedict nodded.

'What for?' But Benedict did not reply, he looked up at the sky. The sun was merely a luminous patch of silver light through the ceiling of sea-fret and wind-driven mist, and still the storm hunted hungrily across the sea.

'It will slow you up on the ladder.' Hugo harped on the shotgun. He wanted very much to part Benedict from it, he wanted it over the side – for its presence aboard would prejudice the plans that Hugo had been forming during the last few hours – plans that took into account the ready market for diamonds in South America, and the undesirability of sharing the proceeds with two partners.

'I'll take it.' Benedict tightened his grip on the stock of the weapon. Without it he would feel naked and vulnerable – and it was part of his own private plans for the future. Benedict's brain had also been busy during the last two hours.

'Suit yourself.' Hugo resigned himself to Benedict's refusal; there would be an opportunity later, during the long passage across the Southern Atlantic. 'You better get up for'ard.'

This time Hugo's approach was neatly executed; in a lull between the colossal swells he touched *Wild Goose*'s bows to the steel side of the factory ship. Benedict stepped across the gap and was up the landing ladder and standing at *Kingfisher*'s rail before the next wave came marching down on them.

He waved Hugo off, then hanging on to the rail, made his way aft to *Kingfisher*'s bridge works.

'Where is Lance?' he demanded of Sergio the moment he stepped on to the bridge, but Sergio glanced significantly at the inquisitively listening helmsman and led Benedict through into his cabin.

'Where is Lance?' Benedict repeated the moment the door was locked.

'He and your sister they are in the conveyor room.'

'The conveyor room?' Benedict was incredulous.

'*Si*. They find out about Kammy's machine. They open the hatch and go inside. I close both doors. Lock them good.'

'They are in there now?' Benedict asked to gain time to reconstruct his plans.

'*Si*. Still there.'

'All right.' Benedict reached his decision. 'Now listen, Caporetti, this is what we are going to do. The whole thing has blown up on us. We are going to wipe out as much of the evidence against us as possible, then we are clearing out. We are going to run for South America in *Wild Goose*. You have got the diamonds – haven't you?'

'*Si*.' Sergio patted the breast of his jacket.

'Give them to me.' Benedict held out his hand, and Sergio grinned.

'I tink I look after them. They keep my heart warm.'

A frown of annoyance narrowed Benedict's eyes, but he let the moment pass.

'All right.' His tone was still friendly. 'Now, what you have to do is get down to the control room and wipe out Kaminikoto's programme. Get his name off that reel. He showed you how to do that?'

'*Si*.' Sergio nodded.

'How long will it take?'

'Half an hour, not longer,' Sergio answered, and Benedict checked his wrist watch, sure that this would give him time enough to do what he had to do.

'Good! Get cracking.'

'Boss.' Sergio hesitated at the cabin door. 'What about my boys, my crew? They good boys, no trouble for them?'

'They're clean,' Benedict pointed out irritably. 'I'll get

them together now, and explain that you have to go ashore. They can keep *Kingfisher* hove to waiting for you to come back. After the storm blows out they are bound to radio base and find out we have disappeared. They'll be all right.'

Sergio nodded his satisfaction.

'I'll call them all to the bridge now. You talk to them.'

The five crew members were gathered on *Kingfisher*'s bridge, and Sergio had disappeared down below.

'Any of you speak English?' Benedict demanded, and two of them affirmed that they did.

'Right,' Benedict addressed them. 'You will have been wondering about all the coming and going in this weather. I want you all to be ready to leave the ship. I want you to get all your valuables – now!'

Quickly they translated to the others, who looked apprehensively at Benedict. He was a strange wild-eyed figure with the shotgun tucked under his arm.

'Right – let's go.' And there was no dissent from any of them as they trooped to the companionway.

Benedict followed them along the passageway towards the crew quarters, and glanced quickly at his wrist watch. Seven minutes had elapsed. He looked at the men ahead of him.

The backs of their heads formed a solid target. He had shot guineafowl like that in Namaqualand when they were on the ground running away from him in a thick file, down on one knee and aim for the thicket of heads, knocking down half the flock with both barrels.

He knew he could take all five of these men with two shots. Just let them get a little further ahead so the shot could spread. But he remembered Ruby and his stomach heaved. The other way was just as sure.

'Stop!' he commanded as the five men came level with

the paint store. They obeyed and turned back to face him. Now he held the shotgun so that there was no mistaking its menace. They stared at the gun fearfully.

'Open that door.' He pointed at the paint store. Nobody moved.

'You.' Benedict picked on one of those who spoke English. Like a man in a trance he went to the steel door and spun the locking handle. He pulled the door open.

'In!' Eloquently Benedict gestured with the shotgun. Reluctantly the five of them filed into the small windowless cubicle, and Benedict slammed the door on them. He spun the lock, throwing all his weight on the handle to set it.

Now he had a clear field, and his wrist watch gave him another twenty minutes. He hurried for'ard, he wanted to keep well clear of the control room and Sergio Caporetti.

Using the forward companionway he dropped down to the working deck, fumbling out his duplicate set of keys.

WARNING. EXPLOSIVES. NO UNAUTHORIZED ENTRY.

He unlocked the door, and laying the shotgun flat on the deck he lifted a twenty-five pound drum of plastique down from its rack.

In his haste he tore a fingernail on the lid of the drum, but hardly felt the pain. He uncoiled a six-foot length of the soft dark toffee-coloured material and slung it around his neck. Next he selected a cardboard box of pencil time fuses. He read the label.

'Fourteen-minute delay. That's about right.'

Blood from his torn nail left brown blotches on the cardboard as he took four of the pencils from the box, picked up the shotgun and hurried aft. The jet engine whine of the cyclone mounted deafeningly as he came closer to it.

Tracey was curled on the bare steel plating of the deck with Johnny's jacket folded under her head. She was in a fatigue-drugged sleep, so deep as to be almost deathlike.

Every few minutes Johnny interrupted his restless patrol of the conveyor room to stand over her and look down on her unconscious form. His worried expression softened a little each time he studied her pale lovely face. Once he stooped over her and tenderly lifted a strand of dark hair from her cheek, before resuming his pacing up and down the narrow cabin.

Each time he reached the door of the conveyor room he glanced through the tiny window. The glass had resisted his attempts to smash it with one of the spanners. He had wanted to open the window to call for help, but his efforts had not marked the thick armoured glass.

There was no way out of the cabin. Johnny had tried every possible outlet. The apertures for the conveyor system were guarded at one end by the furnace, and at the other by moving machinery which would ferociously chew to tatters anyone who became entangled in it. They were caged securely, and Johnny paced his cage.

Again he stopped before the peephole, but this time he flung himself at the door with clenched fists. The steel plate smeared the skin from his knuckles as he hammered on it and the pain sobered him. He pressed his face to the glass and through it watched Benedict van der Byl enter the conveyor room and, without glancing at the window, cross to the cyclone.

Benedict laid aside the shotgun he carried and for a moment stood looking up at the thick steel pipe that carried the gravel down from the deck pumps above. As he lifted the thick rope of plastique from around his neck, Johnny knew exactly what he was going to do.

He watched in fascination as Benedict mounted the steel ladder up the side of the cyclone. Hanging with one hand to the ladder, Benedict reached out with the other and clumsily tied the rope of plastique around the gravel pipe. It hung there like a necklace about the throat of some obscene prehistoric monster.

'You bastard! You murdering bloody swine!' Johnny shouted, and again he beat on the steel door with his fists. But the thickness of the door and the whine of the cyclone drowned his voice. Benedict showed no sign of hearing him – but Tracey sat up and looked about her blearily. Then she came to her feet and staggering to the roll and pitch of the ship she went to Johnny and pressed her face to the window beside her.

Benedict was sticking the time pencils into the soft dark explosive. He used all four fuses, taking no chances on a misfire.

'What's he doing?' Tracey asked after she had recovered from the surprise of recognizing her brother.

'He's going to cut the pipe, and let *Kingfisher* pump herself full of gravel.'

'Sink her?' Tracey's voice was sharp with alarm.

'She'll pump water and gravel into herself at pressures that will tear away all the inner bulkheads.'

'This one?' Tracey patted the steel plate.

'It'll pop like a paper bag. God, you have no idea of the power in those pumps.'

'No.' Tracey shook her head. 'He's my brother. He won't do it, Johnny. He couldn't murder us.'

'By the time he's finished – ' Johnny contradicted her grimly, ' – *Kingfisher* will be lying in 200 feet of water. Her hull will be packed so tightly with gravel that it will be like a block of cement. We, and everything in her, including his little machine, will be so flattened as to be unrecognizable. It would cost millions to salvage *Kingfisher* – and no one will care that much.'

'No, not Benedict.' Tracey was almost pleading. 'He's not that bad.'

Johnny interrupted her brusquely. 'He could get away with it. It's a good try – his best chance. Encase all the evidence against him in concrete, and bury it deep.'

'No, Benedict.' Tracey was watching her brother as he climbed down the cyclone ladder and picked up the shotgun. 'Please, Benedict, don't do it.'

Almost as if he had heard her, Benedict turned suddenly and saw the two faces at the window. The shock of guilt held him rigid for a moment as he stared at them – Tracey's pale lips forming words he could not hear, Johnny's eyes burning with accusation.

Benedict dropped his eyes, he made a gesture that was indecisive, almost pathetic. He looked up at the fused and charged rope of explosive – and then he grinned. A sardonic twitching of the lips, and he stumbled out of the cyclone room and was gone.

'He'll come back,' whispered Tracey. 'He won't let it happen.'

'I wouldn't bet on that – if I were you,' said Johnny.

B enedict reached *Kingfisher*'s rail and clung to it. He looked out to where *Wild Goose* bobbed and hung on the swells. He saw Hugo's face as a white blob behind the wheelhouse window, but as the little trawler began closing in for the pick-up Benedict waved it away. He glanced at his watch again, then looked back anxiously at the bridge.

The long minutes dragged by. Where the hell was the Italian? Benedict could not leave him – not while he still had that diamond; not while he could stop the dredge pumps and release the prisoners locked below.

Again Benedict checked his watch, twelve minutes since he had set the time pencils. He must go back and find Caporetti. He started back along the rail, and at that moment Sergio appeared on the wing of the bridge. He shouted a question at Benedict that was lost in the wind.

'Come on!' Frantically Benedict beckoned to him. 'Come on! Hurry!'

With another last look about the bridge, Sergio ran to the ladder and climbed down to deck level.

'Where my boys?' he shouted at Benedict. 'Why nobody at the con? What you do with them?'

'They are all right,' Benedict assured him. He had turned to the rail and was signalling *Wild Goose* to come alongside.

'Where they?' Sergio demanded. 'Where my boys?'

'I sent them to—' Benedict's reply was cut off as *Kingfisher*'s deck jarred under their feet. The explosion was a dull concussion in her belly, and Sergio's jaw hung open. Benedict backed away from him along the rail.

'Filth!' Sergio's jaw snapped closed, his whole body appeared to swell with anger.

'You kill them, dirty pig. You kill my boys. You kill Johnny – the girl.'

'Keep away from me.' Benedict braced himself against the rail, leaving both hands free to use the gun.

Not even Sergio would advance into the deadly blank eyes of those muzzles. He paused uncertainly.

'I'll blow your guts all over the deck,' Benedict warned him, and his forefinger was hooked around the trigger.

They stared at each other, and the wind fluttered their hair and tore at their clothing.

'Give me those diamonds,' Benedict commanded, and when Sergio stood unmoving, he went on urgently, 'Don't be a hero, Caporetti. I can gun you down and take them anyway. Give them to me – and our deal is still on. You'll come with us. I'll get you out of here. I swear it.'

Sergio's expression of outrage faded. A moment longer he hesitated.

'Come on, Caporetti. We haven't got much time.' It may have been his imagination, but to Benedict it seemed that

Kingfisher's action in the water had altered, she was sluggish to meet the swells and her roll was more pronounced.

'Okay,' said Sergio, and began unbuttoning his jacket. 'You win. I give you.'

Benedict relaxed with relief, and Sergio thrust his hand into his jacket and stepped towards him. He grasped the canvas bag by its neck, and brought it out held like a cosh.

Sergio was close to him, too close for Benedict to swing the shotgun on to him. Sergio's expression became savage, his intentions blazed in his dark eyes as he lifted the canvas bag and poised himself to deliver a blow at Benedict's head, but he had not reckoned with the extraordinary reflexes of the natural athlete he was facing.

As Sergio launched the blow, Benedict rolled his shoulders and head away from it, lifting the butt of the shotgun as a guard. Sergio's wrist struck the seasoned walnut, and he grunted with the pain. His fingers opened nervelessly and the canvas bag flew from his grip, glanced off Benedict's temple and flew on down the deck, sliding to stop against one of the compressed air tanks thirty feet away.

Benedict danced back, dropping the barrels of the shotgun until Sergio looked into the muzzles.

'Hold it, you bastard,' Benedict snarled at him. 'You've made your choice. Now let's see what your guts look like.'

Sergio was hugging his injured wrist to his belly, crouching over it. Benedict was backing away to where the bag lay against the tank. His face was flushed and hectic with anger, but he kept darting side glances at the canvas bag.

At that moment *Kingfisher* took another wave over her bows, and the water came swooshing down the deck, picking up the bag and washing it towards the scuppers.

'Look out!' Sergio shouted. 'The bag! It's going.'

Benedict lunged for it, sprawling full length. With his free hand he grabbed the sodden canvas as it was disappearing over the side. But he was thirty feet away from Sergio, and he still held the shotgun in his other fist. Sergio could not hope to reach him without getting both charges of buckshot in his belly.

Instead Sergio spun round and sprinted back along the deck towards the bridge.

Benedict was on his knees frantically stuffing the bag of diamonds into the side pocket of his jacket and shouting after Sergio.

'Stop! Stop or I'll shoot!'

Sergio did not look back nor check his run, and Benedict had the bag in his pocket and now both hands were free. He lifted the shotgun, and tried to balance himself against *Kingfisher*'s wallowing motion as he aimed.

At the shot, Sergio stumbled slightly but kept on running. He reached the ladder and went up it.

Again Benedict aimed, and the shotgun clapped dully in the wind. This time a spasm of pain shuddered through Sergio's big body, and he froze on the ladder.

Benedict fumbled in his pocket for fresh shells, but before he could reload Sergio had begun climbing again. Benedict broke the gun and thrust the shells into the breech. He snapped the gun closed and looked up just as Sergio disappeared through the storm doors – and the two shots that Benedict loosed after him merely pockmarked the paintwork and starred the glass of the wheelhouse.

'The stupid bastard.' Hugo watched from the wheelhouse of *Wild Goose*. 'He's gone berserk.'

Hugo had heard the explosion and seen the shooting.

'Fifteen years is enough – but not the rope as well.'

He swung the wheel and *Wild Goose* sheered in towards

Kingfisher's side. Peering through the spray and salt-smattered windows, he saw Benedict drag himself to his feet and start after Sergio along the deck.

Hugo snatched the electric loudhailer from its bracket and pulled open the side window of the wheelhouse, holding the hailer to his lips.

'Hey! You stupid bastard, have you gone mad? What the hell you doing?'

Benedict glanced down at the trawler, then ignored it to give all his attention to reloading the gun. He kept going back along the deck, following Sergio to finish him off.

'You'll get us all strung up, you fool,' Hugo called through the loudhailer. 'Leave him. Let's get out of here.'

Benedict kept scrambling and slipping towards *Kingfisher*'s bridge.

'I'm leaving – now! Do you hear me? You can stew in your own pot. I'm getting out.'

Benedict checked and looked down at the trawler. He shouted and pointed at the bridge. Hugo caught one word: 'Diamonds.'

'All right, friend! Do what you like – I'll see you around,' Hugo hailed, and hit the trawler's throttle wide open. The roar of the diesels and the churning of her propeller convinced Benedict.

'Hugo! Wait! Wait for me, I'm coming.' He scampered back to the ladder and started down it.

Hugo throttled back and brought *Wild Goose* in neatly under the ladder.

'Jump!' he shouted through the hailer, and obediently Benedict jumped to hit the foredeck heavily. The shotgun flew from his hands to fall into the water alongside. Benedict cast one longing glance after the gun, then crawled to his feet and limped back to the wheelhouse.

Already *Wild Goose* was plunging away into the wind, but as Benedict entered the wheelhouse Hugo turned on him with his pink albino face set in a snarl of rage.

'What the hell have you done, you bastard? You lied to me. What was that explosion?'

'Explosion – I don't know. What explosion?'

Hugo hit him a stinging open-handed blow across the cheek.

'We agreed no killing – and you put us all on the spot.' Hugo's attention was focused completely on Benedict who had backed into the furthest corner of the wheelhouse. He massaged the dark red finger marks that stained his cheek.

'You set scuttling charges in *Kingfisher* – didn't you, you dirty son of a bitch. God, I hate to think what you've done with Lance and the girl.'

Outside the storm was nearing its climax. A rain squall swept down on *Wild Goose* – a sure sign that the wind must soon drop.

Automatically Hugo switched on the rotating wipers to clear the rain from the screen, as he continued to harangue Benedict.

'I saw you trying to murder the Italian. Christ! What for? He's one of us! Am I next on your list?'

'He had the diamonds,' mumbled Benedict. 'I was trying to get them from him.'

And Hugo's expression changed; he turned away from the wheel and stared at Benedict.

'You haven't got the diamonds? Is that what you're saying?' His tone was almost hurt.

'I tried – he wouldn't—'

And Hugo left the wheel and was across the wheelhouse like a white leopard. He grabbed the front of Benedict's coat, and screamed into his face.

'You left the diamonds! You stick my head in a noose – and I get nothing out of it.'

He was trembling with rage and his pale eyes bulged from their sockets.

Looking into those eyes Benedict realized his own danger. In the time it had taken him to leave *Kingfisher*'s deck and reach the wheelhouse of the trawler he had decided to let Hugo think that Sergio still had the diamonds. Squeamish as Hugo appeared to be about drowning Johnny and Tracey, despite his repeated demands for 'No killing', Benedict knew intuitively that Hugo had no intention of splitting a million pounds' worth of diamonds with him.

Once Hugo was certain that Benedict had the stones on board, Benedict knew there was no chance that he would reach South America alive.

The crossing might take weeks, the crew of the trawler were in Hugo's pay and loyal only to him. Benedict must sleep, and they would take him in the night.

On the other hand, of course, Benedict had no intention of splitting a million pounds' worth of diamonds with Hugo Kramer. He let his voice whine as he cringed in Hugo's grip.

'I tried. Sergio had them. He wouldn't – that's why I shot him.'

Hugo drew back his hand to slap Benedict again. Benedict twisted slightly, and drove his knee into Hugo's crutch, sending him staggering back across the wheelhouse, clutching himself between the legs and whimpering with the pain.

'Right, Kramer,' Benedict spoke softly. 'That's a little lesson for you. Behave yourself, and you'll get your fifty grand on the other side of the Atlantic.'

They stared at each other. Hugo Kramer weak and pale with agony, Benedict standing tall and arrogant again.

'Treat me gently, Kramer. I'm your meal-ticket. Remember it.'

Hugo gaped at him. The positions had reversed so swiftly. He pulled himself upright and his voice was thick with agony, but humble.

'I'm sorry, Mr van der Byl, I lost my temper. It's been a hell of a—'

'Skipper! Ahead!' The warning was shouted by the coloured crewman, Hansie.

Hugo stumbled to the untended wheel, and peered out into the storm.

Wild Goose was shooting down another slope of green water, and just ahead of her bows Hugo saw one of the huge yellow buoys that *Kingfisher* had laid down and then abandoned. It was held captive in the trough of the swells by the anchor cable. The cable was drawn as tight as a rod of steel across the trawler's bows, lifted just above the surface of the water; shivering drops of water flew from it under the tension of the buoy's drag.

'Oh God!' Hugo spun the wheel and threw *Wild Goose*'s engine into reverse – but she was racing down the swell, and her speed was unchecked as the cable scraped harshly along her keel.

Then came the harsh banging and clattering of the drive shaft as the cable fouled the propeller – followed by a crack as the shaft snapped. *Wild Goose*'s engine screamed into overrev as the load was lifted from it.

Hugo shut the throttle, and there was silence in the wheelhouse. *Wild Goose* swung beam on to the seas which came boiling in over her deck. Without her propeller she was transformed from a husky little sea creature to a piece of driftwood at the mercy of each current and the whim of the wind.

Hugo's head swung slowly until he was looking downwind to where the massive shapes of Thunderbolt and Suicide just showed through the rain squall.

'Cover your ears – tight!' Johnny Lance pressed Tracey against the bulkhead as far from the cyclone room as they could get. 'There are twenty-five pounds of plastique in there – it will blow like a volcano. He will have used short fuse, fourteen minutes. We won't have long to wait.'

Johnny set Tracey's shoulders squarely against the steel plating, and crouched over her – trying to shield her with his own body.

They stared into each other's eyes, teeth clenched, the heels of their palms jammed hard over their ears and they cowered away from the blast that must come.

The minutes passed, the longest minutes of Tracey's life. She could not have borne them without screaming hysteria except for that big hard body covering her – even with it she felt her fear mounting steadily during the molasses drip of time.

Suddenly the air lunged at her, driving the breath from her lungs. Johnny was thrown heavily against her. The blast sucked at her eardrums, and burst in her head so that bright lights flashed across her vision and she felt the steel plates heave under her shoulders.

Then her head cleared, and although her eardrums buzzed and sang, she found with a leaping relief that she was still alive.

She reached out for Johnny, but he was gone. In panic she groped, then opened her eyes. He was lurching down the long conveyor room, and when he reached the locked door at the far end he pressed his face to the peephole.

The fumes of the explosion still filled the cyclone room, a swirling bluish fog, but through them Johnny could make out the shambles that was the aftermath.

The huge cyclone had been torn from its mountings, and now sagged against the far bulkhead – crushed. It was worth only a single glance before Johnny froze into rigidity at the true horror.

The gravel pipe had been severed cleanly just below its juncture with the upper deck. It protruded for six feet, but now the force of the jet through it was flicking and whipping it about as though it were not steel but a rubber garden hose.

The jet was a solid eighteen-inch column, a pillar of brown mud and yellow gravel and sea water that beat against the steel plates of the hull with a hollow drumming roar.

In the few seconds since the explosion the cyclone room was already half-filled with a slimy shifting porridge that rushed from wall to wall with the movement of the ship. It was like some monstrous jelly fish which each second gathered weight and strength.

Tracey reached Johnny's side and he placed his arm around her shoulders. She looked through the armoured glass and he felt her body stiffen.

At that moment the yellow monster spread over the window, obscuring it completely. Johnny felt the first straining of the steel plates under his hands. They fluttered and bulged, then began to protest aloud at the intolerable pressure. A seam started, and a fine jet of filthy water hissed from the gap and soaked icily through Johnny's jersey.

'Get back.' Johnny dragged Tracey away from the squeaking, groaning bulkhead. Back along the narrow conveyor room they stumbled, moving with difficulty for the deck beneath their feet was slanting as *Kingfisher* began to lean under the increasing weight in her belly.

Still holding Tracey, he reached the locked door and resisted the futile desire to attack it with his bare hands. Instead he forced his brain to work, tried to anticipate the sequence of events that would lead to the final destruction of *Kingfisher* – and all those aboard her.

Benedict had left the other entrance to the cyclone room wide open. Already that viscous mass of mud and water must be spreading rapidly through the lower levels of the hull, following always the avenue of least resistance – finding the weak spots and bursting through them.

If the walls of the conveyor room held against the pressure, the rest of the hull would be filled and they would be enfolded in the tentacles of that great yellow monster – a small bubble of air trapped within it and taken down with it when it returned to the depths from which it had come.

Would the bulkheads of the conveyor room hold? The answer came almost immediately in the squeal of metal against metal, and the crackle of springing rivets.

The monster had found the weak spot, the aperture through the drying furnace into the conveyor, ripping away the fragile baffles, bursting through the furnace in a cloud of steam, it gushed into the conveyor room bringing with it the sewage stench of deep-sea mud.

Kingfisher made another sluggish roll, so different from her usual spry action, and the mud came racing down the tunnel in a solid knee-high wall.

It slammed both of them back against the steel door with a shocking strength, and the feel of it was cold and loathsome as something long dead and putrefied.

Kingfisher rolled back and the mud slithered away, bunched itself against the far bulkhead then charged at them again.

Waist-deep it struck them, and tried to suck them back with the next roll.

Tracey was screaming now, nerves and muscles reaching their breaking-point. She was clinging to Johnny, coated to the waist in stinking ooze, her eyes and mouth wide open in terror as she watched the mud building up for its next assault.

Johnny groped for some hold to anchor them. They must keep on their feet to survive that next rush of mud. He found the locking handle of the door and braced himself against it, holding Tracey with all his strength.

The mud came again, silently, murderously. It burst over

their heads and punched them with stunning force against the plating.

Then it sucked back once more, and left them down on their knees, anchored only by Johnny's grip on the locking handle.

Tracey was vomiting the foul mud and it filled her ears and eyes and nostrils, clogging them so that it bubbled at her breathing.

Johnny could feel her weakening in his arms, her struggles becoming more feeble as she tried to regain her feet.

His own strength was going. It needed his last reserve to drag them both upright.

The locking handle turned in his fingers, spinning open. The steel door against which he was braced fell away, so that he staggered backwards without support but still clutching Tracey.

There was just a moment to recognize the big, reassuring bulk of Sergio Caporetti beside him and feel an arm like the trunk of a pine tree steady him before the rush of mud down the conveyor room hit them and knocked all three of them down, sending them swirling and rolling end over end before its strength dissipated in the new space beyond the conveyor room door.

Johnny pulled himself up the bulkhead. He had lost Tracey. Dazed but desperate he looked for her, mumbling her name.

He found her swilling aimlessly in the waist-deep mud, floating on her face. He took a handful of muddy hair and lifted her face out, but the mud had hold of his legs, pulling him off balance as it surged back and forth.

'Sergio. Help!' he croaked. 'For God's sake, Sergio.'

And Sergio was there, lifting her like a child in his arms and wading to the ladder that led to the deck above. The mud knocked Johnny down again, and when he surfaced Sergio was climbing steadily up the ladder.

Despite the mud and water that blurred Johnny's eyesight, he could see that Sergio's wide back, from shoulders to hips, was speckled with dozens of punctures as though he had been stabbed repeatedly with a knitting needle. From each tiny wound oozed droplets of blood that spread like brown ink on the blotting-paper of his sodden jacket.

At the head of the companionway Sergio turned, still holding Tracey in his arms; he stood like a Colossus looking down at Johnny wallowing and slipping in the mud below.

'Hey, Lance – go switch off your bloody machine. She drown my ship. I sail her myself now – the right way. No bloody fancy machine.'

Johnny steadied himself against the bulkhead and called up at him:

'Sergio, what happened to Benedict van der Byl, where is he?'

'I think he go with *Wild Goose* – but first he shoot the hell out of me, not half. Fix your machine, no time for talk.'

And he was gone, still carrying Tracey.

Another rush of mud carried Johnny down the flooded passage and threw him against the door to the control room. Already his body seemed to be one aching bruise, and still the battering continued as he tried to unlock and open the control room door.

At last, using the suck of the mud to help him, he yanked it open and went in with a burst of yellow slime following him and flooding the compartment shoulder deep.

Clinging to the console of the computer he reached up and punched the master control buttons.

'Dredge Stop.'

'Dredge engines Stop.'

'Main engines manual.'

'Navigation system manual.'

'All programmes abort.'

Instantly the roar from the severed dredge pipe, which

had echoed through the ship during all their strivings, dwindled as though some vast waterfall had dried. Then there was silence. Though only comparative silence, for the hull still groaned and squeaked at the heart-breaking burden it now carried and the mud slopped and thudded against the plating.

Weak and sick, Johnny clung to the console. He was shivering with cold, and every muscle in his body felt bruised and strained.

Suddenly the ship changed her motion, heaving under his feet like a harpooned whale as she swung broadside to the storm. Johnny roused himself with alarm.

The journey back through the flooded passages to the companionway was an agony of mind and body – for *Kingfisher* was now behaving in a strange and unnatural way.

The scene that awaited Johnny as he dragged himself on to the bridge chilled his soul as the icy mud had chilled his body.

Thunderbolt and Suicide lay less than a furlong off *Kingfisher*'s starboard quarter. Both islands were wreathed in sheets of spray that fumed from the surf that was breaking like cannon-fire on the cliffs.

The maniacal flute of the wind joined with the drum of the surf to produce a symphony fit for the halls of Hell, but above this devil's music Sergio Caporetti bellowed, 'We got no power on port main engine.'

Johnny turned to him. Sergio was hunched over the wheel, and Tracey lay on the deck at his feet like a discarded doll.

'The water, she kill port main.' Sergio was pumping the engine telegraph. Then abandoning the effort, he looked over the side.

The reeking white cliffs were closer now, much closer – as though you could reach out a hand to them. The ship was drifting down rapidly on the wind.

Sergio spun the wheel to full port lock, trying to bring *Kingfisher*'s head round to meet the sea and the wind. She was rolling as no ship was ever meant to roll, hanging over at the limit of each swing, so that the wheelhouse windows seemed but a few feet from the crests of the green waves. She hung like that as though she meant never to come upright again. Then sluggishly, reluctantly, she swung back, speeding up as she reached the perpendicular and the great mass of mud and water in her hull shifted and slammed her over on her other side, pinning her like that for eternal seconds before she could struggle upright again for the cycle to be repeated.

Sergio held the wheel at full lock, but still *Kingfisher* wallowed down towards the cliffs of Thunderbolt and Suicide. The wind had her the way a dog carries a bone in its teeth. Under half power and with her decks awash *Kingfisher* could not break that grip.

Johnny was a helpless spectator, held awe-bound so he could not break away even to succour Tracey who was still lying on the deck. He saw everything with a supernatural clarity – from the dribbling little shot holes in Sergio's back, to the ponderous irresistible rush of the white water up the cliffs that loomed so close alongside.

'She no answer helm. She too sick.' Sergio spoke now in conversational tones which carried with surprising clarity through the uproar of the elements. 'All right then. We go the other way. We take the gap.'

For a moment Johnny did not understand, then he saw it. *Kingfisher*'s bows were coming up to the narrow opening between the two islands.

A passage less than a hundred yards wide at its narrowest point, where the vicious cross-currents met head-on and leapt fifty feet into the air as they collided. Here the surface was obscured by a thick froth of spindrift that heaved and humped up as though the ocean were fighting for breath under the thick cream-coloured blanket.

'No.' Johnny shook his head, staring at that hideous passage. 'We won't make it, Sergio. We won't do it.'

But already Sergio was spinning the wheel from lock to lock, and unbelievably *Kingfisher* was responding. Helped now by the wind she came around slowly, seeming to brush her bows across the white cliff of Thunderbolt, and she steadied her swing and aimed at the gap. It was then Johnny saw it for the first time.

'Christ, there's a boat dead ahead!'

The steep swells had hidden it up to that moment, but now she bobbed up on a crest. It was a tiny trawler, flying a dirty scrap of canvas as a staysail at her stubby mast, and struggling piteously in the granite jaws of Thunderbolt and Suicide.

'*Wild Goose!*' roared Sergio, and he reached for the handle of the foghorn that hung above his head.

'Now we have some fun.' And he yanked the handle. The croaking bellow of the foghorn echoed off the cliffs that were closing on either side of them.

'Kill my boys – hey? Shoot me – hey? Trick me – hey? Now I trick you – but good!' Sergio punctuated his triumphant yells with blasts on the foghorn.

'Christ, no! You can't do it!' Johnny caught urgently at the big Italian's shoulder, but Sergio struck his hand away and steered directly for the trawler as it lay full in the narrow passage.

'I give him plenty warning.' Sergio sent another blast echoing off the cliffs. 'He no give me warning when he shoot me – the bastard.'

There was a group of men on the trawler's foredeck. Johnny could see that they were manhandling an inflatable escape raft, a thick mattress of black rubber, towards the nearest side of the trawler. But now they were frozen by the bellow of *Kingfisher*'s foghorn. They stood looking up at the tall cliff of steel that bore down on them. Their faces were pale blobs in the gloom.

'Sergio. It's murder. Turn away, damn you, you can miss them. Turn away!' Again Johnny lunged across the wheel-house and grabbed at the wheel.

Sergio swung a backhanded blow that cracked against Johnny's temple and sent him reeling back half-stunned against the storm doors.

'*Who* Captain for this bloody ship!' There was blood on Sergio's lips, the shout had torn something inside him.

Kingfisher's bows were lifting and swinging down like an executioner's axe over the trawler. They were close enough now for Johnny to recognize the men on the trawler's deck – but only one of them held his full attention.

Benedict van der Byl cowered against the trawler's rail, gripping it with both hands. His hair fluttered, soft and dark, in the wind. His eyes were big dark holes like those of a skull in the bone-white face, and his lips a pink circle of terror.

Then suddenly the trawler disappeared under *Kingfisher*'s massive bows, and immediately after that came the splintering crunching sound of her timbers shattering. *Kingfisher* bore on down the passage between the cliffs without a check in her speed.

Johnny fumbled with the catch of the storm doors, and the wind tore it open. He staggered out on to the exposed wing of the bridge and reached the rail.

He stood there with the storm clawing at his clothing and looked down at the wreckage that dragged slowly along *Kingfisher*'s hull, and then was left behind.

There were human heads bobbing among the wreckage, and the wash from *Kingfisher*'s propeller pushed them towards the cliffs of Suicide.

A wave picked up one of the men and carried him swiftly on to the cliff, sweeping him high and then swirling back to leave the body stranded on that smooth white slope of granite.

The man was still alive, Johnny saw him clawing at the

smooth rock with his bare fingers, trying to drag himself above the reach of the sea.

It was Hugo Kramer; even through the fog of spray there was no mistaking that head of pale hair and the lithe twisted body.

The next wave reached up and dragged him back over the rock, tearing the nails from his hooked fingers as he tried to find a hold.

He was swirled and flung about in the turbulence below the cliff before another wave lifted him and hurled him on to the granite. One of his legs was broken at the knee by the force of the impact and the lower part of the leg spun loosely like the blade of a windmill as the water tugged at it.

Once again Hugo was left stranded but now he made no movement. His arms were flung wide, and his leg stuck out at a grossly unnatural angle from the knee.

Then from among these mighty waves there rose up a mass of green water which dwarfed all the others.

It reared up with slow majesty and hung poised over the granite cliff before it fell on Hugo's broken body with a boom that seemed to shake the very rock.

When the giant wave drew back, the cliff was washed clean. Hugo was gone.

The same wave that destroyed him came down the passage between the cliffs and, in contrast to its treatment of Hugo Kramer, it was tender as a mother as it lifted *Kingfisher* and carried her out into the open sea beyond the reach of those cruel cliffs.

Looking back into the gap between the islands, the last trace that Johnny saw of the *Wild Goose* was the black rubber escape raft tossing and leaping high on the turmoil of broken water and creamy spindrift.

'They'll have no use for that,' he said aloud. He searched for a sight of any survivor, but there was none. They were chewed to pulp in the jaws of Thunderbolt and Suicide, and swallowed down into the cold green maw of the sea.

Johnny turned away and went back into the wheelhouse. He lifted Tracey from the deck and carried her through to Sergio's cabin.

As he laid her on the bunk he whispered to her, 'I'm glad. I'm glad you didn't see it, my darling.'

At midnight the wind still howled about the ship, hurling sheets of solid rain against the windows of the bridge. Forty minutes later the wind had veered through a hundred and eighty degrees and become a light south-easterly air. The black sky opened like a theatre curtain, and the full moonlight burst through so brightly as to pale the stars. Though the tall black swells still marched in martial ranks from the north, the gentle wind was soothing and lulling them.

'Sergio, you must rest now. I will take the con. Let Tracey dress your back.'

'You take the con!' Sergio snorted scornfully. 'I save the ship – and you sink her for me. Not bloody likely.'

'Listen, Sergio. We don't know how badly you are hurt. You are killing yourself.'

The same argument spluttered and flared intermittently during the long night hours while Sergio clung stubbornly to the helm and coaxed the labouring vessel back towards Cartridge Bay. He insisted on detouring far out to sea to avoid the islands, so that when the bright dawn broke, the land was only a low brown line on the horizon and the mountains of the interior were a distant blue.

An hour after dawn Johnny made radio contact with the very agitated operator at Cartridge Bay.

'Mr Lance, we've been trying to raise you since yesterday evening.'

218

'I've been busy.' Despite his fatigue Johnny grinned at his own understatement. 'Now, listen to me. We are coming into Cartridge Bay. We'll be there in a couple of hours. I want you to have a doctor, Doctor Robin Sutherland, flown up from Cape Town – also I want you to have the police standing by. I want somebody from both the Diamond police and the Robbery and Murder squad – have you got that?'

'The police are here already, Mr Lance. They are looking for Mr Benedict van der Byl. They found his car here – they have a warrant . . .' The operator's voice broke off and Johnny heard the mumble of background voices, then – 'Mr Lance, are you there? Stand by to speak to Inspector Stander of the CID.'

'Negative!' Johnny cut in on his transmission. 'I'm not talking to anybody. He can wait until we get into the Bay. Just you have Doctor Sutherland ready. I've got a badly wounded man on board.'

Johnny leaned over the radio set and shut off the main switch, then he stood and made his way slowly back to the bridge. Every muscle in his body felt stiff and bruised, and he was groggy with tiredness, but he took up the argument with Sergio where they had left off.

'Now listen to me, Sergio. You must lie down. You can take us in over the bar; but now you must get an hour or two's rest.'

Still Sergio would not relinquish the wheel, but he consented to strip to the waist and let Tracey examine his back.

In the expanse of white muscle were little black holes each set in its own purple bruise. Some of the holes had sealed themselves with black clotted blood, from others fluid still oozed – clear or pink in colour – and there was a faint sweetish smell from the wounds.

Johnny and Tracey exchanged worried looks before Tracey reached into the first aid box and set to work.

'How she look, Johnny?' Sergio's jovial tone was belied by his face which was a lump of bread dough touched with greenish blue hues.

'Depends if you like your meat rare.' Johnny matched his tone, and Sergio chuckled but cut it short with a wince.

Johnny put a cheroot between Sergio's lips and held a match for him. As Sergio puffed the tip into a glow, Johnny asked casually, 'What made you change your mind?' And Sergio looked up at him quickly, guiltily, through the cloud of cheroot smoke.

'You had us cold. You could have got away with it – perhaps,' Johnny persisted quietly. 'What made you come back?'

'Listen, Johnny. Me, I've done some damned awful bloody things – but I never killed a man or a woman – ever. He said no killing. Fine, I go along. Then I hear the plastique blow. I know you two in conveyor room. I think the hell with it. Now, I climb off the wagon – but she's going too fast. I get bum full of buckshot.'

They were silent for a while. Tracey was absorbed in patching the shot-wounds with adhesive tape.

Johnny broke the silence. 'Was there a big diamond, Sergio? A big blue diamond?'

'Si.' Sergio sighed. 'Such a diamond you will never see again.'

'Benedict had it?'

'Si. Benedict had it.'

'Did he have it on him?'

'In his coat. He put it in his coat pocket.'

Tracey stepped back. 'That's all we can do for now,' she murmured and caught Johnny's eye, shaking her head slightly and frowning with worry. 'The sooner we get him to a doctor the happier I'll be.'

A little before noon Sergio took *Kingfisher* in through the entrance to Cartridge Bay, handling the mud-filled ship with all the aplomb of the master mariner, but as they approached the first turning in the channel he sagged gently to the deck and the wheel spun out of his hands.

Before Johnny could reach the helm, *Kingfisher* had yawed wearily across the channel. She had so little way on her that when she went up on the sand bank there was only a small jolt and she listed over a few degrees.

Johnny pulled the engine telegraph to 'STOP'.

'Help me, Tracey.' He stooped over Sergio and took him under the armpits. Tracey grasped his ankles. Half dragging, half carrying, they got him through to his cabin and laid him on his bunk.

'Hey, Johnny. Sorry, Johnny,' Sergio was mumbling. 'First time I put ship on bank – ever! Idiot! So close – then *wap*! Sorry, hey, Johnny.'

The motor launch left the jetty and came down the channel towards the sand bank on which *Kingfisher* lay stranded. The launch was crowded, and the whine of the outboard engine raised a storm of water fowl into a whirlwind of frightened wings.

As it drew closer Johnny recognized some of the occupants. Mike Shapiro and with him Robin Sutherland, but there were also two uniformed policemen and another person in civilian clothes who stood up in the launch as it came within hail and cupped his hands about his mouth.

'I am a police officer. I have a warrant for the arrest of Benedict—'

Mike Shapiro touched the man's arm, and spoke softly to him. The officer hesitated and glanced up at Johnny again, before nodding agreement and settling back on to his seat.

'Robin, get up here as quick as you can,' Johnny shouted down at the launch, and when Robin came over the side Johnny hustled him towards the bridge, but Mike Shapiro hurried after them.

'Johnny, I must talk to you.'

'It can wait.'

221

'No, it can't.' Mike Shapiro turned to Tracey. 'Won't you take care of the doctor, please? I must speak to Johnny before the police do.'

Mike led Johnny down the deck and offered him a cigarette, while the three policemen hovered at a discreet distance.

'Johnny, I have some dreadful news. I want to break it to you myself.'

Johnny visibly braced himself. 'Yes?'

'It's about Ruby—'

Johnny made his statement to the police inspector in *Kingfisher*'s guest cabin. It took two hours for him to relate the full story, and during that time one of the uniformed policemen discovered the crew locked in the paint store below decks. They were half poisoned with paint fumes but able to make their statements to the police.

The inspector kept them waiting in the next-door cabin while he finished his interrogation of Johnny.

'Two more questions for now, Mr Lance. In your opinion was the collision between the two ships accidental or deliberate?'

Johnny looked into the steel-grey eyes and lied for the first time.

'It was unavoidable.'

The inspector nodded and made a note on his pad.

'Last question. The survivors from the trawler, what were their chances?'

'In that storm they had none. There was no hope of effecting a rescue with *Kingfisher* almost disabled, and considering the condition of the surf in the passage between the islands.'

'I understand.' The inspector nodded. 'Thank you, Mr Lance. That is all for the present.'

Johnny left the cabin and went quickly to the upper deck. Tracey and Robin were still working over Sergio's bunk, but Robin looked up and came immediately to Johnny as he stood in the doorway.

'How is he, Robin?'

'He hasn't a chance,' Robin replied, keeping his voice low. 'One lung has collapsed, and there appear to be perforations of the bowel and intestine. I suspect a massive peritonitis. I can't move him without risking a secondary haemorrhage.'

'Is he conscious?'

Robin shook his head. 'He's going fast. God knows how he has lasted this long.'

Johnny moved across to the bunk and placed his arm about Tracey's shoulders. She moved closer to him and they stood looking down at Sergio.

His eyes were closed, and a dark pelt of new beard covered the lower part of his face. His breathing sawed and hissed loudly in the quiet cabin, and the fever lit bright spots of colour in his cheeks.

'You magnificent old rogue.' Johnny spoke softly, and Sergio's eyes blinked open.

Quickly Johnny stooped to him.

'Sergio. Your crew – your boys are safe.'

Sergio smiled. He closed those dark gazelle eyes, then opened them again and whispered painfully, 'Johnny, you give me job when I come out of prison?'

'They won't have you in prison – you'd lower the tone of the place.'

Sergio tried to laugh. He managed one strangled chuckle, then he came up on his elbows in the bunk with his eyes bulging, his mouth gaping for breath. He coughed once, a harsh tearing sound, and the blood burst from his lips in thick black clots and a bright red spray of droplets.

He fell back on the pillows, and was dead before Robin reached his side.

Tracey was asleep in the bedroom next door. Robin had sedated her heavily enough to keep her that way for the next twelve hours.

Johnny lay naked on the narrow bunk in the second guest room of the Cartridge Bay depot, and when he switched on the beside lamp his wrist watch showed the time as 2.46 in the morning.

He looked down at his own body. The bruises were dark purple and hot angry red across his ribs and flanks from where the mud had battered him against rough steel plating. He wished now that he had accepted the sleeping pills Robin had offered him, for the ache of his body and the whirl of his thoughts had kept him from sleep all that night.

His mind was trapped on a nightmare roundabout, revolving endlessly the two deaths which Benedict van der Byl must answer for in the dark places to which he had surely gone.

Ruby and Sergio. Ruby and Sergio. One he had seen die, the other he could imagine in all its gruesome detail.

Johnny sat up and lit a cigarette, seeking a distraction from the tortured images with which his overexcited brain bombarded him.

He tried to concentrate on reviewing the practical steps that would be necessary to clear up the aftermath of these last disastrous days.

He had spoken that evening by radio to Larsen, and received from him a promise of complete financial support during the time it would take to clear the mud from *Kingfisher*'s hull and recover the diamonds in the conveyor tunnel, and to tide over the period of salvage and repair before the dredger was ready to begin once more harvesting the rich fields of Thunderbolt and Suicide.

A salvage team would fly in tomorrow to begin the work on *Kingfisher*. He had cabled IBM requesting engineers to check out the computer for water damage.

Six weeks, Johnny estimated, before *Kingfisher* was ready for sea.

Then his unruly imagination leapt suddenly ahead to Ruby's funeral. It was set for Tuesday next week. Johnny rolled restlessly on his bunk, trying to shut his mind against the thoughts that assaulted it – but they crowded forward in a dark host.

Ruby, Benedict, Sergio, the big blue diamond.

He sat up again, stubbed out his cigarette and reached across to switch out the bedside lamp.

He froze like that, as a new thought pressed in on him. He heard Sergio's voice in his memory.

'Such a diamond as you will never see again.'

Now he felt the idea come ghosting along his spine so that the hair at the nape of his neck and on his forearms prickled with excitement.

'The Red Gods!' he exclaimed, almost shouting the name. And again Sergio's voice spoke.

'In his coat. He put it in his coat pocket.'

Johnny swung his legs off the bunk, and reached for his clothes. He felt the pounding of his heart beneath his fingers as he buttoned his shirt. He pulled on slacks and sweater, tied the laces of his shoes and snatched up a sheepskin jacket as he ran from the room.

He was shrugging on the jacket as he entered the deserted radio room and switched on the lights. He crossed quickly to the chart table and pored over it.

He found the name on the map, and repeated it aloud.

'The Red Gods.'

North of Cartridge Bay the coast ran straight and featureless for thirty miles, then abruptly the line of it was broken by the out-thrust of red rock, poking into the sea like an accuser's finger.

Johnny knew it well. It was his job to find and examine any such natural feature that might act as a barrier to the prevailing inshore currents. At such a place diamonds and other seaborne objects would be thrown ashore.

He remembered the red rock cliffs carved by wind and sea into the grotesque natural statues which gave the place its name, but more important he remembered the litter of ocean debris on the beaches beneath the cliff. Driftwood, waterlogged planking, empty bottles, plastic containers, scraps of nylon fishing-net and corks – all of it cast overboard and carried up by the current to be deposited on this promontory.

He ran his finger down the chart and held it on the dots of Thunderbolt and Suicide. He read the laconic notation over the tiny arrows that flew from the islands towards the stark outline of the Red Gods.

'Current sets South South-West. 5 knots.'

Above the chart table the depot keys hung on their little cuphooks, each labelled and numbered.

Johnny selected the two of them marked 'GARAGE' and 'LAND-ROVER'.

The moon was full and high. The night was still and without a trace of wind. Johnny swung the double doors of the garage open and switched on the parking lights of the Land-Rover. By their glow he checked out the vehicle; petrol tank full, the spare five-gallon cans in their brackets full, the can of drinking-water full. He dipped his finger into the neck of the water container and tasted it. It was clean and sweet. He lifted the passenger seat and checked the compartment beneath it. Jack and tyre spanner, first aid kit, flashlight, signal rockets and smoke flares, water bottle, canvas groundsheet, two cans of survival rations, towrope, tool kit, knapsack, knife and compass. The Land-Rover was equipped to meet any of the emergencies of desert travel.

Johnny climbed behind the wheel and started the engine. He drove quietly and slowly past the depot buildings, not wanting to awaken those sleeping inside, but when he hit the firm sand at the edge of the lagoon he switched on the headlights and gunned the engine.

He cut across the sand dunes at the entrance to the bay, and swung northwards on to the beach. The headlights threw solid white beams into the sea mist, and startled sea-birds rose on flapping ghost wings before the rush of the Land-Rover.

The tide was out and the exposed beach was hard and shiny wet, smooth as a tarmac road. He drove fast, and the white beach crabs were blinded by the headlights and crunched crisply beneath the tyres.

The dawn came early, silhouetting the mystic shapes of the dunes against the red sky.

Once he startled a strand wolf, one of the brown hyenas which scavenged this bleak littoral. It galloped, hunch-shouldered, in hideous panic for the safety of the dunes. Even in his urgency Johnny felt a stir of revulsion for the loathsome creature.

The cold damp rush of the wind into his face refreshed Johnny. It cooled the gritty feeling of his eyes, and eased the throb of sleeplessness in his temples.

The sun burst over the horizon, and lit the Red Gods five miles ahead with all the drama of stage lighting. They glowed golden red in the dawn, a procession of huge half-human shapes that marched into the sea.

As Johnny drove towards them the light and shadow played over the cliffs and he saw a hundred-foot tall figure of Neptune stooping to dip his flowing red beard into the sea, while a monstrous hunchback with the head of a wolf pranced beside him. Ranks of Vestal Virgins in long robes of red rock jostled with the throng of weird and fantastic shapes. It was eerie and disquieting. Johnny curbed his fancy and turned his attention to the beaches at the foot of the cliffs.

What he saw started his skin tingling again, and he

227

pressed the accelerator flat against the floorboards, racing across the wet sand to where a white cloud of seabirds circled and dived and hopped about something that lay at the water's edge.

As he drove towards them a gull flew across the front of the Land-Rover. A long ribbon of something wet and fleshy dangled from its beak, and the gull gulped at it greedily in flight. Its crop was distended and engorged with food.

The seabirds scattered raucously and indignantly as the Land-Rover approached, leaving a human body lying in the centre of an area of sand that was dappled by the prints of their webbed feet, and fouled with dropped feathers and excreta.

Johnny braked the Land-Rover and jumped out. He took one long look at the body, then turned quickly away and braced himself against the side of the vehicle.

His gorge rose in a hot flood of nausea, and he gagged it back.

The body was nude but for a few sodden tatters of clothing and a sea boot still laced on to one foot. The birds had attacked every inch of exposed flesh – except for the scalp. The face was unrecognizable. The nose was gone, the eye sockets were empty black holes. There were no lips to cover the grinning teeth.

Above this ruined face the shock of colourless albino hair looked like a wig placed there as an obscene and tasteless joke.

Hugo Kramer had made the long voyage from Thunderbolt and Suicide to the Red Gods.

Johnny took the canvas groundsheet from under the passenger seat of the Land-Rover. Averting his eyes from the task, he wrapped the corpse carefully, tied the whole bundle with lengths of rope cut from the tow line, then laboriously dragged it up the beach well above the high-water mark.

The thick canvas would keep off the birds, but to make doubly certain Johnny collected the driftwood and planking that was scattered thickly along the high-water line and piled it over the corpse.

Some of the planking was freshly broken and the paint on it was still bright and new. Johnny guessed this was part of the wreckage of *Wild Goose*.

He went back to the Land-Rover, and drove on towards the Red Gods which lay only a mile ahead.

The sun was well up by now, and already its heat was uncomfortable. As he drove he struggled out of the sheepskin jacket without interrupting his search of the beach ahead.

He was looking for another gathering of seagulls, but instead he saw a large black object stranded in the angle formed by the red stone cliffs.

He was fifty yards from it before he realized what it was.

He felt his stomach jar violently and then clench at the shock.

It was a black rubber inflatable life raft – and it had been dragged up the beach *above the high-water line*.

As he climbed out of the Land-Rover Johnny felt his legs trembling beneath him, as though he had just climbed a mountain. The hard knotted sensation in his chest shortened his breathing.

He went slowly towards the raft, and there was a story to read in the soft sand.

The smooth drag mark of the raft, and the two sets of footprints. One set made by bare feet; broad, stubby-toed and with flattened arches, the prints of a man who habitually went barefooted.

These tracks had been made by one of the coloured crew of *Wild Goose*, Johnny decided, dismissed them and turned his whole attention to the other set of footprints.

Shod feet, long and narrow, smooth leather soles; the

imprints were sharp-edged suggesting new shoes little worn, the length of the stride and the depth of prints were those of a tall heavily built man.

Johnny realized with mild surprise that his hands were shaking now, and even his lips quivered. He was like a man in high fever; light-headed, weak and shaking. It was Benedict van der Byl. He knew it with complete and utter certainty. Benedict had survived the maelstrom of Thunderbolt and Suicide.

Johnny balled his fists, squeezing hard and he thrust out his jaw, tightening his lips. Still the hatred washed over his mind in dark hot waves.

'Thank God,' he whispered. 'Thank God. Now I can kill him myself.'

The footprints had churned the sand all about the raft. Beside them lay a thick piece of planking which had been used as a lever to rip the emergency water container and the food locker from the floor-boards of the raft.

The food locker had been ransacked and abandoned. They would be carrying the packets of iron rations in their pockets to save weight, but the water container was gone.

The two sets of prints struck out straight for the dunes. Johnny followed them at a run and lost them immediately in the shifting wind-blown sands of the first dune.

Johnny was undismayed. The dunes persisted for only a thousand yards or so, then gave way inland to the plains and salt flats of the interior.

He ran back to the Land-Rover. He had his emotions under control again. His hatred was reduced to a hard indigestible lump below his ribs, and he contemplated for a few seconds lifting the microphone of the Land-Rover's R/T set and calling Cartridge Bay.

Inspector Stander had the police helicopter parked on the landing-strip behind the depot buildings. He could be here in thirty minutes. An hour later they would have Benedict van der Byl.

Johnny dismissed the idea. Officially Benedict was dead, drowned. No one would look for him in a shallow grave in the wastes of the Namib desert.

The crewman with him would be a complication; but he could be bribed and frightened or threatened. Nothing must stand in the path of his vengeance. Nothing.

Johnny opened the Land-Rover's locker and found the knife. He went to the raft and stabbed the blade through the thick material at a dozen places. The air hissed from the holes, and the raft collapsed slowly.

Johnny bundled it into the back of the Land-Rover. He would bury it in the desert; there must be no evidence that Benedict had come ashore.

He started the engine, engaged the four-wheel drive, and followed the spoor to the foot of the dunes.

He picked his way through the valleys and across the knife-backed ridges of sand.

As he descended the last slope of the dunes he felt the oppressive silence and immensity of the land enfold him. Here, only a mile from the sea, the moderating influence of the cold Benguela current did not reach.

The heat was appalling. Johnny felt the sweat prickling from the pores of his skin and drying instantly in the lethal desiccating air.

He swung the Land-Rover parallel to the line of the dunes and crawled along at walking speed, hanging over the side of the vehicle and searching the ground. The bright specks of mica in the sand bounced the heat of the sun into his face.

He cut the spoor again where it came down off the dunes and went away on it, headed arrow straight at the far line of mountains which were already receding into the blue haze as the heat built up towards noon.

Johnny's progress was a series of rushes where the spoor ran true, broken by halts and painstaking casts on the rocky ridges and areas of broken ground. Twice he left the Land-

Rover to work the spoor through difficult terrain, but across one of the flat white salt pans he covered four miles in as many minutes. The prints were strung like the beads of a necklace, cut clearly through the glistening crust of salt.

Beyond the pan they ran into a maze of black rocks, riven by gullies, and guarded by the tall misshapen monoliths.

In one of the gullies he found Hansie, the little old coloured crewman from *Wild Goose*. His skull had been battered in with the blood-caked rock that lay beside him. The blood had dried slick and shiny, and Hansie stared with dry eyeballs at the merciless sky. His expression was of mild surprise.

The story of this new tragedy was written in the sandy bottom of the gully. In an area of milling, confused prints the two men had argued. Johnny could guess that Hansie had wanted to turn back for the coast. He must have known that the road lay beyond the mountains, a hundred miles away. He wanted to abandon the attempt, try for the coast and Cartridge Bay.

The argument ended when he turned his back on Benedict, and returned on his old tracks.

There was a depression in the sand from which Benedict had picked up the rock and followed him.

Standing over Hansie, and looking down at that pathetic crushed head, Johnny realized for the first time that he was following a maniac.

Benedict van der Byl was insane. He was no longer a man, but a raging demented animal.

'I will kill him,' Johnny promised the old white woolly head at his feet. There was no need for subterfuge now.

If he caught up with Benedict and did it, no court in the world would question but that it was self-defence. Benedict had placed himself beyond the laws of man.

Johnny took the deflated rubber raft and spread it over

Hansie. He anchored the edges of the rubber sheet with rocks.

He drove on into the dancing, shimmering walls of heat with a new mood on him; murderous, elated expectation. He knew that at this moment in time he was part animal also, corrupted by the savagery of the man he was hunting. He wanted payment in full from Benedict van der Byl in his own coin. Life for life, and blood for blood.

A mile farther on he found the water container. It had been flung aside violently, skidding across the sand with the force of its rejection; the water had poured from its open mouth, leaving a dried hollow in the thirsty earth.

Johnny stared at it in disbelief. Not even a maniac would condemn himself to such a horrible ending.

Johnny went across to where the five-gallon brass drum lay on its side. He picked it up and shook it, there was the sloshy sound of a pint or so of liquid in it.

'God!' he whispered, awed and feeling a twinge of pity despite himself. 'He won't last long now.'

He lifted the container to his lips and sucked a mouthful. Immediately his nostrils flared with disgust, and he spat violently, dropping the can and wiping at his lips with the back of his hand.

'Sea water!' he mumbled. He hurried to the Land-Rover and washed out his mouth with sweet water.

How the contamination had occurred he would never know. The raft might have lain for years in *Wild Goose* without its stores being checked or renewed.

From that point onwards Benedict must have known he was doomed. His despair was easy to read in the blundering footsteps. He had started running, with panic driving him. Five hundred yards further on he had fallen heavily into the bed of a dry ravine, and lain for a while before dragging himself up the bank.

Now he had lost direction. The spoor began a long curve

to the northwards, running again. It came round full circle, and where it crossed itself, Benedict had sat down. The marks of his buttocks were unmistakable. He must have controlled his panic here because once more the spoor struck out with determination towards the mountains.

However, within half a mile he had tripped and fallen. Now he was staggering off course again, drifting southwards. Once more he had fallen, but here he had lost a shoe.

Johnny picked it up and read the printed gold lettering on the inner sole. 'BALLY OF SWITZERLAND, SPECIALLY MADE FOR HARRODS. That's our boy Benedict, all right. Forty-guinea black kid,' he muttered grimly, and climbed back into the Land-Rover. His excitement was climaxing now. It would be soon, very soon.

Farther on Benedict had wandered down into the bed of an ancient watercourse, and turned to follow it. His right foot was lacerated by the razor flints in the river bed, and at each pace he had left a little dab of brown crumbly blood. He was staggering like a drunkard.

Johnny zigzagged the Land-Rover through the boulders that choked the watercourse. The gully deepened, and spiky cockscomb ridges of black rock hedged it in on either hand. The air in the watercourse was a heavy blanket of heat. It seared the throat, and dried the mucus in Johnny's nostrils brick hard. A small noon breeze came off the mountains, a sluggish stirring of the heavy air, that provided no relief but seemed only to heighten the bite of the sun and the suffocating oppression of the air.

Scattered along the river bed were bushes of stunted scrub. Grotesque little plants, crippled and malformed by the drought of years.

From one of the bushes ahead of the Land-Rover a monstrous black bird flapped its wings lethargically. Johnny screwed up his eyes, uncertain if it was reality or a mirage of the heat and the tortured air.

Suddenly the bird resolved itself into the jacket of a dark

blue suit. It hung in the thorny branches, the breeze stirring the folds of expensive cloth.

'In his coat. He put it in his coat pocket.'

With eyes only for the jacket, recklessly he pressed down the accelerator and the Land-Rover surged forward. Johnny did not see the knee-high boulder of ironstone in his path. He hit it at twenty miles an hour, and the Land-Rover stopped dead with the squeal of tearing metal. Johnny was flung forward against the steering-wheel, the impact driving the breath from his lungs.

He was still doubled up with the pain of it, wheezing for breath, as he hobbled to the jacket and snatched it out of the bush.

He felt the heavy drag of the weighted pocket.

Then the fat canvas bag was in his hands, the contents crunched nuttily as he tore at the drawstring. Nothing else in the world felt like that.

'Such a diamond as you will never see again.'

The drawstring was knotted tightly. Johnny ran back to the Land-Rover. Frantically he scrabbled in the seat locker and found the knife. He cut the drawstring and dumped the contents of the bag on to the bonnet of the Land-Rover.

'Oh God! Oh sweet God!' he whispered through cracked lips. His eyesight blurred, and the big blue diamond glowed mistily, distorted by the tears that flooded his vision.

It was a full minute before he could bring himself to touch it. Then he did so reverently – as though it were a sacred relic.

Johnny Lance had worked all his life to take a stone such as this.

He held it in both hands and sank down into the scrap of shade beside the body of the Land-Rover.

It was another five minutes before the hot cloying smell of engine oil reached the conscious level of his mind.

He turned his head and saw the slowly spreading pool of it beneath the Land-Rover chassis. Quickly he rolled on to his stomach and, still clutching the diamond, crawled under the vehicle. The ironstone boulder had shattered the engine sump. The Land-Rover had bled its lifeblood into the hot sand of the river bed.

He wriggled out from under the body of the Land-Rover and leaned against the front tyre. He looked at his wrist watch and was surprised to see it was already a few minutes after two o'clock in the afternoon.

He was surprised also at the conscious effort it required to focus his eyes on the dial of the watch. Two days and two nights without sleep, the unremitting emotional strain of those days and nights, the battering his body had taken, the long hours in the heat and the soul-corroding desolation of this lunar landscape – all these had taken their toll. He knew he was as light-headed as in the first stages of inebriation, he was beginning to act irrationally. That sudden reckless charge down the boulder-strewn river bed, which had wrecked the Land-Rover, was a symptom of his present instability.

He fondled the great diamond, touching the warm smooth surface to his lips, rubbing it softly between thumb and forefinger, changing it from hand to hand, while every fibre of his muscles and the very marrow of his bones cried out for rest.

A soft and treacherous lethargy spread through his body and reached out to numb his brain. He closed his eyes for a moment to shut out the flare, and when he opened them again with an effort the time was four o'clock. He scrambled to his feet. The shadows in the gully were longer, the breeze had dropped.

Although he moved with the stiffness of an old man, the sleep had cleared his mind and while he wolfed a packet of biscuits spread with meat paste and washed it down with a mugful of lukewarm water he made his decision.

He buried the canvas bag of rough diamonds in the sand

236

beneath the Land-Rover, but he could not bring himself to part with the big blue. He buttoned it securely into the back pocket of his slacks. Into the light knapsack from the seat locker he packed the two-pint water bottle, the first aid kit, a small hand-bearing compass, two of the smoke flares and the knife. He checked his pockets for his cigarette lighter and case.

Then without another glance at the radio set on the dashboard of the Land-Rover he turned away and hobbled up the gully on the spoor of Benedict van der Byl.

Within half a mile he had walked the stiffness out of his body, and he lengthened his stride, going well now. The hatred and hunger for vengeance which had died to smouldering ash since he had found the diamonds now flared up again strongly. It gave power to his legs and sharpened his senses.

The spoor turned abruptly up the side of the gully and he lost it on the black rock of the ridge, but found it again on his first cast.

He was closing fast now. The spoor was running across the grain of the land, and Benedict was clearly weakening rapidly. He had fallen repeatedly, crawled on bloody knees over cruel gravel and rock, he had blundered into the scrub bushes and left threads of his clothing on the hooked red-tipped thorns.

Then the spoor led out of the ridges and scrub into another area of low orange-coloured sandhills and Johnny broke into a jog trot. The sun was sliding down the sky, throwing blue shades in the hollows of the dunes and the heat abated so that Johnny's sweat was able to cool him before drying.

Johnny was intent on the staggering footprints, beginning to worry now that he would find Benedict already dead. The signs were those of a man in extreme distress, and still he was driving himself on.

237

Johnny did not notice the other prints that angled in from the dunes and ran parallel to those of Benedict, until they closed in again and began overlaying the human prints.

Johnny stopped and went down on one knee to examine the broad dog-like pug marks.

'Hyena!' He felt the sick little flutter of revulsion in his stomach as he spoke. He glanced around quickly and saw the other set of prints out on the left.

'A pair of them! They've smelt the blood.'

Johnny began to run on the spoor now. His skin crawled with what he knew could happen when they caught up with a helpless man. The filthiest and most cowardly animals in Africa, but with jaws that could crunch to splinters the thigh bone of a full-grown buffalo, and their thick stubby fangs were coated with such a slime of bacteria from a diet of putrid carrion that their bite was as deadly as that of a black mamba.

'Let me be in time; please God, let me find him in time.'

He heard it then. From beyond the crest of the next dune. The horror of the sound stopped him in mid-stride. It was a shrill giggling gibbering cry that sobbed into silence.

Johnny stood listening, panting wildly from his run.

It came again. The laughter of demons, excited, blood-crazy.

'They've got him.'

Johnny flung himself at the soft slope of sand. He reached the crest and looked down into the saucer-shaped arena formed by the crescent of the dune.

Benedict lay on his back. His white shirt was open to the waist. The blue trousers of his suit were ripped and shredded, exposing his knees. One foot was a bloody lump of sock and congealed dirt.

The pair of hyenas had trampled a path in the sand around his body. They had been circling him for hours, while greed overcame their cowardice.

One hyena sat ten feet from him, squatting obscenely

with its flat snakelike head lowered between humped shoulders. Brown and shaggy, spotted with darker brown, its round ears pricked forward, its black eyes sparkling with greed and excitement as it watched its mate.

The other hyena stood with its front paws on Benedict's chest. Its head was lowered, and its jaws were locked into Benedict's face. It was leaning back, bracing its paws on his chest, tugging viciously as it sought to tear off a mouthful of flesh. Benedict's head was jerking and twitching as the hyena worried it. His legs were kicking weakly, and his hands fluttered on the sand like maimed white birds.

The flesh of his face tore. Johnny heard it distinctly in the utter silence of the desert evening. It tore with the soft sound of silk – and Johnny screamed.

Both hyenas bolted at the scream, scrambling over the far crest of the dune in horrible clownish panic, leaving Benedict lying with a bloody mask for a face.

Looking down at that face Johnny knew he could not kill him now, perhaps could never have killed him. He could not revenge himself on this broken thing with its ruined face and twisted mind.

He dropped on to his knees beside him, and loosened the flap of the knapsack with clumsy fingers.

Benedict's one ear and cheek were hanging over his mouth in a thick flap of torn flesh. The teeth in the side of his jaw were exposed and the blood dribbled and spurted in fine needle jets.

Johnny tore the paper packaging off an absorbent dressing and with it pressed the flap back into place. Holding it there with the full pressure of his spread fingers. The blood soaked through the dressing, but it was slowing at the pressure.

'It's all right, Benedict. I'm here now. You'll be all right,' he whispered hoarsely as he worked. With his free hand he stripped the packaging off another dressing, and substituted it neatly for the sodden one. He maintained the pressure on the clean dressing while he lifted Benedict's head and cradled it in his lap.

'We'll just dry this bleeding up, then we'll give you a drink.' He reached into the first aid kit for a piece of cotton wool and tenderly began to clean the blood and sand from Benedict's nostrils and lips. Benedict's strangled breathing eased a little but still whistled through the black lips. His tongue was swollen, filling his mouth like a fat purple sponge.

'That's better,' Johnny muttered. Still without relaxing pressure on the compress dressing, he got the screw top off the water bottle. Holding his thumb over the opening to regulate the flow, he let a drop of water fall into the dark dry pit of a mouth.

After another ten drops he propped the water bottle in the sand, and massaged Benedict's throat gently to stimulate the swallowing reflex. The unconscious man gulped painfully.

'That's my boy,' Johnny encouraged him, and began again feeding him a drop at a time, crooning softly as he did it.

'You're going to be all right. That's it, swallow it down.'

It took him twenty minutes to administer half a pint of the warm sweet water, and by then the bleeding was negligible. Johnny reached into the kit again and selected two salt and two glucose tablets. He placed them in his own mouth and chewed them to a smooth thin paste then he bent over the mutilated face of the man he had sworn to kill and pressed his own lips against Benedict's swollen dry lips. He injected the solution of salt and glucose into Benedict's mouth, then straightened up and began again dripping the water.

When he had given Benedict another four tablets and half the contents of the water bottle, he stoppered it and returned it to the knapsack. He soaked the compress with bright yellow acriflavine solution, and bandaged it firmly into place. This was a more difficult task than he had anticipated and after a few abortive attempts he passed the bandage under the jaw and over the eyes, swathing Benedict's head completely except for the nose and mouth.

By this time the sun was on the horizon. Johnny stood up and stretched his back and shoulders as he watched the splendid gold and red death of another desert day.

He knew he was delaying his next decision. He reckoned it was five miles to where he had abandoned the Land-Rover in the gully. Five miles of hard going, a round trip of four hours – probably five in the dark. Could he leave Benedict here, get back to the vehicle, radio Cartridge Bay, and return to him?

Johnny swung round and looked up at the dunes. There was his answer. One of the hyenas was squatting on the top of a dune watching him intently. Hunger and the approach of night had made it unnaturally bold.

Johnny shouted an obscenity and made a threatening gesture towards it. The hyena jumped up and loped over the back of the ridge.

'Moon rise at eight tonight. I'll rest until then – and we'll go in the cool,' he decided and lay down on the sand beside Benedict. The lump in his back pocket prodded him, and he took the diamond out and held it in his hand.

In the darkness the hyenas began to cackle and shriek, and when the moon rose it silhouetted their evil shapes on the ridge above the saucer.

'Come on, Benedict. We're going home. There are a couple of nice policemen who want to talk to you.' Johnny lifted him into a sitting position, draped Benedict's arm over his shoulder and came up under him in a fireman's lift.

Johnny stood like that a moment, sinking ankle deep into the soft sand, dismayed at the dead weight of his burden.

'We'll rest every thousand paces,' he promised himself, and began plodding up the dune, counting softly to himself, but knowing that he would not be able to perform that lift again without a rock shelf or some support, against which to brace himself. He had to make it out of the sandhills in one go.

' – Nine hundred and ninety nine. One thousand.' He was counting in his mind only. Husbanding his strength, bowed under the weight, his shoulders and back locked in straining agony, the sand hampering each pace. 'Another five hundred. We'll go another five hundred.'

Behind him padded the two hyenas. They had gobbled the bloody dressings that Johnny had left in the saucer, and the taste of blood was driving them hysterical.

'Right. Just another five hundred.' And Johnny began the third count, and then the fourth, and the fifth.

Johnny felt the drip, drip on the back of his legs. Benedict's head-down position had restarted the bleeding, and the hyenas warbled and wailed at the smell.

'Nearly there, Benedict. Stick it out. Nearly there.'

The first cluster of moon-silver rocks floated towards them and Johnny reeled in amongst them and collapsed face forward. It was a long time before he had regained enough strength to shift Benedict's weight off his shoulders.

He readjusted Benedict's bandages, and fed him a mouthful of water which he swallowed readily. Then Johnny washed down a handful of salt and glucose tablets with two carefully rationed swallows from the water bottle. He rested for twenty minutes by his watch, then using one of the rocks as an anchor he got Benedict across his shoulder again and they went on.

Johnny rested every hour for ten minutes. At one o'clock in the morning they finished the last of the water, and at two o'clock Johnny knew beyond all possibility of doubt that he had missed the watercourse and that he was lost.

He lay against a slab of ironstone, numb with fatigue and despair, and listened to the cackling chorus of death among the rocks nearby. He tried to decide where he had gone wrong. Perhaps the watercourse curved away and he was now moving parallel to it, perhaps he had already crossed it without having recognized it. That was possible, he had heard of others stumbling blindly across a tarmac road without realizing it.

How many ridges had they climbed and descended? He could not remember. There was a place where he had stumbled into a scrub thorn and ripped his legs. Perhaps that was the watercourse.

He crawled across to Benedict.

'Brace up, bucko. We're going back.'

Johnny fell for the last time a little before dawn. When he rolled his head and squinted sideways at his wrist watch it was light enough to read the dial. The time was five o'clock.

He closed his eyes and lay for a long time, he had given up. It had been a good try – but it hadn't worked. In an hour the sun would be up. Then it was finished.

Something was moving near him, soft and stealthy-footed. It was of no interest, he decided. He just wanted to lie here quietly, now that it was finished.

Then he heard the sniffing, the harsh sniffing of a hungry dog. He opened his eyes. The hyena was ten feet away, watching him. Its bottom jaw hung open, and the pink tongue lolled loosely from the side of its mouth. He could smell it, a stink like an animal cage at the zoo, dung and offal and rotting carrion.

Johnny tried to scream, but no sound came from his mouth. His throat was closed, and his tongue filled his mouth. He struggled on to his elbows. The hyena drew back, but without the ludicrous panic of before.

Leisurely it trotted away, and then turned to face him again from a distance of twenty yards. It grinned at him, slurping the pink tongue into its mouth as it gulped saliva.

Johnny dragged himself to where Benedict lay, and looked down at him.

Slowly the blind bandaged head turned towards him, the black lips moved.

'Who's there?' A dry husky whisper. Johnny tried to answer but his voice failed him again. He hawked and chewed painfully, working a trace of moisture into his mouth. Now that Benedict was conscious Johnny's hatred flared again.

'Johnny,' he croaked. 'It's Johnny.'

'Johnny?' Benedict's hand came up and he touched the bandages over his eyes.

'What?'

Johnny reached across, lying on his side, and untied the knot at Benedict's temple. He peeled the bandages from his eyes, and Benedict blinked at him. The light of dawn was stronger now.

'Water?' Benedict asked.

Johnny shook his head.

'Please.'

'None.'

Benedict closed his eyes and then opened them, staring in terror at Johnny.

'Ruby!' Johnny whispered. 'Sergio! Hansie!'

A spasm of guilt twisted Benedict's face, and Johnny leaned closer to him to hiss a single word into his ear.

'Bastard!'

Johnny rested on his elbows, swallowing thickly, then he spoke again.

'Up!' He crawled behind Benedict and pushed and dragged him into a sitting position.

'Look.'

Twenty yards away the two hyenas sat expectantly, leering idiotically and bright-eyed with impatience.

Benedict began to tremble. He made a mewing whimpering sound. Johnny worked him slowly backwards until he had him propped against a rock.

He rested again, leaning on the rock beside Benedict.

'I'm going,' he whispered. 'You stay.'

Benedict made that mewing sound again, shaking his head weakly, staring at the two slobbering animals ahead of him.

Johnny slung the knapsack about his neck. He closed his eyes and called upon the last reserves of his strength. With a heave he got to his knees. Darkness and bright lights obscured his vision. They cleared and he heaved again and he was on his feet. His knees buckled and he grabbed at the rock to steady himself.

'Cheerio!' he whispered. 'Have fun!' And he went lurching and staggering away into the wilderness of black rock.

Behind him the mewing whimper rose to a bubbling scream.

'Johnny. Please, Johnny.'

Johnny closed his mind to that cry, and staggered on.

'Murderer!' screamed Benedict. The accusation checked Johnny. He leaned against another rock for support, and looked back.

Benedict's face was convulsed, and a thin line of bloody froth rimmed his lips. The tears were pouring unashamedly down his cheeks to soak into the blood and antiseptic stained bandages.

'Johnny. My brother. Don't leave me.'

Johnny pushed himself away from the rock. He swayed and almost fell. Then he staggered back to Benedict and slipped down into a sitting position beside him.

From the knapsack he took out the knife and laid it on his lap. Benedict was sobbing and moaning.

'Shut up. Damn you!' whispered Johnny.

The sun was well up now. It was burning directly into Johnny's face. He could feel the skin of his cheeks stretching to bursting-point. The veils of darkness kept passing over his vision, but he blinked them away. The flutter of his eyelids was the only movement he had made in the last hour.

The hyenas had closed in. They were pacing nervously back and forth in front of where Johnny and Benedict sat. Now one of them stopped and stretched its neck out, sniffing eagerly at Benedict's blood-clotted foot, creeping inch by inch closer.

Johnny stirred and the creature jumped back, bobbing its head ingratiatingly, grinning as if in apology.

The time had come to fall back on their last line of defence. Johnny hoped he had not left it too late. He was very weak. His eyes and ears were tricking him, his vision jumped and blurred and there was a humming sound in the silence as though the desert was an orchard filled with bees. He spun the cog-wheel of his cigarette lighter and the flame lit. Carefully he applied it to the fuse of the smoke flare, and it spluttered and caught.

Johnny lobbed the flare towards the hyenas, and as the clouds of pink smoke spewed out, they fled in shrieking terror.

An hour later they were back. Slinking out of the rocks, cautious again. Johnny saw them only in flashes, between the bursts of darkness in his head. The insect drone in his ears was louder, it was confusing, making it difficult for him to think clearly.

It took him ten minutes to light the second flare. His throw was so weak that the flare pitched only a few inches beyond his own feet. The pink smoke spread over them.

Johnny felt the blood humming in his ears as the swirling pink clouds engulfed him. The sulphurous bite of the smoke in his throat choked him. The sound in his ears became a drumming roar, a rushing clattering hissing bellow. Then there was a wild wind in the stillness of the desert. Miraculously the smoke cloud was ripped away by the wind.

Johnny looked up into the sky from which the great wind came. Twenty feet above him, hanging on the glistening dragon-fly wing of its rotor, was the police helicopter.

Tracey's face was framed in the cabin window of the helicopter. He saw her lips form his name before he fainted.

A FALCON FLIES

Once again this book
is for my wife
Danielle Antoinette

1860

Africa crouched low on the horizon, like a lion in ambush, tawny and gold in the early sunlight, seared by the cold of the Benguela Current.

Robyn Ballantyne stood by the ship's rail and stared towards it. She had been standing like that since an hour before dawn, long before the land could be seen. She had known it was there, sensed its vast enigmatic presence in the darkness, detected its breath, warm and spicy dry, over the clammy cold exhalations of the current on which the great ship rode.

It was her cry, not that of the masthead, which brought Captain Mungo St John charging up the companionway from his stern quarters, and the rest of the ship's company crowding to the ship's side to stare and jabber. For seconds only, Mungo St John gripped the teak rail, staring at the land, before whirling to call his orders in the low but piercing tone which seemed to carry to every corner of the ship.

'Stand by to go about!'

Tippoo the mate scattered the crew to their duties with knotted rope-end and clubbed fists. For two weeks, furious winds and low, sullen skies had denied them a glimpse of sun or moon, or of any other heavenly body on which to establish a position. On dead reckoning the tall clipper should have been one hundred nautical miles further west, well clear of this treacherous coast with its uncharted hazards and wild deserted shores.

The Captain was freshly awakened, the thick dark mane

of his hair tangled, rippling now in the wind, his cheeks lightly flushed with sleep, and also with anger and alarm beneath the smooth darkly tanned skin. Yet his eyes were clear, the whites contrasting starkly against the golden-flecked yellow of the iris. Once again, even in this moment of distraction and confusion, Robyn wondered at the sheer physical presence of the man – a dangerous, disturbing quality that at the same time both repelled and attracted her intensely.

His white linen shirt had been stuffed hastily into his breeches, and the front was unfastened. The skin of his chest was dark and smooth also, as if it had been oiled, and the hair upon it was crisp and black, tight whorls of it that made her blush, reminding her too clearly of that morning early in the voyage – the first morning that they had run into the warm blue waters of the Atlantic Ocean below latitude 35° north, the morning which for her had been the subject of much torment and troubled prayer since.

That morning, she had heard the splash and drum of water on the deck above her, and the clank of the ship's pump. She had left the makeshift desk in her tiny cabin on which she was working at her journal, slipped a shawl over her shoulders and gone up on to the maindeck, stepping unsuspectingly into the bright white sunlight and then stopping aghast.

There were two seamen working the pump lustily, and the clear sea water hissed from its throat in a solid jet. Naked, Mungo St John stood beneath it, lifting his face and his arms towards it, the water sleeking his black hair down over his face and neck, flattening his body hair over his chest and the muscled plane of his belly. She had stood and stared, completely frozen, unable to tear her eyes away. The two seamen had turned their heads and grinned lewdly at her while they kept the handles pumping the hissing water.

2

Of course she had seen a man's naked body before, laid out on the dissection table, soft white flesh collapsing over bone, and with belly pouch slit open and the internal organs spilling out of it like butchers' offal, or between the grubby blankets of the fever hospital, sweating and stinking and racked with the convulsions of onrushing death – but never like this, not healthy and vital and overwhelming like this.

This was a marvellous symmetry and balance of trunk to long powerful legs, of broad shoulders to narrow waist. There was a lustre to the skin, even where the sun had not gilded it. This was not an untidy tangle of masculine organs, half-hidden by a bush of coarse hair, shameful and vaguely revolting. This was vibrant manhood, and she had been struck with sudden insight as to the original sin of Eve, the serpent and the apples, here offered again, and she had gasped aloud. He had heard her and stepped from under the thundering jet of water, and flicked the hair from his eyes. He saw her standing near, unable to move or tear her eyes away, and he smiled that lazy, taunting smile, making no move to cover himself, the water still streaming down his body, and sparkling like diamond chips on his skin.

'Good morning, Doctor Ballantyne,' he had murmured. 'Perhaps I am to be the subject of one of your scientific studies?'

Only then had she been able to break the spell, to whirl and rush back to her smelly little cabin. She expected to be greatly disturbed, as she threw herself on the narrow planks of her bunk, waiting to be overwhelmed by a sense of sin and shame, but it did not come. Instead, she was confused by a contraction of her chest and lungs that left her breathless, and a remarkable warmth of her cheeks and the skin of her throat, a prickling of the fine dark hairs at the nape of her neck, and the same warmth of other parts of her body which had so alarmed her that she flung herself hurriedly off the bunk and on to her knees to plead for a

3

proper sense of her own unworthiness and a true understanding of her essential baseness and irretrievable wickedness. It was an exercise she had undertaken a thousand times in her twenty-three years, but seldom with so little success.

For the thirty-eight days of the voyage since then, she had tried to avoid those flecked yellow eyes and that lazy taunting smile, and had taken to eating most of her meals in her cabin, even in the daunting heat of the equator, when the taint of the bucket behind the canvas screen in the corner of the cabin had done little to pique her appetite. Only when she knew that heavy weather would keep him on deck did she join her brother and the others in the ship's small saloon.

Watching him now as he conned his ship off the hostile coast, she felt that disturbing prickle once again, and she turned away quickly to the land that was now swinging across the bows. The tackle roared through the blocks and the yards creaked and crackled, the canvas flogged and then filled again with a crash like cannon.

At the sight of the land she almost overcame those earlier memories and was instead filled with such a sense of awe, that she wondered if it were possible that the land of birth could call so clearly and so undeniably to the blood of its children.

It did not seem possible that nineteen years had passed since as a four-year-old waif, clinging to her mother's long skirts, she had last seen that great flat-topped mountain that guarded the southernmost tip of the continent sink slowly beneath the horizon. It was one of the only clear memories of this land that she retained. She could still almost feel the coarse cheap stuff that a missionary's wife must wear and hear the sobs that her mother tried to stifle and feel them shake her mother's legs beneath the skirts, as she clung closer. Vividly she recalled the fear and confusion of the little girl at her mother's distress, understanding with child-like intuition that their lives were in upheaval, but knowing

4

only that the tall figure that had been up to that time the centre of her small existence was now missing.

'Don't cry, baby,' her mother had whispered. 'We will see Papa again soon. Don't cry, my little one.' But those words had made her doubt that she would ever see her father again, and she had pushed her face into the coarse skirt, too proud even at that age to let the others hear her wail.

As always it had been her brother Morris who had comforted her, three years her senior, a man of seven years, born like her in Africa, on the banks of a far wild river with a strange exotic name, Zouga, which had given him his middle name. Morris Zouga Ballantyne – she liked the Zouga best and always used it, it reminded her of Africa.

She turned her head back towards the quarterdeck, and there he was now, tall but not as tall as Mungo St John, to whom he was speaking excitedly, pointing at the lion-coloured land, his face animated. The features he had inherited from their father were heavy but strong, the nose bony and beaked and the line of the mouth determined, harsh perhaps.

He lifted the glass to his eye again and studied the low coastline, scanning it with the care that he took with any project from the smallest to the greatest, before lowering it and turning back to Mungo St John. They spoke together quietly. An unlikely relationship had developed between the two men, a mutual, though guarded respect each for the other's strengths and accomplishments. But if the truth be told, it was Zouga who pursued the relationship most assiduously. Always one to profit by any opportunity, he had milked Mungo St John of his knowledge and experience. He had done it with an exercise of charm, but since leaving Bristol harbour he had drawn from the Captain most of what he had learned in many years of trading and voyaging along the coasts of this vast savage continent, and Zouga had written all of it down in one of his calf-bound ledgers, storing knowledge against the day.

In addition to this, the Captain had genially undertaken to instruct Zouga in the mystery and art of astronomical navigation. Each day local apparent noon would find the two of them huddled on the sunny side of the quarterdeck with brass sextants poised, waiting for a glimpse of the fiery orb through the layers of cloud, or, when the sky was clear, eagerly sighting it, swaying to the ship's motion, to hold the sun in the field of the lens as they brought it down to the horizon.

At other times they cut the monotony of a long tack with a contest of arms, taking turns at an empty corked brandy bottle thrown over the stern by a crewman, using a magnificent pair of percussion duelling pistols that Mungo St John brought up from his cabin still in their velvet-lined case, and loaded with care on the chart table.

They shouted with laughter, and congratulated each other as the bottles burst in mid-air in an explosion of shards bright as diamond chips in the sunlight.

At other times Zouga brought up the new Sharps breech-loading rifle, a gift from one of the sponsors of the expedition, 'the Ballantyne Africa Expedition', as the *Standard*, that great daily newspaper, had named it.

The Sharps was a magnificent weapon, accurate up to the incredible range of 800 yards, with the power to knock down a bull bison at a thousand. The men who were wiping out the great herds of buffalo from the American prairie at this very time had earned the title 'Sharpshooters' with this weapon.

Mungo St John towed a barrel at the end of an 800-yard length of cable to act as a target, and they shot for a wager of a shilling a bout. Zouga was an accomplished marksman, the best in his regiment, but he had already lost over five guineas to Mungo St John.

Not only were the Americans manufacturing the finest firearms in the world (already John Browning had patented a breechloading repeating rifle that Winchester was evolving into the most formidable weapon known to man), but the Americans were also far and away the finest marksmen. This pointed up the difference between the tradition of the frontiersman with his long rifle, and that of massed British infantry firing smooth-bore muskets in strictly commanded volleys. Mungo St John, an American, handled both the long-barrelled duelling pistol and the Sharps rifle as though they were an extension of his own body.

Now Robyn turned away from the two men, looked back at the land and felt a small dismay to see it already sinking lower into the cold green sea.

She yearned towards it with a quiet desperation, as she had ever since that day of departure so long ago. Her whole life in the intervening years seemed to have been a long preparation for this moment, so many obstacles overcome, obstacles made mountainous by the fact she was a woman; there had been so much struggle against temptation to give in to despair, a struggle that others had read as wilfulness and vaunting pride, as stubbornness and immodesty.

Her education had been gleaned with such toil from the library of her uncle William, despite his active discouragement. 'Too much book learning will only plague you, my dear. It is not a woman's place to trouble herself with certain things. You would do better to assist your mother in the kitchen and learn to sew and knit.'

'I can do both already, Uncle William.'

Later, his reluctant and grumbling assistance changed only slowly into active support when he at last assessed the depth of her intelligence and determination.

Uncle William was her mother's eldest brother, and he had taken in the family when the three of them had returned almost destitute from that far, savage land. They had only their father's stipend from the London Missionary Society, a mere £50 per annum, and William Moffat was not a wealthy man, a physician at King's Lynn with a small practice, hardly sufficient for the ready-made family with which he found himself saddled.

Of course, later – many years later – there had been money, a great deal of money, some said as much as three thousand pounds, the royalties from Robyn's father's books, but it had been Uncle William who had shielded and sustained them through the lean times.

William had somehow found the money to purchase Zouga's commission in his regiment, even selling his two prized hunters and making that humiliating journey to Cheapside and the moneylenders to do so.

With what William could raise, it was perforce not a fashionable regiment, and not even the regular army, but the 13th Regiment of Madras Native Infantry, a line regiment of the East India Company.

It was Uncle William who had instructed Robyn until she was as advanced in formal education as he was himself, and who had then aided and abetted her in the great deception of which she could never bring herself to be ashamed. In 1854 no hospital medical school in all of England would enrol a woman amongst their student body.

With her uncle's help, and active connivance, she had enrolled, using his sponsorship and the assertion that she was his nephew, at St Matthew's Hospital in the east end of London.

It helped that her name needed only changing from Robyn to Robin, that she was tall and small-breasted, that her voice had a depth and huskiness that she could exaggerate. She had kept her thick, dark hair cropped short, and learned to wear trousers with such panache that ever since, the tangle of petticoats and crinolines around her legs had irritated her.

The hospital governors had only discovered the fact that she was a woman after she had obtained her medical qualification from the Royal College of Surgeons at the age of twenty-one. They had immediately petitioned the Royal College to withdraw the honour, and the ensuing scandal had swept the length and breadth of England, made more fascinating by the fact that she was the daughter of Doctor Fuller Ballantyne, the famous African explorer, traveller, medical missionary and author. In the end, the governors of St Matthew's had been forced to retreat, for Robyn Ballantyne and her Uncle William had found a champion in the small, rotund person of Oliver Wicks, editor of the *Standard*.

With a true journalist's eye, Wicks had recognized good copy, and in a scathing editorial had called upon the British tradition of fair play, ridiculed the dark hints of sexual orgies in the operating rooms and pointed up the considerable achievement of this bright and sensitive young girl against almost insurmountable odds. Yet even when her qualification had been confirmed, it was for her only a short step along the road back to Africa, on which she had determined so long ago.

The venerable directors of the London Missionary Society had been considerably alarmed by the offer of the services of a woman. Missionary wives were one thing, were indeed highly desirable to shield the missionaries themselves against physical blandishments and temptations amongst the unclothed heathen, but a lady missionary was another thing entirely.

There was a further complication which weighed heavily against Doctor Robyn Ballantyne's application. Her father was Fuller Ballantyne, who had resigned from the Society six years previously before disappearing once again into the African hinterland; in their eyes he had completely discredited himself. It was clear to them that her father was more interested in exploration and personal aggrandizement than in leading the benighted heathen into the bosom of Jesus Christ. In fact, so far as they were aware, Fuller Ballantyne had made only one convert in all his thousands of miles of African travel, his personal gunbearer.

He seemed to have made himself a crusader against the African slave trade, rather than an emissary of Christ. He had swiftly changed his first missionary station in Africa into a sanctuary for runaway slaves. The station at Koloberg had been on the southern edge of the great Kalahari Desert, a little oasis in the wilderness where a clear, strong spring of water gushed from the ground, and it had been founded with an enormous expenditure of the Society's funds.

Once Fuller had made it a slave refuge, the inevitable had happened. The Trek Boers from the little independent republics which ringed the mission station to the south were the original owners of the slaves to whom Fuller Ballantyne gave sanctuary. They called 'Commando', the medium through which the Trek Boers dispensed frontier justice. They came riding into Koloberg an hour before dawn, dark swift horsemen, a hundred of them, dressed in coarse homespun, bearded and burned by the sun to the colour of Africa's dark earth. The bright flashes of their muzzle-loaders lit the dawn, and then the burning thatch of the buildings of Fuller Ballantyne's mission station made it bright day.

They roped the recaptured slaves together with the station servants and freedmen into long lines, and drove them away southwards, leaving Fuller Ballantyne standing with his family huddled about him, a few pathetic possessions which they had managed to save from the flames scattered at their feet, and the smoke from the smouldering, roofless buildings drifting in eddies about them.

It had confirmed in Fuller Ballantyne his hatred of the institution of slavery, and it had given him the excuse for which he had unwittingly been searching, the excuse to rid himself of the encumbrances which had until then prevented him from answering the call of the vast, empty land to the north.

His wife and two small children were packed off back to England for their own good, and with them went a letter to the directors of the London Missionary Society. God had made his will clear to Fuller Ballantyne. He was bidden to journey to the north, to carry God's word across Africa, a missionary at large, no longer tied to one small station, but with the whole of Africa as his parish.

The directors were greatly troubled by the loss of their station, but they were further dismayed by the prospect of having to mount what seemed to be a costly expedition of exploration into an area which all the world knew was merely a vast desert, unpeopled and unwatered except around the littoral, a burning sand desert which stretched to the Mediterranean Sea four thousand miles northward.

They wrote hurriedly to Fuller Ballantyne, uncertain where exactly the letter should be addressed, but feeling the need to deny all responsibility and to express their deep concern; they ended by stating strongly that no further funds other than his stipend of £50 per annum could be voted for Fuller Ballantyne's highly irregular activities. They need not have expended their energy and emotions, for Fuller Ballantyne had departed. With a handful of porters, his Christian gunbearer, a Colt revolver, a percussion rifle, two boxes of medicines, his journals and navigational instruments, Fuller Ballantyne had disappeared.

He emerged eight years later, down the Zambezi river, appearing at the Portuguese settlement near the mouth of that river, to the great chagrin of the settlers there who, after 200 years of occupation, had pushed no further than 100 miles upstream.

Fuller Ballantyne returned to England and his book *A Missionary in Darkest Africa* created a tremendous sensation. Here was a man who had made the 'Transversa', the over-land passage of Africa from west to east coast, who had seen, where there should be desert, great rivers and lakes, cool pleasant grassy uplands, great herds of game and strange peoples – but most of all he had seen the terrible depreda-tions of the slave-raiders upon the continent, and his

11

revelations rekindled the anti-slavery zeal of Wilberforce in the hearts of the British people.

The London Missionary Society was embarrassed by the instant fame of their prodigal, and they hurried to make amends. Fuller Ballantyne had chosen the sites for future missionary stations in the interior, and at the cost of many thousands of pounds they gathered together groups of devoted men and women and sent them out to the selected sites.

The British Government, prevailed upon by Fuller Ballantyne's description of the Zambezi river as a wide roadway to the rich interior of Africa, nominated Fuller Ballantyne Her Majesty's Consul, and financed an elaborate expedition to open this artery of trade and civilization to the interior.

Fuller had returned to England to write his book, but during this period of reunion with his family, they saw almost as little of the great man as when he was in the depths of Africa. When he was not locked in Uncle William's study writing the epic of his travels, he was in London hounding the Foreign Office or the directors of the L.M.S. And when he had gained from these sources all that he needed for his return to Africa, then he was travelling about England lecturing in Oxford or preaching the sermon from the pulpit of Canterbury cathedral.

Then abruptly he was gone again, taking their mother with him. Robyn would always remember the feel of his spiky whiskers as he stooped to kiss his daughter farewell for the second time. In her mind her father and God were somehow the same person, all-powerful, all-righteous, and her duty to them was blind, accepting adoration.

Years later, when the missionary sites chosen by Fuller Ballantyne had proved to be death-traps, when the surviving missionaries had stumbled back to civilization, their fellows and spouses dead of fever and famine, killed by wild animals and by the wilder men whom they had gone out to save, then Fuller Ballantyne's star had begun to fade.

The Foreign Office expedition to the Zambezi river, led by Ballantyne, had faltered and failed upon the terrible rapids and deep falls of the Kaborra-Bassa gorge through which the Zambezi crashed and roared, dropping a thousand feet in twenty miles. Men wondered how Ballantyne, who had claimed to have followed the Zambezi down from its source to the sea, could have not known of such a formidable obstacle to his dreams. They began to question his other claims, while the British Foreign Office, parsimonious as ever, was considerably miffed by the waste of funds on the abortive expedition and withdrew the title of Consul.

The London Missionary Society wrote another of their lengthy letters to Fuller Ballantyne, requesting him in future to confine his activities to the conversion of the heathen and the propagation of God's word.

Fuller Ballantyne had replied by posting them his resignation, thereby saving the society £50 per annum. At the same time he had penned a letter of encouragement to his two children urging them to show fortitude and faith, and sent the manuscript, in which he vindicated his conduct of the expedition, to his publisher. Then he had taken the few guineas that remained from the huge royalties that his other books had earned and had disappeared once more into the interior of Africa. That was eight years previously and no one had heard from him since.

Now here was this man's daughter, already nearly as notorious as the father, demanding admission to the Society as a working missionary.

Once again, Uncle William had come to Robyn's aid, dear mild bumbling Uncle William with his thick pebble spectacles and wild grey bush of untamable hair. With her he had gone before the board of directors and reminded them that Robyn's grandfather, Robert Moffat, was one of the most successful of all African missionaries, with tens of thousands of conversions to his credit. Indeed the old man was still working at Kuruman and had only recently published his dictionary of the Sechuana language. Robyn herself was dedicated and devout, with medical training and a good knowledge of African languages taught her by her now deceased mother, daughter of the same Robert Moffat, and by virtue of the reverence with which the said Robert Moffat was regarded by even the most warlike African king, Mzilikazi of the Ndebele, or as some people called them, the Matabele, the granddaughter would find immediate acceptance amongst the tribes.

The directors had listened stonily.

Then Uncle William had gone on to suggest that Oliver Wicks, the editor of the *Standard* who had championed the girl against the attempt by the governors of St Matthew's Hospital to deprive her of her medical qualification, would be interested in their reasons for refusing her application to the Society.

The directors sat up and listened with great attention, conferred quietly and accepted Robyn's application. They had then seconded her to another missionary movement who in turn sent her to the industrial slums of northern England.

I t was her brother Zouga who had found the way back to Africa for both of them.

He had returned from India on leave, a man of considerable achievement, already a major in the Indian army, promotion that he had won in the field, with the reputation of being a soldier and military administrator of great promise for one so young.

Despite this, Zouga was every bit as dissatisfied with his lot as was Robyn. Like their father, they were both lone wolves, responding badly to authority and regimentation. In spite of the promising start to his military career, Zouga recognized the fact that he had already made powerful enemies in India, and he had begun to doubt that his future lay on that continent. Like Robyn, he was still a searcher, and they had greeted each other after the parting of years with a warmth that they had seldom displayed during their childhood.

Zouga took her to dinner at the Golden Boar. It was such a change from Robyn's daily surroundings that she accepted a second glass of claret and became gay and sparkling.

'By God, Sissy, you really are a pretty thing, you know,' he had told her at last. He had taken to swearing now, and though it had shocked her at first, she had grown accustomed to it quickly enough. She had heard a lot worse in the slums where she worked. 'You are too good to spend your life amongst those ghastly crones.'

It changed the mood between them instantly, and she was able at last to lean close to her brother and pour out all her frustrations. He listened sympathetically, reaching across the table to squeeze her hand so that she went on quietly but with utter determination.

'Zouga, I have to get back to Africa. I'll die if I don't. I just know it. I will shrivel up and die.'

'Good Lord, Sissy, why Africa?'

'Because I was born there, because my destiny is there – and because Papa is there, somewhere.'

'I was born there also.' Zouga smiled, and when he did so it softened the harsh line of his mouth. 'But I don't know about my destiny. I wouldn't mind going back for the hunting, of course, but as for Father – don't you often think that Papa's main concern was always Fuller Ballantyne? I cannot imagine that you still harbour any great filial love for him.'

'He is different from other men, Zouga, you cannot judge him by the usual yardstick.'

'There are many who might agree with that,' Zouga murmured drily. 'At the L.M.S. and at the Foreign Office – but as a father?'

15

'I love him!' she said defiantly. 'After God, I love him best.'

'He killed mother, you know.' Zouga's mouth hardened into its usual grim line. 'He took her out to the Zambezi in fever season and he killed her as certainly as if he'd put a pistol to her head.'

Robyn conceded after a short, regretful silence, 'He was never a father nor a husband – but as a visionary, a blazer of trails, as a torchbearer . . .'

Zouga laughed and squeezed her hand.

'Really, Sissy!'

'I have read his books, all his letters, every one he ever wrote to mother or to us, and I know that my place is there. In Africa, with Papa.'

Zouga lifted his hand from hers and carefully stroked his thick side whiskers. 'You always had a way of making me feel excited—' Then, seemingly going off at a tangent, 'Did you hear that they have found diamonds on the Orange river?' He lifted his glass and examined the lees in the bottom of it attentively. 'We are so very different, you and I, and yet in some ways so much alike.' He poured fresh wine into his glass and went on casually. 'I am in debt, Sissy.'

The word chilled her. Since her childhood she had been taught a dread of it.

'How much?' she asked at last quietly.

'Two hundred pounds.' He shrugged.

'So much!' she breathed, and then, 'You haven't been gambling, Zouga?'

That was one of the other dread words in Robyn's vocabulary.

'Not gambling?' she repeated.

'As a matter of fact, I have,' Zouga laughed. 'And thank God for that. Without it I would be a thousand guineas under.'

'You mean you gamble – and actually win?' Her horror faded a little, became tinged with fascination.

'Not always, but most of the time.'

She studied him carefully, perhaps for the first time. He was only twenty-six years old, but he had the presence and aplomb of a man ten years older. He was already a hard, professional soldier, tempered in the skirmishes on the border of Afghanistan where his regiment had spent four years. She knew they had been cruel encounters against fierce hill tribes, and that Zouga had distinguished himself. His rapid promotion was proof of that.

'Then how are you in debt, Zouga?' she asked.

'Most of my brother officers, even my juniors, have private fortunes. I am a major now, I have to keep some style. We hunt, we shoot, mess bills, polo ponies—' He shrugged again.

'Will you ever be able to repay it?'

'I could marry a rich wife,' he smiled, 'or find diamonds.'

Zouga sipped his wine, slumped down in his chair, not looking at her, and went on quietly.

'I was reading Cornwallis Harris's book the other day – do you remember the big game we saw when we lived at Koloberg?'

She shook her head.

'No, you were only a baby. But I do. I remember the herds of springbok and wildebeest on the trek down to the Cape. One night there was a lion, I saw it clearly in the light of the campfire. Harris's book described his hunting expeditions up as far as the Limpopo – nobody has been further than that, except Papa, of course. A damned sight better than potting pheasant or black buck. Did you know that Harris made nearly five thousand pounds from his book?'

Zouga pushed his glass away, straightened up in his seat and selected a cigar from his silver case. While he prepared and lit it, he was frowning thoughtfully.

'You want to go to Africa for spiritual reasons. I probably need to go to Africa, for much better reasons, for blood and for money. I make you a proposal. The Ballantyne Expedition!' He lifted his glass to her.

She laughed then, uncertainly, thinking he was joking, but lifted her own glass which was still almost full. 'My word on it. But how? Zouga, how do we get there?'

'What was the name of that newspaper fellow?' Zouga demanded.

'Wicks,' she said, 'Oliver Wicks. But why should he help us?'

'I'll find a good reason why he should.' And Robyn remembered how, even as a child, he had been an eloquent and persuasive pleader of causes.

'You know I rather think you might.'

They drank then, and when she lowered her glass, she had been as happy as she could ever remember being in all her life.

It was another six weeks before she saw Zouga again, striding towards her through the bustle of London Bridge Station as she clambered down from the carriage. He stood tall above the crowd, with the high beaver top hat on his head and the three-quarter length paletot cloak flaring from his shoulders.

'Sissy!' he called, laughing at her, as he lifted her from her feet. 'We are going – we really are going.'

He had a cab waiting for them, and the driver whipped up the horses the moment they were aboard.

'The London Missionary Society were no use at all,' he told her, still with his arm around her shoulders as the cab clattered and lurched over the cobbles. 'I had them down on my list for five hundred iron men, and they nearly had apoplexy. I had the feeling they would rather Papa stayed lost in darkest Africa, and they would pay five hundred to keep him there.'

'You went to the directors?' she demanded.

'Played my losing tricks first,' Zouga smiled. 'The next was Whitehall – actually managed to see the First Secretary. He was damned civil, took me to lunch at the Travellers', and was truly very sorry that they were not able to give financial assistance. They remembered Papa's Zambezi fiasco too clearly, but he did give me letters. A dozen letters, to every conceivable person – to the Governor at the Cape, to Kemp the Admiral at Cape Town, and all the others.'

'Letters won't get us far.'

'Then I went to see your newspaper friend. Extraordinary little man. Smart as a whip. I told him we were going to Africa to find Papa – and he jumped up and clapped his hands like a child at a Punch and Judy show.' Zouga hugged Robyn tighter. 'To tell the truth, I used your name shame-lessly – and it did the trick. He will have all the story rights to our diaries and journals, and the publishing rights to both books.'

'Both books?' Robyn pulled away from him and looked into his face.

'Both.' He grinned at her. 'Yours and mine.'

'I am to write a book?'

'You certainly are. A woman's account of the expedition. I have already signed the contract on your behalf.'

She laughed then, but breathlessly. 'You're going too far, much too fast.'

'Little Wicks was in for five hundred, and the next on the list was the Society for the Extinction of the Slave Trade – they were easy. His Royal Highness is the Society's patron, and he had read Papa's books. We are to report on the state of the trade in the interior of the continent north of the Tropic of Capricorn, and they came up with another 500 guineas.'

'Oh Zouga, you are a magician.'

'Then there was the Worshipful Company of London Merchants Trading into Africa. For the last hundred years they have based all their activities on the west coast, I convinced them that they needed a survey of the east coast. I have been appointed the Worshipful Company's Agent, with instructions to examine the market in palm oil, gum-copal, copper and ivory – and they have come up with the third and last 500 guineas – and a presentation Sharps rifle.'

'One thousand five hundred guineas,' Robyn breathed, and Zouga nodded.

'We are going back home in style.'

'When?'

'I have booked passage on an American trading clipper. We sail from Bristol in six weeks, for Good Hope and Quelimane in Mozambique. I have written asking leave from my regiment for two years – you will have to do the same with the L.M.S.'

I t had all happened with dreamlike rapidity after that. The directors of the L.M.S., perhaps relieved that they were not to pay for her passage nor the expense of relocating Robyn in the African interior, in a flush of extravagance decided to continue her stipend during the period she was away, and made a guarded promise to review the position at the end of that time. If she proved herself capable – then there would be a permanent post in Africa thereafter. It was more than she had ever expected, and she had thrown herself wholeheartedly into assisting Zouga with the preparations for departure.

There was so much to do that six weeks was barely sufficient, and it seemed only days had passed before the mountain of the expedition's equipment was being swayed down into the holds of the big graceful Baltimore clipper.

The *Huron* proved as swift as she looked, another wise choice by Zouga Ballantyne, and under Mungo St John's skilful navigation she made good her westings before attempting to cross the belt of the doldrums at their narrowest point. They were becalmed not a single day and sped across the line at 29° west and immediately Mungo St John put her on the port tack to stand down across the south-east trades. *Huron* clawed her way southwards with the flying fish sailing ahead of her, close on the wind until she broke from their grip at last with Ilha da Trinidade on the horizon. The north-wester came howling at them and *Huron* fled before it, under low, sullen, scudding skies day after day that denied them sight of sun or moon or stars, until she almost hurled herself ashore 200 miles up the west coast of Africa from her destination at the Cape of Good Hope.

'Mr Mate!' Mungo St John called in his clear carrying voice as *Huron* settled down to reach across the wind, pulling swiftly away from the land.

'Captain!' Tippoo bellowed back from the foot of the mainmast, a great blurt of sound from the bull chest.

'Take the name of the masthead watch.'

Tippoo ducked his cannon-ball head on its thick neck like a bare-knuckle fighter taking a punch, and looked up the mast, slitting his eyes in the heavy folds of flesh.

'Twenty minutes more and he would have had us on the beach.' St John's voice was cold, deadly. 'I'll have him on the grating before this day is out, and we will get a peep of his backbone.'

Tippoo licked his thick lips involuntarily, and Robyn, standing near him, felt her stomach heave. There had been three floggings already on this voyage and she knew what to expect. Tippoo was half-Arab, half-African, a honey-coloured giant of a man with a shaven head that was covered with a mesh of tiny pale scars from a thousand violent encounters. He wore a loose embroidered, high-necked tunic over his huge frame, but the forearms that protruded from the wide sleeves were thick as a woman's thighs.

Robyn turned quickly to Zouga as he came down the deck towards her.

'We had a good look at the land, Sissy. Our first definite fix since Ilha da Trinidade. If this wind holds we will be in Table Bay in another five days.'

'Zouga, can't you intercede with the Captain?' she demanded, and Zouga looked startled.

'He is going to flog that poor devil.'

'Damn right too,' Zouga growled. 'The man nearly had us on the rocks.'

'Can't you stop him?'

'I wouldn't dream of interfering with his running of this ship – and nor will I allow you to.'

'Do you have no humanity at all?' she demanded of her brother coldly, but there were bright hot spots of anger on her cheeks, and her eyes snapped clear angry green. 'You call yourself a Christian.'

'When I do, I speak softly though, my dear.' Zouga made the reply he knew would annoy her most. 'And I don't flaunt it at every turn in the conversation.'

Their arguments were always sudden as summer thunderstorms on the African veld, and as spectacular.

Mungo St John sauntered forward to lean his elbow on the railing of his quarterdeck, a long Havana cheroot of coarse black tobacco held between white teeth. He mocked her silently with those flecked yellow eyes, infuriating her further, until she heard her own voice going shrill and she turned from Zouga, and rounded on him.

'The man you had flogged last week could be crippled for life,' she shouted at him.

'Doctor Ballantyne – how would you like Tippoo to carry you down and lock you in your cabin?' Mungo St John asked. 'Until you regain your temper and your manners.'

'You cannot do that,' she flared at him.

'I can, I assure you – that and much more.'

'He's right,' Zouga assured her softly. 'On this ship he can do virtually anything he wants.' He laid his hand on her upper arm. 'Steady now, Sissy. The fellow will be lucky to get away with the loss of a little skin.'

Robyn found she was panting with anger and a sense of helplessness.

'If you are squeamish, Doctor, I will excuse you from witnessing punishment,' Mungo St John mocked her still. 'We must make allowance for the fact that you are a woman.'

'I have never asked for consideration on that score, not once in all my life.' She tried to get her anger back under control, and she shook off her brother's hand and turned away from them.

She walked with stiff back and squared shoulders up into the bows, trying to maintain an aloof dignity, but the ship's motion was awkward and her damned skirts fluttered around her legs. She realized she had thought the word, and she would ask forgiveness later – but not now, and suddenly she said it aloud.

'Damn you, Captain Mungo St John, damn you to hell!'

She stood in the bows, and the wind pulled her hair out of the neat chignon at the nape of her neck and flicked it in her face. It was her mother's thick silken dark hair, shot with tones of russet and chestnut, and as a pale ray of greenish sunlight at last broke through the cloud cover, it turned to a glowing halo around her small neat head.

She stared ahead angrily, hardly noticing the hellish beauty of the scene about her. The cold, green waters, smoked with mist banks, opened and closed about the ship like pearly curtains. Wisps of mist trailed from the sails and yards as though she was on fire.

In patches, the surface of the sea simmered and darkened, for these waters were rich in microscopic sea life, which supported vast shoals of sardine that rose to the surface to feed, to be fed on in turn by flocks of shrieking seabirds that plunged upon them from high above, hitting the water in little cotton puffs of spray.

A thicker bank of mist took the tall ship in its damp cold grip so that when Robyn glanced back she could only just make out the ghostly figures of the Captain and her brother on the quarterdeck.

Then just as suddenly they plunged out again into open sea, and sunny skies. The clouds which had covered them for all these weeks rolled away towards the south while the wind itself increased in strength and veered swiftly into the east, whipping the tops from the waves in graceful ostrich plumes of curling spray.

At the same moment Robyn saw the other ship. It was startlingly close, and she opened her mouth to shout – but a dozen other voices forestalled her.

'Sail oh!'
'Sail fine on the port quarter.'

She was close enough to make out the thin, tall smoke stack between the main and mizzenmasts. Her hull was painted black with a red line below her gunports; five guns each side.

The black hull had a sinister air to it, and the pile of her canvas was not shimmering white as that of *Huron*, but was sullied dirty grey by the belchings from her smoke stack.

Mungo St John played the field of his telescope swiftly over her. Her boilers were unlit, there was not even a tremble of heat from the mouth of her stack. She was under easy canvas only.

'Tippoo!' he called softly, and it seemed that the mate's bulk appeared beside him with the magical speed of a genie.

'Have you seen her before?'

Tippoo grunted and turned his head to spit over the lee rail.

'Lime-juicer,' he said. 'I seen her last in Table Bay eight years back. She called *Black Joke.*'

'Cape Squadron?'

Tippoo grunted, and at that moment the gunboat bore up sharply and at the same time her colours broke out at the masthead. The crisp white and bright scarlet of her ensign shrieked a challenge, a challenge that all the world had learned to heed, and heed swiftly. Only the ships of one nation on earth need not heave to the instant that challenge was flown. The *Huron* was immune, she had only to hoist the Stars and Stripes, and even this importunate representative of the Royal Navy would be forced to respect it.

But Mungo St John was thinking swiftly. Six days before he sailed from Baltimore Harbour, in May, 1860, Abraham Lincoln had been nominated presidential candidate for the United States of America. If elected, as seemed highly likely, he would be invested early in the New Year, and then one of his first actions would surely be to grant to Great Britain the privileges agreed by the Treaty of Brussels, including the right of search of American ships upon the high seas which previous American presidents had so steadfastly denied.

Soon, perhaps sooner than he expected, Mungo St John might have to run his clipper in deadly earnest against one of these ships of the Cape Squadron. It was a heaven-sent opportunity to match his ship, and to observe the capabilities of the other.

He swept one last glance about him, that took in the sea, the wind-driven lines of foam upon it, the piled white pyramids of canvas above him and the evil black hull to leeward – and then his decision was made easier. On the wind came the thud of a gun, and a long feather of gunsmoke spurted from one of the gunboat's bow-chasers, demanding instant obedience.

Mungo St John smiled. 'The insolent bastard!' To Tippoo he said, 'We'll try him on a few points of sailing,' to the helmsman beside him, softly, 'Put the helm down.' And as *Huron* paid off swiftly before the wind, beginning to point directly away from the threatening black ship, 'Shake out all reefs, Mr Mate. Set fore and maintop, hoist studding sails and skysails, crack on the main royal – yes, and flying jib too. By God, we'll show that grubby little coal-guzzling lime-juicer how they build them down Baltimore way!'

Even in her anger, Robyn was thrilled by the manner in which the American worked his ship. With his crew swarming out across the yards to the reefing points, the mainsails swelled out to their full extent, dazzling white in the sunshine, and then high above, seemingly at the very base of the aching blue heaven itself, new unfamiliar-shaped sails popped open like over-ripe cotton pods and the long, graceful hull reacted instantly to the pressures thrust upon her.

'By God, she sails like a witch,' Zouga shouted, laughing with excitement, as she knifed into the crests of the Atlantic rollers and he hustled his sister back from the bows before the first green sheets of water came aboard and swept *Huron*'s decks.

More and still more canvas burst open, and the thick trees of her masts began to arch like drawn longbows under the unbearable pressures of thousands of square feet of spread sails. Now *Huron* seemed to fly, taking off from the crest of each roller and smashing into the face of the next with a crash that shocked her timbers and jarred the teeth of her crew in their skulls.

'A cast of the log, Mr Mate,' Mungo St John called, and when Tippoo bellowed back, 'A touch better than sixteen knots, Cap'n!' the Captain laughed aloud, and strode to the stern rail.

The gunboat was falling astern as though she was standing still, although every inch of her grey canvas was spread. Already she was at extreme cannon range.

Again powder-smoke bloomed briefly on her black bows, and this time it seemed that it was more than merely a warning, for Mungo St John saw the fall of shot. It struck the crest of a roller two cable lengths astern and skipped across the green torn waters, before plunging beneath the surface almost alongside *Huron*'s tall side.

'Captain, you are endangering the lives of your crew and passengers.' The voice arrested him and St John turned to the tall young woman who stood beside him, and he raised one thick black eyebrow in polite enquiry.

'That is a British man-of-war, sir, and we are acting like criminals. They are firing live shot now. You have only to heave to, or at the very least show your colours.'

'I think my sister is right, Captain St John.' Zouga stood beside her. 'I do not understand your behaviour either.'

Huron staggered violently to a larger crest, driven wild by the mountainous press of sail, Robyn lost her balance and fell against the Captain's chest, but instantly pulled away, colouring fiercely at the contact.

'This is the coast of Africa, Major Ballantyne. Nothing is what it appears to be. Here only a fool would accept a strange armed vessel at its face value. Now if you and the good doctor will excuse me, I must attend to my duties.'

He strode forward to gaze down at the maindeck, judging the mood of his crew and the wild abandon of his ship. He unhooked the keyring from his belt and tossed it to Tippoo. 'The arms chest, Mr Mate, a pair of pistols to you and the second mate. Shoot any hand who attempts to interfere with the setting of the sails.' He had recognized the fear which gripped the crew. Most of them had never seen a ship driven like this, there might easily be an attempt to shorten sail rather than have her run herself under.

At that moment *Huron* put her shoulder into the Atlantic

and took it aboard in a solid roaring green wall. One of the topmast men was not quick enough on to the rat-lines. The water plucked him up and flung him down the length of the deck, until he crashed into the side, and lay huddled against the bulwark like a clump of uprooted kelp on a storm-driven beach.

Two of his fellows tried to reach him, but the next wave drove them back as it came pouring aboard waist-deep and then cascading in a roaring white torrent over the side, and when it was gone the fallen topmast man was gone with it and the deck was empty.

'Mr Tippoo, look you to those skysails, they are not drawing as they should.'

Mungo St John turned back to the stern rail, ignoring Robyn Ballantyne's horrified and accusing glare.

Already the British gunboat was hull down and her sails were barely discernible amongst the grey beards of the breaking Atlantic rollers, but suddenly Mungo St John saw something change, and he reached quickly for the telescope in its slot under the chart table. There was a fine black line, as though drawn in Indian ink, extending from the tiny cone of the Englishman's sails for a short way across the bumpy horizon.

'Smoke! She has her boiler fired at last,' he grunted, as Tippoo appeared at his shoulder, with the pistols thrust into his belt.

'One screw. She no catch us.' Tippoo nodded his round shaven head.

'No, not downwind in a full gale,' Mungo St John agreed. 'But I'd like to try her on the wind. We'll harden up now, Mr Mate, on the port tack again. I want to see if I can run up to windward of her and pass her out of cannon shot.'

The unexpected manoeuvre caught the gunboat commander completely off-guard, and he was a few minutes slow in altering course to cut the shorter leg of the triangle and prevent the *Huron* wresting the weather gauge from him.

Huron went streaming past him at extreme cannon shot hard on the wind, her yards sheeted around as close-hauled as she would sail. He tried a shot across her dipping, plunging bows with no visible fall of shot, and then he came around to follow her into the wind, and immediately the flaws in the design and construction of his ship were pointed up as clearly as they had been while both ships were running before the wind.

In order to accommodate the heavy boiler and machinery to drive the big bronze screw under her counter, serious compromise had been made with the design of her masts and the amount of canvas she could carry.

Within five miles it became clear that with all sail set and the boilers belching a solid greasy slug of coal smoke over the stern the *Black Joke* could not point as high into the wind as the beautiful tall ship ahead of her. She was drifting steadily away to leeward, and although the difference in their speeds was not as dramatic as when running before the wind, yet *Huron* was headreaching on her steadily.

The gunboat's commander pointed her higher and higher, trying desperately to hold the bigger ship directly over his bows, but all his sails were shaking and luffing before he could do so.

In a fury of frustration, he took in all his canvas, stripping his poles bare and relying only on the drive of his steam boiler pointed up directly into the very eye of the wind, much much higher than *Huron* could sail. But the gunboat's speed bled away when her screw received no help from her sails. Even though her mast and rigging were bare, the storm whistled and howled head-on through them, acting as a great drogue that slowed her further and *Huron* forged ahead more swiftly.

'A bastard contraption.' Mungo St John watched her battle with all his attention, judging her performance at every point of the wind. 'We are toying with her. As

29

long as there is a breath of wind we'll romp away from her.'

Astern, the gunboat's commander had abandoned his attempt to run the clipper down with steam power alone, and had come back on to a fine reach with all his canvas set, plugging along stubbornly in *Huron*'s streaming wake – until abruptly, with no warning at all, *Huron* sailed into a deep hole in the wind.

The line of the gale was drawn clearly across the surface of the sea. On one side, the water was darkened and furrowed by the talons of the wind, on the other, the humped backs of the rollers in the calm had a polished velvety gloss to them.

As *Huron* crossed that line of demarcation, the clamour of the wind which for week after week had battered their ears, fell to an eerie unnatural silence, and the ship's motion changed from the vital charge of a living, straining sea creature, to the patternless rolling and wallowing of a dead log.

Overhead her canvas volleyed and flapped in the directionless eddies created by her own rolling and pitching, and her tackle crashed and clattered so that it seemed that she might roll her masts clean out of her hull.

Far astern, the black-painted gunboat scrambled on eagerly, swiftly beginning to narrow the distance between them, the black column of coal smoke now rising straight up into the stillness of air, giving her a jubilant and menacing air.

Mungo St John ran to the forward rail of the quarterdeck and stared over his bows. He could see the wind two or three miles ahead clawing at the sea and ruffling it to a sombre shade of indigo, but between them was the oily undulating surface of the calm.

He swung back and the gunboat was closer, sending her smoke spurting high against the bright windswept blue of the sky, so certain of herself now that her gunports were swinging open and the stubby barrels of her 32-pounder cannon protruding from the black sides of her hull, the churning wash of her screw tumbling out from under her counter and sparkling whitely in the sunlight.

With no way upon *Huron*, the helmsman could not hold any course, and the clipper drifted around broadside to the on-rushing man-of-war putting her bows directly into the rollers.

They could make out the individual figures of the three officers on the gunboat's bridge now. Again the bow-chaser fired, and the shell lifted a tall column of water so high and close under the *Huron*'s bows that it collapsed upon the deck and streamed out through the scuppers.

Mungo St John took one last despairing sweep of the horizon, hoping even now for a resurgence of the wind, and then he capitulated.

'Hoist the colours, Mr Tippoo,' he called, and as the gaudy scrap of cloth dropped from the mainyard in the windless air, he watched through the lens of his telescope the consternation it caused upon the gunboat's bridge. That was the last flag they had hoped to see.

They were now close enough to discern the individual expressions of chagrin and alarm and indecision of the naval officers.

'There'll be no prize money for you – not this time around,' Mungo St John murmured with grim satisfaction, and snapped the telescope shut.

The gunboat came on and then rounded up to *Huron*, within easy hail, showing her full broadside, the long 32-pounder cannon gaping menacingly.

The tallest officer on her bridge seemed also the oldest, for his hair was white in the sunlight. He came to the gunboat's near rail and lifted the voice trumpet to his mouth.

'What ship?'

'*Huron*, out of Baltimore and Bristol,' Mungo St John

hailed back. 'With a cargo of trade goods for Good Hope and Quelimane.'

'Why did you not answer my challenge, sir?'

'Because, sir, I do not acknowledge your right to challenge ships of the United States of America on the high seas.'

Both captains knew just what a thorny and controversial question that reply posed, but the Englishman hesitated only a second.

'Do you, sir, accede to my right to satisfy myself as to your nationality and your ship's port of registry?'

'As soon as you run in your guns you may come aboard for that purpose, Captain. But do not send one of your junior officers.'

Mungo St John was making a fine point of humiliating the commander of the *Black Joke*. But inwardly he was still seething at the fluke of wind and weather which had allowed the gunboat to come up with him.

The *Black Joke* launched a longboat on the heavy swell with an immaculate show of seamanship, and it pulled swiftly to *Huron*'s side. While the Captain scrambled up the rope ladder, the boat's crew backed off and rested on their oars.

The naval officer came in through the entry port, so lithe and agile that Mungo St John realized his error in thinking him an elderly man. It was the white-blond hair that had misled him – he was evidently less than thirty years of age. He did not wear a uniform coat, for his ship had been cleared for action, and he was dressed in a plain white linen shirt, breeches and soft boots. There were a pair of pistols in his belt and a naval cutlass in its scabbard on his hip.

'Captain Codrington of Her Majesty's auxiliary cruiser *Black Joke*,' he introduced himself stiffly. His hair was bleached in silver white splashes from the salt and the sun, with darker streaks beneath and it was tied with a leather thong in a short queue at the nape of his neck. His face was weathered to honey-golden brown by the same sun, so that the faded blue of his eyes was in pale contrast.

'Captain St John, owner and master of this vessel.'

Neither man made any move to shake hands, and they seemed to bristle like two dog wolves meeting for the first time.

'I hope you do not intend to detain me longer than is necessary. You can assure yourself that my government will be fully apprised of this incident.'

'May I inspect your papers, Captain?' The young naval officer ignored the threat, and followed St John on to the quarterdeck. There he hesitated for the first time when he caught sight of Robyn and her brother standing together at the far rail, but he recovered immediately, bowed slightly and then turned his full attention to the packet of documents that Mungo St John had ready for his inspection on the chart table.

He stooped over the table, working swiftly through the pile until with a shock of discovery he straightened.

'Damn me – Mungo St John – your reputation precedes you, sir.' The Englishman's expression was strained with strong emotion. 'And what a noble one it is, too.' There was a bitter sting in his voice. 'The first trader ever to carry more than three thousand souls across the middle passage in a single, twelve-month period – small wonder you can afford such a magnificent vessel.'

'You are on dangerous ground, sir,' Mungo St John warned him with that lazy, taunting grin. 'I am fully aware of the lengths to which the officers of your service will go for a few guineas of prize money.'

'Where are you going to pick up your next cargo of human misery, Captain St John?' the Englishman cut in brusquely. 'On such a fine ship you should be able to pack in two thousand.' He had gone pale with unfeigned anger, actually trembling slightly with the force of it.

'If you have finished your investigation—' St John's smile did not slip, but the naval officer went on speaking.

'We have made the west coast a little too hot for you now, have we? Even when you hide behind that pretty piece of silk,' the naval officer glanced up at the flag on the mainyard. 'So you are going to work the east coast now, are you, sir? They tell me you can get a prime slave for two dollars – two for a 10-shilling musket.'

'I must ask you to leave now.' St John took the document from his hand, and when their fingers touched, the Englishman wiped his hand on his own thigh as though to cleanse it of the contact.

'I'd give five years' pay to have the hatches off your holds,' he said bitterly, leaning forward to stare at Mungo St John with those pale fierce eyes.

'Captain Codrington!' Zouga Ballantyne stepped towards the group. 'I am a British subject and an officer in Her Majesty's army. I can assure you that there are no slaves aboard this vessel.' He spoke sharply.

'If you are an Englishman, then you should be ashamed to travel in such company.' Codrington glanced beyond Zouga. 'And that applies equally to you, madam!'

'You overreach yourself, sir,' Zouga told him grimly. 'I have already given you my assurance.'

Codrington's gaze had flicked back to Robyn Ballantyne's face. Her distress was evident, unfeigned. The accusation had shattered her, that she, the daughter of Fuller Ballantyne, the great champion of freedom and sworn adversary of slavery – she, the accredited agent of the Society for the Extinction of the Slave Trade should be travelling aboard a notorious slaver.

She was pale, the green eyes huge and liquid with the shock of it.

'Captain Codrington,' her voice was husky and low, 'my brother is right – I also assure you that there are no slaves aboard this ship.'

34

The Englishman's expression softened, she was not a beautiful woman, but there was a freshness and wholesomeness about her which was difficult to resist.

'I will accept your word, madam.' He inclined his head. 'Indeed, only a madman would carry black ivory towards Africa, but,' and his voice hardened again, 'if only I were able to enter her holds, I'd find enough down there to run her into Table Bay under a prize crew and have her condemned out of hand at the next session of the Court of Mixed Commission.'

Codrington swung on his heel to face Mungo St John again.

'Oh yes, I know that your slave decks will be struck to make way for your trade cargo, but the spare planks are aboard and it won't take you a day to set them up again,' Codrington almost snarled, 'and I'll wager there are open gratings under those hatch covers,' he pointed down at the maindeck, but without taking his eyes from Mungo St John's face, 'that there are shackles in the lower decks to take the chains and leg irons—'

'Captain Codrington, I find your company wearying,' Mungo St John drawled softly. 'You have sixty seconds to leave this ship, before I have my mate assist you over the side.'

Tippoo stepped forward, hairless as an enormous toad, and stood a foot behind Codrington's left shoulder.

With a visible effort, the English captain retained his temper, as he inclined his head towards Mungo St John.

'May God grant we meet again, sir.' He turned back to Robyn and saluted her briefly.

'May I wish you a pleasant continuation of your voyage, madam.'

'Captain Codrington, I think you are mistaken,' she almost pleaded with him. He did not reply but stared at her for a moment longer, the pale eyes were direct and disturbing – the eyes of a prophet or a fanatic – then he turned

35

and went with a gangling boyish stride to *Huron*'s entry port.

Tippoo had stripped off the high-necked tunic and oiled his upper body so it gleamed in the sunlight with the metallic lustre of the skin of some exotic reptile.

He stood stolidly on flat bare feet, balancing effortlessly to the *Huron*'s roll, his thick arms hanging at his side and the lash of the whip coiled on the deck at his feet.

There was a grating fixed at the ship's side and the sailor who had been at the masthead lookout when they had raised the African coast, was spread-eagled upon it like a stranded starfish on a rock exposed at low tide. He twisted his head awkwardly to look back over his shoulder at the mate, and his face was white with terror.

'You were excused witnessing punishment, Doctor Ballantyne,' Mungo St John told her quietly.

'I feel it my duty to suffer this barbaric—'

'As you wish,' he cut her short with a nod, and turned away. 'Twenty, Mr Tippoo.'

'Twenty it is, Cap'n.'

With no expression at all Tippoo stepped up behind the man, hooked his finger into the back of his collar and ripped the shirt down to the belt. The man's back was pale as suet pudding, but studded with fat purple carbuncles, the sailor's affliction, caused by salty, wet clothing and the unhealthy diet.

Tippoo stepped back and flicked out the lash so that it extended to its full length along the seamed oak planking.

'Ship's company!' Mungo St John called. 'The charge is inattention to duty, and endangering the ship's safety.' They shuffled their bare feet, but not one of them looked up at him. 'The sentence is twenty lashes.'

On the grating the man turned his face away and closed his eyes tightly, hunching his shoulders.

'Lay on, Mr Tippoo,' Mungo St John said, and Tippoo squinted carefully at the bare, white skin, through which the knuckles of the spine showed clearly. He reared back, one thickly muscled arm thrown high above his head, and the lash snaked higher, hissing like an angry cobra, then he stepped forward into the stroke, pivoting the full weight and force of his shoulders into it.

The man on the grating shrieked, and his body convulsed in a spasm that smeared the skin from his wrists against the coarse hemp bonds.

The white skin opened in a thin bright scarlet line, from one side of his ribs to the other, and one of the angry purple carbuncles between his shoulder blades erupted in a spurt of yellow matter that ran down the pale skin and soaked into the waistband of his breeches.

'One,' said Mungo St John, and the man on the grating began to sob quietly.

Tippoo stepped back, shook out the lash carefully, squinted at the bloody line across the white shuddering flesh, reared back and grunted as he stepped forward into the next stroke.

'Two,' said Mungo St John. Robyn felt her gorge rise to choke her. She fought it down, and forced herself to watch. She could not allow him to see her weakness.

On the tenth stroke the body on the grating relaxed suddenly, the head lolled sideways and the fists unclenched slowly so that she could see the little bloody half-moon wounds where the nails had been driven into the palms. There was no further sign of life during the rest of the leisurely ritual of punishment.

At the twentieth stroke, she almost flung herself down the ladder to the maindeck and was feeling for the pulse before they could cut the body down from the grating.

'Praise be to God,' she whispered as she felt it fluttering

under her fingers, and then to the seamen who were lifting the man down, 'Gently now!' She saw that Mungo St John had got his wish, for the white porcelain crests of the spinal column were jutting up through the sliced meat of the back muscles.

She had a cotton dressing ready and she placed it across the ruined back as they laid him on to an oak plank and hustled him towards the forecastle.

In the narrow, crowded forecastle, thick with the fumes of cheap pipe tobacco and the almost solid reek of bilges and wet clothing, of unwashed men and mouldering food, they laid the man on the mess table and she worked as best she could in the guttering light of the oil-lamp in its gimbals overhead. She stitched back the flaps and ribbons of mushy, torn flesh with horsehair sutures and then bound up the whole in weak phenol solution, treatment which Joseph Lister had recently pioneered with much success against mortification.

The man was conscious again and whimpering with pain. She gave him five drops of laudanum, and promised to visit him the following day to change the bindings.

As she packed away her instruments and closed her black valise, one of the crew, a little pockmarked bosun, named Nathaniel, picked it up and when she nodded her thanks, he muttered with embarrassment, 'We are beholden, missus.'

It had taken all of them time to accept her ministrations. First it had been only the lancing of carbuncles and sea-boils, calomel for the flux and the grippe, but later, after a dozen successful treatments, which included a fractured humerus, an ulcerated and ruptured eardrum and the banishing of a venereal chancre with mercury, she had become a firm favourite amongst the crew, and her sick-call a regular feature of shipboard life.

The bosun climbed the companionway behind her, carrying the valise, but before they reached the deck, an idea struck her and she stooped to him, placing a hand on his shoulder.

'Nathaniel,' she asked urgently, but in a low voice, 'is there a way of entering the ship's hold without lifting the maindeck hatches?'

The man looked startled, and she shook his shoulder roughly. 'Is there?' she demanded.

'Aye, ma'am, there is.'

'Where? How?'

'Through the lazaretto, below the officers' saloon – there is a hatchway through the forward bulkhead.'

'Is it locked?'

'Aye, ma'am, it is – and Captain St John keeps the keys on his belt.'

'Tell nobody that I asked,' she ordered him, and hastened up on the maindeck.

At the foot of the mainmast, Tippoo was washing down the lash in a bucket of seawater that was already tinged pale pink; he looked up at her, still stripping the water off the leather between thick hairless fingers – and he grinned at Robyn as she passed, squatting down on thick, brown haunches with his loin cloth drawn up into his crotch, swinging his round bald head on its bull neck to follow her.

She found herself panting a little with fear and revulsion, and she swept her skirts aside as she passed him. At the door of her cabin she took the valise from Nathaniel with a word of thanks, and then slumped down upon her bunk.

Her thoughts and her emotions were in uproar, for she had still not recovered from the sudden avalanche of events that had interrupted the leisurely pattern of the voyage.

The boarding by Captain Codrington of the Royal Navy overshadowed even her anger at the flogging or her joy at her first view of Africa in nearly two decades – and now his accusations rankled and disturbed her.

After a few minutes' rest she lifted the lid of her travelling-chest that filled most of the clear space in the tiny cabin, and had to unpack much of it before she found the pamphlets from the anti-slavery society with which she had been armed in London before departure.

She sat down to study them once again, a history of the struggle against the trade up to the present time. As she read, her anger and frustration reawakened at the tale of unenforceable international agreements, all with built-in escape clauses: laws that made it an act of piracy to indulge in the trade north of the equator, but allowed it to flourish unchecked in the southern hemisphere; treaties and agreements signed by all nations, except those most actively engaged in the trade, Portugal, Brazil, Spain. Other great nations – France – using the trade to goad their traditional enemy, Great Britain, shamelessly exploiting Britain's commitment to its extinction, trading political advantage for vague promises of support.

Then there was America, a signatory to the Treaty of Brussels which Britain had engineered, agreeing to the abolition of the trade, but not to the abolition of the institution of slavery itself. America agreed that the transport of human souls into captivity was tantamount to active piracy, and that vessels so engaged were liable to seizure under prize and condemnation by courts of Admiralty or Mixed Commission, agreed also to the equipment clause, that ships equipped for transport of slaves, although not actually with a cargo of slaves on board at the time of seizure, could be taken as prize.

There was America agreeing to all of this – and then denying to the warships of the Royal Navy the right of search. The most America would allow was that British officers could assure themselves of the legality of the claim to American ownership, and if that was proven, they could not search, not even though the stink of slaves rose from her holds to offend the very heavens, or the clank of chains and the half-human cries from her 'tween decks came near to deafening them – still they could not search.

Robyn dropped one pamphlet back into her chest, and selected another publication from the society.

ITEM, in the previous year, 1859, estimated 169,000 slaves had been transported from the coasts of Africa to the mines of Brazil, and the plantations of Cuba, and to those of the Southern States of America.

ITEM, the trade in slaves by the Omani Arabs of Zanzibar could not be estimated except by observation of the numbers passing through the markets of that island. Despite the British Treaty with the Sultan as early as 1822, the British Consul at Zanzibar had counted almost 200,000 slaves landed during the previous twelve-month period. The corpses were not landed, nor were the sick and dying, for the Zanzibar customs dues were payable to the Sultan per capita, live or dead.

The dead and those so enfeebled or diseased as to have little hope of survival were thrown overboard, at the edge of the deep water beyond the coral reef. Here a permanent colony of huge man-eating sharks cruised the area by day and by night. Within minutes of the first body, dead or still living, striking the water, the surface around the dhow was torn into a seething white boil by the great fish. The British Consul estimated a forty per cent mortality rate amongst slaves making the short passage from the mainland to the island.

Robyn dropped that pamphlet and before picking up the next, she reflected a moment on the sheer multitudes involved in the whole grisly business.

'Five million since the turn of the century,' she whispered, 'five million souls. No wonder that they call it the greatest crime against humanity in the history of the world.'

She opened the next pamphlet and skimmed quickly over an examination of the profits that accrued to a successful trader.

In the interior of Africa, up near the lake country where few white men had ever reached, Fuller Ballantyne had discovered – her father's name in print gave her a prickle of pride and of melancholy – Fuller had discovered that a prime slave changed hands for a cupful of porcelain beads, two slaves for an obsolete Tower musket that cost thirteen shillings in London, or a Brown Bess musket that cost two dollars in New York.

At the coast the same slave cost ten dollars, while on the slave market in Brazil he would sell for five hundred dollars. But once he was taken north of the equator, the risks to the trader increased and the price rose dramatically – a thousand dollars in Cuba, fifteen hundred in Louisiana.

Robyn lowered the text, and thought swiftly. The English Captain had challenged that *Huron* could carry 2,000 slaves at a time. Landed in America, they would be worth an unbelievable three million dollars, an amount which would buy fifteen ships like *Huron*. A single voyage would make a man rich beyond mundane dreams of greed, all risks were acceptable to the traders to win such vast wealth.

But had Captain Codrington been justified in his accusations? Robyn knew the counter-accusations that were made against the officers of the Royal Navy, that their zeal arose from the promise of prize money rather than a hatred of the trade and a love of humanity. That every sail they raised was considered a slaver, and that they were swift to apply the Equipment Clause in the widest possible interpretation.

Robyn was searching for the pamphlet that dealt in detail with this Equipment Clause and she found it next on the pile before her.

To enable a ship to be seized as a slaver under the clause, she need only satisfy one of the stipulated conditions. She could be taken if her hatches were equipped with open gratings to ventilate her holds; if there were dividing bulkheads in her holds to facilitate the installation of slave decks; if there were spare planks aboard for laying as slave decks; if she carried shackles and bolts, or leg irons and cuffs; if she carried too many water casks for the number of her crew and passengers; if she had disproportionally large numbers of mess tubs, or her rice boilers were too big, or if she carried unreasonable quantities of rice or farina.

Even if she carried native matting that might be used as bedding for slaves, she could be seized and run in under a prize crew. These were wide powers given to men who could profit financially by seizure.

Was Captain Codrington one of these, were those pale fanatical eyes merely a mask for avarice and a desire for personal gain?

Robyn found herself hoping they were, or at least that in the case of *Huron* he had been mistaken. But then why had Captain St John put down his helm, and run for it the moment he sighted the British cruiser?

Robyn was confused and miserable, haunted with guilt. She needed comfort and she slipped a lace Stuart cape over her head and shoulders before venturing out on to the deck again, for the wind had risen to an icy gale and *Huron* was always a tender ship, she heeled heavily as she beat southwards, flinging spray high into the falling night.

Zouga was in his cabin, dressed in shirt-sleeves and smoking a cigar as he worked over the lists of the expedition's equipment that would still have to be obtained once they reached Good Hope.

He called to her to enter when she knocked, and rose to greet her with a smile.

'Sissy, are you well? It was a most unpleasant business, even though unavoidable. I hope it has not unsettled you.'

'The man will recover,' she said, and Zouga changed the subject as he settled her on his bunk, the only other seating in the cabin.

'I sometimes think we would have been better off with

less money to spend on this expedition. There is always such a temptation to accumulate too much equipment. Papa made the Transversa with only five porter loads, while we will need a hundred porters at the least, each carrying eighty pounds.'

'Zouga, I must speak to you. This is the first opportunity I have had.'

An expression of distaste flickered across the strong, harsh features as though he sensed what she was about to say. But before he could deny her she blurted out, 'Is this ship a slaver, Zouga?'

Zouga removed the cigar from his mouth and inspected the tip minutely before he replied.

'Sissy, a slaver stinks so you can smell it for fifty leagues downwind, and even after the slaves are removed there is no amount of lye that will get rid of the smell. *Huron* does not have the stench of a slaver.'

'This ship is on her first voyage under this ownership,' Robyn reminded him quietly. 'Codrington accused Captain St John of using his profits from previous voyages to purchase her. She is still clean.'

'Mungo St John is a gentleman.' Zouga's tone had an edge of impatience to it now. 'I am convinced of that.'

'The plantation owners of Cuba and Louisiana are amongst the most elegant gentlemen that you could find outside the court of St James,' she reminded him.

'I am prepared to accept his word as a gentleman,' Zouga snapped.

'Are you not a little eager, Zouga?' she asked with a deceptive sweetness, but his tone had kindled sparks in her eyes like the green lights in an emerald. 'Would it not seriously impede your plans to find ourselves shipped on board a slaver?'

'Damn me, woman, I have his word.' Zouga was getting truly angry now. 'St John is engaged in legitimate trade. He hopes for a cargo of ivory and palm oil.'

'Have you asked to inspect the ship's hold?'

'He has given his word.'

'Will you ask him to open the holds?'

Zouga hesitated, his gaze wavered a moment, and then he made his decision.

'No, I will not,' he said flatly. 'That would be an insult to him and quite rightly, he would resent it.'

'And if we found what you are afraid to find, it would discredit the purpose of our expedition,' she agreed.

'As the leader of this mission. I have made the decision—'

'Papa would never let anything stand in his way either, not even Mama or the family—'

'Sissy, if you still feel that way when we reach the Cape, I will arrange for passage on another vessel to Quelimane. Will that satisfy you?'

She did not reply but continued to stare at him with a flat accusing gaze.

'If we did find evidence,' he waved his hands with agitation, 'what could we do about it?'

'We could make a sworn deposition to the Admiralty at Cape Town.'

'Sissy,' he sighed wearily at her intransigence, 'don't you understand? If I were to challenge St John, we could gain nothing. If the accusation is unwarranted we would place ourselves in a damned awkward position, and if in the very unlikely event that this ship is equipped for the trade, we would then be in considerable danger. Do not underestimate that danger. Robyn. St John is a determined man.' He stopped and shook his head decisively, the fashionable curls dangling over his ears. 'I am not going to endanger you, myself or the whole expedition. That is my decision, and I will insist that you abide by it.'

After a long pause, Robyn slowly dropped her gaze to her hands, and inter-meshed her fingers.

'Very well, Zouga.'

His relief was obvious. 'I am grateful for your compliance,

my dear.' He stooped over her and kissed her forehead. 'Let me escort you to dinner.'

She was about to refuse, to tell him she was tired and that, once again, she would dine alone in her cabin, and then an idea struck her, and she nodded.

'Thank you, Zouga,' she told him, and then looked up with one of those sudden smiles so brilliant, so warm and so rare as to disarm him completely. 'I am fortunate to have such a handsome dinner companion.'

She sat between Mungo St John and her brother, and had her brother not known better, he might have suspected her of flirting outrageously with the Captain. She was all smiles and sparkles, leaning forward attentively to listen whenever he spoke, recharging his glass whenever it was less than half-filled with wine and laughing delightedly at his dry sallies.

Zouga was amazed and a little alarmed by the transformation, while St John had never seen her like this. He had covered his original surprise with an amused half-smile. However, in this mood Robyn Ballantyne was an attractive companion. Her stubborn, rather sharp face softened to the edge of prettiness, while her best features, her hair, her perfect skin, her eyes and fine white teeth, gleamed and flashed in the lamplight. Mungo St John's own mood became expansive, he laughed more readily and his interest was clearly piqued. With Robyn plying his glass, he drank more than on any other night of the voyage, and when his steward served a good plum duff he called for a bottle of brandy to wash it down.

Zouga had also been infected by the strangely festive air of the dinner, and he protested as vigorously as St John when suddenly Robyn declared herself to be exhausted and stood up from the board, but she was adamant.

In her cabin she could still hear the occasional shouts of laughter from the saloon, as she went quietly about her preparations. She slid the locking bar into place to assure her privacy. Then she knelt beside her chest and wormed her way down to the bottom layers, from which she retrieved a pair of man's moleskin breeches, a flannel shirt and cravat, with a high-buttoned monkey jacket to go over them, and well-worn half boots.

This had been her uniform and her disguise as a medical student at St Matthew's Hospital. Now she stripped herself naked, and for a moment enjoyed the wicked freedom of the feeling, even indulging herself to the extent of gazing down at her nudity. She was not too certain if it was a sin to enjoy one's own body, but she suspected that it was. Nevertheless she persisted.

Her legs were straight and strong, her hips flared with a graceful curve and then narrowed abruptly into her waist, her belly was almost flat with just an interesting little bulge below the navel. Now here was definitely sinful ground – of this there was no doubt. But still she could not deny the temptation to let her gaze linger a moment. She understood fully the technical purpose and the physical workings of all her body's highly complicated machinery, both visible and concealed. It was only the feelings and emotions which sprang from this source which both confused and worried her, for they had taught her none of this at St Matthew's. She passed on hurriedly to safer ground, lifting her arms to pile the tresses of her hair on to the top of her head and hold them in place with a soft cloth cap.

Her breasts were round and neat as ripening apples, so firm as hardly to change their shape as she moved her arms. Their resemblance to fruit pleased her and she spent a few moments longer than was necessary in adjusting the cloth cap upon her head looking down at them. But there was a limit to self-indulgence, and she swept the flannel shirt over her head, pulled the tails down around her waist, stepped into the breeches – how good they felt again after so long in those hobbling skirts – and then, sitting upon the bunk, she pulled on the half-boots and buckled the ankle straps of the breeches under the arch of her foot, before standing to clinch the belt at her waist.

She opened her black valise, took out the roll of surgical instruments and selected one of the sturdier scalpels, folded out the blade and tested it with her thumb. It was stingingly sharp. She closed the blade and slipped it into her hip pocket. It was the only weapon available to her.

She was ready now, and she closed the shutter on the bull's-eye lantern, darkening the cabin completely before climbing, fully dressed, into her bunk, pulling the rough woollen blanket to her chin and settling down to wait. The laughter from the saloon became more abandoned, and she imagined that the brandy bottle was being cruelly punished by the men. A long while later she heard her brother's heavy, uneven footsteps on the companionway past her cabin and then there was only the creak and pop of the ship's timbers as she heeled to the wind, and far away the regular tapping of some loose piece of equipment.

She was so keyed, with both fear and anticipation, that there was never any danger of her falling asleep. However, the time passed with wearying slowness. Each time that she opened the shutter of the lantern to check her pocket-watch, the hands seemed hardly to have moved. Then, somehow, it was two o'clock in the morning, the hour when the human body and spirit are at their lowest ebb.

She rose quietly from her bunk, picked up the darkened lantern and went to the door of her cabin. The locking bolt clattered like a volley of musketry, but then it was open and she slipped through.

In the saloon a single oil-lamp still burned smokily, throwing agitated shadows against the wooden bulkheads, while the empty brandy bottle had fallen to the deck, and rolled back and forth with the ship's motion. Robyn squatted to pull off her boots and, leaving them at the entrance, she went forward on bare feet, crossed the saloon and stepped into the passageway that led to the stern quarters.

Her breath was short, as though she had run far, and she paused to lift the shutter of the bull's-eye lantern and flash a narrow beam of light into the darkness ahead. The door to Mungo St John's cabin was closed.

She crept towards it, guiding herself with one hand on the bulkhead and at last her fingers closed over the brass door-handle.

'Please God,' she whispered, and achingly slowly twisted the handle. It turned easily, and then the door slid open an inch along its track, enough for her to peep through into the cabin beyond.

There was just light to see, for the deck above was pierced for a repeating compass so that even while in his bunk the master could at a glance tell his ship's heading. The compass was lit by the dull yellow glow of the helmsman's lantern and the reflection allowed Robyn to make out the cabin's central features.

The bunk was screened off by a dark curtain and the rest of the furnishings were simple. The locked doors of the arms chest were to the left, with a row of hooks beyond from which hung a boat cloak and the clothing that St John had been wearing at dinner. Facing the door was a solid teak desk with racks to hold the brass navigational instruments, sextant, straight edge, dividers, and on the bulkhead above it were affixed the barometer and the ship's chronometer.

The Captain had evidently emptied his pockets on to the desk top before undressing. Scattered amongst the charts and ship's papers were a clasp knife, a silver cigar case, a tiny gold inlaid pocket pistol of the type favoured by professional gamblers, a pair of chunky ivory dice – Zouga and St John must have fallen to gaming again after she left them – and then most important, what she had hoped to find, the bunch of ship's keys that St John usually wore on a chain from his belt, lay in the centre of the desk.

An inch at a time Robyn slid the door further open, watching the dark alcove to the right of the cabin. The curtains billowed slightly with each roll of the ship, and she screwed her nerves tighter as she imagined the movement to be that of a man about to leap out at her.

When the door was open enough to allow her to pass through, it required a huge effort of will to take the first step.

Half-way across the cabin she froze; now she was only inches from the bunk. She peered into the narrow gap in the curtaining and saw the gleam of naked flesh, and heard the deep regular breathing of the sleeping man. It reassured her and she went on swiftly to the desk.

She had no way of learning which keys fitted the lazaretto and the hatch to the main hold. She had to take the whole heavy bunch, and realized that it would mean returning to the cabin later. She did not know if she would have the courage to do that, and as she lifted the bunch her hand shook so it jangled sharply. Startled, she clutched it to her bosom and stared fearfully at the alcove. There was no movement beyond the curtains, and she glided back towards the door on silent bare feet.

It was only when the door closed again that the curtains of the alcove were jerked open, and Mungo St John lifted himself on one elbow. He paused only a moment and then swung his legs out of the bunk and stood up. He reached the desk in two quick strides and checked the top.

'The keys!' he hissed, and reached out for his breeches hanging on the rack beside him, pulling them on swiftly and then stooping to open one of the drawers in his desk.

He lifted the lid of the rosewood case and took out the pair of long-barrelled duelling pistols, thrust them into his waistband, and started for the door of the cabin.

Robyn found the correct key to the lazaretto on the third attempt and the door gave reluctantly, dragging on the hinge with a squeal that sounded to her like a bugle call commanding a charge of heavy cavalry.

She locked the door behind her again, feeling a rush of relief to know that nobody could follow her now and she opened the shutter of the lantern and looked about her swiftly.

The lazaretto was no more than a large cupboard used as a pantry for the officers' personal stores. Sides of smoked ham and dried polonies hung from hooks in the deck above, there were fat rounds of cheese in the racks, boxes of tinned goods, racks of black bottles with waxed stoppers, bags of flour and rice, and, facing Robyn, another hatch with the locking bar chained in place by a padlock the size of her doubled fists.

The key, when she found it, was equally massive, as thick as her middle finger, and the hatch so heavy that it took all of her strength to drag it aside. Then she had to double over to get through the low opening.

Behind her, Mungo St John heard the scrape of wood on wood and dropped silently down the steps to the door of the lazaretto. With a cocked pistol in one hand, he laid his ear to the oak planking to listen for a moment before trying the handle.

'God's breath!' he muttered angrily as he found it locked, and then turned away and raced on bare feet up the companionway to the cabin of his first mate.

At the first touch on his thickly muscled shoulder, Tippoo was fully awake, his eyes glistening in the gloom like those of a wild animal.

'Someone has broken into the hold,' St John hissed at him, and Tippoo reared up out of his bunk, a huge dark figure.

'We find him,' he grunted, as he bound the loin cloth around his waist. 'Then we feed fish with him.'

The main hold seemed vast as a cathedral. The beam of the lantern could not reach into its deepest recesses, and a great mass of cargo was piled high, in some places as high as the maindeck fifteen feet above her.

She saw at once that the cargo had been loaded in such a way that it could be unloaded piecemeal, that any single item could be identified and swung up through the hatches without the necessity of unloading the whole. This, of course, must be essential in a ship trading from port to port. She saw also that the goods were carefully packaged and clearly labelled. Flashing the lantern around her she saw that the cases of equipment for their own expedition were packed here, 'Ballantyne Africa Expedition' stencilled in black on the raw, white wood.

She clambered up the rampart of cases and bales and balanced on the peak, turning the lantern upwards towards the square opening of the hatch, trying to see if there was an open grating, and immediately she was frustrated for the white canvas cover had been stretched around the lower surface also. She stood at full stretch to try and touch the hatch, to feel for the shape of a grating under the canvas, but her fingertips were inches short of the hatch and a sudden wild plunge of the ship under her sent her flying backwards into the deep aisle between the piles of cargo. She managed to keep her grip on the lantern, but hot oil from its reservoir splashed over her hand, threatening to blister the skin.

Once again she crawled up and over the mountains of cargo, searching for the evidence that she hoped not to find. There was no dividing bulkhead in the hold, but the main mast pierced the deck above and came through the hold to rest its foot on the keel, and there were stepped wedges fixed to the thick column of Norwegian pine.

Perhaps they were there to hold the slave decks. Robyn knelt beside the mast and sighted across the hold to the outward curved side of the ship. In line with the wedges on the mast were heavy wooden ledges, like shallow shelves, and she thought that these might be supports for the outer edges of the slave decks. She guessed that they were between two and a half and three feet apart, which she had read was the average height of the between decks in a slaver.

She tried to imagine what this hold might look like with those decks in place, tiers of low galleries just high enough for a man to crawl into doubled over. She counted the shelves and there were five of them – five piles of decks, each with its layer of naked black humanity laid out sardine fashion, each one in physical contact with his neighbour on either side, lying there in his own filth and that of the slaves above him which leaked through the seams of the deck. She tried to imagine the heat of the middle passage when the ship lay becalmed in the baking doldrums, she tried to imagine 2,000 of them vomiting and purging with seasickness as the ship reared and plunged in the wild seas where the Mozambique current scoured the Agulhas bank. She tried to imagine an epidemic of cholera or smallpox taking hold of that mass of misery, but her imagination could not rise to the task, and she pushed the hideous images aside and crawled on over the heaving cargo, flashing her lantern into each corner, searching for more solid evidence than the narrow ledges.

If there was planking on board it would be laid flat upon the deck, below this cargo, and there was no way in which Robyn could reach it.

Ahead of her she saw a dozen huge casks bolted to the forward bulkhead. They could be water barrels, or they could be filled with trade rum, or the rum could be replaced with water when the slaves were taken on board. There was no means of checking the contents, but she knocked on the oak with the hilt of her scalpel and the dull tone assured her that the casks contained something.

She squatted down on one of the bales and slit the

stitching with the scalpel, thrusting her hand into the opening she grasped a handful and pulled it out to examine it in the lamp light.

Trade beads, ropes of them strung on cotton threads, a *bitil* of beads was as long as the interval between fingertip and wrist, four bitil made a *khete*. These beads were made of scarlet porcelain, they were the most valued variety called *sam sam*. An African of the more primitive tribes would sell his sister for a *khete* of these, his brother for two *khete*.

Robyn crawled on, examining crates and bales – bolts of cotton cloth from the mills of Salem, called *merkani* in Africa, a corruption of the word 'American' and the chequered cloth from Manchester known as *kaniki*. Then there were long wooden crates marked simply 'Fire Pieces', and she could guess that they contained muskets. However, firearms were common trade goods to the coast, and no proof of the intention to buy slaves – they could just as readily be used to purchase ivory or gum-copal.

She was tired now with the effort of climbing and blundering around over the heaving slopes and peaks of cargo, and with the nervous tension of the search.

She paused to rest a moment, leaning back against one of the bales of *merkani* cloth, and as she did so something dug painfully into her back, forcing her to change her position. Then she realized that cloth should not have hard lumps in it. She shuffled around and once again slit the sacking cover of the bale.

Protruding from between the folded layers of cloth, there was something black and cold to the touch. She pulled it out and it was heavy iron, looped and linked, and she recognized it instantly. In Africa they were called 'the bracelets of death'. Here, at last, was proof, positive and irrefutable, for these steel slave cuffs with the light marching chains were the unmistakable stigmata of the trade.

Robyn tore the bale wider open, there were hundreds of the iron cuffs concealed between layers of cloth. Even if it had been possible, a perfunctory search by a naval boarding party was highly unlikely to have uncovered this sinister hoard.

She selected one set of cuffs with which to confront her brother and she started aft towards the lazaretto, filled suddenly with desire to be out of this dark cavern with its menacing shadows, and back once more in the safety of her own cabin.

She had almost reached the entrance to the lazaretto when suddenly there was a loud scraping sound from the deck above, and she froze with alarm. When the sound was repeated, she had enough of her wits still about her to douse the snuffer of her lantern, and then immediately regretted having done so for the darkness seemed to crush down upon her with a suffocating weight and she felt panic rising up to take possession of her.

With a crash like a cannon shot the main hatch flew open, and as she swung back towards it she saw the white pin-pricks of the stars outlined by the square opening. Then a huge, dark shape dropped through, landing lightly on the piled bales beneath, and at almost the same instant the hatch thudded closed again, blotting out the starlight.

Now Robyn's terror came bubbling to the surface. There was somebody locked in the hold with her, and the know-ledge held her riveted for long, precious seconds, before she plunged back towards the lazaretto hatch, suppressing the scream that choked up into her throat.

The shape she had glimpsed for a moment was unmistak-able. She knew that it was Tippoo in the hold with her, and it spurred her terror. She could imagine the great hairless toadlike figure, moving towards her in the darkness, with repulsive reptilian swiftness, could almost see the pink tongue flickering out over thick cruel lips, and her haste became uncontrolled. She lost her footing in the darkness and fell heavily, tumbling backwards into one of the deep gullies between banks of cargo, cracking the back of her skull on a wooden case, half-stunning herself so that she lost her grip on the extinguished lantern and could not find it when she groped for it. When she scrambled to her knees again, she had lost all sense of direction in the total blackness of the hold.

She knew that her best defence was to remain still and silent until she could place the man who was hunting her, and she crouched down in a crack between two crates. Her pulse beat in her ears like a drum, deafening her – and her heart seemed to have crammed up into her throat so that she must fight for each breath.

It took many minutes and all her determination to bring herself under control, to be able to think again.

She tried to decide in which direction lay the lazaretto hatch, her only means of escape, but she knew that the only way she would be able to find it would be to grope her way to the ship's wooden side and follow it around. The prospect of doing this, with that grotesque creature hunting her in the darkness, was appalling. She shrank down as far as she could into the narrow space and listened.

The hold was filled with small sounds that she had not noticed before, the heave and creak of the ship's timbers, the shifts of the cargo against its retaining ropes and netting – but then she heard the movement of a living thing close behind her and she caught the shriek of terror before it reached her lips, as she lifted her arm to protect her head. Frozen like that, she waited for a blow which never came.

Instead she heard another movement pass behind her turned shoulder, a whisper of sound, yet so chilling that she felt all power of movement drained from her legs. He was here, very close in the blackness, toying with her, cruel as a cat. He had smelt her out. With some sort of animal sense, Tippoo had found her unerringly in the darkness, and now he crouched over her ready to strike and she could only wait.

Something touched her shoulder, and before she could

jerk away it swarmed up over her neck, brushed her face. She flung herself backwards and screamed, striking out wildly with the steel chain and handcuffs which she still held in her right hand.

The thing was furry and quick, and it squealed sharply, like an angry piglet as she struck again and again. Then it was gone, and she heard the scamper of small feet across the rough wood of the crate and she realized that it had been one of the ship's rats, big as a tom cat.

Robyn shuddered, revolted, but with a lift of relief that was short-lived.

There was a flash of light, so unexpected that it almost blinded her, and a lantern beam was thrown in a single swift sweep about the hold, and then extinguished, so that the darkness seemed even more crushing than before.

The hunter had heard her scream, and had flashed his lantern in her direction, had probably seen her, for she had clambered out of her niche between the crates. Now at least she knew in which direction to move, the brief flash of light had orientated her once again, she knew where to find the hatch.

She threw herself over a pile of soft bales, clawing herself towards the hatch, then checking her flight to think a moment. Tippoo must have known how she had entered the hold and would know that she would try to escape in the same direction. She must move with care, with stealth, ready for the moment when he flashed the lantern again, taking care not to rush headlong into the trap he was certainly setting for her.

She changed her grip on the iron chain, only then realizing its potential as a weapon, much more effective than the short-bladed scalpel in her pocket. A weapon! For the first time she was thinking of defending herself, not merely crouching like a chicken before the stoop of the hawk. 'Robbie was always the plucky one.' She could almost hear her mother's voice, troubled but touched with pride, when she had defended herself effectively against the village ruffians, or joined her brother on some of his more hair-raising escapades, and she realized she needed all of that pluck now.

With the chain gripped in her right hand, she started stealthily back towards the hatch, crawling, slithering forward on her belly, pausing to listen every few seconds. It seemed like a complete round of eternity before her out-reaching fingers touched the solid planking of the after bulkhead. She was within feet of the hatch now and that was where he would be waiting for her.

She crouched down, with her back firmly against the planking and waited for the flutter of her heart to abate enough to allow her to hear, but any noise her hunter made was covered by the creak and pop and groan of the working wooden hull, and the thud and hiss of the sea as *Huron* beat up hard to the wind.

Then she wrinkled her nose at a new smell that overlaid the pervading reek of the bilges. It was the hot oil smell of a shuttered lantern burning very close by, and now when she listened for it, she thought she could hear the pinkle and tick of the heated metal of the lantern. He must be close, very close, guarding the hatchway, ready to flash the lantern at the moment he could place her whereabouts accurately.

With the slowness of spreading oil she rose to face into the darkness where he was, she slipped the scalpel into the palm of her left hand, and she drew back her right arm with the chain and iron cuff dangling from her fist, ready to strike.

Then she pitched the scalpel with a short underhanded throw, judging the distance to drop it close enough to force him to move again, but far enough to make him move away from her.

The tiny missile struck something soft, the sound muted, almost lost in other small sounds, but then it slithered softly along the deck, like a hesitant footfall and instantly light flooded the hold.

Tippoo's monstrous shape sprang out of the darkness, huge, menacing, unbelievably close. He held the unhooded lantern high in his left hand, and the yellow light glistened on the bare round dome of his yellow skull, and on the broad plain of his back, the muscle rippling and tensing into valleys and ridges as he swung back the heavy club in his right hand, his head hunched down on the thick corded neck. He was facing away from her, but only for an instant. As he realized that there was nobody in front of him, he reacted with animal swiftness, ducking the great round head on to his chest, beginning to swing away.

She moved entirely by instinct. She swung the heavy iron manacle on the length of chain. It hummed in a glittering circle in the lamp-light, and it caught Tippoo high on the temple with a crack like a branch breaking in a high wind, and the thin layer of yellow scalp opened like the mouth of a purse with a crimson velvet lining. Tippoo swayed drunkenly on straddled legs that started to buckle under him. She pulled back the chain and swung again with all the weight of her body and the strength of her fear behind the stroke. Again a deep red wound bloomed on the polished yellow dome of his skull and he went down slowly on to his knees, an attitude she had seen him adopt so often as he prayed on the quarterdeck, making the Muslim obeisance towards Mecca. Now again, his forehead touched the deck, but this time with his blood dribbling into a puddle under it.

The lantern clattered on to the deck, still burning, and in its light Tippoo rolled heavily on to his side with his breath snoring in his throat and his eyes rolled up into his head, glaring ghastly white and unseeing, his thick legs kicking out convulsively.

Robyn stared at the prostrate giant, aghast at the damage she had done, already feeling the need to administer to any hurt or crippled being – but it lasted for only seconds, as Tippoo's eyes rolled back into their deep sockets, and she saw the pupils beginning to focus. The yellow gleaming body heaved, the movement of limbs was no longer spasmodic but more co-ordinated and determined, the head lifted, still lolling, but swinging questingly from side to side.

Unbelievably, the man was no longer crippled by those two cutting blows, had been stunned for seconds only, in seconds more he would be fully conscious and in his fury more dangerous than ever. With a sob Robyn flew at the open hatch of the lazaretto. As she passed him, cruel fingers hooked at her ankle, pulling her off balance so she almost fell before she could kick herself free and dive through the opening.

Tippoo was on hands and knees in the light of the fallen lantern, creeping towards her as she threw all her weight against the hatch. As it thudded into its jamb, she dropped the locking bar into its seating, and at that moment Tippoo's shoulder crashed into the far side of the hatch with a force that shook it in its frame.

Her frantic fingers were so clumsy that it took three attempts to secure the padlock and chain, and only then could she sink to the deck and sob away her fear until her relief came to buoy her up and give her renewed strength.

When she dragged herself to her feet, she was light-headed, intoxicated with the strange fierce jubilation that she had never felt before. She knew it sprang from having fought herself out of danger, from the unfamiliar experience of inflicting punishment on a hated adversary – and she knew she would feel guilt for it later, but not now.

The keys were still in the door between the lazaretto and saloon where she had left them. She pushed the door open quickly and paused in the opening an instant, feeling the quick flare of alarm overtake her feeling of elation.

She had only an instant of time to realize that somebody had trimmed the saloon lamp, and then fingers seized her wrist from behind and swung her off balance, bearing her down heavily to the deck and holding her there with one arm twisted up between her shoulder-blades.

'Keep still, damn you – or I'll twist your head off your shoulders.' Mungo St John's voice was low and fierce in her ear.

Her arm holding the chain was trapped under her, and now her captor shifted his weight over her, placing his knee in the small of her back and bearing down so painfully that she wanted to cry out with the agony that flared up her curved spine.

'Tippoo flushed you out quickly enough,' St John murmured with grim satisfaction. 'Now let's have a look at you – before we stretch you out on the grating.'

He reached forward and pulled off the cloth cap that covered her head, and she heard his little grunt of surprise as her hair tumbled loose in a slippery shining mass in the lamplight. His grasp slackened and the pressure of his knee into the base of her spine eased. Roughly he grasped her shoulder to turn her on to her back so he could see her face.

She rolled easily towards him and then as her trapped arm came free, she hurled the chain at his face. He threw up both hands to catch the blow. As slippery as an eel she wormed out from under him and flew at the door to the saloon.

He was quick, quicker than she was, and his fingers were steely, they caught in the thin worn flannel of her shirt and it ripped from collar to hem. She turned and lashed at him with the chain again, but now he was ready for her, and he caught her wrist, trapping it.

She kicked wildly at his shins, and succeeded in hooking an ankle behind his heel and they went down together in a tangle on to the wooden deck, and Robyn found herself carried along by a fierce uncaring madness. She was hissing and snarling like a cat as she raked at his eyes with her nails, raising a bloody line down his neck. The tatters of her shirt flapped around her waist and the coarse dark hair of his chest rasped against the tender points of her naked breasts so that dimly in her madness she realized that he wore only breeches, and the man smell of his body filled her nostrils as he tried to smother her wild struggles with his body.

She reached up towards him with her open mouth, trying to sink her teeth into that flushed handsome face, but he caught a handful of her hair at the back of her head and twisted it. The pain seemed to flow down through her body and explode at the base of her belly in a warm soft spasm that took her breath and crammed it down into her lungs.

He held her helplessly and she felt the strength going out of her, to be replaced by spreading languor, and she stared up into his face with a kind of wonder as though she had never seen it before. She saw his teeth were very white, his lips drawn back into a rictus of emotion, and fierce yellow eyes unfocused and smoky with a kind of madness that matched her own.

She made one last feeble effort to repel him, driving her knee upwards, aimed at the fork of his lower body, but he trapped it between his thighs and then holding her thus he reared up over her and looked down at her bared bosom.

'Sweet Mother of God!' he croaked, and she saw the cords strain tight in his throat, saw the hot yellow fire in his eyes and she could not move, not even when he freed his hooked fingers from the twisted tresses of her hair and ran them slowly down her body, cupping first one tight small breast and then the other in the palm of his hand.

She felt his touch had gone, though the memory lingered on her skin like butterflies' wings. Then his fingers were back, tugging demandingly at the fastenings of her breeches. She closed her eyes and refused to let her mind consider what was about to happen. She knew that there was nothing she could do to escape, and cried out softly with the strange elation of the martyr.

But her cry seemed to touch something deep in him, the smoky yellow eyes focused for a moment, the predatory expression of the face became uncertain and then tinged with horror as he looked down at her spread white body. Swiftly he rolled away from her.

'Cover yourself!' he said harshly, and she was overwhelmed with a cold avalanche of loss, followed immediately by as great a rush of shame and of guilt.

She scrambled to her knees, clutching her clothing around her, suddenly shivering, but not with cold.

'You should not have struggled,' he said and though he was obviously fighting to control it his voice shook as hers did.

'I hate you,' she whispered foolishly, and then it became true. She hated him for what he had aroused in her, for the sickness and the guilt that followed it, and for the sense of loss and bereavement with which he had left her.

'I should kill you,' he muttered, not looking at her. 'I should have Tippoo do it.'

She felt no fear at the threat. She had rearranged her clothing as best she could, but still she knelt opposite him.

'Go!' he almost shouted at her. 'Get back to your cabin.'

She rose slowly, hesitated a moment and then turned to the companionway.

'Doctor Ballantyne!' he stopped her, and she turned back. He had risen and now he stood beside the door to the lazaretto, the keys in his one hand, the slave cuff and chain in the other. 'It would be best not to tell your brother of what you discovered tonight.' His voice was controlled now, cold and low. 'I would not have the same scruples with him. We will be at Cape Town in four days,' he went on. 'After that you may do as you will. Until then you will not provoke me again. One chance is already too many.'

She stared at him mutely, feeling small and helpless.

'Goodnight, Doctor Ballantyne.'

She hardly had time to pack her breeches and torn shirt away in the bottom of her chest, inspect and rub a salve on her bruises, pull on her nightdress and climb into her narrow bunk before someone pounded on the cabin door.

'Who is it?' she called huskily and breathless, not yet fully recovered from the night's stresses.

'Sissy, it's me.' Zouga's voice. 'Someone has beaten Tippoo's skull in. He's bleeding all over the deck – can you come?'

Robyn glowed with a fierce spark of pagan glee, which she tried immediately to suppress with less than complete success.

'I'm coming.'

There were three men in the saloon, Zouga, the second mate and Tippoo. Mungo St John was not there. Tippoo sat stolidly on a stool under the oil lamp, naked except for his cotton loin cloth, and his neck and shoulders ran with sheets of dark slick blood.

The second mate held a wad of grubby cotton to his skull and when Robyn lifted it away the wounds began spurting merrily.

'Brandy,' she demanded, and rinsed her hands and her instruments in the spirit – she was an admirer and believer in the teachings of Jenner and Lister – before she probed the points of her forceps into the open wound. She gripped the vessels and twisted them closed. Tippoo made no move, his expression never changed – and she was still carried along on the pagan mood, in direct defiance to the oath of Hippocrates which she had sworn.

'I must clean the wounds,' she told him, and quickly, before her conscience could prevent it, she tipped the raw brandy into the wounds and swabbed them out.

Tippoo sat still as a temple carving of a Hindu devil, making no acknowledgement of the harsh spirit burning open tissue.

Robyn tied off the vessels with silk thread, leaving an end hanging from the wounds, and then she sutured the lips closed, laying precise neat stitches and pulling them up tightly so that the smooth bald scalp came up in a sharp little peak of flesh with each tug.

'I will pull the thread when the vessels mortify,' she told him. 'The stitches will be ready to come out in a week.' She would not deplete her stock of laudanum, she decided, the man obviously was impervious to pain, and she was still in the throes of unchristian spite.

Tippoo lifted the round head. 'You good doctor,' he told her solemnly, and she learned then a lesson that would last her throughout her life – the stronger the purge, the more astringent or foul-tasting the medicine, and the more radical the surgery, then the more impressed with the surgeon's skill was the African patient.

'Yes,' Tippoo nodded gravely, 'you one bloody fine doctor.' And he opened one huge paw. In his palm lay the scalpel that Robyn had lost in *Huron*'s hold. Without expression he placed it in Robyn's own unresisting hand, and with that eerie swiftness was gone from the saloon, leaving her staring after him.

Huron flew southwards, meeting the long South Atlantic rollers and spurning them carelessly, brushing them aside with her shoulder and letting them cream over her rail and then tumble away astern in a long smooth wake.

There were seabirds in company with them now, beautiful gannets with yellow throats and black diamonds painted around their eyes, coming in from the east and soaring above their wake, shrieking and diving for the galley scraps when they were thrown overboard. There were seals too, lifting their whiskered heads high above the surface to stare curiously after the towering clipper as she burst the sea open with her sharp bows in her flight into the south.

Smeared across the brilliant blue water were long serpentine trails of sea-bamboo, torn from the rocky shoreline by the gales and storms of this uneasy and troubled sea.

All these were indications of the land which was always just below the eastern horizon, and Robyn spent many hours of each day alone at the port rail staring towards it, longing for another glimpse of it, smelling the dryness and the spiced aroma of its grass and herbs on the wind, seeing its blown dust in the marvellous reds and glowing gold of the sunsets, but denied sight of it by the offing that St John was making before coming back on to the starboard tack for the final run into Table Bay.

However, as soon as Mungo St John appeared on his quarterdeck Robyn would hurry below without another glance in his direction and she locked herself in her cabin, brooding there alone so that even her brother sensed that something troubled her. He tried a dozen times to draw her out. She sent him away each time, refusing to open the cabin to him.

'I'm all right, Zouga. I just want to be alone.'

And when he tried to join her in her solitary vigils at the ship's rail, she was short and unbending, exasperating him so that he stamped away and let her be.

She was afraid to talk to him, afraid that she would blurt out her discovery of slaving equipment in *Huron*'s hold and put him in deadly danger. She knew her brother well enough not to trust his temper and not to doubt his courage. Neither did she doubt Mungo St John's warning. He would kill Zouga to protect himself – he could do it himself, she had seen him handle a pistol, or he could send Tippoo to do the work in the night. She had to protect Zouga, until they reached Cape Town, or until she did what she had to do.

'Vengeance is mine, saith the Lord.' She had found the passage in her Bible and studied it carefully, and then she had prayed for guidance which had not been given – and she had ended more confused and troubled than she had begun.

She prayed again, kneeling on the bare deck beside her bunk until her knees ached, and slowly her duty became clear to her.

Three thousand souls sold into slavery in a single year – that was what the Royal Naval Captain had accused him of. How many thousands before that, how many thousands more in the years to come if *Huron* and her captain were allowed to continue their depredations, if nobody could prevent them ravaging the east coast of Africa, her land, her people, those peoples whom she was sworn to protect and minister to and to lead into the fold of the Saviour.

Her father, Fuller Ballantyne, was one of the great champions of freedom, the unrelenting adversary of this abominable trade. He had called it 'the running sore on the conscience of the civilized world that must be rooted out with all the means at our disposal'. She was her father's daughter, had made her oath in the sight of God.

This man, this monster, epitomized the sickening evil and monstrous cruelty of the whole filthy business.

'Please show me my duty, oh Lord,' she prayed, and always there was her own guilt and shame. Shame that his eyes had probed her half-naked body, that his hands had touched and fondled her, shame that he had debased her further, by stripping bare his own body. Hastily she thrust the image aside, it was too clear, too over-powering. 'Help me to be strong,' she prayed quickly.

There was shame and there was guilt, a terrible corrosive guilt in the fact that his gaze, his touch, his body, had not revolted and disgusted her, but had filled her instead with a sinful delight. He had tempted her to sin. For the first time in all her twenty-three years she had encountered real sin, and she had not been strong enough. She hated him for that.

'Show me my duty, oh Lord,' she prayed aloud and rose stiffly from her knees to sit on the edge of the bunk. She held her well-worn leather-bound Bible in her lap and whispered again.

'Please give unto your faithful servant guidance.' And she let the book fall open, and with her eyes closed placed her forefinger on the text. When she opened her eyes again she gave a start of surprise, guidance obtained by this little ritual of hers was usually not so unequivocal, for she had chosen Numbers 35: 19. 'The revenger of blood himself shall slay the murderer: when he meeteth him, he shall slay him.'

R obyn had no illusions as to the difficulty she would have in performing the heavy duty placed upon her by God's direct injunction, or how easily roles might be reversed and she become herself the victim rather than the avenger.

The man was as dangerous as he was wicked and time was against her. The accurate observation of the sun that Zouga had made at noon that day placed the ship within a hundred and fifty miles of Table Bay, and the wind stood fair and boisterous. Dawn the next day would reveal that great flat-topped mountain rising out of the sea. She had no time for elaborate planning. Whatever she did must be direct and swift.

There were half a dozen bottles in her medicine chest whose contents would serve – but no, poison was the most disgusting of deaths to inflict. She had seen a man die of strychnine poison when she was at St Matthew's. She would never forget the arching spine as his back muscles convulsed until the man stood on the top of his head and on his heels like a drawn bow.

It must be some other means, there was the big naval colt revolver which Zouga kept in his cabin. He had instructed her in its loading and discharge, or there was the Sharps rifle, but both of those belonged to her brother. She did not want to see him swinging from the gallows on the parade ground below the castle at Cape Town. The more uncomplicated and direct the plan, the greater was its chance of success, she realized, and at the thought she knew just how it must be done.

There was a polite knock on her cabin door, and she started.

'Who is it?'

'Jackson, Doctor.' He was the Captain's steward. 'Dinner is served in the saloon.'

She had not realized how late it had grown.

'I will not be dining tonight.'

'You must keep your strength up, ma'am,' Jackson entreated through the closed door.

'Are you the doctor, then?' she asked tartly, and he went shuffling off down the companionway.

She had not eaten since breakfast but she was not hungry, her stomach muscles were rigid with tension. She lay a while on her bunk, gathering her resolve and then she stood up, selected one of her oldest dresses, in a dark heavy wool. It would be the least loss to her wardrobe, and the dark colour would make it less conspicuous in the shadows.

She left her cabin and went quietly up to the main deck. There was no one but the helmsman on the quarterdeck, his weathered brown face lit faintly by the binnacle lamp.

She moved quietly across to the skylight of the saloon and looked down.

Mungo St John sat at the head of the table, with a joint of steaming salt beef in front of him. He was carving thin slices of meat, laughing across the board at one of Zouga's sallies. One quick glance was enough. Unless there was a call from the lookout or a need to change sail, Mungo St John would not move for another half hour at the least.

Robyn went back down the ladder, past her own cabin

and down the companionway to the after quarters. She reached Mungo St John's quarters and tried the door. Once again it slid open easily, she stepped through and closed it behind her again.

It took her only a few minutes to find the case of pistols in the drawer of the teak desk. She opened it on the desk top and took out one of the beautiful weapons. The mottled Damascus steel barrels were inlaid with bright gold, a hunting scene with horses and hounds and huntsmen.

Robyn sat down on the edge of the bunk, held the weapon muzzle-up between her knees while she unscrewed the silver powder flask and measured the fine powder into the cup. It was a familiar chore, for Zouga had spent hours in her instruction. She rammed the charge down into the long, elegant barrel under the felt wad, and then selected a perfect sphere of lead from the ball compartment of the case, wrapped that in an oiled patch of felt to give it a close fit in the rifled barrel and then rammed it down on top of the powder charge.

Then she reversed the pistol, pointing it down at the deck while she fitted one of the copper percussion caps over the nipple of the breech, drew back the hammer until it clicked at full cock, and laid it on the bunk beside her. She did the same with the other pistol, and when they were both loaded and cocked, she placed them on the edge of the desk, butts towards her, ready for immediate use.

Then she stood and, in the centre of the cabin, lifted her skirts around her waist and loosened the draw string of her drawers. She let them fall to the deck, and the air was cool on her naked buttocks so that she felt the little goose pimples rise on her skin. She dropped her skirts and picked up the cotton under-garment. Holding it across her chest she tore it half through and threw it across the cabin, then she took the fastenings of her bodice in both hands and ripped them down almost to her waist, the hooks and eyes hung on the torn threads of cotton.

She looked at herself in the polished metal mirror on the bulkhead beside the door. There was a lustre in her green eyes and her cheeks were flushed.

For the first time in her life, she thought her image beautiful – not beautiful, she corrected herself, but proud and wild and strong as an avenger should be. She was glad he would see her like this before he died, and she lifted her hand and rearranged one of the thick tresses of hair that had broken free of its retaining ribbon.

She sat back on the bunk and picked up a loaded pistol in each hand, she aimed first the one and then the other at the brass handle of the door, and then laid them in her lap and settled down to wait.

She had left her watch in her own cabin so she could not tell how long she waited. The voices and laughter from the officers' saloon were completely muffled by the closed door, but every time a plank creaked or some part of the ship's gear clattered, her nerves sprang tight and she lifted the pistols to cover the doorway.

Then suddenly she heard his footsteps, there was no mistaking them for any other man aboard. They reminded her somehow of a caged leopard pacing its bars, quick, light and alert. He was on the deck above her, but the footsteps were so close that it seemed he was in the cabin with her. She looked up at the deck, swinging her head slightly to follow his turns from one side of the quarterdeck to the other.

She knew what he was doing, she had watched him on a dozen other nights. First he would talk quietly with the helmsman, checking the slate on which the ship's course was chalked before going back to inspect the log trailing astern. Then he would light one of his thin black Havana cigars and begin pacing the deck, with his hands clasped in the small of his back as he walked, darting quick glances up at the trim of his sails, studying the stars and the clouds for signs of change in the weather, pausing to feel the scend of the sea and the run of the ship under him before pacing on again.

Suddenly the footsteps stopped and Robyn froze, the moment had come. He had paused to flip the stub of his cigar over the rail and watch it fizzle into darkness as it hit the surface of the sea.

There was still time for her to escape, and she felt her resolve weaken. She half rose. She could still reach her own cabin if she moved now, but her legs would not carry her across the cabin. Then she heard his footsteps cross the deck above her with a different tread. He was coming down. It was too late.

Almost choking on her own breath, she sank back on to the bunk and lifted both pistols. They wavered uncertainly and she realized that her hands were shaking. With a tremendous effort she stilled them. The door slammed open and Mungo St John stooped into the cabin, and then stopped as he saw the dark figure and the twin barrels that menaced him.

'They are loaded and cocked,' she said huskily. 'And I will not hesitate.'

'I see.' He straightened slowly, so the dark head just brushed the deck overhead.

'Close the door,' she said, and he pushed it closed with his foot, his arms folded on his chest, and that mocking half-smile on his lips. It made her forget her carefully rehearsed speech, and she stuttered slightly, and was immediately furious with herself.

'You are a slaver,' she blurted, and he inclined his head, still smiling. 'And I have to stop you.'

'How do you propose doing that?' he asked with polite interest.

'I am going to kill you.'

'That should do it,' he admitted, and now he smiled, a flash of white teeth in the gloom. 'Unfortunately they would

72

probably hang you for it, if my crew didn't tear you to pieces before that.'

'You assaulted me,' she said. She glanced at her torn drawers lying near his feet and then with the butt of one pistol touched her torn bodice.

'A rape, by God!' Now he chuckled aloud, and she felt herself blushing vividly at the word.

'It's no laughing matter, Captain St John. You have sold thousands of human souls into the most vile bondage.'

He took one slow pace towards her and she half rose, panic in her voice.

'Don't move! I warn you.'

He took another pace and she thrust both pistols towards him – at the full stretch of her arms.

'I shall fire.'

The smile never wavered on his lips and the yellow flecked eyes held hers steadily as he took another lazy pace closer.

'You have the most beautiful green eyes I have ever seen,' he said, and the pistols shook in her hands.

'Here,' he said gently. 'Give them to me.'

He took the two gold-worked barrels in one hand and turned their muzzles upwards, pointing them at the deck above them. With the other hand he gently began to open her fingers, untangling them from trigger and butt.

'This is not why you came here,' he said, and her fingers went slack. He took the pistols out of her hands and uncocked them before laying them back in their velvet-lined nests within the rosewood case.

His smile was no longer mocking, and his voice was soft, almost tender as he lifted her to her feet.

'I am glad you came.'

She tried to turn her face away, but he took her chin between his fingers and lifted it. As he brought his mouth down to hers, she saw his lips opening, and the warm wet touch was a physical shock.

His mouth tasted slightly salty, perfumed with cigar smoke. She tried to keep her lips closed, but the pressure of his own lips forced them gently open and then his tongue was invading her. His fingers were still on her face, stroking her cheek, smoothing her hair back from her temples, touching lightly her closed eyelids – and she lifted her face higher to his touch.

Even when he slowly unfastened the last hooks of her bodice and eased it down off her shoulders, her only response was to feel the strength go out of her thighs so she had to lean against his hard chest for support.

Then he lifted his mouth from hers, leaving it empty, cooling after the warmth and she opened her eyes. With a sense of disbelief, she saw that his head was bowing to her breast, and she was looking down on the thick dark curls that covered the back of his neck. She knew it must stop now, before he did what she could hardly believe he was about to do.

When she tried to protest, it was only a whimper in her throat. When she tried to seize his head and thrust it away from her, her fingers merely curled into the springing crisp curls the way a cat claws a velvet cushion, and instead of thrusting him away, she drew his head down and arched her back slightly so that her breasts rose to meet him.

Yet she was unprepared for the feel of his mouth. It seemed as though he were about to suck her very soul out through the swollen, aching tips. It was too strong, she tried not to cry out, remembering that the last time she had done so, it had broken the spell – but it was too strong.

It was a sobbing choked-up little cry, and now her legs gave way under her. Still holding his head she sagged backwards on to the low bunk, and he knelt beside the bunk without lifting his mouth from her body. She arched her back and raised her buttocks off the bunk at his touch and allowed him to draw out her billowing skirts from under her and drop them to the deck.

Suddenly, he pulled abruptly away and she almost screamed to him not to go away again – but he had crossed to the door and locked it. Then, as he came back to where she lay, his own clothing seemed to fall away from his body like morning mist from mountain peak, and she came up on one elbow to stare at him openly. She had never seen anything so beautiful, she thought.

'The devil is beautiful also.' A tiny inner voice tried to warn her, but it was far away and so small that she could ignore it. Besides it was too late, far too late to listen to warnings now – for already he was coming over her.

She expected pain, but not the deep splitting incursion that racked her. Her head was flung back and her eyes flooded with the tears of it. Yet even in the stinging agony of it there was never a thought to reject this stretching, tearing invasion and she clung to him with both her arms about his neck. It seemed that he suffered with her, for except for that single swift deep stroke, he had not moved, trying to alleviate her agony by his utter stillness, his body was rigid as hers, she could feel the muscles taut to the point of tearing, and he cradled her in his arms.

Then suddenly she could breathe again, and she took in air with a great rushing sob, and immediately the pain began to change its shape, becoming something she could not describe to herself. It started as a spark of heat, deep within her, and flared slowly so she was forced to meet it with a slow voluptuous movement of her hips. She seemed to break free of earth and rise up through flames, that flickered redly through her clenched eyelids. There was only one reality – and that was the hard body that rocked and plunged above her. The heat seemed to fill her until she could not bear it any longer. Then at the last moment when she thought she might die of it, it burst within her and she felt herself falling, like a tumbling leaf, down, down, at last, to the hard narrow bunk in a half-dark cabin in a tall ship on a wind-driven sea.

When next she could open her eyes his face was very close to hers. He was staring at her with a thoughtful, solemn expression.

She tried to smile, it was a shaky unconvincing effort.

'Please don't look at me like that.' Her voice was even deeper, more husky than it usually was.

'I don't think I ever saw you before,' he whispered and traced the line of her lips with his fingertip. 'You are so different.'

'Different from what?'

'Different from other women.' His reply gave her a pang.

He made the first movement of withdrawing from her, but she tightened her grip on him panic-stricken at the thought of losing him yet.

'We will only have this one night,' she told him, and he did not reply. He lifted one eyebrow, and waited for her to speak again.

'You don't dispute it,' she challenged. There was that mocking little smile beginning to curl his lip again, and it annoyed her.

'No, I was wrong, you are like all other women,' he smiled. 'You have to talk, always you have to talk.'

She let him go, as punishment for those words. But as he slithered free of her she felt a terrible emptiness and she regretted his going fiercely, beginning to hate him for it.

'You have no God,' she accused him.

'Isn't it strange,' he chided her gently, 'that most of the worst crimes in history have been committed by men with God's name upon their lips.'

The truth of it deflated her momentarily, and she struggled into a sitting position.

'You are a slaver.'

'I don't really want to argue with you, you know.' But she would not accept that.

'You buy and sell human beings.'

'What are you trying to prove to me?' He chuckled now, further angering her.

'I'm telling you that there is a void between us that can never be bridged.'

'We have just done so, convincingly,' and she flushed bright scarlet down her neck on to her bosom.

'I have sworn to devote my life to destroy all you stand for,' she said fiercely, pushing her face close to his.

'Woman, you talk too much,' he told her lazily, and covered her mouth with his own, holding her like that while she struggled, gagging her with his lips so her protests were muffled and incomprehensible. Then when her struggles had subsided he pushed her easily backwards on to the bunk and came over her again.

In the morning when she woke, he was gone, but the bolster beside her was indented by his head. She pressed her face into it and the smell of his hair and of his skin still lingered, though the heat of his blood had dissipated and the linen was cool against her cheeks.

The ship was in the grip of intense excitement. She could hear the voices from the deck above as she scurried down the empty passageway to her own cabin, dreading meeting a member of the crew, or more especially meeting her brother. What excuse could she have for being abroad in the dawn, with her cabin unslept in and her clothing torn and rumpled?

Her escape was a matter of seconds only, for as she locked and leaned thankfully against the door of her cabin, Zouga beat upon it with his fist from the far side.

'Robyn, wake up! Get dressed. Land is in sight. Come and see!'

Swiftly she bathed her body with a square of flannel

dipped into the enamelled jug of cold sea water. She was tender, swollen and sensitive and there was a trace of blood on the cloth.

'The trace of shame,' she told herself severely, but it was difficult to sustain the emotion. Instead she felt a soaring sense of physical well-being and a hearty appetite for her breakfast.

Her step was light, almost skipping as she went up on to the main deck and the wind tugged playfully at her skirts.

Her first concern was for the man. He stood at the weather rail, in shirt-sleeves only, and immediately a storm of conflicting feelings and thoughts assailed her, the chief of which was that he was so lean and dark and devil-may-care that he should be kept behind bars as a menace to all womankind.

Then he lowered the telescope, turned and saw her by the companionway and bowed slightly, and she inclined her head an inch in reply, very cool and very dignified. Then Zouga hurried to meet her, laughing and excited, and took her arm as he led her to the rail.

The mountain towered out of the steely green Atlantic, a great grey buttress of solid rock, riven and rent by deep ravines and gullies choked with dark green growth. She had not remembered it so huge, seeming to fill the whole eastern horizon and reaching up into the heavens, for its summit was covered in a thick shimmering white mattress of cloud. The cloud rolled endlessly over the edge of the mountain like a froth of boiling milk pouring over the rim of the pot, but as it sank so it was sucked into nothingness, disappearing miraculously to leave the lower slopes of the mountain clear and close, each detail of the rock-face finely etched and the tiny buildings at its foot as startling white as the wing feathers of the gulls that milled the air about the clipper.

'We'll dine tonight in Cape Town,' Zouga shouted over the wind, and the thought of food flooded Robyn's mouth with saliva.

Jackson, the steward, had the hands spread a tarpaulin to break the wind and they breakfasted under its lee, in the sunshine. It was a festive meal, for Mungo St John called for champagne and they toasted the successful voyage and the good landfall in the bubbling yellow wine.

Then Mungo St John ended it. 'The wind comes through there, funnelled down that break in the mountain.' He pointed ahead, and they saw the surface at the mouth of the bay seething with the rush of it. 'Many a ship has been dismasted by that treacherous blast. We'll be shortening sail in a few minutes.' And he signalled to Jackson to clear away the trestles that carried the remains of their breakfast, excused himself with a bow and went back to his quarterdeck.

Robyn watched him strip the canvas off the upper yards, taking in two reefs in the main and setting a storm jib so that *Huron* met the freak wind readily and ran in for Table Bay, giving Robben Island a good berth to port. When the ship had settled on to its new heading, Robyn went up on to the quarterdeck.

'I must speak with you,' she told him, and St John cocked his eyebrow at her.

'You could not have chosen a better time—' and with the eloquent spread of his hands indicated wind and current and the dangerous shore close under their bows.

'This will be the last opportunity,' she told him quickly. 'My brother and I will be leaving this ship immediately you drop anchor in Table Bay.'

The mocking grin slid slowly from his lips.

'If you are determined, then it seems that we have nothing more to say to each other.'

'I want you to know why.'

'I know why,' he said, 'but I doubt that you do.'

She stared at him, but he turned away to call a change of heading to the helmsman and then to the figure at the foot of the mainmast.

'Mr Tippoo, I'll have another reef on her, if you please.'

He came back to her side, but not looking at her, his head tilted back to watch the miniature figures of his crew on the mainyards high above them.

'Have you ever seen sixteen thousand acres of cotton with the pods ready for plucking?' he asked quietly. 'Have you ever seen the bales going down river on the barges to the mills?'

She did not answer, and he went on without waiting. 'I have seen both, Doctor Ballantyne, and no man dare tell me that the men who work my fields are treated like cattle.'

'You are a cotton-planter?'

'I am, and after this voyage I will have a sugar plantation on the island of Cuba – half my cargo to pay for the land and half of it to work the cane.'

'You are worse than I thought,' she whispered. 'I thought you were merely one of the devil's minions. Now I know you are the devil himself.'

'You are going into the interior.' St John looked down at her now. 'When you get there, if you ever do, you will see true human misery. You will see cruelties that no American slave-owner would dream of. You will see the slaughter of human beings by war and disease and wild beast that will baulk your belief in heaven. Beside this savagery, the barracoons and the slave quarters are an earthly paradise.'

'Do you dare suggest that by catching and chaining these poor creatures you do them favour?' Robyn demanded, aghast at his effrontery.

'Have you ever visited a Louisiana plantation, Doctor?' Then answering his own question, 'No, of course, you have not. I invite you to do so. Come down to Bannerfield as my guest one day and then compare the state of my slaves to the savage blacks you will see in Africa, or even to those damned souls that inhabit the slums and workhouses of your own lovely little green island.'

She remembered those raddled and hopeless human

creatures with whom she had worked in the mission hospital. She was speechless. Then suddenly his grin was wicked again. 'Think of it only as forced enlightenment of the heathen. I lead them out of the darkness into the ways of God and civilization – just as you are determined to do – but my methods are more effective.'

'You are incorrigible, sir.'

'No, ma'am. I am a sea-captain and a planter. I am also a trader in, and an owner of, slaves – and I will fight to the death to defend my right to be all of those things.'

'What *right* is that you speak of?' she demanded.

'The right of the cat over the mouse, of the strong over the weak, Doctor Ballantyne The natural law of existence.'

'Then I can only repeat, Captain St John, that I will leave this ship at the very first opportunity.'

'I am sorry that is your decision.' The hard fierce look in the yellow eyes softened a little. 'I wish it were otherwise.'

'I shall devote my life to fighting you and men like you.'

'And what a waste that will be of a lovely woman.' He shook his head regretfully. 'But then your resolve may give us reason to meet again – I must hope that is so.'

'One final word Captain St John – I shall never forgive last night.'

'And I, Doctor Ballantyne, will never forget it.'

Zouga Ballantyne checked his horse at the side of the road, just before it crossed the narrow neck between the crags of Table Mountain and Signal Hill, one of its satellite promontories.

He swung down out of the saddle to rest his mount, for it had been a hard pull up the steep slope from the town, and he tossed the reins to the Hottentot groom who accompanied him on the second horse. Zouga was sweating lightly and there was a residual pulse of dullness behind his eyes from the wine he had drunk the night before, the magnificent rich sweet wine of Constantia, one of the most highly prized vintages in the world, but capable of delivering as thick a head as any of the cheap and common grogs they sold in the waterfront bars.

In the five days since they had disembarked, the friendliness of the Cape Colony citizens had almost overwhelmed them. They had slept only the first night at a public inn in Buitengracht Street, then Zouga had called upon one of the Cape Colony's more prominent merchants, a Mr Cartwright. He had presented his letters of introduction from the directors of the Worshipful Company of London Merchants Trading into Africa, and Cartwright had immediately placed at their disposal the guest bungalow set in the gardens of his large and gracious home on the mountain slope above the old East India Company's gardens.

Every evening since then had been a gay whirl of dinners and dances. Had Robyn and Zouga not insisted otherwise, the days would have been filled with equivalent frivolity – picnics, sailing and fishing expeditions, riding in the forest, long leisurely lunches on the lawns under spreading oak trees that reminded him so vividly of England.

However, Zouga had avoided these diversions and had managed to accomplish much of the work of the expedition. Firstly there had been the supervising of the unloading of equipment from *Huron* – in itself a major undertaking as the crates had to be swayed up from the hold and lowered into lighters alongside before making the perilous return through the surf to land on the beach at Rogger Bay.

Then he had to arrange temporary warehousing for the cargo. Here again Mr Cartwright had been of assistance. Still Zouga found himself fiercely resenting his sister's insistence that had made all this heavy work necessary.

'Damn me, Sissy, even Papa used to travel in the company of Arab slave-traders when he had to. If this fellow St John is a trader, we would do well to learn all we can from him – his methods and sources of supply. No one could give us better information for our report to the society.'

None of his arguments had prevailed, and only when Robyn had threatened to write home to the Society's directors in London, and to follow that up with a frank talk to the editor of the *Cape Times*, had Zouga acceded, with the worst grace possible, to her demands.

Now he looked down longingly at *Huron*, lying well out from the beach, snubbing around on her anchor cable to point into the rumbustious south-easterly wind. Even under bare poles she seemed on the point of flight.

Zouga guessed that St John would be sailing within days, leaving them to await the next ship that might be bound for the Arab and Portuguese coasts.

Zouga had already presented his letter of introduction from the Foreign Secretary to the Admiral of the Cape Squadron of the Royal Navy, and been promised all consideration. Nevertheless, he spent many hours of each morning visiting the shipping agents and owners in the port in hope of an earlier passage.

'Damn the silly wench,' he muttered aloud, thinking bitterly of his sister and her foibles. 'She could cost us weeks, even months.'

Time was, of course, of the very essence. They had to be clear of the fever-ridden coast before the monsoon struck and risk of malaria became suicidal.

At that moment there was the crack of cannon shot from the slopes of the hill above him, and as he glanced up he saw the feather of gunsmoke drifting away from the lookout station on Signal Hill.

The gunshot was to alert the townsfolk that a ship was entering Table Bay, and Zouga shaded his eyes with his cap as the vessel came into view beyond the point of land. He was not a seaman but he recognized instantly the ugly silhouette and single smoke stack of the Royal Naval gunboat that had pursued *Huron* so doggedly. Was it really two weeks ago, he wondered, the days had passed so swiftly. *Black Joke*, the gunboat, had her boilers fired and a thin banner of dark smoke drifted away downwind as she rounded up into the bay, her yards training around as she pointed through the wind and she passed within half a mile of where *Huron* was already at anchor. The proximity of the two ships raised interesting possibilities of the feud between the two captains being revived, Zouga realized, but his immediate feeling was of intense disappointment. He had hoped that the vessel might have been a trader that could have offered the expedition further passage up the east coast – and he turned away abruptly, took the reins from the groom and swung up easily into the saddle.

'Which way?' he asked the servant, and the little yellow-skinned lad in Cartwright's plum-coloured livery indicated the left branch of the road that forked over the neck and dropped down across the dragon's-back of the Cape peninsula to the ocean shore on the far side.

It was another two hours' ride, the last twenty minutes along a rutted cart track, before they reached the sprawling thatched building hidden away in one of the ravines of the mountain slope behind a grove of milkwood trees. The slopes behind the building were thick with protea bushes in full flower and the long-tailed sugar-birds haggled noisily over the blazing blooms. To one side a waterfall smoked with spray as it fell from the sheer rock face and then formed a deep green pool on which a flotilla of ducks cruised.

The building had a dilapidated air to it, the walls needed white-washing and the thatch hung in untidy clumps from the eaves. Under the milkwood trees were scattered items of ancient equipment, a wagon with one wheel missing and the woodwork almost entirely eaten away by worm, a rusty hand forge in which a red hen was sitting upon a clutch of eggs, and mouldering saddlery and ropes hung from the branches of the trees.

As Zouga swung down from the saddle, a pack of half a

dozen dogs came storming out of the front porch, snarling and barking around Zouga's legs so that he lashed out at them with his riding whip and with his boots, changing the snarls to startled yelps and howls.

'Who the hell are you and what do you want?' A voice carried through the uproar, and Zouga took one more cut at a great shaggy Boerhound with a ridge of coarse hair fully erect between his shoulders, catching him fairly on the snout and forcing him to circle out of range, with fangs still bared and murderous growls rumbling up his throat.

Then he looked up at the man on the stoep of the building. He carried a double-barrelled shotgun in the crook of his arm, and both hammers were at full cock. He was so tall that he had to stoop beneath the angle of the roof, but he was thin as a blue-gum tree, as though the flesh and fat had been burned off his bones by ten thousand tropical suns.

'Do I have the honour of addressing Mr Thomas Harkness?' Zouga called over the clamour of the dog pack.

'I ask the questions here,' the lean giant bellowed back. His beard was as white as the thunderhead clouds of a summer's day on the highveld and it hung to his belt buckle. Hair of the same silver covered his head and flowed down to the collar of his leather jerkin.

His face and his arms were burned to the colour of plug tobacco, and were speckled by the raised blemishes like little moles and freckles, where years of the fierce African sun had destroyed the upper layers of his skin. The pupils of his eyes were black and bright as drops of fresh tar, but the whites were smoky yellow, the colour of the malarial fevers and the pestilences of Africa.

'What is your name, boy?' His voice was strong and deep. Without the beard he might have been fifty years of age, but Zouga knew with certainty that he was seventy-three. He carried his one shoulder higher than the other and the arm on that side hung at an awkward angle to the joint. Zouga knew that a lion had chewed through the shoulder and through the bone of the upper arm, before Harkness had been able to reach his hunting knife on his belt with the other hand and stab it between the forelegs, up into the heart. That had been forty years before and the injury had become the Harkness hallmark.

'Ballantyne, sir.' Zouga shouted to make himself heard above the dogs. 'Morris Zouga Ballantyne.'

The old man whistled once, a fluting double note that stilled the dogs and brought them back around his legs. He had not lowered the shotgun and a frown puckered his sharp features.

'Fuller Ballantyne's pup, is it?'

'That's right, sir.'

'By God, any son of Fuller Ballantyne's is good enough for a charge of my buckshot in the rump. Don't cock your butt when you get back on that horse, boy, for I'm a man who tempts very easily.'

'I've ridden a long way to see you, Mr Harkness.' Zouga smiled that frank and wining smile of his, standing his ground. 'I'm one of your greatest admirers. I've read everything that has ever been written about you and everything you have written yourself.'

'I doubt that,' Harkness growled, 'they burned most of mine. Too strong for their lily livers.' But the hostile glint in his eyes turned to a twinkle and he cocked his head as he studied the young man before him.

'I have no doubt that you're as ignorant and arrogant as your Daddy, but you've got a fairer turn of speech.' And he stared again at the toes of Zouga's boots, and let his gaze move up slowly.

'Priest,' he asked, 'like your Daddy?'

'No, sir, soldier.'

'Regiment?'

'13th Madras Foot.'

'Rank?'

'Major.'

Harkness' expression eased with each reply until his gaze once more locked with Zouga's.

'Teetotal? Like Your Daddy?'

'Perish the thought!' Zouga assured him vehemently and Harkness smiled for the first time as he let the muzzles of the shotgun droop until they pointed at the ground. He tugged at the long spikes of his beard for a moment, then reached a decision.

'Come.' He jerked his head and led the way into the house. There was one huge central room, the high ceiling of dried reed stems kept it cool and the narrow windows kept it gloomy. The floor was of peach-pip shells set into a plaster of mud and cow-dung and the walls were three-foot thick.

Zouga paused in the threshold and blinked with surprise at the collection of strange articles that covered the walls, were piled on every table and chair, and packed to the rafters in the dark corners.

There were books, thousands of books, cloth and leather-bound books, pamphlets and journals, atlases and encyclopaedias. There were weapons, assegai of Zulu, shield of Matabele, bow of Bushman with its quiver of poisoned arrows and, of course, guns – dozens of them in racks or merely propped against the walls. There were hunting trophies, the beautiful zigzag-striped hide of zebra, the dark bush of the lion's mane, the elegant curved horn of harris-buck, teeth of hippopotamus and warthog, and then the long, yellow arcs of ivory thicker than a woman's thigh and taller than a man's head. There were rocks, piles of rocks that glittered and sparkled, crystal rocks of purple and green, metallic nodules, native copper redder than gold, hairy strands of raw asbestos – all of it covered with a fine layer of dust and piled untidily wherever it had fallen.

The room smelled of skins and dogs and damp, of stale brandy and fresh turpentine, and there were stacks of new canvases already stretched in their wooden frames, while other canvases stood on their easels with the subjects sketched in charcoal outline, or partly blocked in with bright oil paint. On the walls were hung some finished pictures.

Zouga crossed to examine one of them while the old man blew into a pair of glass tumblers and polished them on his shirt-tail.

'What do you think of my lions?' he asked, as Zouga studied a huge canvas entitled 'Lion Hunt on the Gariep River. Feb. 1846'.

Zouga made an appreciative sound in the back of his throat. Zouga was himself a dauber and scribbler, but he considered that the meticulous reproduction of the subject was the painter's duty, while these paintings had a guileless, almost childlike joy in every primitive line. The colours also were gay and made no pretence to imitate nature, while the perspectives were wildly improbable. The mounted figure with the flowing beard in the background dominated the pride of lions in the foreground. Yet Zouga knew that these strange creations had remarkable value. Cartwright had paid ten guineas for a fanciful landscape. Zouga could only believe it was a fad amongst the colony's fashionable set.

'They say my lions look like English sheepdogs.' Harkness glowered at them. 'What do you think, Ballantyne?'

'Perhaps,' Zouga started, then saw the old man's expression change. 'But tremendously ferocious sheepdogs!' he added swiftly, and Harkness laughed out loud for the first time.

'By God, you'll do!' He shook his head as he half-filled the tumblers with the dark brown local brandy, the fearsome 'Cape Smoke', and brought one glass to Zouga.

'I like a man who speaks his mind. Rot all hypocrites.' He raised his own glass in a toast. 'Especially hypocritical preachers who don't give a damn for God, for truth or for their fellow men.'

Zouga fancied that he recognized the description, but raised his glass. 'Rot them!' he agreed, and managed to suppress a gasp as the liquor exploded in his throat and sizzled behind his eyes.

'Good,' he said hoarsely, and Harkness wiped his silver moustache, left and right, with his thumb before he demanded,

'Why have you come?'

'I want to find my father, and I think you may be able to tell me where to search.'

'Find him?' fulminated the old man. 'We should all be extremely grateful he is lost, and pray each day that he remains that way.'

'I understand how you feel, sir,' Zouga nodded. 'I read the book that was published after the Zambezi expedition.'

Harkness had accompanied Fuller Ballantyne on that ill-fated venture, acting as second-in-command, expedition manager and recording artist. He had been caught up in the squabbling and blame-fixing that had marred the enterprise from the beginning. Fuller Ballantyne had dismissed him, accusing him of theft of the expedition stores, trading on his own account, artistic incompetence, neglecting his duties to hunt for ivory, and total ignorance of the country-side and its trails, of the tribes and their customs and had included these accusations in his account of the expedition, implying that the blame for the expedition's failure could be laid on Thomas Harkness' uneven shoulders.

Now even mention of that book brought the colour to the sun-raddled face and made the white whiskers twitch.

'I crossed the Limpopo for the first time in the year that Fuller Ballantyne was born. I drew the map that he used to reach Lake Ngami.' Harkness stopped and made a dismissive gesture. 'I might as well try and reason with the baboons barking from the tops of the kopjes.' Then he peered more closely at Zouga.

'What do you know about Fuller? Since he sent you

home to the old country, how often have you seen him? How much time have you spent in his company?'

'He came home once.'

'How much time did he spend with you and your mother?'

'Some months – but he was always in Uncle William's study writing, or he was up at London, Oxford or Birmingham to lecture.'

'But you, nevertheless, conceived a burning filial love and duty for the sainted and celebrated father?'

Zouga shook his head. 'I hated him,' he said quietly. 'I could hardly bear the days until he went away again.'

Harkness tilted his head on one side, surprised, speechless for a moment, and Zouga drank the last few drops of liquor in the glass.

'I never told anybody that before.' He seemed puzzled himself. 'I hardly even admitted it to myself. I hated him for what he did to us, to me and my sister, but especially to my mother.'

Harkness took the empty tumbler from his fingers, refilled it and handed it back. He spoke quietly.

'I also will tell you something that I have never told another man. I met your mother at Kuruman, my God, so long ago. She was sixteen or seventeen and I was nearly forty. She was so pretty, so shy and yet so filled with a special quality of joy. I asked her to marry me. The only woman I ever asked.' Harkness stopped himself, turned away to his painting, and peered at it. 'Damned sheepdogs!' he snapped, and then without turning back to face Zouga, 'So why do you want to find your father? Why have you come out to Africa?'

'Two reasons,' Zouga told him. 'Both good. To make my own reputation and my own fortune.'

Harkness swivelled to face him. 'Damn me, but you can be direct.' There was a tinge of respect in his expression now. 'How do you plan to achieve those very desirable ends?'

Zouga explained swiftly – the newspaper sponsorship, that of the Society for the Extinction of the Slave Trade.

'You'll find much grist for your mill,' Harkness interjected. 'The coast is still rife with the trade, despite what you'll hear in London town.'

'I am also an agent for the Worshipful Company of London Merchants Trading into Africa, but I have my own goods to trade, and 5,000 cartridges for my Sharps rifle.'

Harkness wandered across the dim room and stopped before one of the gigantic elephant tusks propped against the far wall. It was so old and heavy that there was very little taper from root to tip, the point worn blunt and rounded. One third of its length was smooth and a clear lovely butter yellow colour, where it had been buried in the jaw of the beast, the rest of it was stained dark with vegetable juices and scarred from the battles and foragings of sixty years.

'This one weighs one hundred and sixty pounds – its value in London is six shillings a pound.' He slapped it with his open palm. 'There are still bulls like that out there, thousands of them. But take a tip from an old dog – forget your fancy Sharps, and use one of the ten-bore elephant guns. They throw a ball that weighs a quarter of a pound, and though they kick like the devil himself, they drive better than any of these new-fangled rifles.' There was a lightness in the worn old features, a sparkle in the dark eyes. 'Another tip – get in close. Forty paces at the outside, and go for the heart. Forget what you'll hear about the brain shot, go for the heart—' He broke off suddenly and waggled his head, grinning ruefully. 'By God, but it's enough to make a man want to be young again!'

He came back and studied Zouga directly, and a thought occurred to him with a suddenness that struck him like a physical blow, taking him so by surprise that he almost spoke it aloud.

'If Helen had given me a different answer, you could be my son.' But he held back the words, and asked instead:

'How can I help you, then?'

'You can tell me where I can start to search for Fuller Ballantyne.'

Harkness threw up his hands, palms uppermost. 'It's a vast land – so big that you could travel across it for a lifetime.'

'That's why I have come to you.'

Harkness went to the long table of yellow Cape deal that ran nearly the full length of the room and with the twisted arm swept a clear space amongst the books and papers and paint pots.

'Bring a chair,' he instructed, and when they sat facing each other across the cleared space, he recharged both their glasses and placed what remained in the bottle between them.

'Where did Fuller Ballantyne go?' Harkness asked, and took a silver tress from the thick beard and began to twist it around his forefinger. The finger was long and bony and covered with the thick ridges of ancient scar tissue where the recoil of overheated or overcharged firearms had driven the trigger guard to the bone.

'Where did Fuller Ballantyne go?' he repeated, but Zouga realized the question was rhetorical, and he said nothing.

'After the Zambezi expedition, his fortune was exhausted, his reputation all but destroyed – and to a man like Fuller Ballantyne that was unbearable. His entire life had been an endless hunt for glory. No risk, no sacrifice was too great, his own or others. He would steal and lie – aye, even kill for it.'

Zouga looked up sharply, challenging.

'Kill,' Harkness nodded. 'Anybody who stood in his way. I have seen him, but that is another tale. Now we want to know where he went.'

Harkness stretched out and selected a roll of parchment

from the cluttered table-top, checked it quickly and grunted with approval, as he spread it between them.

It was a map of Central Africa, east to west coast, south to the Limpopo, north to the lakes, drawn in India ink and the borders were illuminated by Harkness' characteristic figures and animals.

Instantly Zouga coveted it with his very soul. All that Harkness had accused his father of he felt in his own heart. He had to have this, even if it was necessary to steal – or, by God, kill. He had to have it.

The map was huge – at least five feet square – hand-drawn on the finest quality linen-backed paper. It was unique, the detail enormous; the notations were profuse but succinct, the observations first hand, the details precise, written in a tiny elegant script that needed a reading glass to be deciphered with ease.

'Here heavy concentrations of elephant herds during June to September.'

'Here I sampled gold reef in ancient workings at two ounces to the ton.'

'Here rich copper is worked by Gutus people.'

'Here slave convoys depart for the coast in June.'

There were literally hundreds of these notations, each in a neatly numbered box, that corresponded to the exact location on the map.

Harkness watched Zouga's face, with a sly half-smile on his face, and then handed him a reading-glass to continue his examination.

It took Zouga a few minutes to realize that the pink shaded areas indicated the 'fly corridors' of the high African plateau. The safe areas through which domestic animals could be moved to avoid the tsetse fly belts. The terrible Ngana disease which the fly carried could decimate the herds. Knowledge of these corridors had been gathered by the African tribes over hundreds of years, and here it was faithfully recorded by Thomas Harkness. The value of this knowledge was incalculable.

'Here Mzilikazi's border impis kill all travellers.'

'Here there is no water between May and October.'

'Here dangerous malarial vapours during October to December.'

The areas of greatest hazard were signposted, while the known routes to the interior were clearly marked, though there were few enough of these.

The cities of the African kings were marked, as was the location of their military kraals, the areas of influence of each were defined and the names of the subservient chiefs noted.

'Here concession to hunt elephant must be obtained from Chief Mafa. He is treacherous.'

Harkness watched the young man eagerly poring over the priceless document. His expression was almost fond, and he nodded his head once as a memory passed like a shadow behind his eyes. He spoke at last.

'Your father would be trying to restore his reputation at a single stroke,' Harkness mused. 'He would have to feed that monstrous ego. There are two areas that come immediately to mind. Here!'

He placed his open hand across an enormous area to the north and west of the defined shape of Lake Marawi. In this area the copious and authoritative notations were replaced by meagre, hesitant observations obtained from hearsay or native legend, and by speculation followed by a question mark.

'Sheikh Assab of the Omani Arabs reports River Lualaba runs north and west. Possible flow into Lake Tanganyika.' The dotted outlines of rivers, instead of crisp detail. 'Pemba, the Chief of the Marakan, reports huge lake shaped like butterfly twenty-five days march from Khoto Khota. Called Lomani. Possible source of Luapula and of Herodotus fountain.' The lake was sketched in. 'Question. Is Lake Tanganyika connected to Lake Albert? Question. Is Lake Tanganyika connected to Lake Lomani? If so, Lomani is ultimate source of Nile river?'

Harkness touched the two question marks with his gnarled and bony finger.

'Here,' said Harkness. 'The big question marks. The Nile river. That would attract Fuller. He spoke of it often.' Harkness chuckled. 'Always with the same introductory words, "*Of course, the fame matters not at all to me*".' The old man shook his silver head. 'It mattered not less than the air he breathed. Yes, the source of the Nile river and the fame that it would bring its discoverer – that would fascinate him.'

Harkness stared for a long time at the empty spaces, dreaming perhaps, visions awakening behind the bright black eyes. He aroused himself at last, shaking his shaggy head as if to clear it.

'There would be only one other feat that would attract as much attention, would be greeted with as much acclaim.' Harkness ran his spread hand southwards down the parchment to cover another vast void in the web of mountains and rivers. 'Here,' he said softly. 'The forbidden kingdom of the Monomatapas.'

The name itself had an eerie quality. *Monomatapa*. The sound of it raised the fine hair on the back of Zouga's neck.

'You have heard of it?' Harkness asked.

'Yes,' Zouga nodded. 'They say it is the Ophir of the Bible, where Sheba mined her gold. Have you travelled there?'

Harkness shook his head. 'Twice I started out,' he shrugged. 'No white man has travelled there. Even Mzilikazi's impis have not raided that far east. The Portuguese made one attempt to reach the Emperor Monomatapa. That was in 1569. The party was wiped out, and there were no survivors.' Harkness made a sound of disgust. 'As you could expect of the Ports, they abandoned any further attempt to reach Monomatapa. For the 200 years since then they have been content to sit in their seraglios at Tete and Quelimane, breeding half-castes, and picking up the slaves and ivory that filter down out of the interior.'

'But still there are the legends of the Monomatapa. I heard them from my father. Gold and great walled cities.'

Harkness stood up from the table with the grace of a man half his age and crossed to an iron-bound chest against the wall behind his chair. The chest was not locked but the lid required both the old man's skinny arms to lift it.

He came back with a draw-string bag made of softly tanned leather. It was obviously weighty for he carried it in both hands. He pulled the mouth open, and upended the contents on to the linen map.

There was no mistaking the lovely yellow metal, it had the deep glowing lustre which has bewitched mankind for thousands of years. Zouga could not resist the urge to reach out and touch it. It had a marvellous soapy feeling against his fingertips. The precious metal had been beaten into heavy round beads, each the size of the top joint of Zouga's little finger and the beads had been strung on to animal sinew to form a necklace.

'Fifty-eight ounces,' Harkness told him, 'and the metal is of unusual purity, I have had it assayed.'

The old man lifted the necklace over his own head and let it lie against the snowy fall of his beard. It was only then that Zouga realized that there was a pendant on the string of golden beads.

It was in the shape of a bird, a stylized falcon-like shape with folded wings. It was seated upon a rounded plinth that was decorated with a triangular design, like a row of sharks' teeth. The figure was the size of a man's thumb. The gold metal was polished by the touch of human skin over the ages so that some of the detail had been lost. The eyes of the bird were glassy green chips.

'It was a gift from Mzilikazi. He has no use for gold, nor

for emeralds – yes, the stones are emeralds,' Harkness nodded. 'One of Mzilikazi's warriors killed an old woman in the Burnt Land. They found the leather pouch on her body.'

'Where is the Burnt Land?' Zouga asked.

'I'm sorry.' Harkness fiddled with the little golden bird. 'I should have explained. King Mzilikazi's impis have laid waste to the land along his borders, in some places to a depth of a hundred miles and more. They have killed all who lived there and they maintain it as a buffer strip against any hostile force. The Boer commandos from the south particularly, but from any other hostile invader also. Mzilikazi calls it the Burnt Land, and it was here, to the east of his kingdom, that his border guards killed this solitary old woman. They described her as a very strange old woman, not of any known tribe, speaking a language they did not understand.'

Harkness lifted the necklace from his neck and dropped it carelessly back into the bag, and Zouga felt bereft. He would have liked to feel the full weight and the texture of the metal in his hands for a little longer. Harkness went on quietly:

'Of course you have heard the talk of gold and walled cities, like everybody else. But that is the closest I have ever come to corroboration.'

'Did my father know about the necklace?' Zouga asked, and Harkness nodded. 'Fuller wanted to purchase it, he offered me almost twice its gold value.'

They were both silent for a long while, each brooding on his own thoughts until Zouga stirred.

'How would a man like my father try to reach the Monomatapa?'

'Not from the south nor from the west. Mzilikazi, the Matabele king, will let no man pass through the Burnt Land. I feel that Mzilikazi has some deep superstition attached to the land beyond his eastern border. He does not venture there himself, nor does he allow others to do so.'

Harkness shook his head. 'No, Fuller would have to try from the east, from the Portuguese coast, from one of their settlements.' Harkness began to trace out the possible approach marches on the linen chart. 'Here there are high mountains. I have seen them at a distance and they seemed a formidable barrier.' Outside, night had already fallen, and Harkness interrupted himself to order Zouga, 'Tell your groom to off-saddle the horses and take them to the stables. It is too late to return. You will have to stay overnight.'

When Zouga returned a Malay servant had drawn the curtains, lit the lanterns and laid a meal of yellow rice and chicken cooked in a fiery curry – and Harkness had opened another bottle of the Cape brandy. He went on talking as though there had been no interruption. They ate the meal and pushed the enamelled tin plates away to return to the map, and the hours passed unnoticed by either of them.

In the intimate lantern light, the sense of drama and excitement that gripped them both was heightened by the brandy they drank. Once in a while Harkness would rise to fetch some souvenir of his travels to reinforce a point, a sample of quartz rock in which the seams of native gold were clearly visible in the lamplight.

'If there is visible gold, it's rich,' Harkness told Zouga.

'Why did you never mine the reefs you found?'

'I could never stay long enough in one place,' Harkness grinned ruefully. 'There was always another river to cross, a range of mountain or a lake that I had to reach – or I was following a herd of elephant. There was never time to sink a shaft, or build a house, or raise a herd.'

When the dawn was rising, peeping into the huge gloomy room around the curtains, Zouga exclaimed suddenly, 'Come with me. Come with me to find Monomatapa!' And Harkness laughed.

'I thought it was your father you were intent on finding.'

'You know better!' Zouga laughed with him. Somehow he felt at home with the old man, as if he had known him all his life. 'But can you imagine my father's face when you come to rescue him?'

'It would be worth it,' Harkness admitted, and then the laughter faded, and gave way to an expression of such deep regret, of such consuming sorrow that Zouga felt a compulsion to reach out across the table and touch the misshapen shoulder.

Harkness pulled away from his touch. He had lived alone too long. He would never again be able to take comfort from a fellow man.

'Come with me,' Zouga repeated, letting his hand drop to the table top between them.

'I have made my last journey into the interior,' Harkness said tonelessly. 'Now all I have are my paint pots and my memories.'

He lifted his eyes to the ranks of framed canvas with their brilliant joyous images.

'You are still strong, vital,' Zouga insisted. 'Your mind is so clear.'

'Enough!' Harkness' voice was harsh, bitter. 'I am tired now. You must go. Now, immediately.'

Zouga felt his anger rise hotly to his cheeks at the abrupt rejection, this sudden change of mood, and he stood quickly. For a few seconds he stood looking down at the old man.

'Go!' said Harkness again.

Zouga nodded abruptly. 'Very well.' His eyes slid down to the map. He knew he must have it at any price, though he sensed that there was no price that Harkness would accept. He must plan and scheme for it, but he would have it.

He turned and strode to the front door, and the dogs that had slept around their feet rose and followed him.

'Garniet!' Zouga shouted angrily. 'Bring the horses.' And he stood in the doorway rocking impatiently on his toes and heels, hands clasped behind his back, his shoulders stiff, not looking around at the thin stooped figure who still sat at the table in the lamp-light.

The groom brought the horses at last, and still without turning Zouga called roughly, 'Good day to you, Mr Harkness.'

The reply was in a frail old man's quaver that he hardly recognized.

'Come again. We have more to discuss. Come back – in two days.'

The stiffness went out of Zouga's stance. He started to turn back, but the old man waved him away with a brusque gesture and Zouga stamped down the front steps, vaulted into the saddle and whipped his mount into a gallop along the narrow rutted track.

Harkness sat at the table until long after the hoof beats had faded. Strange that the pain had receded to the very back of his consciousness during the hours that he had sat with the youngster. He had felt young and strong, as though he had suckled upon the vigour and the youth of the other.

Then it had come back with a savage rush at Zouga Ballantyne's invitation, almost as if to remind him that his life was no longer his, that it was already forfeit to the hyena that lived deep in his belly, each day growing stronger, bigger, as it fed upon his vitals. When he closed his eyes he could imagine it, the way that he had seen it so often in the light of a thousand camp fires – up there in the wonderful land that he would never visit again. The thing within him had the same furtive slinking presence and he could taste the fetid breath of it in his throat. Now he gasped as the full strength of the pain returned, as the beast buried its fangs deeper into his gut.

He knocked over the chair in his haste to reach the precious bottle in the back of the cabinet and he gulped a mouthful of the clear pungent liquid without measuring it into the spoon. It was too much, he knew that, but each day he needed more to keep the hyena at bay, and each day the relief took longer to come.

He clung to the corner of the cabinet and waited for it. 'Please,' he whispered, 'please let it end soon.'

There were half a dozen messages and invitations awaiting Zouga on his return to the Cartwright estate that morning, but the one which gave him most excitement was on official Admiralty paper, a polite request to call upon the Hon. Ernest Kemp, Rear-Admiral of the Blue, Officer Commanding the Cape Squadron.

Zouga shaved and changed his clothing, selecting his best jacket for the occasion although it was a long dusty ride to Admiralty House. Despite missing a night's sleep, he felt vital and alert.

The Admiral's Secretary kept him waiting only a few minutes before showing him through, and Admiral Kemp came around from behind his desk to greet Zouga amiably, for the young man came highly recommended and Fuller Ballantyne's name still commanded respect in Africa.

'I have some news which I hope will please you, Major Ballantyne. But first, a glass of Madeira?'

Zouga had to curb his impatience while the Admiral poured the syrupy wine. The Admiral's study was decorated richly, with velvet furnishings and a fashionable profusion of ornaments, small statues, bric-a-brac, stuffed birds in glass showcases, family portraits in ornate frames and pretty ceramics, potted plants and the kind of paintings which Zouga admired.

The Admiral was tall but stooped, as though to accommodate his long frame to the limited headroom between the decks of one of Her Majesty's ships. He seemed old for the responsible appointment he held, guarding the Empire's lifeline to India and the East, but the ageing may have been caused by ill-health rather than years. There were dark-toned pouches of skin below his eyes and other marks of sickness carved around his mouth and evident in the distended blue veins on the back of his hands as he handed a glass of Madeira to Zouga

'Your good health, Major Ballantyne.' And then after he had tasted his own wine, 'I think I have a berth for you. A ship of my squadron anchored in Table Bay yesterday, and as soon as she has replenished her coal bunkers and revictualled I shall detach her for independent duty in the Mozambique channel.'

Zouga knew from his meetings with the directors of the Society for the Extinction of the Slave Trade, that one part of the Admiral's standing orders read:

'You are requested and required to dispose the ships of your squadron in such manner as most expediently to prevent vessels of whatever Christian nation from engaging in the slave trade on the coast of the African continent south of the equator.'

Clearly Kemp intended a sweep of the eastern seaboard with elements of his squadron, and Zouga felt awakening delight as the Admiral went on genially, 'It will not need much of a diversion for my ship to call at Quelimane, and to land you and your party.'

'I cannot thank you sufficiently, Admiral.' His pleasure was transparent, and Admiral Kemp smiled in sympathy. He had put himself out more than his usual wont, for the youngster was attractive and likeable, deserving of encouragement, but now there were other matters awaiting his attention and as he pulled out his gold hunter and consulted it pointedly he went on,

'You should be ready to sail in five days' time.' He returned the watch to the fob of his uniform coat. 'I hope we will see you on Friday? My Secretary did send you an invitation, did he not? Your sister will be with you, I hope.'

'Indeed, sir.' Zouga stood in obedience to his dismissal. 'And my sister and I are honoured.'

In fact Robyn had said, 'I do not waste my evenings, Zouga, and I have no intention of enduring the company of a fleet of tipsy sailors nor of suffering the wagging tongues of their wives.'

The Cape wives were agog with the presence in their midst of the notorious Robyn Ballantyne who had impersonated a man and invaded, successfully, an exclusive masculine preserve. Half of them were deliciously scandalized, and the rest were awed and admiring. However, Zouga was certain that she would pay this price for their passage to Quelimane.

'Very well then.' Admiral Kemp nodded. 'Thank you for calling on me.' And then, as Zouga started for the door, 'Oh, by the way, Ballantyne. The ship is the *Black Joke*, Captain Codrington commanding. My Secretary will give you a letter for him, and I suggest you call upon him to introduce yourself and to learn the date of sailing.'

The name came as a shock, and Zouga checked his stride as he thought quickly of the complication which the choice of ship might bring.

Zouga was sensitive to any threat to the expedition, and Codrington had struck him as being a hot-headed, almost fanatical character. He could not afford any slur to his leadership, and Codrington had seen him sailing in company with a suspected slaver. He could not be sure what Codrington would do.

It was a delicate decision: accept the berth and risk Codrington's denunciation, or refuse the offer of passage and perhaps wait for months in Cape Town before another vessel offered them another.

If they were delayed that long, it would mean missing the cooler and dryer period between the monsoons, they would have to cross the pestilential and fever-ridden coastal lowlands in the most dangerous season.

Zouga made his decision. 'Thank you, Admiral Kemp. I will call on Captain Codrington as soon as possible.'

Thomas Harkness had asked Zouga to return on the second day, and the map was more important even than swift passage to Quelimane.

Zouga sent Garniet, the Cartwrights' groom, down to the beach with a sealed letter addressed to Captain Codrington and with instructions to take one of the water boats out to *Black Joke* and deliver it personally to Codrington. It was a warning, couched in the most polite terms, that Zouga and Robyn would call on the Captain the following morning. Zouga had become aware that his sister had an effect on men quite out of proportion to her physical appearance – even Admiral Kemp had asked for her personally – and he had no compunction in using her to take the edge off Codrington's temper. He would have to warn her to exert her charm, but now there was more important business.

He had mounted on Cartwright's big bay gelding and ridden halfway down the gravelled drive between the oaks, when a thought struck him and he swung the horse's head and cantered back to the guest bungalow again. The Naval Colt revolver was on the top layer of his chest, already fully loaded and with caps on the nipples. He carried it under the tail of his coat while he went back out to the tethered gelding, and then slipped the revolver surreptitiously into the saddlebag as he swung up into the saddle.

He knew he had to have the Harkness map at any price, but he deliberately refused to think what that price might be.

He pushed his mount hard up the steep road to the neck between the peaks, and gave him only a few minutes to blow before starting down the far slope.

The air of dilapidation which hung over the thatched building in the milkwood grove seemed to have deepened. It seemed totally deserted, silent and desolate. He dismounted and threw his reins over a milkwood branch and stooped to ease the girth. Then he quietly unfastened the buckle of the saddlebag and slipped the Colt into his waistband and pulled his coat over it.

As he started towards the stoep, the big ridge-backed Boerhound rose from where it had lain in the shadows and came to meet him. In contrast to its previous ferocity, the animal was subdued, its tail and ears drooping and it whined softly when it recognized Zouga.

He went up on to the stoep and hammered on the front door with his fist, and heard the blows reverberate through the room beyond. Beside him, the Boerhound cocked its head and watched him expectantly, but silence settled again over the old building.

Twice more Zouga beat upon the door, before he tried the handle. It was locked. He rattled the brass lock and put his shoulder to the door; but it was heavy teak in a solid frame. Zouga jumped down off the stoep and circled the house, squinting his eyes at the fierce reflected sunlight from the white-washed walls. The windows were shuttered.

Beyond the farmyard stood the old slave quarters, now used by Harkness' servant, and Zouga called loudly for him, but his room was deserted and the ashes cold in the cooking place. Zouga went back to the main house and stood by the locked kitchen door.

He knew that he should go back to his horse and ride away, but he needed the map, even if just for long enough to make a copy. Harkness was not here, and in three days, perhaps less, he would be sailing out of Table Bay.

There was a pile of broken rusted garden tools in the corner of the stoep. Zouga selected a hand scythe, and carefully probed the metal point of the blade into the crack between jamb and door. The lock was old and worn, the tongue slipped back easily under the blade and he jerked the door open with his free hand.

It was still not too late. He paused in the doorway for many seconds, and then he took a deep breath and stepped quietly into the gloom of the interior.

There was a long passage leading past closed doors towards the front room. Zouga went down it, opening doors quietly as he passed. In one room was a huge fourposter bedstead with the curtains opened and the bed clothes in disorder.

Quickly Zouga passed on to the main room. It was in semi-darkness and he stopped to let his eyes adjust, and immediately was aware of a low sound. The hive-murmur of insects seemed to fill the high room. It was a disturbing, almost menacing sound, and Zouga felt the skin prickle on his forearms.

'Mr Harkness!' he called hoarsely, and the hum rose to a loud buzzing. Something alighted on his cheek and crawled across his skin. He struck it away with a shudder of revulsion and stumbled across to the nearest window. His fingers were clumsy on the fastening of the shutters. A shaft of white sunlight burned into the room as the shutter swung open.

Thomas Harkness sat in one of the carved wingback chairs across the cluttered table, and stared at Zouga impassively.

The flies crawled over him, big metallic blue and green flies that glittered in the sunlight. They swarmed with evident glee upon the deep dark wound in the centre of his chest. The snowy beard was black with clotted blood, and blood had formed a congealed pool beneath his chair.

Zouga was rooted by the shock for many seconds, and then reluctantly he took a step forward. The old man had propped one of his big-bored elephant guns against the table leg, reversing the weapon so the muzzle pressed into his own chest and his hands were still locked around the barrel.

'What did you do that for?' Zouga demanded stupidly, speaking aloud, and Harkness stared back at him.

Harkness had removed the boot from his right foot, and depressed the trigger with his bare toe. The massive impact of the heavy lead ball had driven the chair and the man in it back against the wall, but he had retained his death grip on the barrel.

'That was a stupid thing to do.' Zouga took a cheroot from his case and lit it with a Swan Vesta. The smell of death was in the room, coating the back of his throat and the roof of his mouth. Zouga drew deeply on the tobacco smoke.

There was no reason at all to feel grief. He had known the old man for a single day and night. He had come here for one reason only – to get the map the best way he was able. It was ridiculous now to have the deep ache of sorrow turning his legs leaden and stinging the backs of his eyes. Was he mourning the passing of an era perhaps, rather than the man himself? Harkness and the legends of Africa were interwoven. The man had been history itself.

Slowly Zouga approached the figure in the chair, and then reaching out drew his palm slowly down over the old face, ruined by the elements and by pain, closing the lids down over the staring black eyes.

The old man looked more peaceful that way.

Zouga hooked one leg over the corner of the cluttered table, and smoked the cheroot slowly, in almost companionable silence with Thomas Harkness. Then he dropped the stub in the big copper spittoon beside the chair and went through to the bedroom.

He took one of the blankets off the bed and brought it back.

He brushed the flies away into an angry buzzing circle, and threw the blanket over the seated figure. As he drew it up over the head, he murmured softly, 'Get in close, old man, and go for the heart,' the advice that Harkness had given him as a farewell. Then he turned away briskly to the laden table, and began shuffling through the jumbled pieces of canvas and paper, and slowly his impatience turned to alarm – and then to panic as he hunted through pile after pile without finding the map.

He was panting when at last he straightened up and glared at the blanket-covered figure.

'You knew I was coming for it, didn't you!'

He left the table and went to the chest, lifted the lid against groaning hinges and the leather bag with its golden contents was gone also. He ransacked the chest down to its floorboards, but it was not there. Then he started to search in earnest, going carefully through any possible hiding-place in the crowded room. An hour later he went back and perched on the edge of the table once more.

'Damn you, for a cunning old bastard,' he said quietly. He took one more slow look about the room, making certain that he overlooked nothing. The painting of the lion hunt was no longer on the easel, he noticed.

Suddenly the humour of the situation struck him, and his scowl lightened, he began to chuckle ruefully to himself.

'You had the last joke on the Ballantynes, didn't you? By God, but you always did things your way, Tom Harkness, I'll grant you that.'

He stood up slowly, and placed his hand on the blanket-covered shoulder. 'You win, old man. Take your secrets with you then.' He could feel the twisted old bones through the cloth and he shook him gently and then he went out quickly to his horse for there was much to do. It took him the rest of the day to cross the neck again and reach the magistrate's court, then to get back with the coroner and his assistants.

They buried Thomas Harkness that evening, wrapped in the blanket, under the milkwood grove, for the heat was oppressive in the valley and they could not wait for a coffin to be carted out from the city.

Zouga left the coroner to take charge of the estate, to list the equipment and livestock in the yard and put his seals on the doors of the old house until the contents could be taken in.

Zouga rode home in the golden Cape dusk, his boots

dusty and his shirt sticky with sweat. He was exhausted from the day's exertions, and low in spirits, still oppressed by grief for the old man, and angry with him for the last trick he had played.

The groom took his horse in front of the bungalow.

'Did you deliver the letter to Captain Codrington?' Zouga demanded, and hardly waited for the reply as he went up into the house. He needed a drink now, and while he poured whisky into a cut crystal glass, his sister came into the room, and reached up casually to kiss his cheek, wrinkling her nose at the tickle of his whiskers and the smell of his sweat.

'You had best change. We are dining with the Cartwrights tonight,' Robyn told him. 'I could not avoid it.' And then as an afterthought, 'Oh, Zouga, a coloured servant delivered something for you this morning. Just after you had left. I had it put in the study.'

'Who is it from?'

Robyn shrugged. 'The servant spoke only kitchen Dutch and he seemed terrified. He fled before I could find someone to question him.'

With the whisky glass in his hand Zouga crossed to the door of the study, and stopped there abruptly. His expression changed, and he strode through the doorway.

Minutes later Robyn heard his shout of triumphant laughter, and curiously she crossed to the open door. Zouga stood beside the heavy carved stinkwood desk.

On the desk-top lay a draw-string bag of tanned and stained leather from which spilled a heavy necklace of gleaming gold; beside the bag was spread a magnificently illustrated map on a backing of linen parchment, and Zouga stood with his back to her. He held at arm's length a flamboyant picture in oils in a large frame, a figure on horseback with a band of ferocious wild animals in the foreground, and as she watched, Zouga reversed the picture. There was a message freshly carved into the wood of the frame.

For Zouga Ballantyne. May you find the road to all your Monomatapas – would only that I could have gone with you.

Tom Harkness.

Zouga was laughing still, but there was a strange quality to the laughter and when he turned towards her she realized with a shock that her brother's eyes were bright with tears.

Zouga brushed the crumbs from his lips with the damask table napkin and chuckled as he picked up the sheet of newsprint and shook it open again at the second page.

'Damn me, Sissy, I should have known better than to leave you alone.' He read further and laughed outright. 'Did you really say that to him? Did you really?'

'I cannot remember my exact words,' Robyn told him primly, 'you must remember it was in the heat of battle.'

They sat on the terrace of the bungalow under the pergola of vines, through which the early sun flicked golden coins of light upon the breakfast table.

The previous day the editor of the *Cape Times*, with a speculator's eye to making a profit on Dr Robyn Ballantyne's notoriety, had invited her on a tour of the military hospital at Observatory, and in innocence, Robyn had believed that the visit was at the invitation of the Colony's administration and she had welcomed the opportunity to widen her professional experience.

The visit had succeeded beyond the editor's most extravagant expectations, for the surgeon-general of the Colony had scheduled a tour for the same day and he had walked into the hospital's main operating room, followed by his staff, at the moment that Robyn was expressing herself on the subject of sponges to the hospital matron.

The surgeon's sponges were kept in pails of water, clean water from the galvanized rainwater tanks at the rear of the building. The pails were under the operating table, where the surgeon could reach them readily, and after swabbing away blood and pus and other matter the sponge was dropped into a collection tray, later to be washed out and returned to the original pail of fresh water.

'I assure you, doctor, that my nurses wash the swabs out most thoroughly.' The matron was a formidable figure with the flattened features of a bulldog bitch and the same aggressive thrust to her jaw. She stooped, plunged her hand into the pail, and selected one of the sponges and proffered it to Robyn.

'You can see for yourself how soft and white they are.'

'Just like the soft white germs that swarm in them.' Robyn was angry, with red spots of colour in her cheeks. 'Have none of you here ever heard of Joseph Lister?'

The surgeon-general answered her question from the doorway.

'The answer to that question, Doctor Ballantyne, is NO we have never heard of that person, whoever he may be. We do not have time to concern ourselves with the opinions of every crank or, for that matter, with male impersonators.'

The surgeon-general had a very good idea of the identity of the young woman before him. He had followed the gossip which was the Colony's main recreation, and he did not approve of Robyn.

On the other hand, Robyn had no idea as to the identity of the elderly gentleman with the bushy grey whiskers and beetling brows, though by the dried blood stains on the front of his frock coat she guessed that he was a surgeon of the old school, one who operated in his street clothes and let the stains advertise his profession. Here was a much more worthy adversary than the hospital matron, and she rounded on him with the battle light bright in her eyes.

111

'Then, sir, I am amazed that you admit so readily your ignorance and your bigotry.'

The surgeon-general spluttered for breath and a ready answer.

'By God, madam, you do not truly expect me to look for dangerous poisons in each speck of dust, in each drop of water, on my own hands even.' He held them up for her inspection, shaking them in Robyn's face. There were dark rinds of dried blood under the nails, for he had operated that morning. He pushed his face close to hers, and she drew back a little as his spittle flew angrily.

'Yes, sir,' she told him loudly. 'Look for them there, and on each breath you exhale, on those filthy clothes.'

The editor scribbled delightedly in his shorthand note-book, as the exchanges became more violent, more loaded with personal insult. He had not bargained for anything so spectacular, but the climax came when Robyn had goaded her adversary until he used an oath as potent as his rage.

'Your choice of words is as foul as these lowly little white sponges of yours,' she told him, and let him have the sponge full in the face, hurling it with all her strength so that water flew and dripped from his whiskers on to the front of his frock coat as Robyn marched from the operating room.

'You hit him?' Zouga lowered the newspaper, and stared across the table at his sister. 'Really, Sissy, sometimes you are no lady.'

'True,' Robyn agreed unrepentantly. 'But that is not the first time you have made that observation. Besides I had no idea that he was the surgeon-general.'

Zouga shook his head in mock disapproval and read to her. 'His considered opinion of you, as expressed to the editor, is that you are a fledgling doctor of dubious qualification very recently obtained from an obscure school of medicine, by even more dubious means.'

'Oh, rich!' Robyn clapped her hands. 'He's a better orator than a surgeon.'

'He goes on to say that he is considering going to law to obtain redress.'

'For assault with a sponge.' Robyn laughed lightly as she stood up from the breakfast table. 'A fig for him, but we must hurry if we are to keep our appointment with Captain Codrington.'

Her mood was still gay as she stood beside Zouga in the stern of the water lighter when they came alongside the steel side of the gunboat.

The south-east wind had raked the surface of Table Bay into a cottonfield of white caps, and had spread a thick white table-cloth of cloud upon the flat-topped mountain. The people of the Colony called this wind 'The Cape Doctor', for without it the summers would have been oppressive and enervating. However, it provided a constant hazard to shipping and the bottom of the bay was littered with wrecks. *Black Joke* had two men on her anchor watch as she lunged and fretted against her cable.

As the lighter came alongside, the thick canvas hoses were passed down and a dozen men on the pumps began to transfer the cargo into the gunboat's boiler room tanks, before any attempt was made to take visitors aboard.

As Robyn came up through the entry port to the main deck she looked immediately to the quarterdeck. Codrington was in shirt sleeves, he was a head taller than the group of warrant officers around him, and his sun-bleached blond hair shone in the sunlight like a beacon.

The group was giving its attention to the coal lighter which was secured against the port side of the ship.

'Have the hands secure a tarpaulin over the buckets.' Codrington shouted down to his boatswain in the lighter. 'Else you'll have us looking like a party of chimneysweeps.'

The deck was alive with the purposeful pandemonium of revictualling, bunkering and watering, and with Zouga beside her, Robyn picked her way through the litter. Codrington turned away from the rail and saw them.

He seemed younger than Robyn remembered, for his expression was relaxed and his manner easy. He had an almost boyish air when contrasted to the grizzled and weatherbeaten sailors about him, but the illusion was dispelled the moment he recognized his visitors. Suddenly his features were stern and the line of his mouth altered, the eyes chilled to the hardness of pale sapphires.

'Captain Codrington,' Zouga greeted him with his most studied charm. 'I am Major Ballantyne.'

'We have met before, sir,' Codrington acknowledged, making no effort to return the smile.

Zouga went on unruffled, 'May I present my sister, Doctor Ballantyne?'

Codrington glanced back at Robyn. 'Your servant, ma'am.' It was more a nod than a bow. 'I have read something of your further exploits since our last meeting in this morning's news-sheet.' For a moment the stern expression cracked, and there was a mischievous spark in the blue eyes. 'You have strong views, ma'am, and an even stronger right hand.'

Then he turned back to Zouga. 'I have orders from Admiral Kemp to convey you and your party to Quelimane. No doubt you will find our company dull, after your previous travelling companions.' Deliberately Codrington turned and looked across a half mile of wind-creamed green water to where *Huron* still lay at her anchor, and for the first time Zouga fidgeted uncomfortably as he followed the direction of the Captain's eyes. Codrington went on. 'Be that as it may, I would be grateful if you could present yourselves aboard this ship before noon on the day after tomorrow when I expect a fair tide to leave the bay. Now you must excuse me. I must attend to the management of my ship.' With a nod, not offering to shake hands or make any other civility, Codrington turned away to his waiting warrant officers, and Zouga's charm deserted him. His face darkened and seemed to swell with anger at the abrupt dismissal.

114

'The fellow has a damnable cheek,' he growled fiercely to Robyn. For a moment he hesitated and then with a curt, 'Come, let us leave,' he turned, crossed the deck, and clambered down into the water lighter, but Robyn made no move.

She waited quietly until Codrington had finished the discussion with his boatswain, and looked up again, feigning surprise to see her still there.

'Captain Codrington, we left *Huron* on my insistence. That is why we are now seeking other passage.' She spoke in a low husky voice, but her manner was so intense that his expression wavered.

'You were correct. That ship is a slaver and St John is a slave-master. I proved it.'

'How?' he demanded, his manner altering instantly.

'I cannot speak now. My brother—' She glanced back at the entry port, expecting her brother to reappear at any moment. He had given her strict instructions as to how he expected her to act towards Codrington.

'I will be at the landing place at Rogger Bay this afternoon,' she went on quickly.

'What time?'

'Three o'clock,' she said, and turned away, lifted her skirts above her ankles and hurried to the ship's side.

Admiral Kemp sat impatiently in the immense carved abbot's chair which his junior officers referred to as the 'throne'. The size of it emphasized the thinness of the old man's body. It seemed his shoulders were too narrow to support the mass of gold lace which decorated his blue uniform. He clasped the arms of the chair to keep from fidgeting, for this young officer always made him uneasy.

Clinton Codrington leaned forward towards him and spoke quickly, persuasively, using the finely shaped hands to emphasize each point. The Admiral found this much energy and enthusiasm wearying. He preferred men with less mercurial temperament, who could be relied on to carry out orders to the letter without introducing startling improvisations.

Officers with a reputation for brilliance he viewed with deep suspicion. He had never had that reputation as a young man, in fact his nickname had been 'Slogger' Kemp, and he believed that the word brilliance was a pseudonym for instability.

The nature of duty on this station made it necessary for young men like Codrington to be detached for months at a time on independent service, instead of being kept with the battle fleet under the strict eye of a senior officer, ready with a signal of rebuke to check any hot-headedness.

Kemp had an uneasy conviction that he was going to be seriously embarrassed by this particular officer before his appointment of Commander to the Cape Squadron terminated, and allowed him to collect his knighthood and retire to the peace and beloved seclusion of his Surrey home. That his future plans had not already been prejudiced by young Codrington was only a matter of the utmost good fortune, and Kemp had difficulty keeping his expression neutral when he remembered the Calabash affair.

Codrington had run down on the slave barracoons at Calabash on a clear June morning so that the five Argentinian slave ships had spotted his topsails while he was thirty miles out, and had immediately begun frantically re-landing their cargoes of slaves on the beach.

By the time *Black Joke* reached them, the five captains were grinning smugly, their holds empty, and nearly two thousand miserable slaves in clear view squatting in long lines on the shore. To add to the slavers' complacency they were a good twenty nautical miles south of the equator, and therefore at that time beyond the jurisdiction of the Royal Navy. The barracoons had been sited at Calabash to take full advantage of this provision in the international agreements.

The slavers' complacency turned to indignation when the *Black Joke* ran out her guns, and under their menace sent boats with armed seamen on board them.

The Spanish masters, under their Argentinian flags of convenience, protested vigorously and volubly the presence of armed boarding parties.

'We are not a boarding party,' Codrington explained reasonably to the senior captain. 'We are armed advisers, and our advice is that you begin taking aboard your cargo again – and swiftly.'

The Spaniard continued his protests until the crack of a gun from the *Black Joke* drew his attention to the five nooses already dangling from the gunboat's yardarm. The Spaniard was certain that the nooses could not be put to the use for which they were very obviously intended – then he looked once more into the chilled sapphire eyes of the very young silver-haired English officer and decided not to make any bets on it.

Once the slaves were re-embarked, the Englishman, their self-appointed armed adviser, gave them his next piece of unsolicited advice. That was that the slave fleet up-anchor and set a course which five hours later intercepted the equatorial line.

Here Captain Codrington made a very precise observation of the sun's altitude, consulted his almanac and invited the Spanish captain to check his workings and confirm his finding that they were now in 0°05" North latitude. Then the Englishman immediately arrested him and seized the five vessels; the armed advisers changing their status, without visible pain or discomfort, to that of prize crews.

When Codrington sailed his five prizes into Table Bay, Admiral Kemp listened aghast to the Spaniard's account of his capture, and then immediately retired to his bed with bowel spasms and migraine headache. From his darkened bedroom he dictated first the order confining Codrington to his ship and the ship to its anchorage, and then his horrified report to the First Lord of the Admiralty.

This episode, which might so easily have ended with Codrington court-martialled and beached for life and with the abrupt termination of Admiral Kemp's dogged advance towards his knighthood and retirement, had in fact brought both men riches and advancement.

The sloop carrying Kemp's despatch to the First Lord passed another southbound in mid-ocean, which in its turn bore despatches for the Admiral Commanding the Cape Squadron from not only the First Lord but the Foreign Secretary as well.

Kemp was requested and required in the future to apply the 'equipment clause' to the ships of all Christian nations – with the glaring exception of the United States of America – in all latitudes, both north and south of the equatorial line.

The despatches were dated four days previous to Codrington's raid upon the Calabash barracoons, making his actions not only legal but highly meritorious.

From the very brink of professional disaster, Admiral Kemp had been snatched back, with his knighthood assured and a large sum of prize money paid into his account at Messrs Coutts of the Strand. The five Spaniards were condemned at the next session of the Court of Mixed Commission at Cape Town. Kemp's own share of the prize money had amounted to several thousand pounds, that of his junior captain to nearly twice that amount, and both officers had received personal letters of commendation from the First Lord.

None of this had done anything to increase Kemp's trust or liking for his junior, and now he listened with mounting horror to the suggestion that he sanction the boarding and search of the American trading clipper, which was at present enjoying the hospitality of the port.

For some sickening moments Kemp contemplated his place in history as the officer who had precipitated the second war with the former American colonies. There was nothing equivocal about the view of the American Government as to the sanctity of their shipping, and there were specific sections of Kemp's Admiralty orders covering the subject.

'Admiral Kemp,' Codrington was clearly burning with enthusiasm for the enterprise, 'it is absolutely beyond question that the *Huron* is a slaver, and is equipped for the trade in terms of the act. She is no longer upon the high seas, but lying at anchor within British territorial waters. I can be aboard her within two hours, with impartial witnesses, a Supreme Court judge even.'

Kemp cleared his throat noisily. He had in fact tried to speak, but so appalled was he that the words had not reached his lips. Codrington seemed to take the sound as encouragement.

'This man, St John, is one of the most infamous slavers of modern times. His name is a legend on the coast. They say he carried over 3,000 slaves one year across the middle passage. It's a golden opportunity for us.'

Kemp found his voice at last. 'I dined at Government House on Wednesday. Mr St John was in the company as his Excellency's personal guest. I considered Mr St John to be a gentleman, and I know he is a man of considerable substance and influence in his own country,' he said flatly, no trace of emotion in his voice. His self-control surprised even himself.

'He is a slaver.' Robyn Ballantyne spoke for the first time since she had seated herself at the window of the Admiral's study. The two men had forgotten her existence, but now they both turned to her.

'I have been inside the *Huron*'s main hold and she is fully equipped for the trade,' she said, her voice low but clear, and Kemp felt a sour feeling rising within him. He wondered that he had thought this young woman enchanting at their first meeting. Kemp liked young females and he had personally instructed his Secretary to send the invitations to the Ballantynes, brother and sister, but now he regretted it. He could see, of course, that what he had mistaken for spirit was in fact the mischievousness of the born troublemaker, and that far from being pretty she was in fact downright plain, with a large nose and heavy jaw. He had found her a refreshing change from the simpering and giggling young ladies of the Colony – and realized now that the preference had been unwise. He wondered if he could have his Secretary withdraw the invitation.

'I would think, Admiral Kemp, that it was your duty to send a search party aboard the *Huron*,' Robyn told him – and Kemp leaned back in the big chair and breathed heavily through his mouth. As an Admiral of the Blue it had been some years since anybody had dared point out his duty for him. His grip on his self-control slackened.

He stared at the young person. Did he detect a peculiar venom in her voice? He wondered. She had been a passenger on *Huron*. She had left the ship the instant it reached Table Bay. There was no doubt the woman was 'fast' and that Captain St John was a handsome man.

There was a fine story here, Kemp concluded, as he asked drily, 'Is it true, Miss Ballantyne, that you assaulted the surgeon-general in a fit of ungovernable rage?'

Robyn gaped at him for a moment, the change of direction taking her completely off-balance, and before she could reply he went on,

'You are clearly a highly emotional young lady. I would have to consider very carefully before committing a hostile act against an important citizen of a friendly nation on your unsupported testimony.'

He pulled the gold watch from his fob pocket, and consulted it with his full attention.

'Thank you for calling on me, Miss Ballantyne.' Once again he did not use her professional title. 'We look forward to seeing you tomorrow evening. Perhaps you would allow me a private word with Captain Codrington.'

Robyn felt her cheeks burning as she rose from the window seat.

'Thank you, Admiral, you have been very kind and patient,' she said tightly, and swept from the room.

Kemp was not so mild with Codrington. While the young Captain stood to attention before him, he leaned forward in the throne and the veins stood out like twisted blue ropes in the back of his hands as he gripped the arms of the chair.

'You were misguided in bringing that young person here to discuss navy business,' he snapped.

'Sir, I needed to convince you.'

'That's enough, Codrington. I have heard all you have to say. Now you listen to me.'

'Aye, aye, sir.'

'You are naive not to take into account the changed circumstances in the American administration. Are you not aware that Mr Lincoln is likely to be elected to the presidency?'

'I am, sir.'

'Then even you may be dimly aware that very delicate considerations are in the balance. The Foreign Office is confident that the new administration will have a markedly changed attitude to the trade.'

'Sir!' Codrington agreed stiffly.

'Can you imagine what it will mean to us to have full right of search of American shipping on the high seas?'

'Sir.'

'We will have that – once Mr Lincoln takes the oath, and if no junior officers of this service take independent action to prejudice the attitude of the Americans.'

'Sir.' Codrington stood rigid, staring over the Admiral's head at a painting of a lightly veiled Venus on the panelled wall behind him.

'Codrington,' Kemp spoke now with cold menace, 'you have had one very close shave at Calabash. I swear to you, if you let your wild nature get the better of you once more, I will have you hounded from the service.'

'Sir.'

'You are now under the strictest injunction not to approach closer than a cable's length to the trade clipper *Huron*, and if you should encounter her again at sea, you will pay her passing honours and give her a wide berth. Do I make myself sufficiently clear?'

'Sir.' Only Codrington's lips moved, and the Admiral took two long controlled breaths before going on.

'When will you sail for the Mozambique channel?' he asked, his voice more reasonable.

'I have your orders to take the flood on Saturday, sir.'

'Can you advance that sailing?'

'Yes, sir, but it would mean leaving without fully charged magazines – we expect the powder barge alongside at dawn on Saturday.'

Kemp shook his head, and sighed. 'I would feel better with you at sea,' he muttered. 'But, very well then, I will look to see you flying the Blue Peter at first light on Saturday morning.'

Robyn Ballantyne was waiting for him in the borrowed Cartwright carriage, under the portico of Admiralty House.

Codrington came down the steps, with his cocked hat under his arm and climbed stiffly into the buttoned leather seat beside her.

The Hottentot coachman flicked his whip at the shiny rumps, and they swayed in unison as the carriage jerked away down the tree-lined driveway.

Neither of them spoke until they had left the Admiralty grounds, and were whirling down the hill towards the

122

Liesbeeck bridge with the coachman holding them on a light brake.

'What do we do now?' Robyn asked.

'Nothing,' said Clinton Codrington.

Twenty minutes later as they came around the shoulder of the mountain and looked down at the bay where *Huron* rode at anchor, Robyn spoke again.

'Can't you think of anything to stop this monster?'

'Can you?' he asked sharply, and neither of them spoke again until they reached the landing place.

The fishing-boats were in and beached already, their catch laid out on the sand, a glittering silver and ruby-red pile around which the housewives and their servants bartered and bargained with the brown barelegged fishermen, while the fish horns blared to summon more customers down from the town. The two in the parked carriage watched the commotion with unnatural attention, avoiding each other's eyes.

'You will be at the Admiralty ball tomorrow night? I heard Slogger Kemp say so.'

'No,' Robyn shook her head fiercely. 'I cannot abide the frivolous chatter and silly behaviour of these occasions, and I particularly do not want to be again the guest of that man.'

Codrington turned to her for the first time since they had reached the landing. She was a fine-looking woman, he thought, with that clear lustre to her skin and the thoughtful dark-green eyes under their dark curved brows. He liked a tall strong woman, and he had learned enough of her spirit to accord her respect, a respect that could easily become fascination, he realized.

'Could I prevail upon you to change your mind?' he asked quietly, and she glanced at him, startled. 'I would undertake to provide sober conversation and a dignified dancing partner.'

'I do not dance, Captain.'

'That is a great relief,' he admitted. 'For neither do I,

when I have a choice.' He smiled. She could not remember seeing him smile before. It changed him completely. The coldness went from the pale blue eyes and they darkened with merriment while two deep laughter lines formed at the corner of his mouth and arched up to touch the thin straight nose.

'Slogger Kemp keeps a wonderful chef.' He was wheedling now. 'Fine food and serious conversation.'

His teeth were porcelain white and very regular against the deep-water tan. She felt the corners of her own mouth tugging upwards, and he saw the change in her and pressed his advantage.

'I may have further news, some further plan for *Huron* to discuss with you.'

'That makes it irresistible.' She laughed outright at last, with surprisingly unforced gaiety that made some of the nearest bystanders glance around at her and smile in sympathy.

'I will call for you. Where? When?' He had not realized how attractive she truly was until she laughed.

'No.' She laid her hand on his forearm. 'My brother will accompany me, but if you are there I look forward to our momentous conversation.' She felt the devil in her again, and squeezed his arm, finding pleasure in his immediate response, the way the muscles in his forearm corded under her fingertips.

'Wait,' she told the coachman and watched the slim tall figure go down the beach to where *Black Joke*'s whaler waited for him.

He was wearing his dress uniform for the visit to his commanding Admiral. The gold lace epaulettes emphasized the breadth of his shoulders and his sword belt upped his waist. Abruptly, she wondered whether his body hair was as blond as the queue that was twisted up at the nape of his neck – and then instantly was shocked and agitated at herself. She would never have harboured a thought like that before. Before what? she asked herself, and the answer was clear; before that night on *Huron*. Mungo St John had much to answer for. Comfortably placing blame where it belonged, she pulled her eyes away from the lithe figure of Clinton Codrington and leaned forward to tell the coachman,

'Home, please.'

She would not attend the Admiral's ball, she decided, and then she began to recite silently to herself the Christian articles of faith.

However, Zouga prevailed over her good intentions. The two of them shared the open carriage with the eldest, unmarried Cartwright daughter, for it was one of the balmy nights of a Cape autumn.

Cartwight and his wife followed in the closed carriage, in serious discussion for the entire journey.

'I am sure he is quite taken with Aletta,' Mrs Cartwright had stated.

'My dear, the fellow has no fortune whatsoever.'

'Expectations,' Mrs Cartwright told him benignly. 'I understand he will make thousands from this expedition. He is one of those young men who will get on in life, I am certain of that.'

'I prefer money in the bank, my dear.'

'He has created quite a stir, I assure you – such a serious and sensible young man, and so very attractive. It would be quite a feather in Aletta's cap, and you could make a place for him in the firm.'

Every lamp in Admiralty House was lit and a blaze of golden light welcomed the guests. There were coloured lanterns hung in the trees to illuminate the garden.

The marine band in scarlet and gold had converted the open gazebo into a bandstand, and already there were dancers on the open-air dance floor whirling furiously through the opening waltz, waving and calling greetings to the late-comers in the carriages as they trotted up the curved driveway and took their turn to alight under the portico of the main entrance.

Bewigged footmen in household uniform with silken hose and buckled shoes placed the steps and handed down the ladies on to the welcoming red carpet. At the head of the stairs the major domo announced each group in a low throaty bellow.

'Major and Doctor Ballantyne. Miss Cartwright.'

Robyn had not grown altogether accustomed to the scandalized ripple of feminine interest that followed her entrance to any public gathering here in the Colony. The quick exchange of glances, the nods and murmurs behind covering fans. It still left her with a heightened pulse and a feeling of bitter disdain for all of them.

'Did you bring your sponge, Sissy? They expect you to hit somebody with it,' Zouga murmured, and she shook his arm to silence him, but he went on: 'or to drop your skirts and run up the staircase in your breeches.'

'You are wicked.' She felt the tension go out of her, and she smiled her thanks at him and they were into the throng of uniforms, gold lace on navy blue or dress scarlet, with, here and there, the dead black of evening dress relieved by the high stock and white lace ruffles of silk shirts.

The women's skirts were taffeta and flounced silk over pyramids of petticoats. However, they were at least two years behind the London fashions, for only the most daring ladies had bared their shoulders, and powdered them dead white.

Robyn's own dress made no concession to fashion, for it was years old, the only one that she possessed even vaguely suitable for the occasion. The cloth was wool, the skirt narrow, the bodice adorned with neither seed pearls nor sequins. There were neither ostrich feathers nor diamantines to sparkle in her hair. She should have been dowdy, but instead she was strikingly different.

When she had told Codrington that she did not dance it was because she had never had the occasion, and she regretted it now as she watched Zouga lead Aletta Cartwright on to the floor and swirl away in the graceful dip and turn of the waltz.

She knew that nobody would invite her to dance, and that if they did she would be awkward and untutored. She turned away quickly – seeking a familiar or friendly face. She did not want to be left standing alone in the crowd. She began bitterly to regret that she had not stood by her resolution to stay away.

Her relief came with such a rush that she could have thrown her arms around his neck and embraced him. Instead she said evenly, 'Ah, Captain Codrington. Good evening.'

He really was one of the better-looking men in the room, she thought. She could sense the resentment of some of the younger women, so quite deliberately she took his arm as he offered it, but was surprised that he led her immediately through into the garden.

'He's here!' he said quietly, as soon as they were out of earshot of the others.

She did not have to ask who, and she felt the little jolt of shock that kept her silent for a moment.

'You've seen him?'

'He arrived five minutes before you did – in the Governor's carriage.'

'Where is he now?'

'He went into Slogger's study – with the Governor.' Clinton's face was set and hard. 'The fellow flaunts himself.'

A servant approached them with a silver tray of champagne flutes. Robyn shook her head distractedly but Clinton took a glass and drank it in two gulps.

'The devil of it is that nobody can touch him,' he fumed.

When the evening turned chill, the Marine band moved on to the orchestral balcony above the ballroom floor and punched out the dance tunes with a bouncing martial air that set the dancers spinning and prancing.

One dancer stood out above the others, not merely because of his physical height. When the others hopped and strained with gasping breath and flushed faces, Mungo St John seemed to turn and dip and glide with measured, unhurried grace although he completed a circuit of the ballroom floor more swiftly than any other dancer. Always there was one of the prettiest women in the room swinging in the circle of his arms, laughing up at him, cheeks flushed with excitement, and a dozen others watching him with covert envy over the shoulders of their own partners.

Clinton and Robyn watched him also, from the raised and colonnaded balcony that surrounded the dance floor. They stood in a small circle of Clinton's brother officers and their ladies, making no serious effort to contribute to the light chatter around them.

Robyn found herself hoping that St John would look up at her, would catch her eye so that she could flash her hatred at him; but he never glanced once in her direction.

She even thought of suggesting to Clinton Codrington that they should dance, despite her earlier protestations, but quickly decided against it. She knew that as a dancer, the naval captain would not be able to stand comparison with the elegant American.

When she went in to dinner on Clinton's arm, she saw St John ahead of them. He had a blonde woman on his arm, notoriously the Colony's prettiest, richest and most voracious widow. Her coiffure was an elaborate creation of diamanté and ostrich feathers, her shoulders were bared and she showed more of her bosom than she concealed beneath a brocaded bodice stiff with seed pearls.

Mungo St John wore simple black and white evening

dress with more panache than any of the most elaborate military uniforms around him.

Robyn watched the woman tap his shoulder with her fan to attract his attention and then reach up on tiptoe to whisper in his ear, while St John stooped gravely to listen.

'The woman's a brazen whore,' Robyn hissed, and beside her Clinton covered his shock at the word, and then nodded.

'And he's the very devil himself.'

It was as though St John heard them, for he glanced up and saw them watching him across the room. He bowed and smiled at Robyn.

It was such an intimate, completely knowing smile that she felt as though he had stripped her naked again, the way he had in the stern cabin of *Huron*, and immediately she felt the same feeling of helplessness overwhelm her.

With a huge effort she managed to turn away, but Clinton had been watching her. She could not meet his eyes, she felt he would be able to read it all if she did.

Two hours after midnight the Marine band was playing the less strenuous airs for the lovers and romantics who still circled the ballroom floor, but most of the company had gone up to the cardrooms on the first floor, if not to play themselves, then to crowd around the tables and watch with hushed attention and the occasional bursts of applause at a particularly audacious or successful coup.

In the largest room the game was whist and the players were Slogger Kemp and the older guests. In the second room the younger set were at the light-hearted chemin-de-fer and Zouga smiled at Robyn as she passed. He and Aletta Cartwright were playing a single hand between them, and the girl squealed with glee as she won a handful of silver shillings.

Robyn and Clinton passed on into the third and smallest salon. Here it was the game that had once been popular only in America. Recently, however, it had come into sudden vogue at court when the Queen had found it fascinating, and in consequence it was all the rage around the empire despite the odd name – poker.

In spite of Her Majesty's interest in it, it was still not considered a game for ladies to play in mixed company. Only men sat at the green baize, although the ladies fluttered around them like bright butterflies.

Mungo St John sat facing the doorway so that Robyn saw him the instant she stepped into the room. He lazed in his chair, the waves of his dark hair unruffled and sleeked down as if carved from polished ebony, holding a tight fan of cards low against the snowy lace front of his shirt. There was a long black unlit cheroot between his teeth, and as Robyn watched the blonde widow leaned across his shoulder displaying the creamy cleavage between her breasts and held a Vesta to the cheroot.

St John sucked flame into the tip of the cheroot, blew a long feather of blue smoke and thanked her with a slant of his eyes before he made his next bid.

He was an obvious winner, a careless pile of gold spilled across the table in front of him, each coin embossed with the Grecian style head of Queen Victoria looking much younger than her forty-one years – and as they watched, he won again.

Excitement seemed to exude from the man like a tangible substance, infecting the women about the table so they exclaimed at every wager he made, and sighed with disappointment if he folded his cards and declined to play a hand. The same excitement spread to the five other men at the table. It was obvious in the glitter of the eyes, in the white knuckles of the fingers that held their cards, in the rash calls and the imprudent urge that made them remain in play long after chance and the odds were evidently against them. It was clear that all of them saw St John as the main adversary, and the tension went out of each hand if he was not in play.

Robyn felt herself held by the same fascination, and unconsciously tightened her grip on Clinton's arm as the suspense mounted in each hand and the gold coins tinkled in the centre of the table, and she heard herself gasp with chagrin or relief at the show of cards that ended each hand.

Unconsciously, she had moved closer to the table drawing Clinton with her, so that when one of the players exclaimed with disgust, 'Fifty guineas is enough for one night. Will you excuse me, gentlemen?' gathered his few remaining coins from the table and pushed back his chair, they had to stand back to allow him to leave.

With surprise Robyn felt Clinton disengage her fingers from his arm, and then he slipped quietly into the vacant chair.

'May I join you, gentlemen?'

There were preoccupied grunts of acknowledgement but only St John looked up and asked civilly,

'Are you aware of the stakes, Captain?'

Clinton did not reply but took a roll of five-pound notes from an inside pocket and placed them beside him. The amount surprised Robyn, it could not have been less than one hundred pounds. Then she remembered that Clinton Codrington had been for many years one of the most successful blockade commanders on the slave coasts. Her brother had repeated to her the rumour that during that period he had won prize money in excess of ten thousand pounds, yet somehow she had never thought of him as a rich man.

Then with sudden intuition Robyn realized that by that gesture Clinton had laid down a silent challenge, and with a little smile Mungo St John had accepted it.

Robyn felt a flare of alarm. She was certain that Clinton Codrington had chosen an opponent too experienced and skilled. She remembered that Zouga who relied on gambling to eke out his regimental pay, had been no match for the man, even with moderate stakes, and in his frustration Clinton had been drinking steadily during the evening. She was sure his judgement would be faulty even if he had any knowledge of the game.

Almost immediately St John subtly altered the style of his play, doubling the stakes before the draw, crowding the game, dominating it, playing from the strength and confidence of his already considerable winnings, and Clinton seemed immediately uncertain of himself, hesitant in accepting the doubled stakes, discarding rather than going at risk for more than a few guineas, lacking the nerve to meet St John head on.

Robyn moved slightly to a position from which she could watch both men. Clinton was pale under the deepwater tan, the rims of his nostril bloodless and his lips compressed into a thin line, and she remembered that he had drunk a dozen glasses of champagne during the evening.

He was nervous, indecisive, every watcher could sense it, and their disappointment was evident. They had hoped for a dramatic confrontation when Clinton had made the flamboyant gesture of placing a hundred pounds on the table, but when the amount was slowly whittled away by unexciting over-cautious play, their interest was transferred to a lively exchange between Mungo St John and one of the Cloete sons, the family that owned half the Constantia valley with its fabled vineyards.

They laughed at the light banter with which the two men made their bets, and admired the grace of the loser and the easy manner of the winner. The other players at the table were almost forgotten, picking up an uncontested ante or falling before the aggression of the leading pair.

Robyn could only pity Clinton, nervous and pale – when

he fumbled his cards to expose prematurely one of his few winning hands and suffer the chuckles of the spectators as he picked up odd guineas instead of the fifty that might have been his if he had played it correctly.

Robyn tried to catch his eye, to make him leave before he was further humiliated, but Clinton played on doggedly refusing to look up at her.

Cloete won a hand with four of a kind and as was his right called for a pot to celebrate his good fortune.

'Three of a kind to open the pot and guinea sweeteners,' he announced and grinned across the table at St John. 'To your liking, sir?'

'Very much so,' St John smiled back, and the other players tried to cover their discomfiture. It was a dangerous game – requiring one of the players to be dealt an initial three of a kind to commence the game, but for every unsuccessful deal each player had to contribute a guinea to the pot, and when a lucky player achieved the minimum requirement, he could advance the stakes by as much again as was in the centre of the table. It could amount swiftly to a huge sum, and there was no option of withdrawal – a very dangerous game.

Ten times the deal failed to produce an opening hand, and then with the pot at seventy guineas Mungo St John announced quietly,

'She is open, gentlemen, as wide open as my mother-in-law's mouth.'

Play had stopped at the other tables in the room as St John went on,

'It will cost you another seventy iron men to stay in the game.' He had doubled the pot, and the watchers applauded the bet and then looked eagerly to the other players.

'I am your man,' said Cloete, but at last there was a breathlessness to his voice. He counted out the notes and gold coins and spilled them into the considerable pile in the centre.

Three other players withdrew, dropping their cards with alacrity, clearly relieved to have got out of danger for a mere ten guineas, but Clinton Codrington hunched miserably over his cards and St John had to prod him gently for his decision.

'Please don't hurry, Captain. We have all evening.'

And Clinton looked up at him and nodded jerkily, not trusting his voice, then pushed a sheaf of notes into the centre of the table.

'Three players,' St John said, and swiftly counted the money in the pot. 'Two hundred and ten guineas!'

The next bid could double that amount, and the one after could redouble it. The room was silent now, the players at the other tables in the room had left their seats to watch as the dealer gave two cards to Mungo St John to replace those he tossed into the centre of the table. He was buying honestly, trying to add to the triplets with which he had opened the pot, neither faking a flush nor a full house. Cloete bought three cards, evidently looking for a third to a high pair – and then it was Clinton's turn to request cards.

'One,' he mumbled, and held up a single finger. The finger quivered slightly. The dealer slid the card across to him and he covered it with his hand unable to bring himself to look at it yet. It was all too obvious that he was attempting to find the missing card to a flush or straight.

'The opener to bet,' said the dealer. 'Mr St John.'

There was a pause as St John fanned open his cards, and then he said evenly without a change of expression:

'The bet is doubled.'

'Four hundred and twenty guineas,' said somebody loudly, and this time nobody applauded but every eye swivelled to Cloete as he consulted his cards. Then shook his head abruptly, and let them drop. He had not found another king to go with the original pair.

Now everybody looked to the remaining player. A

transformation had come over Clinton Codrington, it was hard to define it exactly. There was just a touch of colour under the tanned cheeks, his lips were slightly parted and for the first time he was looking directly at St John – but somehow confidence and a barely suppressed eagerness shone from him. There was no mistaking it. The man positively glowed.

'Double again,' he said loudly. 'Eight hundred and forty guineas.' He could hardly contain himself, and every man in the room knew he had found the card he needed to transform his hand from worthlessness to a certain winner.

St John did not have to deliberate more than a few seconds.

'Congratulations,' he smiled. 'You found what you were looking for, I must concede this one to you.'

He dropped his cards and pushed them away from him.

'May we see the cards you required to open the pot?' Clinton asked diffidently.

'I beg your pardon.' St John's tone was lightly ironical, and he flipped his hand, face upwards. There were three sevens and two odd cards.

'Thank you,' said Clinton. His manner had changed again. The trembling eagerness, the nervous indecision, both had disappeared. He was calm, almost icy as he began to gather in the piles of scattered gold and bank notes.

'What cards did he have?' demanded one of the women petulantly.

'He does not have to show them,' her partner explained. 'He beat the others out without a showdown.'

'Oh, I'd die to see them,' she squeaked.

Clinton paused in gathering his winnings and looked up at her.

'I beg of you, madam, not to do so,' Clinton smiled. 'I would not wish to have your life on my conscience.'

He turned his cards face upwards on the green baize, and

it took the company many seconds to realize what they were looking at. There were cards of every suit, and not one of them matched another.

There were delighted exclamations, for the hand was utterly worthless. It could have been beaten by a single pair of sevens, to say nothing of the triple sevens which St John had discarded.

With this valueless hand the young naval captain had beaten the American, out-smarting him for almost 900 guineas. It was a spectacular coup. The company gradually realized how carefully it had been engineered, how Clinton had lured his adversary, how he had made a pretence at fumbling uncertainly until exactly the right moment, the moment when he had struck boldly and decisively. They burst into spontaneous applause, the ladies oohing with admiration and the men calling congratulation.

'Oh jolly good play, sir!'

St John held the smile, but his lips drew tighter with the effort, and there was a savage glitter in his eyes as he stared at the cards and realized how he had been duped.

The applause subsided, some of the spectators were beginning to turn away still discussing the hand, and St John began gathering the cards to shuffle when Clinton Codrington spoke. His voice was low but clear, so that nobody about the tables could miss a word of it.

'Even a black-birder's luck can run out at last,' he said. 'I must admit, I would rather have caught you at your dirty slaver's game than beaten you out of a few guineas tonight.'

The company froze, gaping at Clinton with expressions of horror and comical amazement. The silence in the room seemed impenetrable, the only sound was the ripple and click of the cards as Mungo St John began to shuffle them, splitting the pack with a click and then running them into each other under his thumbs in a blur of movement.

He did not look down at his hands as he rippled the cards. He never removed his gaze from Clinton's face, and his smile had still not slipped – only there was a flush of colour under the dark sun-touched skin.

'You like to live dangerously?' St John smiled the question.

'Oh, no.' Clinton shook his head. 'I am in no danger. In my experience, slavers are all cowards.'

Mungo St John's smile disappeared, extinguished instantly, and his expression was coldly murderous, but his fingers never broke the rhythm of cut and ripple, and the cards flowed under his fingers as Clinton went on evenly, 'I was led to believe that so-called Louisiana gentlemen had some exalted code of honour,' he shrugged. 'I suggest, sir, that you are a living contradiction of that notion.'

Every listener was stunned. Not one of them could doubt what they were hearing; the accusation of slave-dealing. To an Englishman, there could be no worse insult.

The last English duels had been fought in 1840 when Lord Cardigan shot Captain Tuckett, and in 1843 when Munro shot his brother-in-law Colonel Fawcett. In consequence of these encounters, the Queen had made her desire for reform known, and the articles of war were amended in the following year, making duelling an offence. Of course, gentlemen still went abroad, mostly to France, to settle affairs of honour with pistol or sword. But this was Cape Colony, one of the jewels of the Empire and the Naval Captain was one of Her Majesty's commissioned officers. The evening had proved diverting beyond any expectations, and now there was the promise of blood and violent death charging the gaming-room.

'Gentlemen,' an urgent persuasive voice interrupted them. The Admiral's flag-captain had come through from the whist room at the Admiral's orders. 'There has been some misunderstanding.'

But neither of the two men as much as glanced in his direction.

'I don't think there has been any misunderstanding at

all,' Mungo St John said coldly, his gaze still locked with Clinton's. 'Captain Codrington's insults cannot possibly be misinterpreted.'

'Mr St John, may I remind you that you are on British soil, subject to Her Majesty's laws.' The flag-captain was becoming desperate.

'Oh, Mr St John sets little store by laws. He sails his slave ship, fully equipped, into a British harbour.' Clinton stared at the American with cold blue eyes. He would have gone on, but St John interrupted him harshly, speaking to the flag-captain but addressing the words to Clinton Codrington. 'I would not dream of abusing Her Majesty's hospitality. In any event I will sail with the tide before noon this day, and in four days I shall be far beyond Her Majesty's territory, in latitude 31° 38″ south. There is a wide river mouth there, between tall bluffs of stone – a good landing and a wide beach. It is unmistakable.' St John stood up. He had recovered his urbane air, and now he adjusted the ruffles of his shirt front and gave the lovely widow his arm. He paused for a moment to look down at Clinton. 'Who knows but that you and I may meet again, when we must certainly discuss the question of honour once more. Until then I give you good morrow, sir.'

He turned away and the spectators fell back ahead of him and seemed to form a guard of honour as St John and his lady sauntered casually from the room.

The flag-captain flung one furious glare at Clinton.

'The Admiral wishes to speak with you, sir,' and then he hurried after the departing couple, followed them down the curve of the staircase and caught up with them at the double doors of carved teak.

'Mr St John, Admiral Kemp asked me to convey his compliments. He sets no store by the rash accusations of one of his junior captains. If he did, he would be obliged to send a party aboard your ship.'

'None of us would like that,' St John nodded. 'Nor the consequences.'

'Indeed,' the flag-captain assured him. 'Nevertheless the Admiral feels that, in the circumstances, you should take advantage of the next fair wind and tide to proceed on your passage.'

'Please return my compliments to the Admiral – and convey my assurance that I will clear the Bay before noon.'

At that moment the widow's coach came up, and St John nodded distantly to the flag-captain and handed the lady up the steps.

From the deck of *Black Joke* they watched the clipper raise her anchor, her master skilfully backing and filling his topsails to run up on her cable, and break the anchor flukes out of the mud and sand of the bottom of the bay. As soon as it was free, he piled on his canvas, sail after sail bursting out in quick successive explosions of brilliant white, and *Huron* tore eagerly out of Table Bay on the south-east wind.

She would be out of sight beyond the lighthouse at Mouille Point for almost four hours before *Black Joke* was ready to follow her out of the bay. The Admiralty powder barge was alongside, and all the elaborate precautions for taking on explosives were in force. The red swallow-tail warning flag at the masthead, the boiler fires in the engine room extinguished, the crew barefooted, the decks kept wet with a constant stream from the hoses to prevent a chance spark, and each powder barrel carefully inspected for leakage as it came aboard.

While the engineer refired his boilers the last members of the Ballantyne expedition came aboard. Once again, Zouga's letters of introduction had proved invaluable, and together with his persuasive manner had provided him with the most valuable addition so far to his expedition.

Old Tom Harkness had warned him during the long night discussion, 'Don't try to cross the Chimanimani Mountains without a force of trained men. Beyond the narrow coastal belt there is only one law and it is promulgated from the muzzle of a gun.'

On the strength of the letters, the commander of the Cape Town garrison had allowed Zouga to ask for volunteers from his regiment of Hottentot Infantry. 'They are the only natives of Africa who understand the working of a firearm,' Harkness had told him. 'They are the devil with drink and women, but they can fight and march, and most of them are hardened to fever and famine. Pick them carefully and watch them every moment, night as well as day.'

Zouga's request for volunteers had been most enthusiastically received. By reputation the Hottentots could scent plunder or a willing lady from fifty miles, and the pay and rations that Zouga offered were almost thrice that of the British army. They had volunteered to a man and Zouga's difficulty had been in selecting ten of them.

Zouga had taken an instant liking to these wiry little men, with their almost oriental features, slanted eyes and high cheekbones. Despite appearances, they were more African than almost any other breed. They were the original inhabitants that the first navigators had found on the beach at Table Bay – and they had taken readily to the white man's ways, and more than readily to his vices.

Zouga had solved his problem by making one selection only. This was a man with an ageless face, it might have been forty years or eighty, for the skin was the colour and texture of a papyrus parchment, each wrinkle seemed to have been eroded into it by wind and driven dust, but the little peppercorns of hair that covered his skull were untinged with silver.

'I taught Captain Harris to hunt elephant,' he boasted.

'Where was that?' Zouga demanded, for Cornwallis Harris was one of the most famous of the old African hunters. His book *The Wild Sports of Africa* was the great classic of the African chase.

'I took him to the Cashan mountains.' Harris's expedition to the Cashan mountains, which the Boers now called Magaliesberg, was in 1829, thirty-one years previously. That would make the little Hottentot somewhere between fifty and sixty years old, if he were telling the truth.

'Harris did not mention your name,' he said. 'I have read his account carefully.'

'Jan Bloom – that was my name then.' Zouga nodded. Bloom had been one of Harris's most intrepid hunter-retainers.

'Why is your name Jan Cheroot now?' Zouga asked and the dark eyes had twinkled with pixie merriment.

'Sometimes a man gets tired of a name, like he does of a woman, and for his health or his life he changes both.'

The long military-issue Enfield rifle was as tall as Jan Cheroot, but it seemed an extension of his wizened little body.

'Pick nine other men. The best,' Zouga told him, and Sergeant Cheroot brought them aboard while the gunboat was working up a head of steam in her boilers.

Each man carried his Enfield over his shoulder, his worldly possessions in the haversack on his back and fifty rounds in the pouches on his belt

It needed only the 'Rogue's March' to welcome them, Zouga thought wryly, as he watched them come in through the entry port, each one bestowing upon him a beatific grin and a salute so vigorous that it nearly swung the donor off his feet.

Sergeant Cheroot lined them up at the rail. Their original scarlet uniform jackets had suffered strange muta-tions to ten different shades, ranging from sun-faded pink to dusty puce, and each peppercorn head wore its brimless infantry cap cocked at a different angle from all the others.

Thin shanks were bound up with grubby puttees, and brown bare feet slapped the oak planking of the deck in unison as Cheroot brought them to attention, Enfields at the slope and happy grins on each puckish face.

'Very well, Sergeant.' Zouga acknowledged the salute. 'Now let's have the packs open, and the bottles over the side.'

The grins wilted, and they exchanged crestfallen glances – the Major had looked so young and gullible.

'You hear the Major, *julle klomp dom skaape*.' Gleefully Jan Cheroot likened them to 'a herd of stupid sheep' in the kitchen Dutch of the Cape, and as he turned back to Zouga there was for the first time a gleam of respect in the dark eyes.

There are two passages from which a ship may choose when sailing the south-eastern coast of Africa. The master may stay outside the 100-fathom line which marks the edge of the continental shelf, for here the opposing forces of the Mozambique current and the prevailing winds can generate a sea which seamen call with awe the '100 year wave', a wave over 200 foot from crest to trough, which will overwhelm even the sturdiest vessel as though it were a drifting autumn leaf. The alternative and only slightly less hazardous passage lies close inshore, in the shallows where the rocky reefs await a careless navigator.

For the sake of speed Clinton Codrington chose the inshore passage, so that always the land was in sight as they bustled northwards. Day after day the shimmering white beaches and dark rocky headlands unreeled ahead of *Black Joke*'s bows, sometimes almost lost in the smoky blue sea-fret, and at other times brutally clear under the African sun.

Clinton kept steam in his boilers and the single bronze

screw spinning under his counter with every sail set and trained around to glean the smallest puff of the wind, as he drove *Black Joke* on to the rendezvous that Mungo St John had set. His haste was symptom of a compulsion that Robyn Ballantyne only began fully to understand during those days and nights that they drove east and north, for Clinton Codrington sought her company constantly and she spent many hours of each day with him, or all of it that could be spared from the management of the vessel – beginning with the assembly of the ship's company for morning prayers.

Most naval captains went through the motions of divine service once a week, but Captain Codrington held prayers every morning and it did not take Robyn long to realize that his faith and sense of Christian duty was, if anything, greater than her own. He did not seem to experience the terrible doubts and temptations to which she was always such a prey, and if it had not been unchristian to do so she would have felt envy for his sense and secure faith.

'I wanted to go into the church, like my father and my elder brother, Ralph, before me,' he told her.

'Why did you not?'

'The Almighty led me into the path He had chosen for me,' Clinton said simply, and it did not seem pretentious when he said it. 'I know now He meant me to be a shepherd for His flock, here in this land,' and he pointed at the silver beaches and blue mountains. 'I did not realize it at the time, but His ways are wonderful. This is the work He has chosen for me.'

Suddenly she realized how deep was his commitment to the war he was waging against the trade, it was almost a personal crusade. His whole being directed at its destruction, for he truly believed that he was the instrument of God's will.

Yet, like many deeply religious men, he kept his belief closely guarded, never flaunting it in sanctimonious posturing or biblical quotation. The only time he spoke of his God was during the daily prayers and when he was alone with her on his quarterdeck. Quite naturally, he assumed that her belief matched, if not outstripped, his own. She did nothing to disillusion him, for she enjoyed his patent admiration, his deference to the fact that she had been appointed as a missionary, and when she was truthful to herself, which was more and more often these days, she liked the way he looked, the sound of his voice, and even the smell of him. It was a man's smell, like tanned leather or the pelt of an otter she had once had as a pet at King's Lynn.

He was good to be near, a man, as the pale missionary initiates and medical students she had known had not been men. He was the Christian warrior. She found a comfort in his presence, not like the wicked excitement of Mungo St John, but something deeper and more satisfying. She looked upon him as her champion, as though the deadly assignation to which he was hurrying was on her behalf, to wipe out the knowledge of sin and to atone for her disgrace.

On the third day they passed the settlement on the shore of Algoa Bay, where the 5,000 British settlers brought out by Governor Somerset forty years before in 1820 had landed and still eked a hard existence from the unforgiving African earth. The white flecks of painted walls looked pitifully insignificant in that wilderness of water and sky and land, and at last Robyn started to come to some small understanding of the vastness of this continent and how puny were the scratches that man had made upon it. For the first time she felt a small cold dread at her own temerity that had brought her so far, so young and so inexperienced, to venture she was not sure what. She hugged her shawl about her shoulders and shivered in the cutting wind that poured in off the green sea. The Africa she had dreamed of so often seemed harsh and unwelcoming now.

As *Black Joke* closed swiftly with the rendezvous that St John had appointed, Clinton Codrington became quieter, and was more often alone in his cabin. He understood clearly the ordeal that faced him. Zouga Ballantyne had discussed it with him on almost every occasion that presented itself. Zouga was unwavering in his opposition to the meeting.

'You have chosen a formidable opponent, sir,' he told Clinton bluntly. 'And I mean no offence when I say I doubt you are a match for him with either pistol or sword – but he'll choose pistols, you can wager on that.'

'He challenged,' Clinton said quietly. 'My weapon is the naval cutlass. We will fight with those.'

'I cannot support you there.' Zouga shook his head. 'If there was a challenge, and I could make a case against that – but if there was one, it came from you, sir. If you fight, it will be with pistols.'

Day after day he tried to persuade Clinton to miss the rendezvous.

'Damn it, man. Nobody fights duels any more, especially against a man who can split the cheroot in your mouth with either hand, at twenty paces.' Or again, 'There was no challenge, Captain Codrington, I was there, and I would stake my honour on it.' At another time, 'You will lose your commission, sir. You have Admiral Kemp's direct order to avoid the meeting, and it is obvious that Kemp is waiting for an opportunity to haul you before a court martial.' Then again, 'By God, sir, you will serve no one – least of all yourself – by being shot to death on some deserted and Godforsaken shore. If St John is a slaver, then your chance to take him at more favourable odds will come later.'

When Zouga's best arguments made no dent on Clinton's resolve Zouga went to Robyn in her cabin.

'You seem to have some influence on the fellow. Can you not persuade him, Sissy?'

'Zouga, why are you so determined to prevent Captain Codrington defending his honour?'

'He's a likeable enough chap, and I do not wish to see him pipped.'

'And if he were, you might find it difficult to reach Quelimane. Is that not so?' Robyn asked sweetly. 'Your concern is most Christian.'

'St John can choose which of his eyes he will put a ball into. You have seen him shoot.' Zouga ignored the accusation.

'I believe it is Captain Codrington's duty to destroy that monster. God protects the righteous.'

'In my experience he protects only those who shoot fastest and straightest,' Zouga growled with frustration.

'That is blasphemy,' Robyn told him.

'You deserve to hear some real blasphemy for your stubbornness,' Zouga told her curtly and strode out of the cabin. He had learned through long experience when he was wasting his time.

They passed the mouth of the Kei river – the frontier of British influence and beyond it was the wilderness, unclaimed and untamed, peopled by the tribes which had been driven back inexorably by the white advance, and by scattered bands of renegades and bastards, wandering hunters and hardy travellers and traders.

Even the trekking Boers had by-passed this land, going on into the interior, circling out beyond the mountain massif that divided the littoral from the highland plateau.

Far to the north they had turned back and crossed the mountains again and reached the coast, fighting and shattering the impis of the Zulu nation, beginning to settle the fertile coastal strip until the British ships had sailed into Port Natal, following them up from the Cape Colony, from whence they had trekked so long and so hard to avoid British rule. The Boers had loaded their waggons once more, and driving their herds before them, climbed back over the range that they called the Dragon mountains and abandoned the land they had wrested from the Zulu king Dingaan with the flame and smoke of their muskets.

However, this coast along which *Black Joke* now steamed

lay between the English colonies of Cape and Natal, claimed by none, except the wild tribesmen who watched the black-hulled vessel pass almost within arrow shot.

Clinton Codrington had marked the point where latitude 31° 38″ south intersected the coast, and the estuary was shown on the chart with the notation 'St John's river', named, probably, by one of the early Portuguese navigators, but it was ironic that it should bear the same name as the man they were hastening to meet. When *Black Joke* steamed around the last headland the description that St John had given of his namesake was instantly recognizable.

Steep, heavily wooded hills rose almost sheer about a wide lagoon. The forest was very dense dark green, with tall galleries of trees, festooned with lianas. Through the telescope could be made out the troops of little grey vervet monkeys and the brilliant plumage of exotic birds that sported and fluttered through the top branches.

The river came down a deep rocky gorge torn through the barrier of hills, filling the reed-lined lagoon and then flowing out over the bar between the curved white pillows of a sandy beach.

To dispel any doubt that this was the rendezvous, *Huron* lay at anchor a cable's length beyond the first line of breakers, in the deep, where the shoal water turned from pale green to blue.

Clinton Codrington examined her carefully through his telescope, then without a word passed the instrument to Zouga. While he in turn glassed the big clipper, Clinton asked softly, 'Will you act for me?'

Zouga lowered the glass with surprise. 'I expected one of your own officers.'

'I could not ask them.' Clinton shook his head. 'Slogger Kemp would mark their service records if he ever heard of it.'

'You do not have the same qualms about my career,' Zouga pointed out.

'You are on extended absence from your regiment, and you have not been expressly ordered, as I and my officers have been.'

Zouga thought quickly, duelling was not so seriously considered in the army as it was in the Royal Navy, in fact the army manuals still maintained no express prohibitions, and a chance to meet with St John was also a last chance to avert this ridiculous affair that so seriously threatened the continuance of his expedition.

'I accept, then,' Zouga said shortly.

'I am extremely grateful to you, sir,' said Clinton as shortly.

'Let us hope you are as grateful after the business is over,' Zouga told him drily. 'I had best go across to *Huron* right away. It will be dark in an hour.'

Tippoo caught the line as it was thrown from the gunboat's whaler, and held it while Zouga gathered his cloak and jumped the gap of surging green water to the boarding ladder, clambering up before the next swell could soak his boots.

Mungo St John waited for him at the foot of the mainmast. He held himself unsmiling and aloof, until Zouga hurried to him and offered his right hand, then he relaxed and returned the smile.

'Damn it, Mungo, cannot we make an end to this nonsense?'

'Certainly, Zouga,' Mungo St John agreed. 'An apology from your man would settle it.'

'The man is a fool,' Zouga shook his head. 'Why take the risk?'

'I don't consider there is any risk, but let me remind you he called me a coward.'

'There is no chance then?' The two of them had become good friends during the weeks they had spent together and Zouga felt he could press further. 'I admit the fellow is a prig, but if you kill him, you'll make it damned awkward for me, don't you know?'

Mungo St John threw back his head and laughed delightedly. 'You and I could work together, do you know that, Zouga? You are a pragmatist, like I am. I make a prophecy – you'll go a long way in this world.'

'Not very far, if you kill the man who is taking me.' And Mungo St John chuckled again and clapped a friendly hand upon his shoulder.

'I'm sorry, my friend. Not this time,' and Zouga sighed with resignation.

'You have choice of weapons.'

'Pistols,' said Mungo St John.

'Of course,' Zouga nodded. 'Dawn tomorrow on the beach there.' He pointed to the land with his chin. 'Will that suit you?'

'Admirably. Tippoo here will act for me.'

'Does he understand the conventions?' Zouga asked doubtfully, as he glanced at the half-naked figure that waited near at hand.

'He understands enough to blow Codrington's head off at the shoulders if he levels his pistol a moment before the signal.' Mungo St John flashed that cruel white smile. 'And that's all he needs to know, as far as I am concerned.'

Robyn Ballantyne slept not a minute during the night and it still lacked two hours of dawn when she bathed and dressed. On an impulse she chose her old moleskin breeches and man's woollen jacket. There would be the need to disembark through the surf from the ship's boat and skirts would hamper her, added to which the morning was damp and chill and her jacket was of good thick Scottish tweed.

She laid out her black leather bag, and checked its contents, making certain she had everything she needed to cleanse and staunch a bullet wound, to bind up torn flesh or hold together shattered bone, and to reduce the agony of either man.

All of them had taken it without question that Robyn would be on the beach that morning. The gunboat did not rate a surgeon, and neither did *Huron*. She was ready with an hour to wait, and she opened her journal and began making the previous day's entry, when there was a light tap on her door.

When she opened it, Clinton Codrington stood in the opening, his face pale and strained in the smoky lamplight and she knew intuitively that he had slept as little as she had. He recovered swiftly from the first shock of seeing her in breeches, dragging his eyes up to her face again.

'I hoped I might speak with you,' he muttered shyly. 'It will be the last opportunity before . . .'

She took his arm and drew him into the cabin. 'You have not breakfasted?' she asked sternly.

'No, ma'am.' He shook his head and his eyes dropped to her trousered legs, and then jerked up guiltily to her face again.

'The medicine worked?' she asked.

He nodded, too embarrassed to reply. She had administered a purge the evening before, for as a surgeon she could dread the effects of a pistol ball through a full bowel or through a belly loaded with breakfast.

She touched his forehead. 'You are warm, you have not taken a chill?' She felt protective towards him, like a mother almost, for he seemed once again so young and untried.

'I wondered if we might pray together.' His voice was so low that she barely caught the words, and she felt a warm, almost suffocating rush of affection for him.

'Come,' she whispered, and she took his hand.

They knelt together on the bare deck of the tiny cabin, still holding hands, and she spoke for both of them, and he made the responses in a soft but firm voice.

When they rose stiffly at last, he kept her hand in his for a while longer.

'Miss Ballantyne – I mean, Doctor Ballantyne – I cannot tell you now what a profound effect meeting you has had on my life.'

She felt herself blushing and tried feebly to disengage her hand, but he clung to it.

'I would like to have your permission to talk to you again in this vein after,' he paused, 'if this morning goes as we hope it will.'

'Oh, it will,' she said fiercely. 'It will – I know it will.' Hardly knowing what she was doing she pressed herself swiftly to him and reaching up kissed him full on the mouth. For a moment he froze, and then clumsily he crushed her to him so that the brass buttons of his coat dug into her bosom and his teeth crushed her lips until she felt them bruising.

'My darling,' he whispered. 'Oh, my darling.'

The strength of his reaction startled her, but almost immediately she found she was enjoying the strength of his embrace, and she tried to free her arms to return it – but he misunderstood her movements and released her hurriedly.

'Forgive me,' he blurted out. 'I don't know what came over me.'

Her disappointment was sharp enough to turn instantly to annoyance at his timidity. Buttons and teeth notwithstanding, it had felt very pleasant indeed.

B oth boats left the two ships at the same time, and they converged through the thin pearly morning mist as their crews pulled for the low lines of breaking surf and the pale outline of the beach in the dawn.

They landed within a hundred yards of each other, surfing in on the crest of the same low, green wave and the oarsmen leapt out waist-deep to run the boats high up the white sand.

Both parties moved separately over the crest of the sand bar and then down to the edge of the lagoon, screened from the boat crews by the intervening dunes and the stands of tall fluffy-headed reeds. There was a level area of firm damp sand at one edge of the reeds.

Mungo St John and Tippoo halted at one end, and Mungo lit a cheroot and stood with both hands on his hips staring out at the crests of the hills, ignoring the activity about him. He was dressed in black tight-fitting breeches and a white silk shirt with full sleeves, open at the throat to reveal the dark curls of his body hair. The white shirt would give his opponent a fair aiming point, he was observing the conventions scrupulously.

Robyn watched him covertly as she stood beside Clinton Codrington at the further end of the clearing. She tried to capture the hatred she felt for St John, to hold on to her outrage at the way he had abused her, but it was a difficult emotion to sustain. Rather, she was excited and with a strange sense of elation, the satanic presence of this man heightened the feeling. She caught herself staring openly and dragged her eyes off him.

Beside her Clinton stood very erect. He wore his blue uniform jacket with the gold lace of his rank gleaming even in the soft pink light of early dawn. He had scraped the sun-bleached hair back from his forehead and temples and bound it at the nape of his neck, leaving clean the purposeful line of his jaw.

Zouga went forward to meet Tippoo who carried under his arm the rosewood case of pistols. When they met in the centre of the level ground, he opened the case and proffered it, standing straddle-legged and attentive, while Zouga took each weapon from its velvet nest and loaded it with a carefully measured charge of black powder before ramming home the dark-blue leaden ball and setting the cap on the nipple.

The sight of the long-barrelled weapons reminded Robyn

forcefully of that night aboard *Huron*, and she bit her lip and shifted uncomfortably.

'Do not fret yourself, Miss Ballantyne.' Clinton mistook her emotion, and whispered soothingly to comfort her while he unbuttoned his jacket and shrugged out of it. Beneath it he also wore a plain white shirt to give St John a fair aim. He handed her the jacket and would have spoken again but Zouga called.

'Will the principals come forward.'

And Clinton gave her another tightly strained smile before he strode out, his heels leaving deep prints in the damp yellow sand.

He faced Mungo St John, holding his gaze steadily, both of them completely expressionless.

'Gentlemen, I appeal to both of you to settle this affair without bloodshed.' Zouga went through the ritual attempt at reconciliation. 'Captain Codrington, as challenger, will you tender an apology?'

Clinton shook his head once, curtly.

'Mr St John, is there any other way in which we can avoid bloodshed?'

'I think not, sir,' St John drawled as he carefully tapped half an inch of grey ash from his cheroot.

'Very well,' Zouga nodded and went on immediately to set the conditions of the meeting. 'At the command "Proceed" each of you gentlemen will take ten paces, which I will count aloud. Immediately after the count of ten I will give the command "Fire" upon which you will be at liberty to turn and discharge your weapon.'

Zouga paused and glanced at Tippoo, there was a long-barrelled muzzle-loading pistol thrust into the waistband of his baggy breeches.

'Both seconds are armed.' Zouga laid his right hand on the butt of the Colt pistol in his own belt. 'If either principal attempts to fire before the command to do so, then he will immediately be shot down by the seconds.'

He paused again looking from one to the other. 'Is that clearly understood, gentlemen?' They both nodded. 'Do either of you have a question?' Zouga waited in silence for a few seconds then went on. 'Very well, we will proceed. Mr St John you have first choice of weapon.'

Mungo St John dropped the cheroot and ground it into the sand with his heel, before stepping forward. Tippoo offered him the rosewood case and after a momentary hesitation St John lifted out one of the beautifully inlaid weapons. He pointed at the sky and cocked the hammer with a sweep of his free hand.

Clinton took the remaining pistol and weighed it experimentally in his hand, settling the butt deeply into the vee formed by thumb and forefinger, turning half away and lifting the pistol at full stretch of his arm to aim it at one of the black and bright yellow bishop birds that chattered in the reeds nearby.

With relief Robyn watched the familiar ease with which her champion handled the weapon, and she felt completely certain of the outcome now. Good must triumph, and she started to pray again, silently, only her lips moving as she recited the twenty-third psalm.

'Though I walk through the valley of the shadow of death.'

'Take your positions, gentlemen.' Zouga stepped back and gestured to Robyn. Still praying, she hurried to where Zouga stood, and fell in a few paces behind him, well out on the flank from the lines of fire of the two principals.

Beside Zouga, Tippoo drew the long, clumsy-looking pistol from the sash around his waist and cocked the big ornate hammer, the bore of the barrel gaped like a cannon as he lifted it into the 'present' position. Zouga drew his own Colt revolver and stood quietly while the two principals walked the last few paces towards each other and then turned back to back.

Beyond them the early sun was gilding the hilltops with bright gold, but leaving the lagoon still in shadow so the still dark waters steamed with wisps of mist. In the silence a ghostly grey heron croaked hoarsely and then launched into flight from the edge of the reedbank with slow wingbeats, its neck drawn back into a snakelike 'S' to balance the long beak.

'Proceed!' called Zouga, so loudly that Robyn started violently.

The two men stepped out, away from each other, with deliberate strides, heeling heavily in the yielding sand, pacing to the count that Zouga called.

'Five.' Mungo St John was smiling softly, as though at some secret joke, and his white silk sleeve fluttered like a moth's wing around the uplifted arm that held the slim steel-blue barrel pointed at the dawn sky.

'Six.' Clinton leaned forward, stepping out with long legs clad in white uniform breeches. His face was set, pale as a mask, his lips drawn into a thin determined line.

'Seven.' Robyn felt the beat of her heart crescendo, thumping painfully against the cage of her ribs so that she could not breathe.

'Eight.' She noticed for the first time the patches of sweat that had soaked through the armpits of Clinton's white shirt, despite the chill of the dawn air.

'Nine.' Suddenly she was deadly afraid for him, all her faith dissolved in a sudden premonition of disaster rushing down upon her.

'Ten!' She wanted to scream to them to stop. She wanted to rush forward and throw herself into the space between the two men. She didn't want them to die, either of them. She tried to fill her lungs but her throat was closed and dry, she tried to drive her legs forward but they were locked rigidly under her, beyond her control.

'Fire!' shouted Zouga, his voice cracking with the tension

that gripped all the watchers and out on the damp dark-yellow sand both men turned, like a pair of dancers in a meticulously rehearsed ballet of death.

Their right arms out-flung towards each other in a gesture like that of parted lovers, the left hand on the hip to balance, the classic stance of the expert marksman.

Time seemed suspended, the movements of the two men graceful but measured, without the urgency of onrushing death.

The silence was complete, there was no wind to rustle the reeds, no bird nor animal called from the looming forest across the lagoon, the footfalls of the two men were deadened by the yielding sand – the world seemed to hold its breath.

And then the crash of pistol fire awoke the echoes and sent them bounding and booming across the gorge, leaping from cliff to cliff, startling the birds into raucous flight.

The two shots were within one hundredth of a second of each other, so that they blended into a single blurt of sound. From the levelling blue barrels the dead white powder smoke spurted, and then the barrels were flung upwards in unison with the shock of discharge.

Both men reeled backwards, keeping their feet, but Robyn had seen the smoke fly from the muzzle of Mungo St John's pistol a fraction of a second earlier, and then an instant later Mungo's big dark head flinched as though he had been struck with an open hand across the cheek.

After that one staggered pace backwards, he steadied, drawing himself to his full height, the pistol still smoking in his raised hand, staring at his adversary, and Robyn felt a rush of relief. Mungo St John was unscathed. She wanted to run to him, and then suddenly the joy withered, a dark red snake of blood slithered from the thick hair at his temple down the smoothly shaven olive skin of his cheek and dripped with slow sullen drops on to the white silk of his shirt.

She lifted her hand to her mouth to cover the cry that rose within her, and then her attention was distracted by another movement in the corner of her vision and she swung her head with a jerky movement towards Clinton Codrington.

He also had been standing erect, with almost military bearing, but now he began to bow slowly forward at the waist. The right hand holding the pistol hung at his side and now his fingers opened and the ornate weapon dropped into the sand at his feet.

He lifted the empty hand and placed it across his chest with fingers outspread in a gesture that seemed reverential and slowly his body bent forward and now his legs gave way under him and he dropped to his knees, as though in prayer. Kneeling, he lifted his hand away from his chest and examined with an expression of mild surprise the small smear of blood that coated his fingers and then he pitched forward face down on to the sand.

At last Robyn could move. She raced forward and dropped to her knees beside Clinton's fallen body, and with strength of panic rolled him on to his back. The front of the white linen shirt was damp with a little blood around the neat puncture in the cloth six inches to the left of the line of mother of pearl buttons.

He had been half-turned to fire and the ball had taken him low and left, at the level of the lungs, she saw instantly. The lungs! She felt despair overwhelm her. It would mean death, no less certain because it was slow and agonizing. She would have to watch this man drown inexorably in his own blood.

Sand crunched beside her and she looked up.

Mungo St John stood over her, his shirt a mess of wet blood. He was holding a silk kerchief to his temple to staunch the copious flow where the pistol ball had stripped a long ribbon of scalp off his skull above the ear.

His eyes were bleak, his expression forbidding and his voice cold and distant as he told her quietly.

'I trust you will be satisfied at last, madam.' Then he turned abruptly up the white dune towards the beach. She wanted to run after him, to restrain him, to explain – to explain she knew not what, but her duty was here, with the man more gravely stricken. Her fingers shook as she unbuttoned the front of Clinton's shirt and saw the dark blue hole punched into the pale flesh from which a little thick slow blood oozed. So little blood at the mouth of the wound – it was the worst indication – the bleeding would be inside, deep inside the chest cavity.

'Zouga, my bag,' she called sharply.

Zouga came to her, carrying the bag and went down on one knee beside her.

'I am but lightly struck,' murmured Clinton. 'I have no pain. Just a feeling of numbness here.'

Zouga did not reply. He had seen a multitude of gunshot wounds in India, pain was no indication of the severity. A ball through the hand or foot was unbearable agony, another through both lungs was only mildly discomforting.

Only one thing puzzled him and that was why Mungo St John's shot had been so wide. At twenty paces he would surely have taken the head shot, aiming between the eyes with an expectation of the ball deviating less than an inch from the point of aim, yet the shot had taken Clinton low in the chest.

While Robyn pressed a dressing over the wicked little blue mouth of the wound, Zouga picked the pistol out of the sand. The barrel was still warm and there was the peppery whiff of burned powder as he examined it and saw instantly why Mungo's shot had struck wide.

There was a bluish smear of bright new lead on the steel trigger guard.

Mungo St John had indeed aimed at the head, but

Clinton had lifted the pistol to his eye at the same instant directly in the line of sight. St John's ball had struck the metal guard and been deflected downwards.

That would account for the fact that Clinton's own ball had been high, for as a less expert marksman he would surely have aimed for St John's chest. The strike of the ball had thrown his weapon upwards at the moment of discharge.

Zouga looked up and handed the pistol to Tippoo who waited impassively close at hand. Without a word, Tippoo took the weapon, turned away and followed his master over the dunes.

By the time four seamen from the gunboat could carry Clinton Codrington down the beach using his boat cloak as a hammock, St John was climbing up from the *Huron*'s whaler on to his main deck, and before they could rig a block and tackle to lift the prostrate form of her captain into *Black Joke* – *Huron* had broken out her anchor and was spreading sail before the south-westerly breeze and bearing away with the sunrise transforming her into a vessel etched in golden fire.

For twenty-four hours Clinton Codrington surprised Robyn with the strength of his recovery. She looked to see blood on his lips, and she expected him to experience the agony of breathing as the damaged lung collapsed. Every few hours she listened with her sounding trumpet to his chest, stooping over the bunk in his cabin to catch the hiss and saw of his breathing, listening for the bubbling sound of blood, or for the dry rubbing of the lung against the rib cage, and was puzzled when none of these symptoms occurred.

Indeed Clinton was unaccountably resilient for a patient with ball through the chest cavity. He complained only of stiffness reaching up into his left armpit and semi-paralysis of that arm – and he was vociferous in his advice to his surgeon.

'You will bleed me, of course?' he asked.

'I will not,' Robyn told him shortly as she cleaned the area around the wound and then lifted him into a sitting position to bandage his chest.

'You should take at least a pint,' Clinton insisted.

'Have you not bled enough?' Robyn asked witheringly, but he was undaunted.

'There is black rotten blood that must be taken off.' Clinton indicated the massive bruise that was spreading around his chest like some dark parasitic plant around the smooth pale trunk.

'You must bleed me,' Clinton insisted, for all his adult life he had been exposed to the ministrations of naval surgeons. 'If you don't, fever is sure to follow.'

He offered Robyn the inner curve of his elbow where the thin white scars over the blue blood vessels marked where he had been bled before.

'We no longer live in the dark ages,' Robyn told him tartly. 'This is 1860,' and she pushed him down on the bolster and covered him with a grey ship's blanket against the shivering and chilling nausea which she knew must soon accompany such a wound. It did not come, and for the next twenty hours he continued to manage the ship from his bunk, and he chafed at the restraint she placed on him. However, Robyn knew the pistol ball was still in there and that there must be drastic consequences. She wished that there was some technique that could enable a surgeon to locate the whereabouts of a foreign body accurately, and then allow him to enter the rib cage and remove it.

That evening she fell asleep in the rope chair beside his bunk, awakening once to hold the enamel cup of water to his lips when he complained of thirst and noting the dryness

160

and heat of his skin, and in the morning all her fears were confirmed.

He was only semi-conscious, and the pain was fierce. He moaned and cried out at the smallest movement. His eyes were sunken into plum-coloured cavities, his tongue was thickly coated with white and his lips were dried and cracked. He pleaded for drink and his skin was hot, the heat increasing every hour until it seemed to be burning out the core of his being, and he was restless and flushed, tossing in the narrow bunk, fighting off the blankets with which she tried to keep him covered, and whimpering in his delirium at the agony of movement. His breath was sawing painfully in his swollen and bruised chest, his eyes were glittering bright, and when Robyn unwound the bandage to sponge his body with cool water there was only a little pale fluid staining the dressing, but her nostrils flared as she smelled it. It was so horribly familiar, she always thought of that stench as the fetid breath of death itself.

The wound had shrunk, but the crust that had formed over it was so thin that it cracked at one of Clinton's restless movements and through it rose a thick droplet of custard-coloured matter. Immediately the smell was stronger. This was not the benign pus of healing, but the malignant pus that she so dreaded to see in a wound.

She swabbed it away carefully, and then with cold sea-water sponged his chest and the hard hot swollen flesh below his armpit. The bruising was extensive and it had changed colour, dark blue as storm clouds, tinged with the yellow of sulphur and the virulent rose of some flower from the gardens of Hell itself.

There was one area just below the point of his shoulder blade particularly sensitive for he screamed when she touched it, and a sparse prickle of sweat broke out across his forehead and amongst the fine golden bristles of his unshaven cheeks.

She replaced the bandage with a fresh dressing and then forced between his dry lips four grams of laudanum mixed with a warm draught of calomel. She watched while he fell into a restless drug-induced sleep.

'Another twenty-four hours,' she whispered aloud, watching him toss and mutter. She had seen it so often. Soon the pus would suffuse his whole body, building up steadily around the ball deep in his chest. She was helpless. No surgeon could enter the rib cage, it had never been done before.

She looked up as Zouga stooped into the cabin. He was grave and quiet, standing beside her chair for a moment and placing a hand on her shoulder comfortingly.

'He improves?' he asked softly.

She shook her head, and he nodded as though he had expected the answer.

'You must eat.' He offered her the pannikin. 'I brought you some pea soup. It has bacon in it, it's very good.'

She had not realized how hungry she was and she ate gratefully, breaking the hard ship's bread into the broth and Zouga went on quietly,

'I loaded the pistols with less than a full charge, as little as I dared.' He shook his head irritably. 'Damned bad luck. After Mungo's ball hit the trigger guard, I'd not have expected it to have entered the chest – it must have lost most of its power.'

She looked up quickly. 'It struck the trigger guard – you did not tell me.'

He shrugged. 'It's not important now. But the ball was deflected.'

She sat very still in her chair for ten minutes after he was gone, and then purposefully she stood over the bunk and stripped back the blanket, unwound the bandage and examined the wound again.

Very carefully she began to sound the ribs beneath it, pressing in with thumb gently to feel for the give of shattered bone. The ribs were all firm, yet that was no proof that the ball had not driven between two of them.

She pressed her thumb into the swelling towards the outside of his chest, and although he thrashed around weakly, she thought she felt the rasp of bone against bone – as though the rib had been chipped or even as if a long splinter had been cracked off it.

She felt a little flutter of excitement, and extended her examination gradually backwards, guided by his delirious cries of pain, until once again she reached the lower point of his shoulder and he came upright in the bunk with another wild yell of agony, breaking out once more into the burning sweat of high fever. But with the tip of her finger she thought she had felt something, something that was neither bone nor knotted muscle.

The excitement quickened her breathing. From the angle at which Clinton had stood, half-turned away from Mungo St John, she now believed that it was possible that the ball had followed a different path from that which she had assumed.

If the pistol had been undercharged with powder, and if the ball had struck the trigger guard, it was just possible that it had not had the velocity to penetrate the rib cage, it had been turned by the bone and ploughed along under the skin, skidding along the groove between two ribs following the track that she had just probed, and lodging at last in the thick bed of the *latissimus dorsi* and *tenes major* muscle.

She stood back from the bunk. She could be very wrong, she realized, but if she was he would die anyway, and that very soon.

'I will cut for it,' she decided, the decision direct and swift. She glanced up at the skylight in the roof of the cabin. There was only an hour or two of good daylight left.

'Zouga!' she called as she raced from the cabin. 'Zouga! Come quickly!'

Robyn administered another five grams of laudanum before they moved Clinton. It was as large a dose as she dared, for he had taken nearly fifteen grams in the preceding thirty-six hours. She waited as long as she could in the failing light for the drug to begin taking effect. Then she passed the word to Lieutenant Denham to shorten sail, reduce engine revolutions and make the ship's motion as easy as possible.

Zouga had chosen two seamen to assist them. One was the boatswain, a burly and greying sailor, the other was the officers' steward who had impressed Robyn with his quiet, controlled manner.

Now as the three of them half-lifted Clinton and rolled him on to his side, the steward spread a sheet of fresh white canvas on the bunk under him to receive the spilled blood, then Zouga swiftly knotted the lengths of soft cotton rope around Clinton's wrists and ankles. He had chosen cotton in preference to the coarse hemp which would tear the skin, and he made the bowline knots that would not slip under pressure.

The boatswain helped him to tie down the ends at the head and foot of the bunk, stretching out the almost naked white body so that for an instant it reminded Robyn of the painting of the crucifixion that stood in Uncle William's study at King's Lynn – the Roman legionaries spread-eagling Christ upon the cross before driving home the nails. She shook her head irritably, driving the memory from her mind, concentrating all her attention on the task ahead.

'Wash your hands!' she ordered Zouga, indicating the bucket of almost boiling water and yellow lye soap the steward had provided.

'Why?'

'Do it,' she snapped, in no mood to argue. Her own hands were pink with the heat of the water and tingled with the harsh soap, as she wiped her instruments with the cloth dipped in a mug of the pungent ship's rum, laying them out on the shelf above the bunk, and then with the same cloth swabbed the hot discoloured flesh at the base of Clinton's shoulder blade. He jerked against his bonds and muttered an incoherent protest, but she ignored it and nodded to the boatswain.

He seized Clinton's head, drawing it back slightly, and thrust a thick pad of felt, a piece of wadding for the main-deck cannon, between his teeth, holding it in place.

'Zouga!'

He took Clinton's shoulders and locked them in his powerful fingers, preventing Clinton from rolling on to his belly.

'Good.'

Robyn took one of the razor-sharp scalpels from the shelf, and then with the forefinger of her other hand probed brutally down to where she had felt that hard foreign body.

Clinton's back arched and he let out a shattering cry that was muffled by the wadding, but Robyn felt it clearly this time, hard and unyielding in the swollen flesh.

She cut swiftly, without hesitation, opening the skin cleanly, following the direction of the muscle fibre beneath, dissecting down through layer after layer of muscle, separating the bluish membrane capsules that covered each of the muscles with the hilt of the scalpel, probing deeper and deeper with her fingers towards that elusive lump in the flesh.

Clinton writhed and heaved at his bonds, his breathing rasping in the back of his throat, his teeth locked into the felt wadding in a fierce rictus so that cords of muscle stood out along his jawline and white spittle frothed upon his lips.

His wild lunges made her task that much more difficult, and her fingers were slippery with blood in his hot flesh, but she found the rubbery pulsing snake of the lateral thoracic artery and worked past it gently, pinching off the smaller spurting blood vessels with the forceps, and tying them with a loop of catgut, torn between the need for speed and the danger of doing greater damage.

Finally she had to use the blade again, and she reversed the scalpel and paused a moment to locate the lump with the tip of her forefinger.

She could feel the sweat sliding down her cheek, and she was aware of the strained tense faces of the men who held Clinton as they watched her work.

She guided the scalpel down into the open wound, and then cut down firmly and a sudden bright yellow fountain spouted up through her fingers, and the cloying stomach-twisting stench of corruption that filled the hot little cabin made her gasp.

That sharp rush of pus lasted only a second, and then there was something black and sodden blocking the wound. She picked it out with the forceps, releasing another lesser welling up of the thick custardy matter.

'The wad,' Zouga grunted with the effort of holding down the struggling naked body. They all stared at the soft rotten object in the teeth of her forceps.

The patch of cloth had been carried deeply into the flesh by the passage of the ball, and Robyn felt a surging lift of relief – she had been right.

Quickly she went back to work, running her finger into the tunnel cut by the ball until she felt it with her fingertip.

'There it is!' She spoke for the first time since she had cut, but the metal pellet was slippery and heavy, she could not prise it loose and she had to cut again and then lock the teeth of the pair of bone forceps over it. It came out with reluctant suck of clinging tissue, and she dropped it impatiently on the shelf. It made a heavy clunk against the wood. There was a temptation to come out immediately, to sew up and bind up – but she resisted it and took an extra ten seconds to probe the wound thoroughly. She was almost immediately justified, there was another rotting and stinking tatter of cloth in the wound track.

'A piece of the shirt.' She identified the white shreds, and Zouga's face mirrored his disgust. 'Now we can come out,' Robyn went on complacently.

She left a bristle in the wound to allow the remaining pus to drain off. It stuck out stiffly between the stitches with which she closed up.

When she stood back at last she was smugly satisfied with her work. There had been nobody at St Matthew's who could lay down stitches so neatly and regularly, not even the senior surgeons could match her.

Clinton had collapsed with the shock of deep surgery. His body was wet and slick with his sweat, and the skin at his wrists and ankles had been smeared away where he had fought against his bonds.

'Let him loose,' she said softly. She felt a vast pride, almost of ownership, in him now, as though he were her special creation, for she had dragged him back almost bodily from the abyss. Pride was a sin, she knew, but it did not make the sensation any the less agreeable, and in the circumstances, she decided, she had earned the pleasure of a little sin.

Clinton's recovery was almost miraculous. By the next morning he was fully conscious, and the fever had abated to leave him pale and shaky, with just enough strength to argue bitterly when she had him carried up into the sunshine and laid behind the canvas wind shelter that the carpenter had rigged under the poop.

'Cold air is bad for gunshot wounds, everybody knows that.'

'And I suppose I should bleed you before closing you up

in that hot little hell hole you call a cabin,' Robyn asked tartly.

'A navy surgeon would do so,' he muttered.

'Then thank your Maker that I am not one.'

On the second day he was sitting up unaided and eating voraciously, by the third he was managing the ship from his litter, and on the fourth day he was on his quarterdeck once more, although his arm was in a sling and he was still pale and gaunt where the fever had wasted away the flesh from his face, but strong enough to keep on his feet for an hour at a time before resorting to the rope chair the carpenter had rigged at the rail. That day Robyn withdrew the bristle from the wound and was relieved at the tiny quantity of benign pus that followed it out. They watched the little town of Port Natal come up ahead of them, the primitive buildings huddled under the whale-backed mountain they called the Bluff like chickens under the wing of the hen. *Black Joke* did not call, even though this was the furthest outpost of the British Empire on this coast, but steamed on briskly into the north, each day becoming perceptively warmer with the sun standing higher at noon and the sea changing to the darker azure of the tropics beneath *Black Joke*'s bows, and once again the flying fish sported ahead of them on filmy silver wings.

The evening before they reached the Portuguese settlement of Lourenço Marques on the deep bight of Delagoa Bay, Robyn dressed the stitches in Clinton's side, making small cooing and clucking sounds of satisfaction and approval as she saw how cleanly they were healing.

When she helped him into his shirt and then buttoned it for him, like a mother dressing her child, he told her gravely, 'I am aware you have saved my life.'

'Even though you do not approve of my methods?' she asked with a twinkle of a smile.

'I ask your forgiveness for my impertinence.' He dropped his eyes. 'You have proved yourself to be a brilliant physician.'

She made a modest murmur of denial, but when he insisted, 'No, I truly mean that. I think you are gifted,' Robyn protested no further, but moved slightly to make it easier for him to reach her with his good arm, but his declaration of faith in her skills seemed to have exhausted his courage for the moment.

That evening she vented some of her frustration by confiding to her journal that, 'Captain Codrington is clearly a man that can be trusted by a woman, in any circumstances – though a little more boldness would make him a great deal more attractive.'

She was about to close the journal and lock it away in her chest, when another thought occurred to her, and she thumbed back through the preceding pages each crammed with her small neat script, until she reached the single sheet that had become a milestone in her life. The entry for the day before *Huron* reached Cape Town she had left blank. What words were there to describe it? Each moment of it would be engraved for ever on her memory. She stared for many minutes at the empty sheet, and then she made a silent calculation, subtracting one date from another. When she had the answer she felt a chill of foreboding, and went over the calculation again, reaching the same answer.

She closed the journal slowly and stared at the lantern flame.

She had missed her lunar courses by almost a week, and with a prickle of dread lilting the fine hair at the nape of her neck she laid her hand upon her own belly as if there was something to feel there like the pistol ball in Clinton's flesh.

Black Joke called at Lourenço Marques for coal bunkers, despite the town's notoriety as a fever port. The swamps and mangroves that half-circled the town to the southwards spread the miasmic airs across the port.

Although Robyn had only very limited first-hand experience of the peculiar fevers of Africa, she had made a close study of all the writings on the subject – the most notable of these being probably those of her own father. Fuller Ballantyne had written a long paper for the British Medical Association, in which he recognized four distinct types of African fever – the recurrent fevers with a definite cycle, he divided into three categories – quotidian, tertian and quartan – by the length of the cycle. These he called the malarial fevers. The fourth type was the black vomit, or the yellow jack.

In his own unmistakable style, Fuller Ballantyne had shown that these diseases were neither contagious nor infectious. He had done so by a horrifying but courageous demonstration to a group of sceptical brother physicians at the military hospital at Algoa Bay.

While the other physicians watched, he had collected a wine glass of fresh vomit from a victim of the yellow jack. Fuller had toasted them with the awful draught and then drunk it down at a single swallow. His colleagues had waited, with a certain keen anticipation for his demise and had some difficulty in hiding their disappointment when he showed no ill effects and set off a week later to walk across Africa. Fuller Ballantyne was a man it was easier to admire than to like. The episode had become part of the legend that surrounded him.

In his writings her father insisted that the disease could only be contracted by breathing the night air of tropical areas, particularly those airs released by swamps or other large bodies of stagnant water. However, some individuals most certainly had a natural resistance to the disease, and this resistance was probably hereditary. He cited the African

170

tribes who lived in known malarial areas, and his own family and that of his wife who had lived and worked in Africa for sixty years with only mild afflictions.

Fuller wrote of the 'seasoning fever' – the first of which either killed or gave partial immunity to the subject. He used the example of the high mortality rate amongst newly arrived Europeans in Africa.

He quoted the case of Nathaniel Isaacs who in 1832 left Port Natal in a party with twenty-one newly arrived white men to hunt hippopotamus in the estuary and swamps of St Lucia river. Within four weeks nineteen had perished while Isaacs and the other survivor were so wracked with the disease that they were invalids for a year thereafter.

These losses were unnecessary, Fuller Ballantyne pointed out. There was a preventative and a cure that had been known for hundreds of years though under various names, Peruvian bark, or Chinchona bark, more recently called essence of quinine when manufactured in powder form by the Quaker brothers Luke and John Howard. Taken at the dosage of five grains daily it was a highly effective preventative, for even if the disease was subsequently contracted, it was in such mild form as to be no more dangerous than a common cold and responded immediately to a more massive dose of twenty-five grains of quinine.

Of course, Robyn had heard the accusation that her father played down the dangers of the disease to further his grand design. Fuller Ballantyne had a vision of Africa settled by colonists of British stock, bringing to the savage continent the true God, and all the benefits of British justice and ingenuity. His catastrophic expedition to the Zambezi had been in pursuit of this vision, for the great river was to have been his highway to the high and healthy plateau of the interior where his Englishmen would settle, driving out the slavers, bringing the warring and godless tribes to order, and taming and cultivating the savage earth.

Part of his vision had died upon the terrible torrents and rapids of the Kaborra-Bassa gorge.

With a sneaking sensation of disloyalty Robyn admitted to herself that there was probably some shred of justification in the accusation, for she had as a child seen her father in the grip of a malarial fever brought out by the cold of an English winter. It had not seemed as mild as a common cold then. Despite this, there was no one in the medical profession who doubted that Fuller Ballantyne was probably one of the world's leading authorities on the disease and that he had a real talent in diagnosing and treating it. So she followed his dictates faithfully, administering the daily five grains of quinine to herself, to Zouga and under protest to Captain Codrington. She had no success with Zouga's Hottentot musketeers, however. At the first dose Jan Cheroot had begun staggering in circles, clutching his throat and rolling his eyes horribly, crying to all his Hottentot gods that he had been poisoned. Only a tot of ship's rum saved him, but none of the other Hottentots would touch the white powder after that. Not even the thought of a tot of rum would tempt them, which was a measure of their opposition to the cure. Robyn could only hope that they possessed the resistance to the fever that her father spoke of.

Her store of quinine was meant to last for the duration of the expedition, possibly as long as two years, so she had reluctantly to refrain from pressing any of it on *Black Joke*'s seamen. She stilled her conscience with the fact that none of them would be spending a night ashore, therefore they would not be exposed to the dangerous airs. She prevailed upon Clinton Codrington to anchor in the outer roads where the onshore breeze kept the air sweet and, as an added attraction, the distance offshore prevented the swarms of mosquitoes and other flying insects from coming on board during the night.

The first night at anchor, the sound of music, of drunken

laughter and the shrill cries of women at play and at work carried across the still waters to the nine Hottentot musketeers in their corner of the forecastle, and the lights of the bordels and bars along the waterfront were as irresistible to the nine as a candle to a hawk moth. Temptation was made unbearable by the weight and heat of the golden sovereign that each of them carried in some secret place upon his person, the princely advance that Major Ballantyne had made against their salaries.

Sergeant Cheroot woke Zouga a little before midnight, his features a mask of outrage.

'They are gone.' He was shaking with anger.

'Where?' Zouga was still more asleep than awake.

'They swim like rats,' stormed Cheroot. 'They all go drinking and a-whoring.' The thought of it was insupportable. 'We must catch them. They will burn their brains out on smoke and pox themselves—' His rage was mixed with an equal portion of raw envy, and once they were ashore his enthusiasm for the chase was almost a frenzy. Cheroot had an unerring instinct that led him directly to the lowest dives on the waterfront.

'You go in, Master,' he told Zouga. 'I'll wait around the back,' and he twitched the short oaken club in his hand with gleeful anticipation.

The tobacco smoke and the fumes of cheap rum and gin were like a solid wall, but the musketeers saw Zouga the moment he stooped through the doorway into the yellow lantern light. There were four of them. They overturned two tables and smashed a dozen bottles in their eagerness to depart, jamming in a solid knot in the doorway to the rear alley before bursting out into the night beyond.

It took Zouga half a minute to push and wrestle his way through the crowd. The women of a dozen rich shades between gold and ebony reached out to pluck shamelessly at the more private areas of his anatomy as he passed, forcing him to defend himself, and the men deliberately blocked his path until he drew the Colt revolver from under his coat tails – only then they sullenly opened a way for him to pass. When he reached the back door Sergeant Cheroot had the four Hottentots laid out in a row in the filth and dust of the alley.

'You haven't killed them?' Zouga asked anxiously.

'*Nee wat*! They got heads of solid bone.' Cheroot tucked the club back in his belt, and stooped to pick up one of the bodies. The strength of his wiry little body was out of all proportion to its size. He carried them down to the beach one at a time as though they were bags of straw, and dumped them head first into the waiting whaler.

'Now we find the others.'

They ferreted them out, singly and in pairs from the fan-tan parlours and the gin hells, tracking down the ninth and last to the embrace of an enormous naked Somali lady in one of the shacks of mud walls and corrugated iron roofs behind the waterfront.

It was almost dawn when Zouga climbed wearily out of the whaler on to *Black Joke*'s deck and booted the ninth Hottentot down the forecastle ladder. He started for his own cabin, red-eyed and irritable, aching with fatigue when it occurred to him that he had not noticed Sergeant Cheroot amongst the dark figures in the whaler, and his penetrating voice and biting sarcasm had been silent on the return from the beach.

Zouga's mood was murderous as he landed once again, and picked his way through the narrow filth-choked alleys to the mud and iron shack. The woman made up four of Jan Cheroot. She was a mountain of polished dark flesh, gleaming with oil, each of her widespread thighs thicker than his waist, her great mammaries each as large as his head, and Jan Cheroot's head was buried between them as though he was drowning himself in exotic and abundant flesh so that his ecstatic cries were almost smothered.

The woman looked down at him fondly, chuckling to

herself as she watched Sergeant Cheroot's upended buttocks. They were skinny and a delicate shade of buttercup yellow, but they seemed to blur with the speed of movement, and the shock waves they created were transferred into the mountain of flesh beneath him, creating ripples and waves that undulated through the woman's belly and elephantine haunches, travelling up to agitate the pendulous folds that hung from her upper arms, and at last breaking in a wobbling heaving surf of gleaming black flesh around Sergeant Cheroot's head.

On the final return to the gunboat, Sergeant Cheroot sat, a small dejected figure, in the bows of the whaler. His post coital tristesse considerably enhanced by the buzzing in his ears and the ache in his head. Only Englishmen had the alarming habit of bunching up a hand suddenly, and then hitting with more effect than a man wielding a club or hurling a brick. Sergeant Cheroot found his respect for his new master increasing daily.

'You should be an example to the men,' Zouga growled at him as he hoisted him up the ladder by the collar of his uniform jacket.

'I know that, Master,' Cheroot agreed miserably. 'But I was in love.'

'Are you still in love?' Zouga demanded harshly.

'No, Master, with me love don't last too long,' Cheroot assured him hurriedly.

'I am a modestly wealthy man,' Clinton Codrington told Robyn seriously. 'Since my days as a midshipman I have saved as much of my pay as I did not need to live by, and of recent years I have been fortunate in the matter of prize money. This, together with the legacy of my mother, would enable me to care very comfortably for a wife.'

They had lunched with the Portuguese Governor at his

invitation and the *vinho verde* that had accompanied the meal of succulent seafood and tasteless stringy beef had given Clinton a flush of courage.

Rather than returning immediately to the ship after the meal, he had suggested a tour of the principal city of the Portuguese possessions on the east coast of the African continent.

The Governor's dilapidated carriage rumbled over the rutted roads and splashed through the puddles formed by the overflow from the open sewers. A raucous flock of ragged child-beggars followed them, dancing in their dirty rags to keep pace with the bony, sway-backed mule that drew the carriage, and holding up their tiny pink-palmed hands for alms. The sun was fierce but not as fierce as the smells.

It was not the appropriate setting for what Clinton Codrington had in mind, and with relief he handed Robyn down from the carriage, scattered the beggars by hurling a handful of copper coins down the dusty street, and hurried Robyn into the cool gloom of the Roman Catholic cathedral. The cathedral was the most magnificent building in the city, its towers and spires rising high above the hovels and shacks that surrounded it.

However, Robyn had difficulty in concentrating on Clinton's declaration in these popish surroundings, amongst the gaudy idols, saints and virgins in scarlet and gold leaf. The reek of incense and the flickering of the massed banks of candles distracted her even though what he was saying was what she wanted to hear, she wished he had chosen some other place to say it.

That very morning she had been taken by a sudden spell of vomiting, and a mild nausea persisted even now. As a physician she knew exactly what that heralded.

Before the courtesy visit to the Portuguese Governor's mouldering palace, she had tentatively decided that she would have to take the initiative. That attack of vapours had convinced her of the urgency of the situation, and she had pondered how she could induce Clinton Codrington to stake some sort of claim to the burden she was convinced she was carrying.

When Zouga had still lived at King's Lynn with Uncle William, she had discovered a cheaply printed novel of a most disreputable type concealed amongst the military texts on Zouga's desk. From a furtive study of this publication she had learned that it was possible for a woman to seduce a man, as well as the other way around. Unfortunately, the author had not provided a detailed description of the procedures. She had not even been certain if it were possible in a carriage, or whether anything should be said during the process, but now Clinton was obviating the necessity for experiment by a straightforward declaration. Her relief was tinged with shades of disappointment, after having been forced into the decision to carry out his seduction, she had found herself looking forward to the experience.

Now, however, she forced herself to assume an attentive expression and, when he hesitated, to encourage him with a nod or a gesture.

'Even though I am without powerful friends in the service, yet my record is such that I would never expect a half pay appointment, and although it might sound immodest I would confidently look forward to hoisting my own broad pennant before I am fifty years of age.'

It was typical of him that he was already thinking twenty-five years ahead. It required an effort to prevent her irritation showing, for Robyn preferred to live in the present, or at least the immediately foreseeable future.

'I should point out that an Admiral's wife enjoys a great deal of social prestige,' he went on comfortably, and her irritation flared higher. Prestige was something she had always intended to win at first hand – crusader against the slave trade, celebrated pioneer in tropical medicine, writer of admired books on African travel.

She could not contain it longer, but her voice was sweet

and demure. 'A woman can have a career as well as being a wife.'

Clinton drew himself up stiffly. 'A wife's place is in the home,' he intoned, and she opened her mouth, then slowly closed it again. She knew she was bargaining from weakness, and when she was silent Clinton was encouraged. 'To begin with a comfortable little house, near the harbour in Portsmouth. Of course, once there are children one would have to seek larger premises—'

'You would want many children?' she asked still sweetly, but with colour mounting in her cheeks.

'Oh yes, indeed. One a year,' and Robyn recalled those pale drabs with whom she had worked, women with brats hanging from both breasts and every limb, with another one always in the belly. She shuddered, and he was immediately concerned.

'Are you cold?'

'No. No, please go on.' She felt trapped, and not for the first time resented the role that her sex had forced upon her.

'Miss Ballantyne – Doctor Ballantyne – what I have been trying to say to you – is that I would be greatly honoured if you could find it in yourself to consent to become my wife.'

Now when it came she was not really ready for it, and her confusion was genuine.

'Captain Codrington, this comes as such a surprise—'

'I do not see why. My admiration for you must be apparent, and the other day you led me to believe—' He hesitated, and then with a rush, 'you even allowed me to embrace you.'

Suddenly she was overcome with the urge to burst out laughing, if only he had known her further intentions towards him – but she skittered away from the subject, her expression as solemn as his.

'When would we be able to marry?' she asked instead.

'Well, on my return to—'

'There is a British consul at Zanzibar, and you are bound

there, are you not?' she interrupted quickly. 'He could perform the ceremony.'

Clinton's face lit with slow, deep joy. 'Oh Miss Ballantyne, does that mean – can I take it that—' He took a pace towards her, and she had a vivid image of the tiny house in Portsmouth bursting at the seams with little blond replicas of himself, and she took a quick pace backwards and went on hurriedly.

'I need time to think.'

He stopped, joy faded and he said heavily, 'Of course.'

'It means such a change in my life, I would have to abandon all my plans. The expedition – it's such a big decision.'

'I could wait a year, longer if necessary. Until after the expedition, as long as you wished,' he told her earnestly, and she felt a flutter of panic deep in her belly.

'No, I mean I need a few days, that is all,' and she laid her hand on his forearm. 'I will give you an answer before we reach Quelimane. I promise you that.'

Sheikh Yussuf was a worried man. For eight days the big dhow-rigged vessel had lain within sight of land, the single, huge lateen sail drooped from the long yard, the sea about her was velvet smooth during the day and afire with phosphorus during the long moonless, windless nights.

So deep and utter was the calm that not the slightest swell moved the surface. The dhow lay so still that she might have been hard aground.

The Sheikh was a master mariner who owned a fleet of trading vessels and who for forty years had threaded the seaways of the Indian Ocean. He knew intimately each island, each headland and the tricks of the tides that swirled about them. He knew the great roads that the currents cut across the waters the way a post coachman knows each turn and dip of the road between his stages, and he could run them without compass or sextant, steering only by the heavenly bodies a thousand miles and more across open water, making his unerring landfalls on the great horn of Africa, on the coast of India and back again on the island of Zanzibar.

In forty years he had never known the monsoon wind to fail for eight successive days at this season of the year. All his calculations had been based on the wind standing steady out of the south-east, day and night, hour after hour, day after day.

He had taken on his cargo with that expectation, calculating that he could discharge again on Zanzibar Island within six days of loading. Naturally, a man expected losses, they were an integral part of his calculations. Ten per cent losses was the very least, twenty was more likely, thirty was acceptable, forty was always possible and even losses of fifty per cent would still leave the voyage in profit.

But not this. He looked up at the stubby foremast from which drooped the fifteen-foot-long scarlet banner of the Sultan of Zanzibar, beloved of Allah, ruler of all the Omani Arabs and overlord of vast tracts of eastern Africa. The banner was as faded and soiled as the lateen sail, both of them veterans of fifty such voyages, of calms and hurricanes, of baking sun and the driving torrential rains of the high monsoon. The golden Arabic script that covered the banner was barely legible now, and he had lost count of the number of times it had been taken down from the masthead and carried at the head of his column of armed men deep into the interior of that brooding land on the horizon.

How many times had that banner wafted out proudly, long and sinuous as a serpent on the breeze, as he brought his vessel up under the fort at Zanzibar Island. Sheikh Yussuf caught himself dreaming again. It was an old man's failing. He straightened up on the pile of cushions and precious rugs of silk and gold thread, and looked down from his command position on the poop deck. His crew lay like dead men in the shade of the sail, their grubby robes folded up over their heads against the heat. Let them lie, he decided, there was nothing mortal man could do now, except wait. It was in the hands of Allah now. 'There is one God,' he murmured. 'And Mohammed is his prophet.' It did not occur to him to question his fate, to rail or pray against it. It was God's will, and God is great.

Yet he could not help feeling regret. It was thirty years since he had taken such a fine cargo as this, and at prices that compared with those of thirty years before.

Three hundred and thirty black pearls, each one perfectly formed, young, by Allah, not one of them over sixteen years of age. They were of a people he had never encountered before, for he had never before traded so far south. It was only in this last season that he had heard of the new source of black pearl from beyond the Djinn Mountains, that forbidden land from which no man returned.

A new people, well favoured and beautifully formed, strong and tall, sturdy limbs, not those stickline legs of the people from beyond the lakes; these had full moon faces and good strong white teeth. Sheikh Yussuf nodded over his pipe, the water bubbled softly in the bowl of the hookah at each inhalation and he let the smoke trickle out softly between his lips. It had stained his white beard pale yellow at the corners of his mouth, and at each lungful he felt the delightful lethargy steal through his old veins, and take the edge off the cold frosts of age which seemed always now to chill his blood.

There was suddenly a higher pitched shriek, that rose above the low hubbub which enveloped the dhow. The sound was part of the ship, day and night it came up from the slave hold below the dhow's main deck.

Sheikh Yussuf removed the mouthpiece from his lips and cocked his head to listen, combing his fingers through his scraggling white beard – but the shriek was not repeated. It was perhaps the final cry from one of his fine black pearls.

Sheikh Yussuf sighed, the din from below decks had slowly decreased in volume while the dhow lay becalmed, and he was able to judge with great accuracy how high his losses were by that volume. He knew he had already lost half of them. Another quarter at the very least would perish before he could reach Zanzibar, many more would go even after they were landed, only the very hardiest would be fit for the market, and then only after careful convalescence.

Another indication of his losses, though not as accurate, was the smell. Some of them must have gone on the very first day of the calm and without the wind the heat had been blinding. It would be even worse in the holds, the corpses would be swollen to twice life-size. The smell was bad, he could not recall a worse stench in all his forty years. It was a pity that there was no way in which to remove the bodies, but this could only be done in port.

Sheikh Yussuf dealt only in young females. They were smaller and much hardier than males of the same age, and could be loaded more densely. He had been able to reduce the clearance between each deck by six inches, which meant an entire extra deck could be built into the hold.

Females had a remarkable ability to go without water for longer periods than the males, like the camel of the desert they seemed able to exist on the accumulated fat in their thighs, buttocks and bosoms, and to make the Mozambique passage even in the fairest conditions of wind and tide required five days without water.

Another consideration was the loss of males destined for China and the Far East by the necessary surgery. The Chinese buyers insisted that all male slaves be castrated before they would purchase. It was a logical precaution to prevent breeding with local populations, but one that entailed additional losses to the trader who must perform the operation.

The final reason that Sheikh Yussuf dealt only in comely young females was that they commanded a price almost twice that of a young male slave in the Zanzibar market.

Before Sheikh Yussuf loaded his wares, he allowed them to fatten for at least a week in his barracoons, with as much to eat and drink as they could force down their throats. Then they were stripped naked, except for light chains, and at low tide taken out to the dhow where it lay high and dry on the shoaling beach.

The first girls aboard were made to lie on the bare boards of the hull in the bottom of the hold, each on her left side with her knees raised slightly so that the knees of the girl behind her could fit against the back of her legs, the front of her pelvis against buttocks, belly against back, like a row of spoons in a rack.

At intervals the chains were snapped into the ring bolts already set into the deck. This was not only for security but also to prevent the layer of human bodies sliding about in rough weather, piling up in heaps and crushing those beneath.

Once the bottom of the hold was covered with a layer of humanity, the next deck was placed in position over them, so close that they could not attempt to sit upright nor roll over. The next layer of girls was laid over them, and the next deck over them again.

To reach the lower decks meant laboriously unchaining and unloading each layer of humanity, and lifting the intervening decks. It could not even be attempted at sea. However, with the trade wind standing fair, it was a straight run down the channel, and the wind blowing in through the canvas scoops and ports, kept the air below decks breathable, and the heat bearable.

Sheikh Yussuf sighed again and lifted his rheumy eyes to the unbroken blue line of the eastern horizon.

'This will be my last voyage,' he decided, whispering aloud in the way of old men. 'Allah has been good and I am a rich man with many strong sons. Perhaps this is his sign to me. This will indeed be my last voyage.'

It was almost as though he had been overheard, for the scarlet banner moved lazily, like an adder emerging from long hibernation, then slowly reared its head, and Sheikh Yussuf felt the wind on his seared and wrinkled cheek.

He stood up suddenly, quick and supple as a man half his age, and stamped his bare foot on the deck.

'Up', he cried. 'Up, my children. Here is the wind at last.' And while his crew scrambled to their feet, he took the long tiller under his arm and threw back his head to watch the sail bulge outwards, and the thick clumsy pole of the mainmast heel slowly across a horizon suddenly dark with the scurry of the trade wind.

Clinton Codrington caught the first whiff of it during the night. It woke him from a nightmare that he had lived through on many other nights, but when he lay sweating in his narrow wooden bunk the smell persisted and he threw a boat cloak over his bare shoulders and hurried up on deck.

It came in gusts out of the darkness, for minutes at a time the warm sweet rush of the trade wind brought only the iodine and salt smell of the sea, then suddenly there was another curdled whiff of it. It was a smell that Clinton would never forget, like the smell of a cage full of carnivorous beasts that had never been cleaned, the stench of excrement and rotten flesh, and his nightmare came rushing back upon him in full strength.

Ten years before, when Clinton had been a very junior midshipman aboard the old *Widgeon*, one of the very first gunboats of the anti-slavery squadron, they had taken a slaver in northern latitudes. She was a schooner of 300 tons burden, out of Lisbon, but flying a Brazilian flag of convenience, with the unlikely name of *Hirondelle Blanche*, the *White Swallow*. Clinton had been ordered into her as prize-master with orders to run her into the nearest Portuguese port and deliver her to the Courts of Mixed Commission to be condemned as a prize.

They had made the capture a hundred nautical miles off the Brazilian coast, after the *Hirondelle Blanche* and her five hundred black slaves had almost completed the dreaded middle passage. Under orders, Clinton had turned the schooner and sailed her back to the Cape Verde Islands, crossing the equator to do so and lying three days becalmed in the doldrums before breaking from their suffocating grip.

In the harbour of Praia, on the principal island of São Tiago, Clinton had been refused permission to land any of his slaves, and they had lain sixteen days waiting for the Portuguese president of the Court of Mixed Commission to reach a decision. Finally, the president had decided, after strenuous representation by the owners of the *Hirondelle Blanche* that he had no jurisdiction in the case, and ordered Clinton to sail her back to Brazil and submit the vessel to the Brazilian courts.

However, Clinton knew very well what a Brazilian court would decide, and instead set a course for the British naval station on St Helena Island, once more crossing the equator with his burden of human misery.

By the time he dropped anchor in Jamestown roads, the surviving slaves aboard had made three consecutive crossings of the terrible middle passage. There were only twenty-six of them still alive, and the smell of a slaver had become part of the nightmares which still plagued Clinton ten years later.

Now he stood on the darkened deck and flared his nostrils, the same smell coming to him out of the tropical night, horrible and unmistakable. He had to drag himself away with a physical effort to give the orders to fire *Black Joke*'s furnace and work up a head of steam in her boilers, ready for the dawn.

Sheikh Yussuf recognized the dark shape with a sense of utter disbelief, and the dismay of one finally deserted by Allah.

She was still five miles distant, indistinct in the dusty pink light of the dawn, but coming up swiftly, a thick column of black smoke smearing away on her beam, carried low over the green waters of the inshore channel by the boisterous trade wind. The same wind spread out her ensign to full view from the poop of the dhow, and in the field of Sheikh Yussuf's ancient brass and leatherbound telescope it snapped and flickered, the snowy white field crossed by bold bright scarlet.

How he hated that flag, the symbol of an arrogant, bullying people, tyrants of the oceans, captors of continents. He had seen gunboats like this one in Aden and Calcutta, he had seen that same flag flying in every far corner of every sea he had ever sailed. Very clearly he knew what it all meant.

He put up the helm, in that gesture acknowledging the final end to a disastrous voyage, the dhow came around reluctantly, creaking in every timber, the great sail flagging before it could be trimmed to take the wind over the stern.

There had seemed to be so little risk, he thought with weary resignation. Of course, the treaty that the Sultan had made with the Zanzibar consul of these dangerous infidels allowed his subjects to trade in the black pearls, between any of the Sultan's possessions, with the proviso that only Omani Arabs loyal to the Sultan could indulge in the lucrative traffic. No person of Christian European extraction, not even a converted Muslim, could sail under the Sultan's flag, and not even an Omani Arab might trade beyond the borders of the Sultan's possessions.

The Sultan's African possessions had been very carefully defined in the treaty, and here was he, Sheikh Yussuf, with a cargo of three hundred and thirty living, dying and dead slaves at least one hundred and fifty miles south of the furthest of the Sultan's borders with a British gunboat bearing down upon him. Truly the ways of Allah were wonderful, passing the comprehension of man, Sheikh Yussuf thought with only the slightest taste of bitterness in his throat, as he hung grimly on to the tiller and made his run for the land.

A gun thumped from the gunboat's bows, powder smoke flew bright as a seabird's wing in the first rays of the low sun, and Sheikh Yussuf spat passionately over the lee bulwark and said aloud,

'El Sheetan, the Devil,' using for the first time the name with which, in time, Captain Clinton Codrington would be known throughout the length of the Mozambique channel, and as far north as the great Horn of Africa.

The bronze screw under *Black Joke*'s counter thrashed out a long wide wake behind her. She still had main and jib set, but Codrington would shorten to 'fighting sail' just as soon as he had made the adjustment to counter the dhow's turn away towards the land.

Zouga and Robyn were on the quarterdeck to watch the chase, infected by the restrained businesslike excitement which gripped the vessel so that Zouga laughed aloud and called, 'Gone away! Tally-ho!' as the dhow turned, and Clinton glanced at him with a conspiratory grin.

'She's a slaver. Apart from the stink, that turn-away proves it beyond doubt.'

Robyn strained forward to watch the filthy little vessel, with its discoloured and patched sail, the unpainted timbers of the hull striped with zebra stripes of human excrement and other wastes. It was her first view of a slaver actually carrying on its grisly trade, and she felt herself filled with a new purpose; she had come so far for this moment, and she tried to capture every detail of it all for her journal.

'Mr Denham, give him a gun, if you please,' ordered Clinton.

The bow-chaser thudded, but the dhow held to her new course.

'Be ready to round up and send the boat away on the instant.' Clinton's excitement had given way to obvious anxiety. He turned to look across at his boarding party. They had been issued with cutlass and pistol, and waited now in the waist under the command of a young ensign.

Clinton would dearly have liked to command the boarding-party himself but his arm was still in its sling and the stitches still in the wound. To go aboard a dhow in a rough sea, and fight its crew required both hands and the agility which his wound denied him. Reluctantly he had put Ferris in charge of the boarders.

Now he looked back at the dhow, and his expression was grim.

'He is going to beach.'

They were all silent now, staring ahead, watching the slaver run in towards the land.

'But there is a coral reef.' Robyn spoke for them all, pointing to the black points that broke the surface a quarter of a mile short of the land itself, they looked like a necklace of sharks' teeth, and the surf broke and swirled about them as it was driven in upon the trade wind.

'Yes,' Clinton agreed. 'They will run it up on the coral and escape across the lagoon.'

'But, what about the slaves?' Robyn asked, horrified, and nobody answered her.

Black Joke rushed on purposefully, but with the wind almost dead astern the dhow trained her long boom around to go on to her best point of sailing. The boom was longer than the hull itself, and the huge triangular mainsail bulged far out, almost touching the surface of the water as she hurled down upon the reef.

'We may just cut her off,' Zouga said loudly, but he did

not have the seaman's eye for bearing and speed, and Clinton Codrington shook his head angrily.

'Not this time.'

But it was very finely run. Clinton held his course up until the very last moment, and the dhow passed a mere two hundred yards ahead of his bows. So close that they could clearly make out the features of the helmsman on the vessel's poop deck, a skinny old Arab in long, flowing robe and with the tasselled fez on his head that declared that he had made the pilgrimage to Mecca. On his belt glittered the gold filigree hilt of the short curved dagger of a sheikh, and his long scraggling white beard fluttered in the wind as he leaned on the long tiller and turned his head to watch the high black hull bearing down on him.

'I could put a bullet through the bastard,' Zouga growled.

'It's too late for that,' Clinton told him, for the dhow had passed beneath their bows and the gun boat was as close to the menacing fangs of coral as she dared go. Clinton called to his quartermaster at the gunboat's wheel, 'Heave to! And bring her head to wind.' Then, spinning on his heel, 'Away, the boarding-party!'

There was a squeal of davits as the crowded whaler dropped out of sight towards the choppy green sea alongside, but already the dhow was pitching wildly in the lines of seething white surf that guarded the reef.

It was two days since that dead flat calm had broken, and since then the trades had worked up a goodly swell. It came sweeping in across the inshore channel, in low green humps with dark wind-scarred backs, but as soon as they felt the tilt of the land, they peaked up eagerly, the crests turning opaque as green jelly, shivering and wobbling, and then collapsing on themselves and surging in tumultuous white water up on to the black fangs of the reef.

The dhow caught one of the taller swells, threw up her stern and went racing down upon it like a surf boat, with the skinny old Arab at the tiller prancing like a trained monkey on the stick of the tiller to hold her in the wave, but the dhow was not built for this work, and she dug her shoulder rebelliously into the sliding, roaring chute of green water, breaching so fiercely to the wave that the water poured aboard her in a green wall and she wallowed broadside, half swamped before she took the reef with a force that snapped off her single mast at deck level and sent yard and sail and rigging crashing over the side.

In an instant she was transformed into a broken hulk, and clearly the watchers on *Black Joke*'s deck could hear the crackle and rending of her bottom timbers.

'There they go!' Clinton muttered angrily, as the dhow's crew began to abandon her, leaping over the side and using the swells to carry them over the reef into the quieter waters of the lagoon, thrashing and kicking until the beach shelved up beneath them.

They saw the old Arab steersman amongst the survivors. He waded ashore, beard and robe plastered against his body with seawater, and then lifted his robes to his waist, exposing skinny legs and shrunken buttocks as he scampered up the white beach with the agility of a goat and disappeared amongst the palm groves.

Black Joke's whaler pulled swiftly into the first line of breakers, the Ensign in the stern peering over his shoulder to judge the surf, and then catching his wave taking her in with a rush, swinging sharply into the lee of the dhow's stranded hull where there was calmer water.

They watched the Ensign and four of his men go up over the side, pistols and cutlasses drawn, but by this time the last of the Arab crew were staggering up the beach and into the sanctuary of the palm grove a quarter of a mile away across the lagoon.

The Ensign led his men below decks, and they waited on *Black Joke*'s quarterdeck, watching the abandoned dhow through the telescope. A minute passed before the Ensign appeared on deck again. He crossed quickly to the dhow's rail and leaned against it to vomit over the side, then straightened up and wiped his mouth on the back of his sleeve, before shouting an order down to the oarsmen in the whaler.

Immediately the whaler shot out from the lee of the hull, and pulled lustily back through the surf towards *Black Joke*.

The boatswain came in through the entry port, and knuckled his forehead to his captain.

'Mr Ferris's compliments, sir, and he needs a carpenter to get the slave decks open, and two good men with bolt cutters for the chains.' He had gabbled this out on a single breath, and he paused to refill his lungs. 'Mr Ferris says as how it's fierce bad below decks, and some of them is trapped – and he needs the doctor—'

'I'm ready to go,' Robyn cut in.

'Wait,' Clinton snapped, but Robyn had gathered her skirts and run.

'If my sister goes, I'm going too.'

'Very well, then, Ballantyne, I'm obliged for your assistance,' Clinton nodded. 'Tell Ferris we have an incoming tide, full moon tonight, so it will make springs. There is a twenty-two foot tidal fall on this coast. He will have less than an hour in which to work.'

Robyn appeared on deck again, lugging her black leather valise, and she had exchanged skirts for breeches once more. The seamen on deck gawked at her legs curiously, but she ignored them and hurried to the ship's side. The boatswain gave her a hand and she scrambled down into the whaler with Zouga carrying her valise behind her.

The ride in through the surf was terrifying and exhilarating at the same time; the whaler tilted forward at an alarming angle, the water hissing and creaming alongside, with a belly-swooping rush that ended alongside the dhow's heavily canted side.

The deck was running with water and listed so steeply that Robyn had to crawl up it on her hands and knees, and each time a wave struck the hull, it quivered and shook and more water came streaming down over the deck.

The Ensign and his boarding party had ripped the hatches off the main hold and as she reached them, Robyn gagged and choked with the solid stench that came out of that square opening. She had believed herself hardened to the smell of death and corruption, but never had she experienced anything like this.

'Did you bring the bolt cutters?' the Ensign demanded, white-faced with nausea and horror.

The bolt cutters were heavy duty shears, used for cutting the shrouds and halyards from a dismasted vessel. Two men wielded them now, as they lifted a bunch of small black bodies through the hatch, all of them fastened together at wrist and ankle by the clanking black steel links. It reminded Robyn of the cut-out paper dolls she had amused herself with as a child, fashioning with scissors a single figure from the folded sheet, and then pulling out a chain of identical dolls. The cutters crunched through the light chain and the limp little bodies fell apart.

'They are children,' she cried out aloud, and the men around her worked in grim silence, dragging them out of the hatch, cutting them free and dropping them on to the tilted wet deck.

Robyn seized the first of them, a skeletal stick figure, crusted and streaked with dried filth, vomit and faeces, head lolling as she lifted it into her lap.

'No,' There was no life. The eyeballs had dried already. She let the head drop and a seaman dragged it away.

'No,' and 'No,' and 'No again.' Some of them were already in an advanced state of putrefaction; at a word from the Ensign, the seamen began dropping the wasted corpses over the side, to make room for those still coming up from below.

Robyn found her first live one, there was feeble pulse and fluttering breath, but it did not need a physician's instinct to tell that the hold on life was tenuous. She worked swiftly, apportioning her time to where the chance of life seemed greater.

Another taller wave struck the dhow, and it tilted sharply, the timbers crackled and snapped deep inside her.

'Tides flooding. Work faster,' shouted the Ensign. They were into the hull now. Robyn could hear the thud of sledge hammers and the rip of irons as they began to tear the slave decks out of her.

Zouga was in there, stripped to the waist, leading the attack on the wooden barricades. He was an officer, with the easy way of command and his natural leadership was swiftly acknowledged by the seamen around him.

The hubbub reminded Robyn of a rookery at sundown, the shrieks of the returning birds and the answering cries from the chicks in the nests. The mass of black girls were aroused from the lethargy of approaching death by the dhow's wild antics, the crash of breaking timbers and the flood of cold salt water into the hold.

Some of those lying in the bilges were already drowning as the hold flooded, and some of them had realized that there was a rescue team aboard, and cried aloud with the last strength of waning hope.

Alongside the slaver, the whaler was moored, and she was almost filled with the wasted bodies in which some life still burned, while on the surface of the lagoon bobbed a hundred or more corpses with gas-swollen bellies like the corks on a fishing-net.

'Take them to the ship,' the Ensign shouted down to his oarsmen in the whaler, 'and come back for more.' As he spoke, another white-crested wave struck the dhow solidly and she heeled, but she was pinned down by the spikes of coral driven through her bottom timbers, otherwise she would have turned turtle.

'Robyn!' Zouga shouted at her from the hatch. 'We need you!'

For a moment she did not even glance at him, but shook her head at the sailor beside her.

'No, she's gone.' Expressionlessly the sailor picked up the frail body and dumped it over the side.

Then Robyn scrambled up to the hatch and dropped through it.

It was a descent into the pit, dark after the brilliant noontide, so that she paused for a moment to let her vision adjust.

The tilting deck beneath her feet was slippery with human waste, so that she had to cling for a handhold.

The air was so thick that for a minute she nearly panicked, as though she were being smothered by a stinking damp pillow. She almost fought her way back into the sweet sunlight, but then she forced herself to take a gulp of it, and though her guts heaved and she tasted the bitter flooding into the back of her throat, she kept it down.

Then her own discomfort was forgotten as she made out the chaos about her.

'That last wave,' Zouga grunted as he steadied her with an arm about her shoulder, 'the decks have collapsed.'

Like a house of cards they had folded in upon each other, raw splinters of timber thrusting up out of the gloom, baulks crossed like the blades of a scissors, with small dark bodies trapped in the jaws, others crushed by the fallen beams, squashed so that they were no longer recognizable as human, others dangling head down on their ankle-chains, suspended in space, writhing weakly as crippled insects, or hanging quietly, swinging to the dhow's movements.

'Oh, sweet Mother of God, where to begin,' Robyn whispered. She let go her handhold, starting forward and her feet shot out from under her on the slick coated deck, and she went plunging down into the hold.

She struck heavily and pain flared in her back and lower body, but she dragged herself on to her knees. Her own pain seemed insignificant in this terrible prison.

'Are you all right?' Zouga asked anxiously, but she shrugged his hands away.

There was one girl screaming. Robyn crawled to her. Her legs were crushed below the knees, trapped under a baulk of hand-hewn timber.

'Can you move that?' she asked Zouga.

'No, she's done for. Come, there are others—'

'No.' Robyn crawled back to where her bag had fallen. The pain was bad, but she forced it below the surface of her consciousness.

She had only seen a leg amputation performed once before. When she started the girl threw off the seamen who held her and attacked Robyn like a tortured wild cat. Her nails raked skin from Robyn's cheek, but by the time Robyn had freed the first leg below the knee, the child was silent and limp. She died before Robyn had reached the bone of the second leg, and Robyn was weeping chokingly to herself as she left the body still hanging in the grip of the timber baulk.

She scrubbed her hands together, they were bloodied to the elbows, her palms sticky against each other. She felt consumed by guilt at her failure, without the strength to move. Dully, she stared about her now.

The hold was more than half flooded, the tide pounded remorselessly at the ship's hull.

'We have to get out,' Zouga called to her urgently, and when she did not turn her head, he seized her shoulder and shook her roughly. 'There's nothing more we can do. It will capsize at any moment.'

Robyn was staring down into the stinking black waters which sloshed from one end of the hold to the other, a single hand broke through the surface directly below her, a child's hand with soft pink palm and pretty tapered fingers spread in a gesture of appeal. The iron cuff seemed too large for the narrow wrist, and weighted it down so the hand sank gently disappearing from her sight. She stared after it with infinite regret, then Zouga hauled her roughly to her feet.

'Come on, damn it!' His face was savage, haunted with the horrors he had experienced in this fetid, half flooded hull.

The next wave hit the swamped hull, and this time it broke the grip of the coral. Timber squealed as it twisted and tore, and the dhow began to roll, the filthy waters rose up out of the darkness into a steep black wave and burst about them, shoulder deep.

The damaged slave decks broke free, sliding down over each other, tumultuous and lethal, releasing a fresh layer of tightly packed black bodies to tumble loosely into the flooded depths.

'Robyn! We'll be trapped.' He dragged her up towards the square of brilliant sunlight, clambering over timbers and bodies.

'We can't leave them,' Robyn resisted.

'They're finished, damn it. The whole thing is going. We have to get out.'

She pulled her arm free, stumbled, collapsed backwards, hit something so that pain flared through her lower body again and she cried out with the strength of it. She was lying on her side, couched on a pile of linked bodies, and there was a face a foot from hers. It was alive, she had never seen such eyes, cat fierce, falcon bright, the colour of boiling honey.

'This one is strong enough!' Robyn thought, and then she shouted. 'Help me Zouga.'

'For God's sake, Robyn.'

She crawled forward and reached the black child, and the deck tilted viciously under her, fresh cold water bursting into the hull.

'Leave her,' shouted Zouga.

The fresh flood swirled up around Robyn's head, and the chained girl disappeared below it.

Robyn lunged for her, groping blindly below the surface, feeling panic rise in her when she could not find the child.

Ducking her head under, choking as the pain in her belly made her gasp and she swallowed water, she at last got a grip on the girl's shoulder, feeling her struggling as desperately as she.

Together they came out above the surface, coughing and gasping weakly, Robyn holding the girl's mouth just clear of the surface, but when she tried to lift her further the chain anchored them both and she screamed.

'Zouga, help me!'

Another surge of water, smelling like raw sewage, filled her mouth and both of them went under once more.

She thought she would never come up again, but stubbornly she held on, sliding one arm under the girl's armpit, and with the other hand gripping her chin, forcing it up so that when they broke out again the girl's face was lifted to take another precious breath of the stinking air, and Zouga was with them.

He took a double turn of the chain around his wrists and threw all his weight on to it. In the gloom of the hull he towered over them, the light from the hatch highlighting the bulging wet muscles in his arms and shoulders as he strained at the chain, his mouth opened in a silent scream of effort, sinewy cords standing out of his throat.

Another wave hissed over them, and this time Robyn was not ready for it, she felt the burn of it in her lungs and knew she was drowning. She need only release her hold on the girl's head and shoulders and she would be free to breathe, but she held on stubbornly, determined suddenly that she would never let this little soul go. She had seen it in the girl's eyes, the fierce will to live. This one she could save, this one out of three hundred or more was the only one she could be certain of saving. She had to have her.

The wave receded and Zouga was still there, his hair streaming with water, slicked down over his face and into his eyes, and now he shifted slightly, jamming his legs against one of the heavy timbers and he reared back once more against the chain, and a low bellow broke from his throat in the agony of effort.

The ring bolt that held the loop of chain to the deck ripped out cleanly, and Zouga dragged both women clear of the water, the chain slithering after them for ten feet or more before coming up hard against the next ring bolt.

Robyn had never suspected Zouga capable of such strength, she had never seen his naked upper body – not since he was a child – had not realized that he had the lean hard muscle of a prizefighter. But even so, he could not repeat the effort, and the girl was still chained. They had won only a temporary respite. Zouga was bellowing now, and the young naval Ensign scrambled down through the open hatch. To re-enter the doomed hull was an act of courage in itself, Robyn realized, as she saw that the Ensign carried the cutting shears, lugging the heavy tool with him as he floundered towards the struggling group in the bottom of the hold.

The hull rolled through another five degrees, the water swirling higher towards them hungrily, it sucked at their bodies. Had not Zouga given them the extra few feet of chain they would be far below the surface now.

Zouga stooped over her and helped her hold the black girl's head above the water, while the Ensign groped for the chain links and fed them into the jaws of the shears. But the blades had been blunted and chipped by the heavy work they had already done, and the Ensign was still only a lad. Zouga pushed him aside.

Again muscle bunched in his shoulders and upper arms and the chain parted with a metallic clunk. Zouga cut twice at ankle and at wrist, then he dropped the shears, picked the frail naked body up against his chest and climbed frantically up towards the hatch.

Robyn tried to follow him, but something tore deep in

her belly, she felt it go, tearing like brittle parchment, and the pain was a lance that transfixed her. She doubled over it, clutching herself, unable to move, and the wave hit her, knocking her down, swirling her over the broken timbers into the dark waters, and the darkness began to fill her head. There was temptation to let go now, to let the water and darkness take her, it would be so easy, but she gathered her anger and her obstinacy to her and went on fighting. She was still fighting when Zouga reached her, and dragged her up towards the light.

As they crawled out through the hatch into the sunlight, so the dhow rolled all the way, flinging them as though from a catapult over the side into the shocking cold of the green waters.

As the dhow capsized so the last faint cries from within her were extinguished, and the hull began to break up under the remorseless hammer of the sea. When Robyn and Zouga surfaced, still clinging together, the whaler was hovering over them, the Ensign risking all to come in over the reef for the pick-up.

Strong hands reached down, and the overladen boat heeled dangerously as they were pulled aboard. Then the Ensign swung the bows to meet the next boiling line of surf and they climbed its steep side and crashed over the top, the seamen pulling frantically to hold her bows on.

Robyn crawled to where the black girl lay on a heap of other bodies in the bottom of the boat, her relief at finding her aboard and still alive outweighing the pain of her sodden lungs and the deep ache in her belly.

Robyn rolled the girl on to her back, and lifted her lolling head to cushion it from the pounding of the whaler's hull over the steep swells that threatened to crack her skull against the floorboards.

She saw immediately that the girl was older than she had imagined, although the body was desiccated, dried out and wasted. Yet her pelvis had the breadth of maturity. She would be sixteen years old at least, Robyn thought, and pulled a corner of the tarpaulin over her body to screen her from the men's gaze.

The girl opened her eyes again, and stared at Robyn solemnly. Those eyes were still the colour of dark honey, but the ferocity had dimmed to some other emotion as she looked up into Robyn's face.

'*Ngi ya bonga*,' the girl whispered, and with a shock Robyn realized that she understood the words. She was transported in an instant to another land and another woman, her mother, Helen Ballantyne, teaching her those same words, repeating them to her until Robyn had them perfected.

'*Ngi ya bonga*, I praise you!'

Robyn tried to find a reply, but her mind was as battered as her body, and it had been so long ago that she had learned the language, the circumstances so different that the words came only haltingly.

'*Velapi wena*, who are you and from where do you come?'

The black girl's eyes flew wide with shock.

'You!' she whispered. 'You speak the language of the people.'

They had taken on board twenty-eight living black girls. By the time *Black Joke* got under way again and turned from the land, towards the open sea, the dhow's hull had burst open and the planking and timbers swung and pitched end over end as they sawed across the exposed reef.

A squawking raucous flock of seabirds squabbled over the reef, hovering above the gruesome remnants that were mixed in the floating debris of the wreck, dropping to seize a tidbit and rise again on delicate fans of pearly wings.

In the deeper water along the seaward side of the reef,

the shark packs were gathering, lashing themselves into a frenzy, the stubby rounded triangles of their dorsal fins crisscrossing the green sweep of the current, while every few seconds a long torpedo-shaped body would break clear of the surface in an ecstasy of greed, falling back heavily with a boom like distant cannon as it struck the surface.

Twenty-eight from three hundred and more was no great haul, Robyn thought, as she hobbled along the line of barely living bodies, her own bruised limbs protesting every step, and her despair deepened as she realized how far gone they were. It was easy to see which of them had already lost the will to resist. She had read her father's treatise on the sick African, and she knew how important this will to resist was in treating a primitive people. A perfectly healthy man could will himself to die, and once he did so there was nothing that could save him.

That night, despite Robyn's constant attention, twenty-two of the girls died and were carried aft to be dropped over *Black Joke*'s stern. By morning all the others were sinking into the coma and fever of renal failure, their kidneys, shrivelled and atrophied by lack of fluid, were no longer filtering the urinary wastes from the bodies' systems. There was only one treatment and that was to force the patient to drink.

The little Nguni girl resisted strongly. Robyn knew that she belonged to the Nguni group of peoples, although she was uncertain of which tribe, for many of them spoke variations of the original Zulu, and the girl's accent and pronunciation had been strange to Robyn's ear.

Robyn had tried to keep her talking, keep her conscious and keep the will to resist burning in her. She had conceived an almost maternal possessiveness for the child, and though she tried to spread her attention fairly amongst the other survivors, she always returned to where the girl lay under a strip of tarpaulin and held the pannikin of weak sugar solution to her lips.

They shared only a few hundred words in which to converse, as the girl rested between each painful sip of fluid.

'I am called Juba,' the child whispered, in answer to Robyn's question. Even the sound of it brought back to Robyn the memory of the cooing of the plump blue-grey ring-necked doves in the wild fig trees that grew above the mission cottage in which she had been born.

'Little Dove. It is a pretty name.' And the girl smiled shyly, and went on in that dry tortured whisper. Much of it Robyn could not follow, but she listened and nodded, realizing with a pang that the sense of it was going, Juba was sinking into delirium, that she was talking to phantoms from her past. Now she tried to resist when Robyn forced her to drink, muttering and crying out in fear or anger, gagging on the tiny mouthfuls of liquid.

'You must rest yourself,' Zouga told Robyn brusquely. 'You have been with her for almost two days without sleeping. You're killing yourself.'

'I am quite well, thank you,' Robyn told him, but her face was gaunt and white with fatigue and pain.

'At least let me take you down to your cabin.'

By this time Juba was the only black girl still alive, all the others had gone over the stern to feed the following shark pack.

'Very well,' Robyn agreed, and Zouga carried the child down from the makeshift shelter on the aft deck which Robyn had used as a surgery.

The steward brought a canvas pallet filled with straw and laid it on the deck of Robyn's tiny cabin. There was only just room for it, and Zouga laid the naked body upon it.

Robyn was tempted to stretch out on her own narrow bunk to rest for a while, but she knew that if she let go now, even for a moment, she would fall into a deathlike sleep, and her patient would die of such neglect.

Alone in the cabin, she sat cross-legged on the straw

pallet, wedged her back against the sea-chest, and lifted Juba into her lap. Doggedly she went on with the task of forcing liquid between the girl's lips, drop by drop, hour after hour.

Through the single port, the light turned to a ruby glow at the short tropical sunset and then it swiftly faded. It was almost completely dark in the cabin when suddenly Robyn felt a copious warm flood soak through her skirts into her lap, and she smelt the strong ammoniacal taint of the girl's urine.

'Thank you, God,' she whispered. 'Oh, thank you, God!' The girl's kidneys were functioning again, she was safe. Robyn rocked the girl in her lap, feeling no revulsion from her soaked skirts, welcoming them as the promise of life.

'You did it,' she whispered. 'You did, with sheer pluck, my little dove.'

She had just enough strength left herself to wipe down the child's body with a cloth soaked in sea water, then she stepped out of her soiled dress and collapsed face down on her hard narrow bunk.

Robyn slept for ten hours, and then the cramps woke her groaning. Her knees were drawn up against her chest by the severity of the pain, and her belly muscles were hard as stone and it felt as though she had been clubbed across the back, a deep bruised sensation that alarmed her seriously.

For many minutes after waking she believed herself seriously stricken, and then with a rush of relief and joy that was far stronger than the pain she realized what was happening to her. She dragged herself across the cabin, doubled over with the pain, and bathed in the bucket of cold sea water. Then she knelt beside Juba on the pallet.

The girl's fever had abated. The skin felt cooler to the touch. Her continued recovery added to Robyn's sense of deep pleasure and relief. Now she would have to find the right moment to tell Clinton Codrington that she would not marry him, and the vision of the little house above the Portsmouth harbour receded. Despite the pain, she felt free, light of body, like a bird poised on the point of flight.

She filled the pannikin with water and lifted Juba's head.

'We will be all right now,' she told the girl, and Juba opened her eyes.

'We'll both be all right now,' she repeated, watching the girl drink thirstily, smiling happily to herself.

Juba's recovery was swift. Soon she ate with a robust appetite. Her body filled almost before Robyn's eyes, her skin took on the lustre of health and youth again, her eyes regained the sparkle of high spirits – and Robyn realized with proprietorial approval that she was a pretty girl, no, more than that, she had natural grace and poise, the voluptuous curve of bosom and buttock which ladies of high fashion tried to achieve with bustle and padded bodice. She possessed also a sweet moon-face, the big wide-set eyes and full sculptured lips that were exotic and strangely beautiful.

Juba could not understand Robyn's concern with having her cover her breasts and legs, but Robyn had seen the seamen's eyes when the girl followed her up on to the deck with only a scrap of canvas covering the most vital point of her anatomy and showing no concern at all when the wind lifted the canvas and fluttered it like a beckoning flag. Robyn commandeered one of Zouga's oldest shirts. It hung to Juba's knees and she belted it at the waist with a bright ribbon that had the child cooing with the eternal feminine delight in pretty things.

She followed Robyn about like a puppy, and Robyn's ear tuned to the Nguni language. Her vocabulary expanded swiftly, and the two of them chatted late every night, sitting side by side on the straw pallet.

Clinton Codrington began showing acute signs of jealousy. He had become used to having more of her company, and Robyn was using the girl as an excuse to taper off their relationship, preparing him for the news that she must deliver before they reached Quelimane.

Zouga also disapproved of her growing intimacy with the girl.

'Sissy, you must remember that she is a native. It never pays to let them get too familiar,' Zouga told her gravely. 'I've seen that happen too often in India. One has to keep one's reserve. After all, you are an English woman.'

'And she is a Matabele of Zanzi blood, which makes her an aristocrat, for her family came up with Mzilikazi from the south. Her father was a famous general and she can trace her bloodline back to Senzangakhona, the King of the Zulu, and the father of Chaka himself. We, on the other hand, can trace our family as far as great-grandfather, who was a cattle herd.'

Zouga's expression stiffened. He did not enjoy discussing the family origins.

'We are English. The greatest and most civilized people in the world's history.'

'Grandfather Moffat knows Mzilikazi,' Robyn pointed out, 'and thinks him a great gentleman.'

'You are being foolish,' Zouga snapped. 'How can you compare the English race to these blood-thirsty savages.' But he stooped out of the cabin for he did not wish to continue the discussion. As usual, Robyn had her facts correct and her logic was infuriating.

His own grandfather, Robert Moffat, had first met Mzilikazi back in '29 and over the years the two men had become firm and trusted friends. The King relied on Moffat, whom he called Tshedi, for counsel in his dealings with the world beyond his borders and for medical treatment for the gout which plagued him as he grew older.

The route northwards to the land of the Matabele always passed through Robert Moffat's mission station at Kuruman.

A prudent traveller would ask for a safe conduct from the old missionary, and the Matabele impis guarding the Burnt Land along the border would honour that safe conduct.

Indeed, the ease with which Fuller Ballantyne had moved through the wild, untamed tribes along the Zambezi river as far west as Lake Ngami, unmolested and unharmed, was in great part due to his relationship with Robert 'Tshedi' Moffat. The mantle of protection which the Matabele King spread over his old friend extended to his immediate family, and was recognized by all the tribes within range of the Matabele's long arm, an arm that wielded the assegai, the terrible stabbing spear which King Chaka of the Zulu had first conceived, and with which he had conquered his known world.

In his pique at Robyn's comparison of the ancestry of his family and that of the pretty half-naked black girl, Zouga at first missed the significance of what she had told him. When it struck him, he hurried back to Robyn's cabin.

'Sissy,' he burst out excitedly. 'If she comes from Mzilikazi country – why! that's almost a thousand miles due west of Quelimane. She must have passed through the land of Monomatapa to reach this coast. Get her to tell us about it.'

He regretted his childhood inattention to the language when his mother had taught them. He concentrated savagely now as the two girls chatted animatedly, and began recognizing some of the words, but it needed Robyn to translate the full sense for him.

Juba's father was a famous Induna, a great warrior who had fought the Boers at Mosega, and a hundred other battles since then, his shield had been thick with the tassels of cow tails, black and white, each of which signified a heroic deed.

He had been granted the headring of the Induna when he was still a young man of less than thirty summers, and had become one of the highest elders in the council of the nation. He had fifty wives, many of them of pure Zanzi blood like his own, a hundred and twelve sons and uncounted daughters. Although all the cattle of the nation belonged to the King, yet over five thousand were put in the charge of Juba's father, a mark of the King's high favour.

He was a great man – perhaps too great for his own safety. Somebody whispered the word 'treason' in the King's ear, and the King's executioners had surrounded the kraal in the dawn, and called out Juba's father.

He had stooped out through the low entrance of his thatched beehive hut, naked from the embrace of his favourite wife.

'Who calls?' he cried into the dawn, and then he saw the ring of black figures, tall in their feather headdresses, but standing motionless, silent and menacing.

'In the King's name,' a voice answered him, and out of the ranks stepped a figure he recognized immediately. It was one of the King's Indunas also, a man named Bopa a short powerful man, with a deep-muscled bare chest and a head so heavy that the broad features seemed to have been carved out of a chunk of granite from the kopjes across the Nyati River.

There was no appeal, no escape, not that either consideration even passed briefly through the old Induna's mind.

'In the King's name.' That was sufficient. Slowly he drew himself to his full height. Despite the grey cap of his hair, he was still a finely built warrior with broad rangy shoulders and the ridged battle scars which crawled across his chest and flanks like live serpents.

'The Black Elephant,' he began to recite the praise names of his King. 'Bayete! The Thunder of the Heavens. The Shaker of the Earth. Bayete!'

Still calling the King's names he went down on one knee, and the King's executioner stepped up behind him.

The wives and elder children had crawled from their huts now and huddled together in dread, watching from the shadows, their voices blending in a single cry of horror and sorrow as the executioner drove his short thick-bladed assegai between the Induna's shoulder blades, and two hands breadth out of his chest. As he withdrew the blade, there was the crude sucking sound of steel leaving flesh and the old Induna's life blood spurted head high as he fell forward on to his face.

With his smeared red blade the King's executioner commanded his warriors forward, for the sentence of death included the old man's wives, every one of their sons and daughters, the household slaves and their children, every inhabitant of the large village, three hundred or more souls.

The executioners worked swiftly, but there was a change in the ancient ritual of death. The old women, the grey-headed slaves died swiftly, not honoured with the blade but clubbed to death with the heavy knobkerries that each warrior carried. The infants, and unweaned toddlers were snatched up by the ankles, and their brains were dashed from their skulls against the trunk of a tree, against the heavy poles of the cattle enclosure or against a convenient rock. It was very swift, for the warriors were highly trained and disciplined troops and this was something they had done many times before.

Yet this time there was a difference, the younger women, the adolescent children, even those on the verge of puberty, were hustled forward and the King's executioner glanced at them appraisingly, and with a gesture of the bloody spear sent them left or right.

On the left hand was swift death, while those who were sent right were forced into a trot and led away towards the east, towards the sunrise as the girl Juba explained it to Robyn.

'Many days we travelled,' her voice sinking, the horror of it still in the dark brown eyes. 'I do not know how long it was. Those who fell were left where they lay, and we went on.'

'Ask her what she remembers of the country,' Zouga demanded.

'There were rivers,' the girl replied. 'Many rivers and great mountains.' Her memories were confused, she could make no estimates of distance, they had encountered no other people, no villages nor towns, they had seen no cattle nor standing crops. Juba shook her head to each of Zouga's questions, and when he showed her the Harkness map in a forlorn hope that she might be able to point out features upon it, the child giggled in confusion. Drawn symbols on parchment were beyond her comprehension, she could not begin to relate them to features of landscape.

'Tell her to go on,' he ordered Robyn impatiently.

'At the end we passed through deep gorges in high mountains where the slopes were covered with tall trees, and the rivers fell with white spray, until at last we came to where the *bunu* – the white men – waited.'

'The white men?' Robyn demanded.

'Men of your people,' the girl nodded. 'With a pale skin and pale eyes. There were many men, some white and others brown or black men, but dressed as the white men were dressed, and armed with the *isibamu*, with guns.' The Matabele people knew well the power and effect of firearms, they had encountered enemies armed with them at least thirty years before. Even some of the Matabele Indunas carried muskets, although they always handed these to a servant to carry when there was serious fighting in the offing.

'These people had built kraals, such as we build for our cattle, but these were filled with people, a great multitude of people and with them we were bound with the *insimbi*, the links of iron.' She rubbed her wrists instinctively at the memory, and the calluses raised by the slave cuffs still blemished the skin of her forearms.

'Each day that we stayed at this place in the mountains, more people came. Sometimes only as many as the fingers on both your hands, on other days there were great numbers so we could hear their lamentations at a distance. And always there were warriors guarding them.

'Then before the sun one morning, at the time of the horns,' Robyn recalled the expression for the time of dawning when the horns of the cattle first show against the morning sky, 'they led us from the kraals, wearing the *insimbi*, and we formed a great snake of people so long that the head was out of sight ahead of me in the forest while the tail was still up in the clouds of the mountains when we came down the Hyena Road.'

'The Hyena Road, *Ndlele umfisi.*' It was the first time that Robyn had heard the name spoken. It conjured up an image of a dark trail through the forests, beaten by tens of thousands of bare feet, with the loathsome eaters of dead flesh slinking along beside it, chuckling and shrieking their inane chorus.

'Those that died, and those that fell and could not rise again were released from their chains and dragged to one side. The *fisi* have grown so bold along the road, that they rushed out of the bushes and devoured the bodies while we passed in full sight. It was worst when the body still lived.'

Juba broke off and stared unseeingly at the bulkhead across the cabin. Slowly her eyes filled with tears, and Robyn took her hand and held it in her lap.

'I know not how long we followed the Hyena Road,' Juba went on, 'for each day became as the one before it, and as the one that followed it, until at last we came to the sea.'

Afterwards Zouga and Robyn discussed the girl's story.

'She must have gone through the kingdom of Monomatapa, and yet she says there were no towns nor signs of occupation.'

'The slavers might have avoided contact with Monomatapa's people.'

'I wish she had seen and remembered more.'

'She was in a slave caravan,' Robyn pointed out. 'Survival was her only concern.'

'If only these damned people could even read a map.'

'It's a different culture, Zouga,' and he saw the spark in her eye, and sensed the drift of the conversation and turned it aside swiftly.

'Perhaps the legend of Monomatapa is only a myth, perhaps there are no gold mines.'

'The important thing about Juba's story is that the Matabele are dealing in slaves, they have never done so before.'

'Nonsense!' grunted Zouga. 'They are the greatest predators since Genghis Khan! They and all the Zulu splinter tribes – the Shangaan, the Angoni and the Matabele. War is their way of life, and plunder is their main crop. Their whole nation was built on a system of slave-taking.'

'But they have never sold them before,' said Robyn mildly. 'At least as far as we know from all that grandfather and Harris and the others have written.'

'The Matabele never found a market before,' Zouga replied reasonably. 'Now they have at last made contact with the slavemasters, and found an opening to the coast. That was all it lacked before.'

'We must witness this, Zouga,' Robyn spoke with quiet determination. 'We have to bear witness to this terrible crime against humanity and carry word back to London.'

'If only the child had seen evidence of the Monomatapa, or the gold mines,' muttered Zouga. 'You must ask her if there were elephant.' He pored over the Harkness map, lamenting the blank spaces. 'I cannot believe that it does not exist. There is too much evidence.' Zouga looked up at his sister. 'One other thing – I seem to have forgotten almost every word of the language that mother taught us, except some of the nursery rhymes and lullabies.

'*Munya, mabili zinthatu, Yolala. umdade wethu*,'* he

* 'One, two, three,
Go to sleep my little sister,'

211

recited, and then chuckled and shook his head. 'I shall have to study it again, you and Juba will have to help me.'

The Zambezi comes to the sea through a delta of vast swampland, and a hundred confused shallow channels spread out for thirty miles down the low featureless coastline.

Floating islands of papyrus grass break free from the main pastures which blanket the waters of the delta and are carried out to sea on the dirty brown water. Some of these islands are many acres in extent and the roots of the plants so entangled that they can support the weight of a heavy animal. On occasions small herds of buffalo are trapped and carried twenty miles out to sea before the action of the waves smashes up the islands and plunges the great bovine animals into the water, prey for the big sharks which cruise the tainted estuary waters for just such a prize.

The muddy smell of the swamps carries far from the land when the wind is right, and the same wind carries strange insects with it. There is a tiny spider no bigger than the head of a wax Vesta which lives in the papyrus banks of the delta. It spins a gossamer web on which it launches itself into the breeze in such numbers that the gossamer fills the sky in clouds, like the smoke from a raging bushfire, rising many hundreds of feet and eddying and swirling in misty columns that are touched by the sunset into shades of pink and lovely pale mauve.

The river pours a muddy brown effluent into the sea, silt enriched with the bodies of drowned animals and birds, and the huge Zambezi crocodiles join the shark packs at the feast.

Black Joke found the first of these hideous creatures ten miles from land, wallowing in the low swells like a log, the rough scales glittering wetly in the sunlight until the gunboat approached too closely and the reptile dived with a lash and swirl of its powerful ridged tail.

Black Joke steamed obliquely across the multiple mouths of the river, none of which offered passage for a vessel of her size. She was headed further north for the Congone channel which was the only passage upriver to the town of Quelimane.

Clinton Codrington planned to enter it the following morning, after having lain hove-to during the night off the mouth. Robyn knew that she must remove the stitches from the wound under his armpit, though she would have liked to leave them a few days longer. They must come out before she left the gunboat at Quelimane.

She decided to use the same opportunity to give him the answer for which he had waited so patiently. She knew it was going to be painful for him to hear that she would not marry him, and she felt guilty that she had so encouraged him. It was alien to her nature to inflict suffering on another, and she would try to tell him as gently as possible.

She ordered him to her cabin for the removal of the horsehair stitches, seating him on the narrow bunk naked to the waist, with his arm raised. She was delighted with the way the wound had healed, and proud of the neatness of her work. She cooed over each knot as she snipped it with the pointed scissors, then seized it with the forceps and gently tugged and worried it free of his flesh. The stitches left twin punctures, one on each side of the raised purple welt of the scar and they were clean and dry. Only one of them leaked a single drop of blood which she swabbed away gently.

Robyn was training Juba as her assistant, teaching her to hold the tray of instruments, and receive the discarded and soiled dressings or instruments. Now she stood back and appraised the healed wound, without looking at Juba.

'You may go now,' she said quietly. 'I will call you when I need you.'

Juba smiled like a conspirator, and murmured, 'He is truly beautiful, so white and smooth,' and Robyn blushed pinkly, for that was exact!y what she had been thinking. Clinton's body, unlike that of Mungo St John, was hairless as a girl's but finely muscled, and the skin had an almost marble sheen to it.

'His eyes are like two moons when he looks at you, Nomusa,' Juba went on with relish, and Robyn tried to frown at her but her lips kept puckering into a smile.

'Go swiftly,' she snapped, and Juba giggled.

'There is a time to be alone,' and she rolled her eyes lewdly. 'I shall guard the door, and hardly listen at all, Nomusa.' Robyn found it impossible to be angry when the child used that name, for it meant 'the daughter of mercy', and Robyn found it highly acceptable. She would have had difficulty picking a better name for herself, and she was smiling as she hurried Juba from the cabin with a gentle slap and a push.

Clinton must have had some idea of the exchange, for he was buttoning his shirt as she turned back to him, and looking embarrassed.

She drew a deep breath, folded her arms and began.

'Captain Codrington, I have thought unceasingly of the great honour you have done me by inviting me to be your wife.'

'However,' Clinton forestalled her, and she faltered, the prepared speech forgotten, for her next word would indeed have been, 'However'.

'Miss Ballantyne, I mean Doctor Ballantyne, I would rather you did not say the rest of it.' His face was pale and intense, he really was beautiful now, she thought, with a pang. 'That way I can still cherish hope.'

She shook her head vehemently, but he lifted a hand.

'I have come to realize that you have a duty, to your father and the poor unhappy people of this land. I under-stand and deeply admire that.'

Robyn felt her heart go out to him, he was so good and so perceptive to have understood that about her.

'However, I feel sure that one day, you and I shall . . .'

She wanted to spare him pain.

'Captain,' she began, shaking her head again.

'No,' he said. 'Nothing you say will ever make me abandon hope. I am a very patient man, and I realize that now is not the time. But I know in the depths of my soul that our destiny binds us together, even if I must wait ten or fifty years.'

A time-span of that magnitude no longer frightened Robyn. She relaxed visibly.

'I love you, my dear Doctor Ballantyne, nothing will ever alter that, and in the meantime I ask only your good opinion, and friendship.'

'You have both,' she said, with truth and relief. It had been a great deal easier than she had expected, yet strange that a shadow of regret lingered.

There was no further opportunity to speak privately, for Clinton was fully occupied with bringing *Black Joke* into the treacherous channel, with its shifting banks and uncharted shoals guarding the mouth. The channel meandered twenty miles through the mangrove forests to the port of Quelimane on the northern bank.

The heat in the delta was rendered scarcely bearable by the humid effluxion of mud and rotting vegetation that rose from the mangrove forests. The weird shapes of the mangroves fascinated Robyn and she stood by the rail and watched them slide past. Each tree stood clear of the slick chocolate-coloured mud on its pyramid of roots, like the multiple legs of a grotesque insect reaching up to join the thick pulpy stem which in turn extended upwards to the roof of poisonous green foliage. Amongst the roots skittered the purple and yellow fiddler crabs, each of them holding aloft a single disproportionately huge claw, and waving it in menace or ponderous greeting at the passing vessel.

Black Joke's wake spread across the channel, flopping wavelets on the mud banks and startling the small green and purple night herons into laborious flight.

Around a bend in the channel the decaying buildings of Quelimane came into view, dominated by the square towers of the stucco church. The plaster was falling away in unsightly chunks and the whitewash was streaked and splotched with grey and green mould, like a ripening cheese.

This port had once been one of the most busy slave ports on the entire African coast. The Zambezi river had acted as a highway to the interior for the slave-masters, and the Shire river, its major tributary, led directly to Lake Marawi and the highlands which had been the mother lode from which hundreds and thousands of black slaves had poured.

When the Portuguese, under British pressure, had signed the Brussels Agreement, the barracoons at Quelimane and Lourenço Marques and Mozambique Island had been closed down. However, the slaving dhow that *Black Joke* had intercepted proved that the abominable trade still flourished covertly along the Portuguese coast. That was typical of these people, Clinton Codrington thought.

Clinton curled his lip with distaste. In the many hundred years since their great navigators had opened up this coast, the Portuguese had clung to the narrow unhealthy strip of the littoral, making only one half-hearted effort to penetrate the interior and since then, lying here like their disintegrating buildings and crumbling empire, content with the bribes and extortions of petty officialdom and their seraglios of women, tolerant of any crimes or evil as long as there was a little dash or profit in it.

As he worked *Black Joke* in towards the quay, he could see them gathering already, gaudy vultures, in their fool's motley of uniform, tarnished gold braid and ornate swords sported by even the lowliest customs officer.

There would be endless forms and declarations unless he was firm, and always the open palm and the leering wink. Well, this time there would be none of that. This was a ship of the Queen's Navy.

'Mr Denham,' Clinton called sharply, 'issue pistols and cutlasses to the anchor watch, and nobody comes aboard without the express permission of the officer of the watch.'

He turned away to shake hands briefly with Zouga; they had found little in common during the voyage and the parting was cool.

'Never thank you enough, sir,' said Zouga briskly.

'Only my duty, Major.' But already Zouga's eyes were following Sergeant Cheroot as he assembled his men on the foredeck. They were in full marching order, eager to be ashore after the tedious voyage.

'I must see to my men, Captain,' Zouga excused himself and hurried forward.

Clinton turned to Robyn and looked steadily into her green eyes.

'I beg a small token of remembrance,' he said quietly.

In response to his request she reached up and took one of the cheap paste earrings from her lobe. As they shook hands, she slipped the little ornament into his palm, and he touched it briefly to his lips before slipping it into his pocket.

'I will wait,' he repeated, 'ten or even fifty years.'

Black Joke had come up-channel on the flood, unloaded the mountainous stores of the Ballantyne Africa Expedition on to the stone quay during slack water, and two hours later thrown off her mooring ropes and swung sharply across the ebb, pointing her high bows down the channel.

From his position on the quarterdeck, Clinton Codrington stared across the widening gap at the slim, tall figure in long skirts standing on the very edge of the quay. Beyond her, her brother did not look up from his lists as he checked the stores and equipment. Sergeant Cheroot stood armed guard with his little pug-featured Hottentots, and the idlers and watchers kept well clear.

The Portuguese officials had treated the red wax seals and ribbons which decorated Zouga's letters of authority from the Portuguese ambassador in London with great respect. However, even more important was the fact that Zouga was an officer of Queen Victoria's army, that he had arrived in a Royal Navy gunboat, and lastly that there was every reason to believe that the same gunboat would remain in the area for the foreseeable future.

The Governor of Portuguese East Africa himself would not have commanded greater respect. Already minor officials were scampering about the squalid little town arranging the best accommodation, securing warehousing for the stores, commandeering river transport for the next leg of the journey up-river to Tete, the last outpost of Portuguese empire on the Zambezi, drafting orders to have bearers and guides meet the expedition at Tete, and doing everything else that the young British officer casually demanded as though it was his God-given right.

In this turmoil of activity Robyn Ballantyne stood alone, staring after the blue-clad figure on *Black Joke*'s quarterdeck. How tall he was, and his hair caught the sunlight in a flash of white gold as he lifted his hand in farewell. She waved until *Black Joke* disappeared behind a palisade of mangrove, though her masts and fuming smoke stack stayed in view for a long time after. She watched until they, too, dwindled to nothingness, and only the smear of black smoke lay low over the tops of the green mangrove.

C linton Codrington stood on his deck, hands clasped loosely at the small of his back, and an expression of near rapture in the pale blue eyes. In this temper the knight-errant of old must have ridden out at chance, Clinton thought.

He did not find the notion at all melodramatic. He felt truly ennobled by his love, sensing somehow that he must earn something so precious, and that the opportunity to do so lay ahead of him. The earring that Robyn had given him was suspended by a thread around his neck, lying under his shirt against his skin. He touched it now, peering impatiently ahead down the channel. It seemed to him that for the first time he had a steady direction in his life, constant as the pole star to the navigator.

This gallant mood was still strong five days later when *Black Joke* rounded the headland of Ras Elat and steamed into the anchorage. There were eight large dhows keeled over on the exposed sand bar at low tide. The tidal fall on this coast at full springs was twenty-two feet. These craft were designed to take the ground readily, and it facilitated loading. The long ranks of chained slaves were being goaded out to the stranded vessels, slipping and splashing through the shallow tidal pools, to await patiently their turn to climb the ladder up the side of the dhow.

Black Joke's unannounced arrival caused pandemonium, and the beach was alive with running stumbling figures, the screams and shrieks of the slaves, the pop of the kurbash whips and the frantic cries of the slave-masters carried clearly to *Black Joke*'s deck as she dropped her anchor just beyond the reef and rounded up to the wind.

Clinton Codrington stared longingly at the heeled vessels and the concourse of panicky humanity, the way a slum-child stares at the display in the window of a food shop.

His orders were clear, had been spelled out by Admiral Kemp with painful attention to detail. The Admiral remembered with lingering horror his young Captain's capture of the slaving fleet at Calabash after forcing the masters to load their cargoes and sail north of the equator. He wanted no repetition of this type of risky action on this patrol.

Black Joke's commander was strictly adjured to respect the territorial integrity of the Sultan of the Omani Arabs, and the exact letter of the treaty that the British Consul had negotiated at Zanzibar.

Clinton Codrington was strictly forbidden to interfere with any subject of the Sultan who was engaged in trade between any of the Sultan's dominions. He was denied even the right of search of any vessel flying the red-and-gold flag of Omani on any of the Sultan's recognized trade routes, and these were carefully defined for Captain Codrington's benefit.

He was to confine his patrol to intercepting only vessels that did not belong to the Sultan, particularly vessels of the European powers. Naturally no American vessel might be searched on the high seas. Within these limits Captain Codrington had powers of independent action.

Far from being allowed to seize or search the Sultan's vessels, Clinton was ordered to use the first opportunity to make a courtesy call on the port of Zanzibar. There he would take counsel from the British Consul as how best to use his influence to reinforce the existing treaties, and especially to remind the Sultan of his own obligations under those treaties.

So now Clinton paced his deck like a caged lion at feeding time, and glowered helplessly, through the pass in the coral reef, at the slaving fleet of Omani engaged in legitimate trade, for the Gulf of Elat was very much part of the Sultan's possessions, and had so been recognized by Her Majesty's Government.

After the first wild panic ashore, the beach and dhows were now deserted, but Clinton was aware of the thousands of watchful eyes upon him from the mud-walled town and the shadows of the coconut groves.

The thought of hauling his anchor and sailing away filled him with bitter chagrin, and he stood bare-headed and stared with cold hungry blue eyes at the prize spread before him.

The palace of the Sheikh of Elat, Mohamed Bin Salim, was an unpainted mud-walled building in the centre of the town. The only opening in the parapeted wall was the gate closed by thick, brass-studded double doors in carved teak, which led through to the dusty central courtyard.

In this courtyard, under the spreading branches of an ancient takamaka tree the Sheikh was in earnest conclave with his senior advisers and the emissaries of his supreme sovereign – the Sultan of Zanzibar. They were discussing a matter, literally, of life or death.

Sheikh Mohamed Bin Salim had the plump smooth body of the bon vivant, the bright red lips of the sensualist, and the hooded eyes of a falcon.

He was a very worried man, for his ambition had led him into dire danger. His ambition had been quite simply to accumulate the sum of one million gold rupees in his treasury, and he had very nearly satisfied that reasonable goal, when his overlord, the omnipotent Sultan of Zanzibar, had sent his emissaries to call the Sheikh to account.

Sheikh Mohamed had begun to satisfy his ambition ten years previously by very gently mulching the Sultan's tithe, and each year since then he had increased his depredations. Like all greedy men, one successful coup was the signal for the next. The Sultan had known this, for though he was old, he was also exceedingly cunning. He knew that the missing tithes were safely stored for him in the Sheikh's treasure house, to be collected whenever he felt inclined. He need only benignly feign ignorance of the Sheikh's manipulations, until he was so deeply in the trap that no squirming or squealing would get him out again. After ten years that moment had arrived. The Sultan would collect not only his due but the Sheikh's own accumulations.

Further retribution would be a lengthy business. It would begin with a beating on the soles of the Sheikh's bare feet, until all those delicate little bones were cracked or fractured making it extremely painful for the Sheikh to be marched into the Sultan's presence. There, the final judgement would be read, and it would end with the knotted strip of buffalo hide wound up tighter and tighter around the forehead, until first the Sheikh's falcon eyes popped from their sockets and then his skull collapsed like a bursting melon. The Sultan truly enjoyed these spectacles – and had been looking forward to this particular one for ten years.

Both men knew the ritual, and it had begun with the polite visit of the Sultan's emissaries who even now were sitting opposite the Sheikh under the takamaka, sipping thick black coffee from the brass thimbles, munching the yellow and pink coconut sweetmeats, and smiling at the Sheikh with cold passionless eyes.

It was into this chilling atmosphere that the messengers from the harbour came running to fling themselves prostrate and gabble out the news of the British warship, whose great guns threatened the harbour and the town.

The Sheikh listened quietly and then dismissed the messengers, before turning back to his distinguished guests.

'This is a serious business,' he began, relieved to be able to change the subject under discussion. 'It would be wise to view this strange vessel.'

'The Ferengi have a treaty with our master,' pronounced one greybeard, 'and they set great store by these pieces of paper.'

They all nodded, none of them showing the agitation which filled all of their breasts. Although this coast had only received passing attention from these brash northern people, still it had been enough to engender fear and apprehension.

The Sheikh deliberated for a few minutes, stroking his thick curly beard, hooding his eyes as the ideas began to flow. His mind had been almost paralysed with the extent of the disaster which had overtaken him, but now it began to work again.

'I must go out to this warship,' he announced. There was an immediate hubbub of protest, but he held up his hand to silence them. He was still the Sheikh of Elat, and they had, perforce, to hear him out.

'It is my duty to ascertain the intentions of the commander and to send word immediately to our master.'

Clinton Codrington had almost resigned himself to give the order to weigh anchor. There had been no sign of life on the beach for many hours, and there was nothing he could accomplish here. His hope that he might catch a European slaver, actually taking on slaves in the anchorage, had proved forlorn. He should have sailed hours ago, the sun was half way down the sky already and he did not want to run the dangers of the inshore channel in darkness, but some instinct had kept him here.

He kept returning to the starboard rail, and glassing the flat-roofed mud buildings that just showed amongst the palm trees. Each time his junior officers stiffened expectantly, then relaxed as he turned away without a word or change of expression.

This time Clinton saw movement, the flash of white robes in the deserted, single street of the town, and as he watched through the telescope he felt a prickle of excitement and a lift of self-congratulation. A small deputation was emerging from the grove and coming down the beach.

'Pass the word to my steward to lay out my number ones and sword,' he ordered without lowering his glass. The party on the beach was led by a portly figure in blindingly white robes and a full headdress that gleamed with gold. Behind him a bearer carried the long floating banner, scarlet and gold, of the Sultan.

'We'll treat him as a Governor,' Clinton decided. 'And give him four guns.' With that he turned on his heel and went to his cabin to change his uniform.

The Arab climbed out of the little felucca and came in through the entry port puffing for breath, assisted by two house-slaves. As his foot touched the deck, the first gun of the salute crashed out unexpectedly, and the Sheikh let out a whinny like a wild stallion and leapt two feet straight up in the air, the high colour flying from his cheeks leaving them ashen and trembling.

Clinton stepped forward, resplendent in cocked hat, blue and gold jacket, white breeches and sword and took the Sheikh's arm to steady him through the rest of the salute, and to prevent him stampeding back into the crazily rocking felucca where the oarsmen were in equal terror.

'Will you step this way, Your Excellency,' Clinton murmured and, without releasing his iron grip on the Sheikh's pudgy arm, marched him briskly down to his cabin.

Translation was a problem, but one of the Sheikh's entourage had a smattering of French and some English. It was almost dark before Clinton was able to see through the flowery verbiage and the atrocious rendering of his mother tongue. When it came it was like a great light filling the cabin, and Clinton found himself buoyed up with a savage, warlike glee.

The fat Sheikh, Governor of Elat, with his soft, red lips, was asking for the protection of her Britannic Majesty against the injustices and tyrannies of the Sultan of Zanzibar.

'*Dites lui je ne peux pas* – oh, damn it, tell him I can only protect him if he declares Elat free of the Sultan's dominions, *comprenez vous?*'

'*Je m'excuse, je ne comprends pas.*'

It was tedious, especially in view of Clinton's eagerness

to remove the province of Elat from under the sway of Zanzibar.

The Foreign Office had provided all commanders of the Atlantic anti-slavery squadrons with blank treaty forms, drawn up with deference to protocol and in correct legal terminology. These were for signature of any indigenous chiefs, warlords, petty princes and native kings who could be induced to place their mark upon them.

These documents started with a declaration of mutual recognition between Her Majesty's government and the signatory, went on in vague terms to promise protection and free trade, and ended in very specific terms with a round condemnation of the slave trade and the granting of rights to Her Majesty's government to search, seize and destroy all ships engaged in such trade within the signatory's territorial waters. Further it granted rights to Her Majesty's Navy to land troops, destroy barracoons, free slaves, arrest slave-masters and do any such other act as should be deemed necessary to the extinction of the trade in all the signatory's lands and possessions.

Admiral Kemp in Cape Town had overlooked the fact that Captain Codrington had a good supply of these documents in his possession. They had been intended for use entirely on the west African coast north of the equator. The good Admiral would have been a very worried man indeed if he had realized that he had detached his most brilliant but mercurial junior on independent patrol armed with anything so explosive.

'He must sign here,' Clinton explained briskly, 'and I will give him an order on the British treasury for a hundred guineas.' The treaty made provision for annual tribute to be paid to the signatory. Clinton considered a hundred guineas sufficient. He was not sure of what authority he had to write treasury orders, but Sheikh Mohamed was delighted. He had negotiated for life alone, and received not only the protection of this fine warship but the promise of good gold as well. He was smiling happily, pursing his red lips as he signed his long signature under his new title 'Prince and Supreme Ruler of the sovereign possessions of Elat and Ras Telfa.'

'Good,' said Clinton briskly, rolling his copy of the treaty and slipping the retaining ribbon over it as he hurried to the door of his cabin.

'Mr Denham,' he bellowed up the companionway. 'I want a landing-party, muskets, pistols, cutlasses and carrying combustibles, forty men ready to go ashore at first light tomorrow!' He was grinning as he turned back and told the Sheikh's translator, 'It would be best if His Excellency remained on board tonight. We will see him safely installed at noon tomorrow.' And for the first time the Sheikh felt a thrill of apprehension. This Ferengi had the cold blue merciless eyes of a devil. 'El Sheetan,' he thought, 'the very devil.' And made the sign against the evil eye.

'Sir, may I speak?' Mr Denham, *Black Joke*'s First Lieutenant looked puzzled in the light of the binnacle. It was an hour short of sunrise and he glanced down at the ranks of armed seamen squatting on the foredeck.

'Speak your heart,' Clinton invited him magnanimously. Lieutenant Denham was not accustomed to this jovial mood from his captain, and he expressed himself cautiously. In essence Lieutenant Denham's views came very close to those of the Admiral in Cape Town.

'If you would like to make a protest against my orders, Lieutenant,' Clinton interrupted him cheerfully, 'I will be pleased to enter it in the ship's log.'

Thus absolved of responsibility for having been party to an act of war on the territory of a foreign ruler, Lieutenant Denham was so relieved that when Clinton told him, 'I am taking command of the landing party. You will command the ship in my absence,' he shook Clinton's hand impulsively.

'Good luck, sir,' he blurted.

They went ashore in two boats, the whaler leading through the pass in the reef and the gig following two lengths astern. The moment the keel touched, Clinton sprang knee-deep into the blood-warm water and the rush of armed men followed him ashore. He drew his cutlass and his shoes squelched as he led his team of five men to the nearest dhow at a dead run.

As he jumped down from the ladder on to the dhow's heavily canted deck, an Arab watchman ducked out of the stern cabin and aimed a long jezail at Clinton's head. The range was point blank and Clinton struck out instinctively in an underhand parry, just as the gun's lock clicked and smoke and spark shot from the pan under the steel and flint.

His blade clashed against the steel barrel, deflecting it upwards as the jezail roared an instant after the snap of the lock, and a blinding billow of smoke and burning powder struck his face and singed his eyebrows, but the chunk of beaten potleg howled inches over his head. When his vision cleared the watchman had thrown his empty weapon aside, leaped over the side of the dhow and was hopping and hobbling across the sand towards the grove of palms.

'Search her, and then put fire into her,' Clinton ordered brusquely.

It was the first chance he had had to look across at the other dhows of the fleet. One of them was already on fire, the flames bright in the early light, rising straight up with little smoke. The furled mainsail was blackening like a dried leaf, and he could hear the crackle of the tinder-dry timbers of the hull and stern cabin. His seamen were spilling out of her and straggling across to the next vessel.

'She's aflame, sir,' his boatswain panted, and a hot gust of air struck Clinton's cheek at that moment and a quiver of heat hung over the main hatch.

'We'd best be getting on,' he said mildly, and scrambled down the ladder on to the packed damp sand. Behind him the flames roared like a cageful of wild animals.

The biggest dhow, a two-hundred tonner, lay ahead of them and Clinton reached it fifty paces ahead of his men.

'Make sure there is nobody below,' he ordered, and one of the seamen came back on deck carrying a rolled silk prayer rug under his arm.

'Belay that!' snapped Clinton. 'There'll be no looting.'

Reluctantly the seaman dropped the precious burden back into the hatchway, and the flames sucked up in a hot breath to accept it as though it was an offering to Baal.

By the time they reached the tree line, all eight of the stranded vessels were burning fiercely, the stubby masts collapsing as they burned through at the base, the furled sails disappearing in bright explosions of flame. In one of the burning hulls a keg of powder went up with a thunderous crash of sound, and a tall column of dark smoke hovered over the beach for a few seconds, shaped like a gigantic grey octopus before it drifted slowly out across the reef, leaving the dhow shattered, its timbers scattered across the sand, the flames extinguished by the shock wave of the explosion.

'Was there anybody aboard her?' Clinton demanded quietly.

'No, sir.' His boatswain was panting beside him, redfaced with excitement, and with a bared cutlass in his hand. 'All accounted for.'

Clinton hid his relief behind a cool nod, and spent a few precious minutes drawing his men into an orderly formation, giving them time to regain their breath, and getting them well in hand again.

'Check your muskets,' he ordered, and there was the click of the locks. 'Fix bayonets.' Metal rattled on metal as the long blades were fitted to the barrels of the Enfield rifles. 'If there is resistance we'll find it in the town, I fancy,' and he ran an eye down the uneven ranks. They were neither marines nor lobster-backs, he thought with quick affection. They might not be perfect in drill, but they were men with spirit and initiative, not parade-ground automatons.

'Come along then.' He waved them forward into the dusty street between the mud-brown flat-roofed buildings. The town smelled of wood smoke and raw sewage, of rice cooked with saffron and of ghee, clarified butter.

'Shall we burn 'em?' His boatswain jerked a thumb at the buildings that flanked the deserted street.

'No, we are here to protect them,' Clinton told him stiffly. 'They belong to our new ally, the Sheikh.'

'I see, sir,' the boatswain grunted, looking mildly perplexed, and Clinton took pity on him.

'We are after the barracoons,' he explained, as they trotted up the street in compact formation. They halted where the road branched left and right.

The heat was oppressive and the silence menacing. There was no wind and the coconut groves had stilled the eternal clatter of their fronds. From the beach far behind them, came the faint popping of burning timbers, and overhead the ubiquitous pied crows of Africa circled and cawed raucously, but the buildings and dense coconut groves were deserted.

'I don't like this,' one of the men croaked behind Clinton. He could understand the man's point of view. A seaman always felt awkward when parted from his ship, and there were a mere forty of them, out of sight of the beach and surrounded by thousands of unseen but none the less savage warriors. Clinton knew he must keep the momentum of surprise rolling through the town, yet he hesitated a moment longer until he realized that the amorphous sacklike shape lying on the edge of the right hand street was a human body, naked and black and very dead. One of the slaves trampled in the previous day's panic and left where he had fallen.

That way must lie the barracoons, he decided. 'Quiet!'

he cautioned his men and, cocking his head, listened with all his attention to the faint sussuration on the still air. It might have been the wind except there was no wind, or the flames, except that the flames were behind them. It was the distant sound of human voices, he decided, many voices, thousands of voices.

'This way. Follow me.' They went forward at a full run, taking the right fork and running immediately into the ambush which had been so carefully prepared for them.

The volley of musket fire crashed out from both sides of the narrowing track, and powder smoke rolled out towards them and hung like a thick, pearly curtain amongst the palm boles and the cashew nut trees.

Through the smoke danced the ethereal robed figures of the attackers, brandishing the long-barrelled jezails or swinging the half-moon-bladed scimitars, with wild shrieks of 'Allah Akbar, Allah is great!'

They rushed down on the little band of seamen, caught in enfilade on the narrow track. There were at least a hundred of them, Clinton judged instantly, and they were pressing in determinedly. Those scimitars were glittering – bright bare steel has a particularly chilling effect.

'Close up,' Clinton shouted. 'We'll give them a volley then take the bayonet to them, through the smoke.'

The first rank of racing Arabs were almost on top of the levelled Enfields. Incongruously, Clinton noticed that many of them had tucked up the skirts of their robes, leaving their legs bared to the thighs. Their skins varied from the colour of ivory to tobacco, and there were wrinkled grey beards in the front rank, screaming and howling with rage and battle-lust. They had just seen their livelihood burned to heaps of ash upon the beach. All that remained to them of their wealth was the contents of the barracoons set back amongst the groves of cashew nut and coconut trees.

'Fire!' roared Clinton, and the solid blast of sound deafened him for a moment. The gun smoke wiped out all vision ahead of him and then hung on the windless morning in an impenetrable fog bank.

'Forward!' howled Clinton and led the charge into the smoke. He stumbled over the body of an Arab. The man's turban had unwound and come down over his eyes, soaked with blood like the sultan's scarlet banner which Clinton could see waving ahead of him above the smoke.

A figure loomed ahead of him, and he heard the fluting whistle of a scimitar blade, like wild goose wings overhead. He ducked. The sharp breeze puffed a loose strand of hair into his eyes, as the blade passed an inch from his forehead, and Clinton straightened from the knees and put his whole body into the counter-lunge.

The point of his blade went in with a dead, soggy feel, sliding grudgingly through flesh until the point grated on bone. The Arab dropped his scimitar and clutched the cutlass blade with bare hands. Clinton leaned back and jerked the cutlass free of flesh. As the blade slid through the Arab's nerveless fingers the tendons parted with a faint popping sound, and the man went down on his knees, holding his multilated hands up in front of his eyes with a look of amazement on his face.

Clinton ran on to catch up with his seamen, and found them scattered in little groups amongst the grove, laughing and shouting with excitement.

'They've run like steeplechasers, sir,' the boatswain called. 'Grand National, ten to one the field!' He snatched up the fallen banner of the Sultan and waved it furiously over his head, completely overtaken by excitement.

'Did we lose anybody?' Clinton demanded. He also felt the dizzy euphoria of battle. The killing of the Arab, far from sickening him, had elated him. In that moment he was quite capable of turning back and taking the man's scalp. However, the question sobered them.

'Jedrow caught one in the belly, but he can walk. Wilson got a sword cut in the arm.'

'Send them back to the beach. They can escort each other. The rest of you, come on!'

They found the barracoons a quarter of a mile further on. The guards had fled.

The slave-pens stretched out for a half mile along both banks of a small stream that provided both drinking water and sewage disposal for the inmates.

They were unlike the barracoons that Clinton had captured and sacked on the west coast, for those had been built by European traders with the white man's orderly eye. There was no resemblance in these sprawling compounds built of rough, unbarked forest poles, bound together with rope made from plaited palm fronds. Behind the outer barricades were open godowns with thatched roofs in which the chained slaves could find some shelter from sun and rain. The only thing the same was the smell. An epidemic of tropical dysentery had swept through the barracoons and most of the sheds contained the decomposing bodies of the victims. The crows and buzzards and vultures were waiting patiently in the palms and cashew trees, misshapen, dark silhouettes against the hard bright blue of the morning sky.

Clinton met the new ruler of the state of Elat, Sheikh Mohamed, at the water's edge and escorted him up the beach. The incoming tide was dousing the piles of smouldering ash that marked the last resting places of eight fine dhows, and the Sheikh tottered uncertainly, like a man in deep shock, relying for support on the sturdy shoulder of one of his house-slaves, looking about him with lugubrious disbelief at the carnage that had overtaken him. The Sheikh owned one third shares in every one of those smoking piles of ash.

He had to rest when they reached the tree line above the

beach. A slave placed a carved wooden stool in the shade, and another waved a fan of plaited palm fronds over his head to keep off the flies and to cool his heated brow that was dewed with the heavy sweat of despair.

His misery was completed by the lecture in broken French and pidgin English which 'El Sheetan', the mad British sea captain with the devil's eyes, was relaying to him through the shocked and incredulous interpreter. Such things could only be repeated in a hoarse whisper, and the Sheikh greeted each new revelation with a soft cry of 'Waai!' and the upturning of his eyes to heaven.

He learned that the village blacksmiths had been dragged out of the bushes and were already knocking the fetters off long rows of perplexed slaves.

'Waai!' wailed the Sheikh. 'Does the devil not realize that those slaves have already been purchased and that the tax has been collected.'

Comfortably Clinton explained that once the slaves were freed, they would be marched back into the interior, and the Sheikh would send guards with them to see them safely home, and to warn any slave caravans that they encountered on the down-route that all the ports of Elat were now closed to the trade.

'Waai!' This time the Sheikh's eyes actually brimmed with tears. 'He will beggar me. My wives and children will starve.'

'El Sheetan counsels you to enlarge the trade in gum-copal and copra,' explained the interpreter in a sepulchral voice. 'And as your closest ally, he promises to call upon you regularly with his great ship of many guns, to make certain that you heed this advice.'

'Waai!' The Sheikh plucked at his beard, so that long curly hairs came out between his fingers. 'This ally makes one long for ordinary enemies.'

Twenty-four hours later *Black Joke* sailed into Telfa bay, forty miles further up the coast. Nobody had thought to warn the slaving fleet that was anchored there of the new policies of the state of Elat to which the territory now belonged.

The five dhows in the outer anchorage managed to cut their anchor cables and slip away into the maze of shallow coral channels and shoals to the north of the bay, where *Black Joke* could not follow.

However, there were another six smaller vessels on the beach and four magnificent double-decked ocean-going dhows lying in the inner anchorage. Clinton Codrington burned two of them and seized the four newest and biggest vessels, put prize crews into them and sent them south to the nearest British base at Port Natal.

Two days later, off the beach at Kilwa, Clinton Codrington exercised his ship at gunnery practice. Running out his thirty-two pounders, and firing them in broadsides which set the surface of the lagoon seething and dancing with foam and white fountains of spray. The thunder of gunfire burst against the far hills and rumbled back across the sky like cannon-balls rolled across a wooden deck.

The Sultan's local Governor was reduced to a quivering jelly of terror by this display of might, and had to be carried bodily into *Black Joke*'s whaler to be rowed out to a conclave with the gunboat's Captain. Clinton had the treaty forms already filled out and ready for signature when they carried the Governor aboard to learn that he was heir to a kingdom to which he had never aspired, and a title which he knew was too grandiose not to bring with it certain retribution from somebody whose name he did not dare to breathe aloud.

Admiral Kemp, sitting in his study in the magnificent mansion of Admiralty House, overlooking the wide smoky-blue haze of the Cape flats to the far mountains of the Hottentots Holland, hopefully dismissed the first reports as the wild imaginings of some crazed subordinate who had served too long in the godforsaken outpost of Port Natal, and who was suffering from the bush madness of 'El Cafard' that sometimes affects a person so isolated.

Then the details began to arrive with every despatch from the north, and they were too graphic to be lightly dismissed. An armada of captured prizes was arriving in the bay of Port Natal, twenty-six sizable dhows to date, some of them loaded with slaves.

The Lieutenant-Governor of Port Natal was desperate for the Admiral's advice as to what should be done with the dhows. The slaves had been taken ashore, released and immediately been contracted as indentured labourers to the hardy and hopeful gentlemen who were attempting to raise cotton and sugarcane in the wilderness of the Umgeni valley. The shortage of labour was critical, the local Zulu tribesmen much preferred cattle raiding and beer drinking to agricultural labour, so the Governor would be delighted to receive as many freed slaves as the Royal Navy wished to send him. (The Admiral was not entirely certain of the difference between indentured labourers and slaves.) However, what was the Lieutenant-Governor to do with twenty-six – no, the latest figure was thirty-two captured dhows. A further flotilla of six vessels had arrived as the Governor was dictating his report.

Two weeks later, one of the captured dhows, which had been purchased into the Colonial Service by the Lieutenant-Governor, arrived in Table Bay bearing a further batch of despatches.

One of these was from Sir John Bannerman, H.M. Consul on the island of Zanzibar. Another was from the Sultan of Zanzibar in person, with copies to the Foreign Secretary in

London and, quite remarkably, to the Governor-General in Calcutta. The Sultan evidently believed that as the representative of the Queen of England the Governor-General would have some jurisdiction in the Indian Ocean, which was virtually his front garden.

Admiral Kemp split the seals on both packets with a queasy feeling of impending doom.

'Good God!' he groaned, as he began reading, and then, 'Oh sweet merciful Jesus, no!' And later, 'It's too much, it's like some sort of nightmare!'

Captain Codrington, one of the most junior post-captains on the Admiralty list, seemed to have taken powers unto himself which would have made a Wellington or a Bonaparte pause.

He had annexed to the British Crown vast African territories, which hitherto had formed part of the Sultan's dominions. With a high hand he had negotiated with various local chiefs and dignitaries of dubious title and authority, pledging recognition and good British gold. 'Good God!' the Admiral cried again in real anguish, 'What will that blighter Palmerston have to say.' As a staunch Tory, Kemp had no great opinion of the new Whig prime minister.

Since the troubles in India, the sepoy risings of a few years previously, the British government was very wary of accepting further responsibility for overseas territory and backward peoples. Their orders were specific, and Captain Codrington's recent activity went far from the essence of those orders.

The scramble for Africa was still in the future, and the spirit of the Little Englanders motivated British foreign policy – of this Admiral Kemp was very painfully aware. Daunting as this was, yet it was far from the entire story, Kemp realized, as he read on into the Consul's despatch, his breath rasping hoarsely, his colour rising steadily, and his eyes behind the gold-wire framed reading-glasses swimming with tears of rage and frustration.

'When I get my hands on that puppy—' he promised himself.

Captain Codrington seemed to have declared single-handed war upon the Sultan. Yet even in his outrage the Admiral felt a prickle of professional appreciation for the scope of his subordinate's operations.

There was a formidable list of over thirty separate incidents recited by H.M. Consul. The puppy had stormed fortified castles, raided ashore to burn and destroy barra-coons, released tens of thousands of slaves, seized slaving vessels on the high seas, burned others at their moorings, and wreaked the kind of chaos worthy of a marauding Nelson himself.

The Admiral's reluctant admiration for Codrington's technical conduct of the campaign in no way lessened his determination to exact vengeance for the disruption of his life and career that those actions presaged.

'Nothing can save him this time. Nothing!' the Admiral rumbled, as he turned to study the Sultan's protest. This was obviously the work of a professional letter-writer, and every paragraph began and ended with incongruous and flowery enquiries after Kemp's health, between which were sand-wiched cries of anguish, screams of outrage, and bitter protest against the broken promises and treaties of Her Majesty's government.

At the very end the letter-writer had not been able to resist adding a prayer for the Admiral and the Queen's prosperity and health in this life, and happiness in the one to follow. This detracted a little from the tone of injury in which the protests and demands had been couched.

The Sultan assessed his losses at over fourteen lakhs of rupees, almost a million of sterling, in plundered shipping and released slaves, and that did not take into account the irreplaceable damage to his prestige, nor the break-down of the entire trade along the coast. The confusion was such that some ports might never again be opened to the trade.

The system of gathering slaves in the interior of the continent and the network of routes to the coastal ports had been so sadly disrupted that they might take years to re-open, to say nothing of the gross shortage of shipping resulting from the depredations of 'El Sheetan'. Those ports still open to the trade were swamped with patient slaves, waiting for the dhows which were already scattered wrecks upon the reefs and beaches of the Mozambique channel, or sailing southwards under prize crews.

'Nothing can save him,' repeated Admiral Kemp, and then paused. His own career was finished also. He realized that, and he felt the deep injustice of it. For forty years he had put not a single foot wrong, and his retirement was so close, so very close. He shook off the lethargy of despair, and began to draft his orders.

The first was to all ships of his squadron, to detach immediately and to steam in search of *Black Joke*. In despair he realized that it might take as much as six weeks for his orders to reach his commanders, for they were scattered across two oceans. It might also take as long again for them to search out the errant gunboat in the maze of islands and bays along the Mozambique channel.

However, when they did so, Captain Codrington was to be relieved of his command with immediate effect. Lieutenant Denham was to take over as temporary commander, with orders to bring *Black Joke* into Table Bay as soon as possible.

Admiral Kemp was confident that he could assemble sufficient senior officers on the Cape Station to convene an immediate court martial. It might help his own position a little if he could report to the First Lord that a savage sentence had already been handed down to Codrington.

Then there was a despatch to H.M. Consul in Zanzibar, suggesting he keep the Sultan reassured and quiescent until the situation could once more be brought under control – and until instructions could be forwarded from the Foreign

Office in London regarding possible redress and compensation, although naturally at this stage, no promises or commitments were to be given the Sultan, beyond expressions of good faith and commiseration.

Then there was the onerous task of making his report to the Admiralty. There were no words to soften the actions of his subordinate, and his own responsibility. Besides he had been a serving officer too long to make any such attempt. Yet when the bare facts were stated, even in the beloved unemotional jargon of the navy, they seemed so magnified that Admiral Kemp was himself utterly appalled, all over again. The packet-boat was delayed five hours while the Admiral completed, sealed and addressed this missive. It would be in London in less than a month.

His last despatch was addressed to the officer commanding Her Majesty's ship *Black Joke* in person. And in it Admiral Kemp allowed himself to give expression to some of his own bitterness, taking a sour sadistic pleasure in weighing the relative effectiveness of such words as 'corsair' and 'pirate' – or 'malicious' and 'irresponsible'. He had his little masterpiece of venom written out in five copies to be disseminated in every direction and by every available means that might most speedily bring the puppy to heel. Yet when they were sent, all he could do was wait – and that was the worst part of the affair. Uncertainty and inaction seemed to corrode his very soul.

He dreaded each new arrival in Table Bay, and whenever the signal gun on the hill above the town boomed its brief feather of gunsmoke, his spirit quailed and that sour ache of dread stabbed him in his guts.

Each new despatch lengthened the toll of destruction and depredation, until at last there was a report from the culprit himself, sewn up in a package of canvas and addressed to Admiral Kemp, delivered by the prize crew of a particularly valuable dhow over eighty feet long and of a hundred-ton burden.

The tone in which Captain Codrington listed his achievements infuriated the Admiral as much as the deeds themselves. In an almost casual opening paragraph, Captain Codrington recorded the addition of some million square miles of Africa to the Empire.

He had the grace to admit that his action may have exceeded his orders, and he explained away the discrepancy winningly. 'It had been my firm intention to avoid scrupulously, whilst on this service, every act of a political nature. However, I was forced to accept the cession of the kingdoms of Elat and Telfa by the entreaties of the Sheikh and the Imam – together with that of their people, who seek refuge from the inimical and savage acts of the Sultan of Zanzibar.'

This was hard fare to serve their Lordships, especially the First Lord, Lord Somerset, who had always grudged the use of his men and ships to fight against slavery. However, much worse was to follow. Captain Codrington went on to lecture the Admiral and to deliver a few homilies for their Lordships' instruction.

'By God's providence, an Englishman with no other force than the character of his noble nation has brought to these poor people salvation. Their Lordships must pardon me for using an unfashionable argument,' a sneer at the Little Englanders, 'however, it is as clear to me as the African sun that God has prepared this continent for the only nation on earth that has the public virtue sufficient to govern it for its own benefit, and for the only people who take the revealed word for their moral law.'

Admiral Kemp gulped as he read it, swallowed the wrong way, and was prostrated with a coughing fit from which he recovered some minutes later to read on.

'In all the foregoing I have been influenced by no personal motive or interest, by no desire of vainglory, but my endeavours have solely been to use the powers granted me to the honour of my God, my Queen and to the benefit of my country and all mankind.'

The Admiral removed his reading glasses and stared at the glass case of stuffed songbirds on the wall opposite him.

'To write that, he is either the world's greatest fool – or a brave man or both,' he decided at last.

Admiral Kemp was wrong in his estimate. In fact, Clinton Codrington was having an attack of cockiness and self-importance occasioned by the sense of limitless power which this command had given him. He had been wielding this power for many months, and his judgement and good sense had warped. Yet he still truly believed that he was fulfilling, in this order, the will of God, his patriotic duty and the spirit of his orders from the Lords of the Admiralty.

He was also fully aware that he had demonstrated superior professional ability in carrying out a series of land and sea actions, nearly always against superior numbers, without a single reverse and with the loss of only three men killed in action and less than a dozen wounded. There was only one area in which Clinton had any reservations regarding the success of the patrol up to this time.

Huron was still on the coast, he had intelligence of her from a dozen sources. Mungo St John was trading, paying top prices for only the best merchandise, hand-picked by either himself or his mate, the bald yellow giant, Tippoo. They were taking only healthy, mature men and women fit to withstand the long voyage back around Good Hope and across the middle passage, and there was a shortage of this type of merchandise.

For Clinton every dawn brought the hope that *Black Joke* would once again raise that towering pyramid of beautiful white sails, but each day the hope faded slowly with the passage of the fierce tropical sun across the heavens, to be extinguished each night as the huge red orb plunged into the sea, and to be resurrected again with the next dawn.

Once they missed the big American clipper by a single day. She had slipped out of the bay at Lindi twenty-four hours before *Black Joke*'s arrival, after taking on fifty prime slaves. The watchers from the beach could not be certain if she had turned north or south, for she had made her offing below the horizon, and had been lost to sight from the shore before coming around on to her intended course.

Clinton guessed *Huron* would go south, and had steamed in pursuit for three days, over an empty sea, down a seemingly deserted coast with barren anchorages before admitting that St John had sailed away from him again, and he was forced to abandon the hunt.

At the very least he knew that if *Huron* had turned north, St John was still trading, and there was always a chance of another encounter. He prayed for it every evening. It was all that he needed to make this patrol a clean sweep, so that he could fly a broom from his masthead when he sailed into Table Bay again.

This time he had the sworn and witnessed depositions of the men who had sold St John his slaves as proof positive that *Huron* was trading. He did not have to rely on the equipment clause, or the dubious rights of search. He had his proof and somehow he knew his chance would come.

The tide was running at full flow for Clinton. He was imbued with a new sense of worthiness, a new, iron-clad confidence in himself and his own good fortune. He carried himself differently, chin higher, shoulders squarer, and if his gait was not yet a swagger, it was at least an assured stride. He smiled more often and when he did there was a wicked curl to his upper lip and a devilish twinkle in the pale blue eyes. He had even grown a full moustache, golden and curling that gave him a piratical air, and his crew, who had always respected his cold precise management of the ship, but felt little affection for him, now cheered him when he came back on board from one of his forays ashore.

'Good old Tongs!' It was his new nickname from 'Hammer and Tongs'. They had never had a pet name for him before, but now they were a proud ship, proud of themselves, and proud of their twenty-seven-year-old 'Old Man'.

'Give 'em hell, Tongs!' they cheered behind him, as he led them, naked cutlass in hand, over the outer stockade of a barracoon with musket smoke swirling around his lanky figure.

'At 'em, the Jokers!' they cheered themselves, as they leaped the gap between *Black Joke*'s rail and the deck of a slaving dhow, swinging their cutlasses, pistols popping as they drove the slavers down into their own holds and battened the hatches down on them, or chased them over the side where the sharks were cruising.

They knew they were creating a legend. Tongs and his Jokers sweeping the slavers from the Mozambique channel, a hell of a story to tell the nippers back home, and a good slice of prize money to prove the tale.

It was in this mood that *Black Joke* sailed into Zanzibar harbour, the stronghold of the Omani Sultan, little Daniel into the lion's den. The gunners on the parapets of the fort, though they stood with the slow match burning in their hands, could not bring themselves to touch them to the huge bronze cannon as the ugly little gunboat came fussing up the Zanzibar roads.

Black Joke had her yards manned with neatly uniformed seamen. A spectacular display, geometrical white ranks of men against the backdrop of tropical anvil-headed thunder-clouds.

Her officers were in cocked hats and full ceremonial dress, uniform, swords, white gloves and white breeches, and as she made her turn into the inner harbour, *Black Joke* began firing her courtesy salute – which was a signal for most of the population to head for the hills, jamming the narrow alleyways of the old city with a lamenting torrent of refugees.

The Sultan himself fled his palace, and with most of his court took refuge in the British consulate, overlooking the harbour.

'I am not a coward,' the Sultan explained bitterly to Sir John Bannerman, 'but the Captain of that ship is a madman. Allah himself does not know what he will do next.'

Sir John was a large man, of large appetite. He possessed a noble belly like the glacis of a mediaeval castle and full mutton-chop whiskers around a florid face, but the clear intelligent eyes, and the wide friendly mouth of a man of humanity and humour. He was a noted oriental scholar, and had written books of travel and of religious and political appraisal of the East, as well as a dozen translations of minor Arabic poets.

He was also a confirmed opponent of the slave trade, for the Zanzibar markets were held in the square below the windows of his residence and from his bedroom terrace he could watch on any morning the slaving dhows unloading their pitiful cargoes on the stone wharf they called, with cruel humour, the 'Pearl Gate'.

For seven years he had patiently negotiated a series of treaties with the Sultan, each one nipping a few more twigs off the flourishing growth which he detested, but found almost impossible to prune effectively, let alone root out entirely.

In all the Sultan's territories Sir John had absolute jurisdiction only over the community of Hindu traders on the island, for they were British subjects, and Sir John published a bulletin requiring them to free all their slaves forthwith, against a penalty of £100 for non-compliance.

His bulletin made no mention of compensation, so the most influential of the merchants sent Sir John a defiance which was the Pushtu equivalent of 'The hell with you and your bulletins'.

Sir John, with his one good foot, personally kicked in the merchant's door, dragged him out from under his charpoy bed, dropped him to his knees with a full-blooded round-house punch, chained him around the neck and marched him through the city streets to the consulate and locked him in the wine cellar until the fine was paid and the slaves' manumission papers signed. There had been no further defiance and no takers for the Hindu merchant's subsequent, privately circulated offer to pay another £100 to anybody who would stick a knife between Sir John's ribs during one of his evening promenades through the old city. Thus it was that Sir John was still bluff and hale as he stood on his terrace puffing a cigar – his only indisposition was the gouty foot thrust into a carpet slipper. He watched the little black-hulled gunboat coming up the channel.

'She behaves like a flagship,' he smiled indulgently, and beside him, Said the Sultan of Zanzibar, hissed like a faulty steam valve.

'El Sheetan!' His wrinkled turkey neck turned bright red with impotent anger, his bony nose beaked like that of an unhappy parrot. 'He sails here, into my harbour, and my gunners stand by their cannon like dead men. He who has beggared me, who has plunged my empire into ruins – what does he dare here?'

The answer that Clinton 'Tongs' Codrington would have given him was quite simple. He was carrying out to the letter the orders given him in Cape Town many months previously by Admiral Kemp, the Commander of the South Atlantic and Indian Ocean Squadron.

'You are further requested and required to take advantage of the first opportunity to call into the harbour of Zanzibar, where you will accord to his Royal Highness the Sultan of Zanzibar full honours, while taking the advice of Her Majesty's Consul, Sir John Bannerman, as to reinforcing existing treaties between His Royal Highness and her Britannic Majesty's Government.'

Which, being translated, was an instruction to show the Union Jack against a background of thirty-two pounder cannon, and by doing so remind the Sultan of his commitments under the various treaties.

'To teach the naughty old beggar to mind his P's and Q's,' as Clinton explained cheerfully to Lieutenant Denham with a twirl of his new golden moustache.

'I would have thought, sir, that the lesson had already been given,' Denham answered darkly.

'Not at all,' Clinton demurred. 'The treaties with the new sultans on the mainland no longer affect the Zanzibar fellow. We still have to ginger this old boy up a little.'

Sir John Bannerman limped up on to *Black Joke*'s deck, favouring his gouty foot and cocked a lively eye at the young naval officer who stepped forward to greet him.

'Well, sir, you have been busy indeed,' he murmured. My God, the fellow was little more than a boy, a fresh-faced youngster, despite the cocked hat and moustache. It was difficult to believe that he had created such havoc with this tiny ship.

They shook hands, and Bannerman found himself liking the boy, despite the turmoil that he brought into the Consul's normally tranquil existence.

'A glass of madeira, sir?' Clinton suggested.

'Damned decent of you, I must say.'

In the small cabin, Bannerman mopped his streaming face, and came directly to business.

'By God, you've put the cat amongst the pigeons,' he wagged his big head.

'I don't see . . .'

'Now, listen to me,' Bannerman snapped, 'and I'll explain to you the facts of life as they apply to eastern Africa in general, and Zanzibar in particular.'

Half an hour later, Clinton had lost much of his new-found bumptiousness.

'What should we do?' he asked.

'Do?' Bannerman asked. 'What we do is take full advantage of the situation which you have precipitated, before these idiots in Whitehall come stumbling in. Thanks to you the Sultan is at last in a mood to sign the treaty I have been after for five years. I'll trade a handful of these utterly illegal, untenable treaties that you have made with non-existent states and mythical princes for one that will truss the old goat up the way I've wanted him for years.'

'Excuse me, Sir John,' Clinton looked slightly perplexed, 'from what you said earlier, I understood that you heartily disapproved of my recent actions.'

'On the contrary,' Sir John grinned at him expansively, 'you have stirred my blood, and made me proud to be an Englishman again. I say, do you have a little more of the madeira?'

He raised the glass to Clinton. 'My hearty congratulations, Captain Codrington. I only wish that I could do something to save you from the fate that so certainly awaits you, once the Admiralty and Lord Palmerston catch up with you.' Sir John drank half the glass, smacked his lips, 'Jolly good stuff,' he nodded, set the glass aside and went on briskly, 'Now, we have to work fast and get the Sultan to sign an iron-clad treaty – before Whitehall rushes in with apologies and protestations of good faith which will put to naught all the fine work you have done to date. Something tells me that won't be very long,' he added lugubriously, and then more brightly, 'You could have your ship's guns run out whilst we are ashore. Do wear your sword. Oh, and one other thing, don't take your eyes off the old goat while I do the talking. There is already talk about your eyes – that extraordinary colour of blue, don't you know – and the Sultan has heard about them already. As you probably know, they now call you "El Sheetan" on this coast, and the Sultan is a man who sets great store by djinns and the occult.'

Sir John's predictions as to the imminent arrival of tidings from higher sources was almost clairvoyant, for as he spoke H.M. sloop *Penguin*, with urgent despatches on board for Sir John Bannerman, for the Sultan, and for Captain Codrington, was on a fair wind, which, if it held, would bring her into Zanzibar harbour within the next two days. Time was shorter than even Sir John believed.

With some trepidation, the Sultan had moved back into his palace. He had only half believed Sir John's assurances, but, on the other hand, the palace was half a mile from the harbour where that evil black ship was displaying its formidable broadside of cannonades, while the consulate was on the harbour front – or the front line of fire, depending on how one looked at it.

On Sir John's advice Clinton had come ashore with a bodyguard of a dozen picked seamen, who could be trusted to resist the temptations of the old city's red-light area, the grog and the women that seamen dream of. It was dusk when the party plunged into the labyrinth of narrow alleys, where the balconies almost met overhead, led by Sir John who despite his game leg set a good pace, picking his way around heaps of noisome garbage and avoiding the puddles in the uneven paving that looked like a cold minestrone soup and smelled a great deal higher.

He chatted affably with Clinton, pointing out the various sites and buildings of interest, relating the island's history and giving a quick perceptive character-sketch of the Sultan and the more important men in his empire, including those unfortunate new princes who had signed Clinton's blank treaty forms.

'That's one thing, Sir John. I wouldn't want anything to happen to them,' Clinton cut in for the first time. 'I hope they won't be victimized for having, well, how can one say, for having seceded from the Sultan's empire—'

'Forlorn hope.' Sir John waggled his head. 'Not one of them will be alive by Ramadan. The old goat has a nasty streak.'

'Couldn't we put a clause in the new treaty to protect them?'

'We could, but it would be a waste of paper and ink.' Sir John clapped his shoulder. 'Your concern is misplaced. The finest collection of ruffians, rogues and assassins south of the equator, or north of it, for that matter. One of the side benefits of the whole business, getting rid of that lot. Old goat will have a lovely time, compensate him for the loss of face when he straps their heads or hands them the cup of datura tea. Ghastly death, datura poisoning. Oh, by the way, you must look at these gates.' They had reached the front of the palace. 'One of the most magnificent examples of craftsmanship on the island.'

The massive teak doors were fifteen feet high, intricately carved, but in accordance with Muslim law the carvings depicted neither human nor animal figures. They were the only impressive feature of the drab square building with its blank walls relieved by the wooden balconies high above street level, shuttered against the night air and the gaze of the curious.

The gates swung open at their approach, and the palace guards armed with ancient jezails were the first living beings they had seen since leaving the harbour. The city was still deserted, and cowering under the menace of *Black Joke*'s guns.

Clinton noticed, since Sir John had mentioned it, that the guards averted their gaze as he passed, one of them actually covering his face with the loose tail of his turban. So the business of the eyes was true. He was not sure whether to feel insulted or amused.

'You must see these.' Sir John stopped him in the cavernous ante-chamber lit with guttering oil lamps suspended in heavy brass chandeliers from a ceiling lost in the gloom. 'The heaviest recorded specimens in the world, one of them over three-hundred-pounds weight.'

They were a pair of African elephant tusks, suspended on the stone wall with retaining bands of copper, two incredible curves of ancient ivory, as thick as a girl's waist, taller than a man could reach, with hardly any taper from hilt to blunt tip, gleaming with the lustre of precious porcelain.

'You have not hunted these beasts?'

Clinton shook his head, he had never even seen one of them, but the huge teeth impressed him none the less.

'Before my foot, I shot them in India and in Africa. No other sport to beat that, incredible animals.' He patted one of the tusks. 'The Sultan killed this one when he was a young man, with a jezail! But there aren't any monsters like that around any more, more's the pity. Come along, mustn't keep the old goat waiting.'

They went on through half a dozen chambers, each of them Aladdin's caves of rare treasures, carved jade, beautifully worked ivory carvings, a palm tree and suspended moon, the symbol of Mohamed, in solid gold, carpets of silk and thread of gold and silver, a collection of fifty priceless Korans in silver and golden containers set with precious and semi-precious stones.

'Look at that shiner.' Sir John stopped again, and pointed out a native-cut diamond in the hilt of an Arabian scimitar. The diamond was cushion-shaped and a little out of true, but burning with a weird blue and icy fire even in the semi-dark. 'Legend says the sword was Saladin's, I doubt it, but the stone is one hundred and fifty-five carats. I weighed it myself.' And then as he took Clinton's arm and stumped off again, 'Old goat is rich as Croesus. He has been milking rupees out of the mainland for forty years, and his father for fifty years before that. Ten rupees for every slave, ten for every kilo of ivory, God knows how much for copra and gum-copal concessions.'

Clinton saw instantly why Sir John called him the old goat. The resemblance was startling, from the white, pointed beard and square yellow teeth to the mournful Roman nose and elongated ears.

He took one look at Clinton, catching his eye for a split part of a second, before he looked away hurriedly, blanching visibly, as he waved his two visitors to the piles of velvet and silk cushions.

'Keep the old beady eye on him,' Sir John counselled aside, 'and don't eat anything.' He indicated the display of sweetmeats and sugared cakes on the bronze trays. 'If they aren't poisoned, they'll probably turn your stomach anyway. It's going to be a long night.'

The prediction was accurate, the talk went on hour after tedious hour, in flowing Arabic hyperbole and flowery diplomacy, that concealed the hard bargaining. Clinton understood not a word. He forced himself not to fidget, though his buttocks and legs soon lost all feeling from the unusual position on the cushions, yet he maintained a stern expression and kept his gaze fixed on the Sultan's wrinkled and whiskered visage. Sir John assured him later that it had helped greatly to shorten the negotiations, yet it seemed a full round of eternity before Sir John and the Sultan were exchanging polite fixed smiles and deep bows of agreement.

There was a triumphant gleam in Sir John's eye, as he strode out of the palace, and he took Clinton's arm affectionately.

'Whatever happens to you, my dear fellow, generations unborn will have reason to bless your name. We have done it, you and I. The old goat has agreed. The trade will wither and die out within the next few years now.'

On the walk back through the narrow streets, Sir John was as lively and cheerful as a man returning from a convivial party rather than the bargaining table. His servants were still waiting his coming, and all the lamps in the consulate were burning.

Clinton would have liked immediately to go back aboard his ship, but Sir John restrained him with an arm about his shoulder, as he called for his Hindu butler to bring champagne. On the silver tray with the green bottle and crystal glasses was a small package in stitched and sealed canvas. While the uniformed butler poured the champagne, Sir John handed Clinton the package.

'This came in earlier on a trading dhow. I did not have the opportunity to deliver it to you before we left for the palace.'

Clinton accepted it warily, and read the address: – 'Captain Clinton Codrington, Officer Commanding Her Majesty's Ship Black Joke. Please forward to H.M. Consul at Zanzibar to await collection.'

The address was repeated in French, and Clinton felt a quick thrill kindle his blood as he recognized the bold round script in which the package was addressed. It took an effort to restrain himself from ripping the package open immediately.

However, Sir John was handing him a glass of wine, and Clinton suffered through the toasts, the loyal toast to the Queen, and that ironical one to the Sultan and the new treaty, before he blurted out, 'Excuse me, Sir John, I believe this to be a communication of importance,' and the Consul waved him into his study and closed the door after him to give Clinton privacy.

On the leather top of the marquetry desk, Clinton slit the seals and stitching of the package with a silver knife from the Consul's desk set. From it fell a thick sheaf of closely written notepaper, and a woman's earring of paste and silver, the twin to the one that Clinton wore under his shirt against his chest.

*B*lack Joke groped her way out through the dark, unbuoyed channel an hour before the first flush of dawn in the eastern sky. Turning southwards she set all canvas and worked up swiftly to her best speed.

She was making eleven knots when she passed the sloop *Penguin* a little before midnight the following night. *Penguin* bearing her urgent dispatches was hull down on the eastern horizon and her running lights were obscured by a heavy tropical deluge, the first fanfare of the coming monsoon that passed between the two vessels hiding them from each other's lookouts.

By dawn the two ships were fifty nautical miles apart, and rapidly widening the gap, while Clinton Codrington paced his quarterdeck impatiently, stopping at every turn to peer impatiently into the south.

He was hurrying to answer the most poignant appeal, the most pressing duty of a dutiful man, the call for succour from the woman he loved, a woman in terrible and pressing jeopardy.

*T*he flow of the Zambezi had a majesty that Zouga Ballantyne had seen on no other great river, neither the Thames, nor the Rhine, nor the Ganges.

The water was the almost iridescent green of molten slag pouring down the side of a steel-yard dump, and it formed powerful, slowly turning vortices in the angles of the broad bends, while in the shallows it seemed to roll upon itself as though the leviathan of all the world sported below its dark mysterious surface. Here the main channel was more than a mile across, though there were other lesser channels, and other narrower mouths beyond the waving banks of papyrus and cotton-headed reeds.

The small flotilla of boats hardly seemed to move against

the current. In the lead was the steam launch *Helen*, named after Zouga's mother.

Fuller Ballantyne had designed the vessel and had it manufactured in Scotland for the disastrous Zambezi expedition which had penetrated only as far as the Kaborra-Bassa gorge. The launch was almost ten years old now, and for most of that time had been the victim of the engineering prowess of the Portuguese trader who had purchased her from Fuller Ballantyne when the expedition was abandoned.

The launch's steam engine creaked and thudded, leaked steam from every pipe and joint, and sprayed sparks and thick black smoke from her wood-burning furnace, exerting herself far beyond the dictates of her age and maker's specifications as she towed the three deeply laden barges against the flow of the mighty river. They were making good a mere fifteen miles a day, and it was more than two hundred up river from Quelimane to Tete.

Zouga had chartered the launch and her barges to carry the expedition upstream to the jump-off point at Tete. He and Robyn rode in the first barge, together with the most valuable and delicate equipment: the medical stores, the navigational equipment, sextants, barometers and chronometers, the ammunition and firearms, and the personal camping gear.

In the third and last barge, under the bright and restless eye of Sergeant Cheroot, were the few porters that had been recruited at Quelimane. Zouga was assured that the additional hundred porters that he needed could be procured at Tete, but it had seemed prudent to sign on these healthy and vigorous men, as they became available. So far there had been no desertions, which was something unusual for the beginning of a long safari, when the proximity of home and hearth could be expected to exert sudden irresistible attractions on the weaker souls.

In the middle barge, on the tow-line directly behind Zouga, were the bulkier stores. In the main these were trade goods, cloth and beads, knives and axes, some cheap muskets and lead bars for ball, bags of black powder and flints. These were essential commodities with which to buy fresh provisions, to bribe local headmen for right to passage, to purchase concessions to hunt and prospect, and generally sustain the expedition's objects.

In charge of this middle barge was Zouga's newest and most dubious acquisition, who had been hired as guide, translator and camp manager. His slight admixture of blood showed in his skin, a smooth dark olive and his hair, thick and lustrous as a woman's. His teeth were very white and he flashed them in a perpetually ready smile. Yet, even when he smiled, his eyes were cold and black as those of an angry mamba.

The Governor in Quelimane had assured Zouga that this man was the most famous elephant hunter and traveller in all the Portuguese territories. He had ventured further into the interior than any other living Portuguese, and he spoke a dozen of the local dialects and understood the customs of the local tribes.

'You cannot travel without him,' the Governor assured Zouga. 'It would be madness to do so. Even your own father, the famous Dr Fuller Ballantyne, made use of his services. It was he who showed your sainted father the way to reach Lake Marawi.'

Zouga had raised an eyebrow. 'My father was the first man to reach Lake Marawi.'

'The first *white* man,' the Governor corrected him delicately, and Zouga smiled as he realized that it was one of the subtle distinctions which Fuller Ballantyne used to protect the value of his discoveries and explorations. Of course, there had been men living on the shores of the lake for at least two thousand years, and the Arabs and mulattos had traded there for two hundred years, but they were not *white* men. That made an enormous difference.

Zouga had at last acceded to the Governor's suggestions

when he had realized that this paragon was also the Governor's nephew, and that the further course of the expedition would be much smoothed by employing somebody so well connected.

He had reason to reconsider this opinion within the first few days. The man was a braggart and a bore. He had an endless fund of tales, of which he was always the hero, and the evident disregard for the truth that these demonstrated made all his facts and information suspect.

Zouga was uncertain just how well the man spoke the tribal dialects. He seemed to prefer to communicate with the toe of his boot or the *sjambok* of cured hippopotamus hide which he always carried. As for his hunting prowess, he certainly expended a great deal of powder and shot.

Zouga was sprawled on the barge's afterdeck, in the shade of the canvas awning, and he was sketching on the board he held on his knees. It was a pastime he had taken up in India, and though he knew that he had no great talent, yet it filled the idle hours of camp life and served as a useful record of places and persons, of events and animals. Zouga intended incorporating some of the sketches and water colours in the book describing the expedition. The book which would make his fortune and reputation.

He was trying to capture on paper the river's immensity, and the tallness of that aching blue sky set with the afternoon's building thunderheads, when there was the sharp crack of a rifle shot, and he looked up frowning with annoyance.

'He is at it again.' Robyn dropped her book into her lap and glanced back at the second barge.

Camacho Nuño Alvares Pereira sat high on the barge's cargo, reloading the rifle, ramrodding the charge down the long barrel. The high beaver hat sat on his head like a chimney stack and the bunch of white ostrich feathers plumed out above the crown like smoke from the furnace. Zouga could not see what he had fired at, but he guessed what would be his next target, for the steamer was being pushed out by the current to the outside of a broad bend in the river, and it was forced to steer between two low sandbanks.

The sand shone in the sunlight with the peculiar brilliance of an alpine snowfield, contrasting with the dark shapes upon it that looked like rounded granite boulders.

As the steamer slowly closed the gap, the shapes resolved into a troop of sprawling somnolent hippopotami. There were a dozen of them, one a huge scarred bull, lying on his side and exposing the expanse of his belly.

Zouga glanced back from the huge sleeping animals to the figure of Camacho Pereira on the second barge. Camacho lifted the plumed beaver and waved it in jovial salute. His teeth flashing like a semaphore even at that distance.

'You chose him,' said Robyn sweetly, following the direction of his gaze.

'That's a great comfort.' Zouga glanced at his sister. 'They told me he was the greatest sportsman and guide on the east coast.'

They both watched Camacho finish loading the rifle and setting the cap on its nipple.

The sleeping hippopotami suddenly became aware of the approaching vessels. They scrambled upright with amazing alacrity for such clumsy-looking animals, and galloped over the white sand, scattering clouds of it under their huge feet and then entered the water in a high crashing cascade of thrown spray, disappearing swiftly, and leaving the water churned and flecked with foam. Standing in the bows of the first barge, Zouga could clearly see the dark shapes below the surface of the water, galloping in comical slow motion, their movements inhibited by the water. They were silhouetted against the lighter-coloured sandbanks, and as he watched them, the ungainly creatures evoked his sympathy and amusement. He remembered a nursery rhyme that his

Uncle William had recited to him as a child that began 'A hippo – what – amus?'

Zouga was still smiling as the bull hippo surfaced fifty paces from the barge's side. The bulky grey head broke clear, the flaps of flesh that sealed the nostrils flared open as he breathed and the small round ears fluttered like the wings of a bird as he cleared them of water.

For a moment he stared at the strange vessels through pinkly inflamed, piggy blue eyes. Then he opened the full gape of his jaws, a cavern the colour and the texture of a pink rose. The tusks were yellow and curved to murderous cutting edges, quite capable of biting a bullock cleanly in half, and he no longer seemed fat and comical. Instead he looked exactly what he was, the most dangerous of all African big game.

Zouga knew that the hippopotamus had killed more human beings than all the elephant and lion and buffalo together. With ease they could crunch in the fragile hull of a dugout canoe, the ubiquitous *makoro* of Africa, and then cut in half the terrified swimmers. They would readily leave the water to chase and kill any human who they believed threatened a calf, and in areas where they had been hunted they would attack without provocation. However, the steel hulls of the barges were invulnerable even to the jaws and tusks of the massive creature, and Zouga could afford to watch with complete objectivity.

From the bull's gaping pink jaws came a challenging series of bellows, each mounting in volume and menace as he moved closer to drive off the intruders who threatened his females and their young. Camacho put his hand up behind his head, and tilted the beaver hat at a jaunty angle over one eye. As always, he was smiling as he swung up the rifle and fired.

Zouga saw the strike of the bullet deep in the animal's throat, it severed an artery and instantly bright crimson blood gushed against the roof of the open mouth, discolouring the gleaming tusks, and pouring in a quick flood over the rubbery, bewhiskered lips. The bull's bellow rose into a piercing scream of agony, and he lunged half clear of the surface in a burst of white water.

'I keel heem!' roared Camacho, and his shout of laughter filled the sudden void of silence as the bull dived below the surface, leaving his blood to swirl away down the current.

Robyn had jumped up and was clinging to the barge's rail, a flush overlaying her sunbronzed cheeks and throat.

'That was callous butchery,' she said quietly.

'No point in it,' Zouga agreed. 'The animal will die below the surface and be washed out to sea.'

But he was wrong, for the bull surfaced again, closer to the barge. His jaws still gaping and streaming gouts of blood, he thrashed and lunged in maddened circles, his bellows distorted by blood and water, as his death frenzy rose to a crescendo. Perhaps the bullet had damaged his brain, making it impossible for him to close his jaws or to control his limbs.

'I keel heem!' roared Camacho, dancing with excitement on the foredeck of the second barge, pouring shot after shot into the immense grey body, grabbing a rifle from his black gunbearer or from his second loader as soon as it was primed.

His two black servants worked with the expertise of long practice, so that it seemed that Camacho always had a loaded rifle ready to snatch and waiting hands ready to take the smoking weapon from him the moment he had fired.

Slowly the string of barges drew away upstream, leaving the stricken animal wallowing with increasing feebleness in its own expanding circle of blood-tinged waters, until at last it rolled belly upwards, all four stubby legs sticking up towards the sky for a moment before it sank at last and the blood was diluted and swept away downstream.

'That was sickening,' whispered Robyn.

'Yes, but he has trained those gunbearers of his damned well,' said Zouga thoughtfully. 'If one is going to hunt elephant, that is the way to do it.'

Two hours before sunset the *Helen* edged in towards the south bank. For the first time since leaving Quelimane there was some feature on the shore, beyond the endless reed swamps and sandbanks.

The bank was steeper here, rising ten feet above the river, and game paths had been cut into the grey earth by thousands of sharp hooves, and polished to shiny clay by the sliding wet bellies of the long lizard-like shapes of the big crocodiles that came tobogganing down the almost vertical slope when they were disturbed by *Helen*'s churning propeller. The heavily armoured reptiles with the staring yellow eyes set on a hard horny scale atop the long saurian head repulsed Robyn, the first African animal to do so.

There were trees on the bank now, not just waving stands of papyrus. Chief of these were the graceful palms with stems sculptured like a claret bottle.

'Ivory palms,' Zouga told her. 'The fruit has a kernel like a ball of ivory.'

Then far beyond the palms, low against the ruddy evening sky, they could make out the first silhouette of hills and kopjes. They were leaving the delta at last, and that night the company would camp on firm ground, instead of soft white sand, and burn heavy logs on the camp fires rather than the pulpy papyrus stems.

Zouga checked the sentries that Sergeant Cheroot had placed over the irreplaceable cargoes in the barges on which the whole expedition depended, then he supervised the siting of the tents before taking the Sharps rifle and starting out into the open forest and grassland beyond the camp site.

'I come weeth you,' offered Camacho. 'We keel something.'

'Your job is to make camp,' Zouga told him coolly, and the Portuguese flashed his smile and shrugged.

'I make one damn fine camp – you see.'

But as Zouga disappeared amongst the trees, the smile slid off his face, and he hawked in his throat and spat in the dust. Then he turned back into the turmoil of men raising canvas on poles, or dragging in branches of freshly cut thorns to build the *scherm* against marauding lions or scavenging hyena.

Camacho lashed out at a bare black back. 'Hurry, you one mother, twenty-seven fathers.' The man cried out at the pain of the cured hippo-hide whip, redoubling his efforts as a purple welt, thick as a man's little finger, rose across his sweat-oiled muscles.

Camacho strode on towards the small grove of trees which Zouga had picked as the site for the tents of his sister and himself, and he saw that the tents had already been erected and that the woman was busy with the evening muster when she treated the ailments of the camp.

She had been seated at the collapsible camp table, but as Camacho approached, she rose and stooped to examine the foot of one of the bearers whose axe had slipped and almost severed a toe.

The Portuguese stopped abruptly, and his throat dried out as he watched her. As soon as they had left Quelimane, the woman had taken to wearing men's breeches. Camacho found them more provocative than naked flesh itself. It was the first time he had ever seen a white woman dressed like this, and he found it hard to take his eyes off her. Whenever she was in sight, he would watch her surreptitiously, waiting hungrily for the moment when she stooped or leaned forward and the moleskin stretched over her buttocks, as it was now. It lasted too short a time, for the woman straightened up and began speaking to the little black girl who seemed more of a companion to her than a servant.

Still Camacho leaned against the bole of one of the tall umsivu trees and watched her with those black eyes gone velvet and swimming with desire. He was carefully weighing the consequences of what he had dreamed about every night since they had left Quelimane. He had imagined every detail, every expression, every word, each movement and each sigh or cry.

It was not as improbable as it seemed at first. She was an English woman, of course, daughter of a famous man of God, both facts should have been prohibitive to his intentions. However, Camacho had a canny instinct when it came to women, there was a sensuality about her eye and in the full soft lips, and she moved with animal awareness of her body. Camacho stirred restlessly and thrust his hands into his pockets, kneading and tugging gently at himself.

He was fully aware that he was a magnificent specimen of masculinity – those thick tresses of black hair, the gypsy feyness of eyes, the blazing smile, and powerful and well-proportioned body. He was attractive, perhaps irresistibly so, for more than once he had intercepted a quizzical appraising look from the woman. Often the admixture of his blood was attractive to white women, it was an exoticism, the attraction of the forbidden and dangerous, and he sensed in this woman a rebellious disregard for the rules of society. It was possible, no more than possible, Camacho decided, and there was unlikely to be a better opportunity than now. The cold stiff English brother was out of camp, would be so for another hour or more, and the woman had finished attending the little group of sick bearers. A servant had brought a kettle of boiling water to her tent, and she was closing the fly.

Camacho had watched this little ritual every evening. Once the oil lamp had cast her shadow upon the canvas, and he had watched her silhouette lowering those tantalizing breeches, and then using the sponge to – he shuddered deliciously at the memory, and pushed himself away from the tree trunk.

Robyn mixed the hot water from the kettle into the enamel basin. It was still scalding hot, but she liked it so that it reddened the skin and left her feeling glowing with cleanliness. She began to unbutton the flannel shirt, sighing with pleasant weariness, when something scratched on the fly of the tent.

'Who is it?' she called sharply, and felt a faint stir of alarm as she recognized the low voice. 'What do you want?'

'I want talk you, missus.' Camacho's tone was conspiratorial.

'Not now, I am busy.' The man repelled her, and yet in a contrary manner, fascinated her as well. She had found herself staring at him more than once – as she would at a beautiful but poisonous insect. She was annoyed that he had noticed and was vaguely aware that it was unwise to show even the vaguest interest in a man like that.

'Come back tomorrow.' It suddenly occurred to her that Zouga was not in the camp, and she had sent little Juba on an errand.

'I cannot wait. I am sick.'

That was one appeal she could not deny.

'Oh very well. Wait,' she called, and buttoned her shirt, and then as an excuse perhaps to delay the moment turned her attention to her instruments still spread on the table that had been carried into the tent. It reassured her to touch them, and rearrange the bottles and pots of medicines.

'Enter,' she called at last, and faced the entrance of the tent.

Camacho stooped through the entrance, and for the first time she realized how tall he was. His presence in the small tent was almost overpowering, and his smile seemed to light up the interior. His teeth were startlingly white and perfect, she found herself staring again, like a chicken at the dancing cobra perhaps. He was beautiful in a decadent, overblown way, he was bare-headed, all dark dancing hair and scalding eyes.

'What is the trouble?' she asked, trying to sound brisk and businesslike.

'I show you.'

'Very well,' she nodded, and he unbuttoned his shirt. His skin had the sheen of wet marble, but was deep olive in colour, and his body hair grew in crisply springing whorls. His belly was moulded like that of a greyhound and his waist narrow as a girl's. She ran her eyes down his body, quite certain that her gaze was level and professional, but there was no denying the fact that he was an extraordinary animal.

'Where is it?' she asked, and with a single movement he had unclinched and lowered the light duck trousers that were all he wore on his lower body.

'Where?' she asked, and realized that her voice croaked, and she could not go on with the question, for suddenly it dawned upon her that she had been the victim of a carefully planned ruse, and she was in a potentially dangerous position.

'Is that where it hurts?' she found her voice was still a husky whisper.

'Yes,' his voice was a whisper also, and he made a slow stroking movement. 'You can fix, maybe.' He took a step towards her.

'I can fix, certainly,' she said softly, and her hand dropped on to the array of surgical instruments. She actually experienced a twinge of real regret, for it was a superior example of nature's art, and afterwards she was relieved that she had selected a needle probe – and not one of the razorlike scalpels that she had reached for.

The instant before she stabbed, he realized what was about to happen and an expression of utter terror blanched his swarthy, handsome face. He tried frantically to return it to whence it had come, but fear had slowed his hand.

He screamed like a teenage girl as the probe plunged into him, and kept on screaming as he spun around on the same spot as though one foot had been nailed to the ground. Now he was using both hands to hold himself, and once again, with cool professional interest, Robyn noticed the quite miraculous change that had taken place.

She advanced the probe once more into the ready position, and Camacho could no longer stand his ground. He snatched up his trousers, and with a last terrified howl, launched himself head-first into the tent pole. The collision checked him only a moment, and then he was gone, and Robyn found herself trembling violently and yet she was strangely elated. It had been an unusual and instructive experience. However, she would have to use her own personal code when describing it in her journal.

From that evening onwards the Portuguese kept well away from Robyn, and she was relieved not to have those hot dark eyes caressing her wherever she turned. She thought of telling Zouga of the incident, but decided that the embarrassment to both of them and the difficulty of finding the correct words was not worth it. Then there would be the extreme reaction which Zouga would almost certainly have, or that she expected he would have. She had learned never to expect the obvious reaction from her brother, behind the cool and reserved exterior she suspected there existed mysterious passions and dark emotions. After all, they were full-blooded brother and sister, and if she was so afflicted, why should he not be also?

On the other hand, she suspected that, like a cornered wild animal, the Portuguese could be a grave danger even to an experienced soldier and man of action like Zouga. She had a horror of forcing her brother into a position which might lead, if not to his death, then at least to serious injury. Besides which, she had effectively taken care of the man herself. He would be no further trouble, she decided comfortably, and she dismissed Camacho from her mind, and concentrated instead on the pleasure of the last few leisurely days of the voyage upstream.

The river had narrowed, and the flow was swifter, so that the rate of the convoy's advance slowed even further. The banks provided an ever-changing panorama. Sitting under the awning, with Zouga sketching or writing beside her, she was able to call his attention to the new birds and trees and animals and to have the benefit of his knowledge, gathered to be sure mostly from books, but still wide-reaching and interesting.

The hills of the escarpment rose in a series of cockscombs, so two-dimensional that they seemed to be cut out from thin sheets of some opaque material that allowed the colours of the sunrise to glow through with a weird luminosity. As the sun rose higher, the colours washed out to ethereal eggshell blues, and finally faded altogether in the heat haze of midday, to reappear in the late afternoon in a new suit of totally different colours – pale pinks and ash of roses, ripe plum and delicate apricot.

The hills formed a backdrop to the forests that now ran in a narrow belt along the river banks. Tall galleries of trees, with spreading upper branches in which the troops of vervet monkeys frolicked. The trunks of these trees were daubed with multi-coloured lichens, sulphurous yellows, burnt oranges and the blues and greens of a summer sea. The tangled ropes of lianas, which as a child Robyn had called 'monkey ropes', dangled down from the upper branches to touch the surface of the river or cascade into the dense dark greens of the undergrowth.

Beyond this narrow strip of vegetation, there were

occasional glimpses of a different forest on the higher, drier ground, and Robyn saw again with a sharp nostalgic pang the ugly and bloated baobab tree with its scrubby bare little branches topping the huge swollen stem. The African legends that her mother had repeated to her so often explained how the Nkulu-kulu, the great great one, had planted the baobab upside down, with its roots in the air.

Nearly every baobab had a nest of one of the big birds of prey in its bare branches, each a shaggy mass of dried twigs and small branches looking like a small, air-borne haystack. Often the birds were at the nest site, sitting on a lookout branch, with that typical raptorial stillness, or soaring above in wide circles, with only an occasional lazy flap of the spread wings, and the stiff tip-feathers feeling the air currents like the fingers of a concert pianist upon the ivory keys.

There was very little game along this part of the river, and the rare antelope rushed back into cover at the first distant approach – a pale blur of movement, with a mere fleeting glimpse of the tall corkscrew horns of a greater kudu, or the flirt of the white, powder-puff underside of a reedbuck's tail.

The game close to the river had been heavily hunted, if not by the Portuguese themselves then by their armed servants, for nearly two hundred years.

When Zouga asked Camacho, 'Do you ever find elephant on this part of the river?' the Portuguese had flashed his smile and declared, 'If I find heem, I keel heem.'

A sentiment that was probably shared by nearly every traveller along this busy waterway, and which accounted for the timidity and scarcity of game in the area.

Camacho was reduced to firing at the roosting fish eagles on their fishing perches overhanging the water. These handsome birds had the same snowy white head, breast and shoulders of the famous American bald eagle, and a body of lovely dark russet and glistening black. When a shout of Camacho's laughter signalled a hit, a bird would tumble untidily over its disproportionately large wings as it fell into the green water, reduced from imperial dignity to awkward and ungainly death by the strike of the lead bullet.

Within a few days Camacho had recovered from the peculiar, bow-legged and deliberate gait, with which he favoured the injury that Robyn had inflicted on him, and his laughter regained its ringing timbre. But there were other injuries that did not heal so readily, those to his pride and his masculinity. His lust had been changed on the instant to burning hatred, and the more he brooded upon it the more corrosive it became and the deeper his craving for vengeance.

However, his personal considerations would have to wait. There was still much important work for him to do. His uncle, the Governor of Quelimane, had placed great trust in him by assigning him to this task – and his uncle would be unforgiving of any failure. The family fortune was involved, and to a lesser extent the family honour, although this last was a commodity that through constant attrition had lost much of its lustre. However, the family fortunes had suffered considerably since Portugal had been forced to heed the Brussels Treaty. What was left to the family had to be protected. Gold before honour, and honour only when it does not affect the profits – this might have been the family motto.

His uncle had been perceptive, as always, in recognizing in this English expedition a further threat to their interests. It was, after all, headed by the son of a notorious trouble-maker who could be expected to aggravate the enormous damage done by the father. Furthermore, nobody could be sure of the real objects of the expedition.

Major Ballantyne's assertion that it was an expedition to find his missing father was, of course, utterly absurd. That explanation was much too simple and direct, and the English were never simple or direct. This elaborate expedition must have cost many thousands of English pounds, a huge sum of money, far beyond the means of a junior army officer, or the family of a missionary whose futile effort to navigate the Zambezi had ended in disgrace and ridicule, a sick old man who must have perished years ago in the uncharted wilderness.

No, there was another motive for all this activity – and the Governor wanted to know what it was.

It was, of course, possible that this was a clandestine reconnaissance by an officer of the British army ordered by his overbearing government. Who knew what outrageous designs they had upon the sovereign territory of the glorious Portuguese empire? The avarice of this impudent race of shopkeepers and tradesmen was scarcely to be believed. The Governor did not trust them, despite their traditional alliance with Portugal.

On the other hand, it might indeed be a private expedition, but the Governor never lost sight of the fact that it was led by the son of that notorious old busybody who had possessed the scavenging eye of a vulture. Who knew what the old devil had stumbled upon out there in the unknown land, a mountain of gold or silver, the fabled lost city of Monomatapa with all its treasures intact: anything was possible. Of course, the old missionary would have sent news of the discovery to his own son. If there was a mountain of gold out there, then the Governor would be very pleased to know about it.

Even if there were no new treasures to discover, there were certainly old ones to protect. It would be Camacho Pereira's duty to steer the expedition away from certain areas, to prevent it stumbling on secrets known not even to the Governor's masters in Lisbon.

Camacho's orders were clear: distract the Englishman by accounts of the insurmountable difficulty of travel in certain directions, the swamps, the mountain ranges, the disease, the savage animals and even more savage men, and contrast that with the friendly people in pleasant lands, rich with ivory, that lay in other directions.

If this was unsuccessful – and Major Ballantyne had all the earmarks of arrogance and intractability peculiar to his nation – then Camacho was to use what other means of persuasion came to hand. This was a euphemism perfectly understood by both the Governor and his nephew.

Camacho had almost convinced himself that this was really the only sensible course of action. Beyond Tete there was no law, except that of the knife, and Camacho had always lived by that law. Now he savoured the thought. He had found the Englishman's unconcealed contempt as galling as the woman's rejection had been painful.

He had convinced himself that the reason for the attitude towards him of both brother and sister was his mulatto blood. This was a sensitive area of Camacho's self-esteem, for even in Portuguese territories where miscegenation was almost universal practice, mixed blood still carried a stigma. He would enjoy the work ahead, for not only would it wipe out the insults he had suffered, but it would bring rich pickings, and even after they had been shared out with his uncle, and others, there would still be much profit in it for himself.

The equipment that the expedition carried represented, in Camacho's view, a vast fortune. There were barge-loads of excellent trade goods. Camacho had taken the first opportunity to check secretly the contents of the packs. There were firearms, and valuable instruments, chronometers and sextants, and there was a forged-steel field-safe that the Englishman kept locked and guarded. The merciful God alone knew how many golden English sovereigns it contained, and if He did not know – then his good uncle the Governor knew less. It would make the division of spoils more in Camacho's favour. The more he brooded upon it, the more he looked forward to the arrival at Tete, and the jump-off into the unexplored territory beyond.

To Robyn the tiny town of Tete marked her real arrival in Africa, and her return to the world for which she had trained so assiduously and yearned so deeply.

She was secretly glad that Zouga had used the unloading of the barges as an excuse not to accompany her.

'You find the place, Sissy, and we'll go there together tomorrow.'

She had changed back into skirts, for small and isolated as it was, Tete was still a backwater of civilized behaviour and there was no point in giving offence to the local inhabitants. Though she found the heavy folds about her legs annoying, she soon forgot them as she walked the single, dusty street of the village where her father and mother must have walked together for the last time, and peered at the mud-walled trading stores built haphazardly along a rough line with the bank of the river.

She stopped at one of these little *dukas* and found that the storekeeper could understand a mixture of her basic Swahili, English and Nguni language, enough anyway to direct her on to where the village street petered out in a mere footpath that meandered off into the acacia forest.

The forest was hushed in the heat of the noon, even the birds were silent and the mood weighed on Robyn, depressing her and awakening the memories of long-ago mourning.

She saw a flash of white amongst the trees ahead, and stopped, reluctant to go on to what she knew she would find. For a moment she was transported to girlhood again, to a grey November day standing beside her Uncle William waving upwards at the passenger decks of the departing ship, her eyes so dimmed with tears that she could not make out at the crowded rail the beloved face for which she searched, while the distance between ship and quay opened like the gulf between life and death.

Robyn shook the memory away and went on. There were six graves amongst the trees, she had not expected that, but then she recalled that there had been heavy mortality amongst the members of her father's Kaborra-Bassa expedition, four of disease, one drowned and a suicide.

The grave for which she searched stood a little apart from the others. It was demarcated by a square of white-washed river stones and at the head was a cross built of mortar. It also had been whitewashed. Unlike the other graves, it had been kept cleaned of grass and weed, and the cross and stones freshly painted. There was even a small bunch of wilted wild flowers standing in a cheap blue china vase. They were not more than a few days old. That surprised Robyn.

Standing at the foot of the grave she read the still fully legible lettering on the plaster cross:

In loving memory of
Helen
beloved wife of Fuller Morris Ballantyne.
Born August 4th 1814. Died of fever December 16th
1852.
God's will be done.

Robyn closed her eyes and waited for the tears to come up from deep inside, but there were no tears, they had been shed long ago. Instead there were only the memories.

Little fragments of memory played over and over in her mind – the smell of strawberries as they gathered them together in Uncle William's garden, standing on tiptoe to place one of the lush red fruit between her mother's white teeth and then eating the half that was left especially for her; lying cuddled under her bedclothes as she listened drowsily to her mother's voice reading aloud to her in the candlelight; the lessons at the kitchen table in winter, under the elm trees in summer and her own eagerness to learn and to please; her first pony ride, her mother's hands holding her in the saddle, her legs too short for the stirrups; the feel of the soapy sponge down her back as her mother stooped over the iron hipbath; the sound of her mother's laughter, and then at night the sound of her weeping beyond the thin partition beside her cot; then the final memory of the smell of violets and lavender as she pressed her face to her mother's bodice.

'Why must you go, Mama?'

'Because your father needs me. Because your father has sent for me, at last.'

And Robyn's own terrible consuming jealousy at the words, mingled with the sense of impending loss.

Robyn knelt in the soft cushion of dust beside the grave, and began to pray, and as she whispered, the memories came crowding back, happy ones and sad ones together, and she had not felt so close to her mother in all the intervening years.

She did not know how long she had knelt there, it seemed an eternity, when a shadow fell across the earth in front of her and she looked up, jerked back to the present with a little gasp of surprise and alarm.

A woman and child stood near her, a black woman, with a pleasant, even pretty, face. Not young, in her middle thirties possibly, though it was always difficult to guess an African's age. She wore European-style clothing, cast-offs probably, for they were so faded that the original pattern was hardly visible, but starched and fastidiously clean. Robyn sensed that they had been donned for the occasion.

Although the child wore the brief leather kilt of the local Shangaan tribe, he was clearly not a full-blooded African. He could not have been more than seven or eight years of age, a sturdy boy, with a head of dusty-coloured curls and oddly pale-coloured eyes. There was something vaguely familiar about him that made Robyn stare.

He carried a small bunch of the yellow acacia flowers in his hands, and smiled shyly at Robyn before hanging his head and shuffling his feet in the dust. The woman said something to him and tugged at his hand, and he came hesitantly to Robyn and handed her the flowers.

'Thank you,' she said automatically, and raised the bouquet to her nose. They were faintly, but sweetly perfumed.

The woman hiked her skirts and squatted beside the grave, removed the wilted flowers and then handed the little blue china vase to the boy. He scampered away towards the river-bank.

While he was gone the woman plucked out the first green sprouts of weeds from the mound of the grave and then rearranged the whitewashed stones carefully. The familiar manner in which she performed the chore left no doubt in Robyn's mind that she was responsible for the upkeep of her mother's grave.

Both women maintained a friendly, comfortable silence, but when their eyes met they smiled and Robyn nodded her thanks. The child came trotting back, muddy to the knees and slopping water from the vase, but puffed up with self-importance. He had clearly performed this task before.

The woman took the vase from him and set it carefully on the grave, then both of them looked expectantly towards Robyn and watched her while she arranged the acacia flowers in the vase.

'Your mother?' said the woman softly, and Robyn was startled to hear her speak English.

'Yes,' she tried to hide her surprise. 'My mother.'

'Good lady.'

'You knew her?'

'Please?'

After the valiant opening, the woman had very little English, and their communication was halting, until Robyn, out of the habit of talking to little Juba, said something in Matabele. The woman's face lit with pleasure and she answered swiftly in a language which was obviously one of

274

the Nguni group, and whose inflection and vocabulary differed very little from that to which Robyn was accustomed.

'You are Matabele?' Robyn demanded.

'I am Angoni,' the woman put in hastily, for there was rivalry and hostility between even the closely related tribes of the Nguni.

Her tribe, the Angoni, had swept northwards from their origins in the grassy hills of Zululand, and crossed the Zambezi river thirty years before, she explained in her lilting musical dialect. They had conquered the land along the northern shores of Lake Marawi. It was from there that the woman had been sold to one of the Omani slave-masters, and had come down the Shire river in chains.

Unable to keep up with the slave caravan, reduced by starvation, and the fevers and hardships of the long journey, she had been freed of her chains and left for the hyenas beside the slave road. It was there that Fuller Ballantyne had found her and taken her into his own small camp.

She had responded to his rough nursing and when she was recovered, Fuller had baptized her with the Christian name of Sarah.

'So my father's detractors are mistaken,' Robyn laughed, and spoke in English. 'He made more than one convert.'

Sarah did not understand but laughed in sympathy. By now it was almost dusk and the two women, followed by the half-naked child, left the little cemetery and started back along the footpath, with Sarah still telling how when Robyn's mother, summoned at last by Fuller Ballantyne, arrived in Tete with other members of the Kaborra-Bassa expedition, Sarah had been presented to her by Fuller as a personal servant.

By now they had stopped at a fork in the path, and after a moment of hesitation Sarah invited Robyn to her village which was only a short way off the path. Robyn glanced up at the sun and shook her head, it would be dark in an hour and Zouga would be certain to turn out the camp to search for her if she had not returned by then.

She had enjoyed the hours spent with the young woman and the bright sweet child, and when she saw Sarah's obvious disappointment, she said quickly, 'Although I must go – I will return tomorrow at the same time. I wish to hear all you can tell me of my mother and my father.'

Sarah sent the little boy with her as far as the buildings of the village and after the first few yards Robyn quite naturally took the boy's hand, and he skipped along beside her, chattering gay childish nonsense, which helped to lift her sombre mood until Robyn laughed and chattered with him.

Before they reached the outskirts of Tete, Robyn's fears were confirmed. They met Zouga and Sergeant Cheroot. Zouga was armed with the Sharps rifle and angry with relief the moment he saw her.

'Damn me, Sissy, but you have had us all beside ourselves. You've been missing for five hours.'

The child stared at Zouga with wide eyes. He had never seen anything like this tall lordly man with the imperious manner and sharp commanding voice. He must be a great chief, and he slipped his hand out of Robyn's, retreated two paces, then turned and darted away like a sparrow from the circling hawk.

Some of Zouga's anger left him as he watched the child go, and a small smile touched his lips.

'For a moment I thought you'd picked up another stray.'

'Zouga, I found Mama's grave.' Robyn hurried to him and took his arm. 'It's only a mile or so.'

Zouga's expression changed again and he glanced up at the sun that was already on the tops of the acacia trees and turning deep smouldering red.

'We'll come back tomorrow,' he said. 'I don't like to leave the camp after dark, there are too many jackals lurking

276

about – two legged jackals.' Firmly, he led her back towards the village, continuing his explanation as they walked.

'We are still having a great deal of difficulty obtaining porters, despite the fact that the Governor in Quelimane assured me they would be readily available, and the good Lord knows there are any amount of able-bodied men hereabouts. Yet that strutting poppinjay Pereira finds obstacles at every turn.' The frown made him look much older than his years as did the full beard which he had allowed to grow since disembarking from *Black Joke*. 'He says that the porters refuse to contract until they know the direction and duration of the safari.'

'That seems logical,' Robyn agreed. 'I know I wouldn't carry one of those huge packs, unless I knew where I was going.'

'I don't think at all that it's the porters – there is no reason why the destination should worry them. I am offering top wages, and not a single man has come forward.'

'What is it, then?'

'Pereira has been trying to wheedle our intentions out of me, ever since we left the coast. I think this is a form of blackmail, no porters until I tell him.'

'Then why don't you tell him?' Robyn asked, and Zouga shrugged.

'Because he is too damned insistent. It's not a casual interest, and instinct warns me not to trust him with any information which it is not essential for him to know.'

They walked on in silence until they reached the perimeter of the camp. Zouga had laid it out on the lines of a military base, with an outer stockade of acacia thorn branches, a Hottentot guard at the gate and the *boma* for the porters separated from the stores depot by the tent lines.

'It looks like home already,' Robyn congratulated him, and would have left him for her own tent when Camacho Pereira hurried forward.

'Ah! Major, I wait for you with good news.'

'That's a pleasant change,' Zouga murmured drily.

'I find man who has seen your father, not eight months ago.'

Robyn turned back instantly, her excitement matching that of the flamboyant Portuguese and she spoke directly to him for the first time since the incident in her tent.

'Where is he? Oh, this is wonderful news.'

'If it's true,' qualified Zouga, with considerably less enthusiasm.

'I bring the man, damned quick – you see!' Camacho promised, and hurried away towards the porters' *boma*, shouting as he went.

Within ten minutes he returned dragging with him a skinny old man dressed in greasy tatters of animal skins, and with his eyes rolling up into his head with terror.

As soon as Camacho released him, the old man prostrated himself before Zouga who sat in one of the canvas camp chairs under the awning of the dining tent, and gabbled replies to the queries that Camacho shouted at him in hectoring tones.

'What dialect is that he speaks?' Zouga interrupted within the first few seconds.

'Chichewa,' Camacho replied. 'He no speak other.'

Zouga glanced at Robyn, but she shook her head. They had to rely entirely on Camacho's rendition of the old man's replies.

It seemed that the old man had seen 'Manali', the man with the red shirt, at Zimi on the Lualaba river. Manali had been camped there with a dozen porters, and the old man had seen him with his own eyes.

'How does he know it was my father?' Zouga asked.

Everybody knew 'Manali', the old man explained, he was a living legend from the coast to 'Chona langa', the land where the sun sets.

'When did he see Manali?'

One moon before the coming of the last rains, which made it in October of the previous year, as Camacho had said, about eight months previously.

Zouga sat lost in thought, but his gaze fixed with such ferocity on the unfortunate who grovelled before him that the old man suddenly burst out on a plaintive note that made Camacho's handsome face darken with anger and he touched the skeletal ribs with the toe of his boot, a threatening gesture that quieted that old man instantly.

'What did he say?' Robyn demanded.

'He swears he speak the truth only,' Camacho assured her, resurrecting his smile with an effort.

'What else does he know of Manali?' Zouga asked.

'He speak with the porters of Manali, they say they go follow the Lualaba river.'

It made some sense, Zouga thought. If Fuller Ballantyne was indeed seeking the source of the Nile river to recover his lost reputation, then that is where he would have gone. The Lualaba, which was reputed to run directly northward, was one of obvious choices for the source river.

Camacho questioned the old man for another ten minutes, and would have taken the hippo-hide whip to jog his memory, but Zouga stopped him with a gesture of annoyance. It was obvious that there was nothing further to learn from him.

'Give him a bolt of *merkani* cloth and a *khete* of beads – and let him go,' Zouga ordered and the old man's gratitude was pathetic to watch.

Zouga and Robyn sat later than usual beside the camp fire, while it collapsed slowly upon itself in spasmodic torrents of rising sparks and the murmur of sleepy voices from the porters' *boma* died into silence.

'If we go north,' Robyn mused, watching her brother's face, 'we will be going into the stronghold of the slave trade, from Lake Marawi northwards. From that area into which no white man, not even Pater, has ever ventured must come all the slaves for the markets of Zanzibar and the Omani Arabs—'

'What about the evidence of the trade to the south,' Zouga glanced across the clearing at the silent figure of Juba, waiting patiently by the entrance to Robyn's tent. 'That girl is the living proof that a new trade is flourishing south of the Zambezi.'

'Yes, but it seems to be nothing compared to the activity north of here.'

'The northern trade has been fully documented. Father reached Marawi and followed the slave caravans down to the coast fifteen years ago, and Bannerman at Zanzibar has written a dozen reports on the Zanzibar market,' Zouga pointed out, nursing a precious tumbler of his fast-dwindling supply of whisky, and staring into the ashes of the fire. 'Whereas nobody knows anything about the trade with the Monomatapa and the Matabele south of here.'

'Yes, I acknowledge that,' Robyn admitted reluctantly. 'However, in his *Missionary Travels* father wrote that the Lualaba was the source of the Nile and he would one day prove it by following it from its headwaters. Besides which, he has been seen in the north.'

'Has he, though?' Zouga asked mildly.

'That old man . . .'

'. . . Was lying. Somebody put him up to it, and I don't need more than one guess,' Zouga finished.

'How do you know he was lying?' Robyn demanded.

'If you live long enough in India you develop an instinct for the lie,' Zouga smiled at her. 'Besides why would father wait eight years *after* he disappeared to explore the Lualaba river. He would have gone there directly – if he had gone north.'

'My dear brother,' Robyn's voice was stinging, 'it would not be the legend of Monomatapa that makes you so stubbornly determined to go south of the river, would it? Is that gleam in your eye the gold?'

'That is a mean thought,' Zouga smiled again. 'But what does intrigue me is the determination of that great guide and explorer, Camacho Pereira, to discourage any journey to the south, and instead to lead us northwards.'

Long after Robyn had disappeared into her tent and the lantern within was extinguished Zouga sat on beside the fire, nursing the whisky in the tumbler and staring into the fading coals. When he reached his decision he drained the glass and stood up abruptly. He strode down the lines to where Camacho Pereira's tent stood at the furthest end of the camp.

There was a lantern burning within even at this late hour, and when Zouga called out, a squeak of alarm in feminine tones was quickly hushed with a man's low growl and a few minutes later Camacho Pereira pulled the fly aside and peered out at Zouga warily.

He had thrown a blanket over his shoulders to cover his nudity, but in one hand he carried a pistol and relaxed only slightly as he recognized Zouga.

'I have decided,' Zouga told him brusquely, 'that we'll go north, up the Shire river to Lake Marawi – and then on to the Lualaba river.'

Camacho's face shone like the full moon as he smiled.

'That is very good. Very good – much ivory, we find your father – you see, we find him damn soon.'

Before noon the following day Camacho, with a great deal of shouting and swishing of the kurbash, marched a hundred strong healthy men into the camp. 'I find you porters,' he announced. 'Plenty porters – damn good, hey?'

The Christian girl Sarah was waiting beside the grave again when Robyn came down through the acacia forest the following afternoon.

The child saw her first and ran to greet her, he was laughing with pleasure, and Robyn was struck once again by the familiarity of his face. It was something she had known was so forcible that she stopped dead and stared at him, but could not recapture the memory before the boy took her hand and led her to where his mother waited.

They went through the little ritual of changing the flowers on the grave and then settled side by side on a fallen acacia branch. It was cooler in the shade and in the branches above their heads a pair of shrikes hunted little green caterpillars. The birds were black and white across the back and wings, but their breasts were a striking shade of crimson that glowed like the blood of a dying gladiator, and Robyn watched them with rare pleasure while she and Sarah talked quietly.

Sarah was telling her about her mother, how brave and uncomplaining she had been in the terrible heat of the Kaborra-Bassa where the black ironstone cliffs turned the gorge into a furnace.

'It was the bad season,' Sarah explained. 'The hot season before the rains break.' Robyn recalled her father's written account of the expedition in which he had laid the blame for the delays upon his subordinates, old Harkness and Commander Stone, so that they had missed the cool season, and entered the gorge in the suicide month of November.

'Then when the rains came, the fever came with it,' Sarah went on. 'It was very bad. The white men and your mother became sick very quickly.' Perhaps her mother had lost much of her immunity to malaria during the years in England while she waited for her husband's summons. 'Even Manali himself became sick. It was the first time I had seen him sick of the fever. He was filled with the devils for many days,' the expression described vividly the delirium of malarial fever, Robyn thought. 'So he did not know when your mother died.'

They were silent again. The child, bored by the interminable talk of the two women, threw a stone at the birds in the acacia branches above their heads, and with flash of their marvellous crimson breasts the two shrikes winged away towards the river, and again the child engaged Robyn's attention. It was as if she had known that face all her life.

'My mother?' Robyn asked, still watching the child.

'Her water turned black,' said Sarah simply. The blackwater fever – Robyn felt her skin prickle. When malaria changed its course, attacking the kidneys and transforming them into thin-walled sacks of clotted black blood that could rupture at the patient's smallest movement. The blackwater fever, when the urine changed to dark mulberry-coloured blood, and few, very few victims, ever recovered.

'She was strong,' Sarah went on quietly. 'She was the last of them to go.' She turned her head towards the other neglected graves. The curly pods of the acacia were scattered thickly over the unadorned mounds. 'We buried her here, while Manali was still with his devils. But later, when he could walk he came with the book and said the words for her. He built the cross with his own hands.'

'Then he went away again?' Robyn asked.

'No, he was very sick, and new devils came to him. He wept for your mother.' The thought of her father weeping was something so completely alien that Robyn could not imagine it. 'He spoke often of the river that had destroyed him.'

Through the acacia trees there were glimpses of the wide green river, and both their heads turned towards it naturally.

'He came to hate that river as though it was a living enemy that had denied him a road to his dreams. He was like a man demented, for the fever came and went. At times he would battle with his devils, shouting his defiance the way a warrior *giyas* at the enemy host.' The *giya* was a

283

challenge dance with which the Nguni warrior baits his adversary. 'At other times he would speak wildly of machines that would tame his enemy, of walls that he would build across the waters to carry men and ships up above the gorge.' Sarah broke off, her lovely dark moon face stricken with the memory, and the child sensed her distress and came to her, kneeling on the earth and laying his dusty little head in her lap. She stroked the tight cap of curls with an absentminded caress.

With a sudden little chill of shock Robyn recognized the child. Her expression changed so drastically that Sarah followed the direction of her gaze, looking down with all her attention at the head in her lap, then up again to meet Robyn's eyes. It did not really need words to pass, the question was posed and answered with silent exchange of feminine understanding, and Sarah drew the child towards her with a protective gesture.

'It was only after your mother . . .' Sarah began to explain and then fell silent, and Robyn went on staring at the little boy. It was Zouga at the same age, a dusky miniature Zouga. It was only the colour of his skin which had prevented her from seeing it immediately.

Robyn felt as though the earth had lurched beneath her feet, then it steadied again and she felt a strange sense of release. Fuller Ballantyne was no longer the god-like figure hewn from unforgiving, unbending granite that had over-shadowed her entire life.

She held out both her hands to the child and he went to her unhesitatingly, trustingly. Robyn embraced him, and his skin was smooth and warm as she kissed him. He wriggled against her like a puppy, and she felt a deep glow of affection and of gratitude to the child.

'He was very sick,' said Sarah softly, 'and alone. They had all gone or died, and he was sad, so that I feared for his life.'

Robyn nodded understanding. 'And you loved him?'

'There was no sin in it, for he was a God,' said Sarah simply.

'No,' thought Robyn with intense relief. 'He was a man, and I, his daughter, am a woman.'

In that moment she knew that she never need again feel shame and guilt for her body and the demands and desires which sprang from it. She hugged the child who was proof of her father's humanity, and Sarah smiled with relief.

For the first time in her life Robyn was able to face the fact that she loved her father, and she understood part of the compulsion that over the years had grown stronger rather than dwindling.

The longing she had felt for the father had been submerged completely by the awe and majesty of the legend. Now she knew why she was here, on the banks of this majestic river, on the very frontier of the known world. She had come not to find Fuller Ballantyne, but to discover rather the father and the self that she had never known before.

'Where is he, Sarah, where is my father? Which way did he go?' she demanded eagerly, but the woman dropped her eyes.

'I do not know,' she whispered. 'I woke one morning and he was gone. I do not know where he is, but I will wait for him, until he returns to me and his son.' She looked up quickly. 'He will come back?' she asked pathetically. 'If not to me then for the child?'

'Yes,' Robyn answered with certainty that she did not feel. 'Of course he will come back.'

The selection of porters was a lengthy business, and after Zouga had signalled his choice with a slap on the shoulder, the men were sent to Robyn's tent to be examined for signs of disease or infirmity that might prevent them performing their duties.

Then came the allocation of packs.

Although Zouga had already made up and weighed each pack, making sure that not one of them exceeded the stipulated eighty pounds weight, the newly engaged porters had to watch the loads reweighed publicly, and then there was interminable haggling over the size and balance of the burden that each of them would carry for months, perhaps even years ahead.

Although Zouga brusquely forbade Pereira to hasten the selection process with his kurbash, and entered good-naturedly into the spirit of banter and bargain, he was, in fact, using the occasion to assess the spirit of his men, to pick out the malcontents who would sour that spirit in the hardships ahead, and also to select the natural leaders to whom the others turned instinctively for decision.

The following day when planning the order of march, Zouga used the knowledge he had gained in this way. To begin with, seven of the more obvious trouble-makers were given a *khete* of beads each and ordered out of the camp without explanation or apology. Then Zouga called out five of the brightest and best and made them captains of divisions of twenty porters each.

They would be responsible for maintaining the pace of the march, for preventing pilfering of the loads, making and breaking camp, distributing rations, and acting as the spokesmen of each division, presenting complaints to and transmitting orders from Zouga.

When the roll was complete there were one hundred and twenty-six names upon it, including Sergeant Cheroot's Hottentots, the porters who had come up from Quelimane,

Camacho Pereira and the two principals – Robyn and Zouga himself.

It would be a slow and unwieldy caravan unless properly organized, that was bad enough – but on the march it would also be very vulnerable. Zouga gave much thought to defence of the column, and he and Sergeant Cheroot shared the last quarter bottle of whisky as they pooled their experience and planned the order of march.

Zouga, with a small party of local guides and personal bearers, planned to travel independently of the main caravan, reconnoitring the terrain ahead of the march, and making himself free to prospect and hunt as the opportunity arose. He would return most nights to rejoin the caravan, but would be equipped to spend many days out of contact.

Camacho Pereira with five of the Hottentot musketeers would lead the van of the main column, and even when Robyn chaffed him lightly, Zouga saw nothing ludicrous in Camacho marching under the Union Jack.

'It's an English expedition, and we will carry the flag,' Zouga replied stiffly.

'Rule, Britannia,' Robyn laughed irreverently, and Zouga ignored her and went on describing the order of march.

The divisions of porters would remain separate but closed up, and Sergeant Cheroot with the remainder of his musketeers would form the rearguard of the column.

There was a simple system of signals to control the movements of the column, a pre-arranged series of blasts on the kudu-horn trumpets would sound the 'march' or 'halt', the 'close-up or 'form square'.

For four days Zouga exercised the column in these evolutions and though proficiency would only come much later, at last he felt that they were ready to make a start, and he told Robyn so.

'But how are we to cross the river?' she asked looking across at the north bank.

The river was half a mile wide, and the heavy rainfalls over millions of square miles had drained into it. The flow was swift and powerful. If they were going northwards to the Shire river and Lake Marawi they would need a flotilla of dugout canoes and many days to make the crossing to the north bank.

The steam launch *Helen* had long ago departed on the flood of the river, making good at least twenty knots with the current pushing her, so she would already be back in Quelimane.

'All the arrangements have been made,' Zouga told her, and she had to be satisfied with that.

O n the last day Robyn allowed Juba to accompany her to the cemetery for the first time, and both of them were laden with gifts. Bolts of trade cloth and a thirty-five-pound bag of ceramic beads, the most sought-after scarlet *sam-sam* variety.

It was as much as she dare ask from the expedition's stores without arousing Zouga's ire and interest.

She had thought of telling Zouga of Sarah and the child, but had wisely decided against it.

Zouga's reaction to finding a half-brother of mixed blood was too terrible to contemplate. Zouga had acquired his opinions on caste and colour in the hard school of the Indian army, and to find that his own father had trespassed against these iron rules would be too much of a shock. Instead, Robyn had mentioned that she had met one of her father's former servants, who tended their mother's grave over the years. The gift would have to be in proportion to these services.

Sarah and her child were waiting by the grave, and she accepted the gifts with a gracious little curtsy and her palms held together at the level of her eyes.

'We leave tomorrow,' Robyn explained, and saw the immediate regret in Sarah's eyes, followed by acceptance.

'It is God's will,' and Robyn could almost hear her father saying it.

Juba and the child soon became involved in collecting the pods from a kaffir boom tree nearby and stringing together the pretty scarlet lucky beans, each with its little black eye at the end to make necklaces and bracelets. The two of them, girl and boy, made a delightfully uninhibited pair, their laughter and shrill happy voices a pleasant background to the talk under the acacia tree.

Robyn and Sarah had become friends in the short time they had known each other. Her father had written in his *Missionary Travels* that he preferred the company of black people to white, and certainly all the evidence seemed to point that way. It seemed that Fuller Ballantyne had done nothing but squabble with his own kind. Contact with other white men seemed to bring out in him all the pettiness, suspicion and jealousy of his complex nature; while he had spent the greater part of his life with black men, and received from them trust and honour and lasting friendships. His relationship with Sarah was only a natural extension of those feelings, Robyn realized. She contrasted these feelings with those of her brother, and knew that he could never cross the dividing line. A black man could possibly earn Zouga's liking and even his respect, but the gulf was too wide. For Zouga they would always be 'those people', and she guessed that he could never change in that. If he lived on in Africa for another fifty years, he would never learn to understand them, while she, within weeks, had made real friendships. She wondered if, like her father, she would come one day to prefer them to her own kind. It didn't seem possible now, but she recognized in herself the capacity for adjustment and change.

Beside her, Sarah was speaking, so softly, so shyly that

Robyn had to make an effort to tear herself away from her own thoughts and ask:

'What was that you said?'

'Your father, Manali, will you tell him about the boy when you find him?'

'He did not know?' Robyn was stunned, and Sarah shook her head.

'Why did you not go with him then?' Robyn demanded.

'He did not wish it. He said the journey would be too hard, but in reality, he is like an old bull elephant who does not like to stay too long with his cows but always must follow the wind.'

Camacho Pereira towered above the wiry little tribesmen, checking off their names against the camp register. This evening he wore a jerkin of kudu skin that was decorated with fancy stitching and trade beads, and unfastened down the front to expose bulging hairy chest and the flat belly with its ridged muscles like the patterns in the sand of a windswept beach.

'We feed them too much,' he told Zouga. 'Fat nigger is lazy nigger.' He chuckled when he saw Zouga's expression, for that word had already been the cause of dissension. Zouga had forbidden him to use it, particularly in front of the expedition's black servants, for some of them it was the only word of English they understood. 'Feed small, kick heavy and they work hard,' Camacho went on with relish.

Zouga ignored these gems of philosophy. He had heard them before, many times. Instead he turned to the captains of divisions, and watched them finish the rationing.

Two of them, their arms floury to the elbows, dipped into the bags of meal and doled it out to the shrinking line of porters. Each man had his calabash or chipped enamel basin to receive the scoop of stone-ground red-brown grain. Then one of the other captains slapped a split and smoked river fish on top of the pile. The fish looked like Scotch Kippers, but their odour punched like a prize-fighter. Weevils, maggots and all, they were a delicacy that the porters would miss once they left the river.

Pereira suddenly hauled a man out of the line and hit him a lusty clout across the back of the head with the stock of his kurbash.

'He come twice, try for double,' he explained cheerfully, and took a playful kick at the man as he ran. If it had landed it would have knocked him off his feet, but all the porters had come to anticipate Camacho's flying boots.

Zouga waited until the last porter had been rationed, then he called to the captains.

'*Indaba*! Tell the men, *indaba*.'

It was the call to council, to discuss affairs of consequence, and the whole camp left the cooking fires and came hurrying, agog with excitement.

Zouga paced up and down before the ranks of intent squatting black men, drawing out the moment, for he had come quickly to realize the African love of the theatrical. Most of them could understand the basic Nguni which Zouga now spoke with some fluency, for many of them were Shangaans or Angoni.

Now he spread his arms to his audience, paused for a second and then announced portentously.

'*Kusasa isufari*, tomorrow the march begins!'

There was a hive murmur of comment and excitement, and then one of the captains rose from the front rank.

'*Phi*? *Phi*? Where? Which way?'

Zouga lowered his arms, and let the suspense build up for a few moments, and then he stabbed out towards the far blue southern hills with a bunched fist.

'*Laphaya*! That way!'

They roared with approval, just as they would have done if Zouga had pointed north or west for that matter. They were ready to go now. The direction was not important. The captains of divisions, the indunas, were shouting out the translation to those who had not understood and the first roar of the crowd settled to a boisterous rumble of comment and speculation, but it died away suddenly, and Zouga turned quickly.

Camacho Pereira had stepped beside Zouga, and his face was swollen and dark with rage. This was the first time he had heard Zouga's intention to go southwards so that when he started to speak, it was so forcefully that droplets of spittle flew from his lips. He was using one of the local dialects, and speaking so swiftly that Zouga understood only a word here and there. The sense was unmistakable, however, and he saw the shock on the faces of the men who squatted before them.

Camacho was warning them of the dangers beyond the southern hills. Zouga heard the word 'Monomatapa' and knew that he was speaking of the terrible armies of the legendary empire, merciless legions whose favourite sport was to cut off a man's genitals and force him to eat them himself. The shock of the listening black faces was changing swiftly to terror and Camacho had been speaking for only a few seconds, a minute more and nothing would induce the caravan to march, two minutes more and most of his porters would have deserted by morning.

There was nothing to be gained by arguing with the Portuguese except an unseemly shouting match, which would be watched with interest by the entire assembled camp. One thing that Zouga had learned was that the Africans, like the Asians he had come to know well in India, were immensely respectful of a victor and impressed by success. He could demonstrate neither qualities by becoming embroiled in an undignified wrangle in a language that none of the spectators could understand.

'Pereira!' he snapped, in a tone that cut through the Portuguese's torrent of words, and for an instant stilled them. Zouga had the Englishman's peculiar sense of fair play which made him warn an enemy before an attack. As the Portuguese turned to face him, Zouga swung in towards him with two light steps and he flicked his left hand at Camacho's eyes, forcing him to throw up both hands to protect his face. As he did so, Zouga slammed his fist into his belly, just under the ribs, with a force that doubled him in the middle, his breath whooshing out of his gaping mouth in an agonized explosion of sound and his hand dropping to cover his injured belly, leaving his face open for the next blow.

It was a short chopping left-handed shot that took Camacho cleanly under the right ear, on the hinge of the jaw. The plumed beaver hat spun off his head. His eyes rolled up into his skull, leaving the whites glaring madly, and Camacho's knees gave way under him. He pitched forward, making no effort to cushion his fall and dropped face down on to the grey sandy soil.

The silence lasted only a second, and then a shout went up from the watchers. Most of them had felt Camacho's boots or his kurbash, and they hugged one another happily. The trepidation that Camacho's little speech had raised was completely lost in wonder at the swiftness and the effect of those two blows. Most of them had never seen a man strike with a bunched hand, and the novelty of this form of combat impressed and delighted them.

Casually, Zouga turned his back on the prone figure. Not a trace of anger showed on his face, in fact he was smiling slightly as he strolled down the front rank of men and lifted one hand to quieten them.

'There are soldiers who travel with us,' he told them, in a low voice that yet reached clearly to every one of them, 'and you have seen them shoot.' He had made sure they had, and that news of the prowess in weapons of Sergeant Cheroot's men would travel ahead of them.

'You see that flag?' He flung one hand dramatically at the red and white and blue jack floating above the main tent on its improvised flag pole. 'No man, no chief nor warrior dare . . .'

'Zouga!' shrieked Robyn. To the terrible urgency of her tone Zouga reacted instantly, spinning aside with two dancing paces, and the crowd exploded with a single word, a deep drawn out 'Jee!'

It is a blood-chilling sound, for it is the cry with which an African warrior encourages himself or another in the fatal moments of a battle to the death.

Camacho's stroke had been aimed at the small of Zouga's back. He was a man who had fought with the knife many times before, and he had not taken the more tempting target between the shoulder blades, where the point could turn against the ribs. He had gone for the soft area above the kidneys, and even with Robyn's warning, Zouga was not quite quick enough. The point raked his hip, slitting the cloth of his breeches in a six-inch rent, beneath which skin and flesh opened cleanly and the bright blood spread swiftly to the knee.

'Jee!' The deep sonorous chant of the watchers as Camacho reversed the stroke of his extended right hand, cutting sideways at Zouga's belly. The blade twinkled and hissed like an angry cobra, ten inches of tempered steel, and Zouga threw up his hands and sucked in his belly muscles as he jumped back. There was a sharp tug as the point caught in his shirt, but it did not touch the skin.

'Jee!' again as Camacho lunged. His face was bloated and mottled with purple and white, the eyes squinting with rage and the after-effects of the blow to his jaw. Zouga felt the sting of the wound on his hip pulling open as he swayed back out of the path of the blade, and the stronger flood of warm blood down his leg.

He paused out of range of the knife for Zouga had heard the snap of a weapon being cocked and from the corner of his eye he saw Sergeant Cheroot levelling the Enfield, waiting for a clean shot at the Portuguese.

'No! Don't shoot!' Zouga called urgently. He did not relish a bullet in his own belly, for he and Camacho were dancing close together, with the weaving point of the knife seeming to bind them to each other.

'Don't shoot, Sergeant!' There was another reason why he could afford no interference. There were a hundred men judging him now, men with whom he would march and work in the months and years ahead. He needed their respect.

'Jee!' sang the watchers, and Camacho was panting with rage. Again the blade in his right hand whispered like the wing of a swallow in flight, and this time Zouga over-reacted, blundering back half a dozen paces, and then losing his balance for a moment he dropped on one knee and put a hand to the ground to steady himself. But as Camacho charged again, he rebounded to his feet and arched his hips aside, the way a matador swings out of the line of the bull's run. In the hand that had touched the ground, Zouga held a handful of the coarse grey sand.

His eyes were locked to those of the Portuguese, it was the eyes not the knife hand that would signal Camacho's intentions. They flickered left, while the hand feigned the other way, and Zouga moved in past the blade, and was ready again when Camacho rounded.

They faced each other, shuffling in a slow cycle that stirred wisps of pale dust around their feet. Camacho kept the knife low, and stirred it gently as though he was conducting a slow passage of music, but Zouga studying his eyes saw the first small nervous flickers of uncertainty.

He jumped in, launching himself off the right foot.

'Jee!' roared the watchers, and for the first time Camacho broke ground, falling back and then turning hurriedly as Zouga checked and feinted to his open side.

Twice more Zouga drove him back with threats, until it needed only a feint with his upper body to make Camacho scramble away. The watchers were laughing now, mocking shouts of glee every time he gave ground, and the rage that had flushed Camacho's face had given way to fear, the angry purple mottling had chilled to white. Zouga was still watching his eyes, as they darted from side to side seeking an escape – but the knife kept weaving between them, bright and razor sharp, broad as three fingers and grooved along its length to break the suction of clinging wet flesh once it was buried.

Camacho's eyes flickered away once more and Zouga moved, pulling the knife hand around as he crossed the man's front, holding out his empty hand for the eyes to follow, keeping the other low and moving in as close to the knife as he dared, then at the moment that Camacho lunged, using the momentum of his avoiding turn, he hurled the handful of coarse sand into Camacho's eyes – blinding him, and still in the same movement reversing his direction, and going straight in on the knife, chancing it all on locking the wrist before the man could see again.

'Jee!' the crowd roared as Camacho's wrist slapped into Zouga's palm, and he locked it down with all his strength.

Tears were already streaming from Camacho's eyes, and his lids fluttered, grinding the sharp grains across the unseeing eyeballs. He could not judge nor meet Zouga's weight as, still locked grimly to the wrist, he threw him off-balance. As Camacho went over, Zouga reared back, resisting with all his strength, holding the knife arm against the fall. Something went with a loud rubbery popping sound in Camacho's shoulder and he screamed, as he sprawled again, face down, with the arm twisted up behind him.

Once again Zouga jerked viciously, and this time Camacho screeched like a girl and the knife dropped from his fingers. He made a feeble effort to snatch it with the other hand, but Zouga trod down on the blade with a booted foot, then scooped it up, released the damaged arm and stepped back holding the heavy weapon in his right hand.

'*Bulala*!' chanted the watchers. '*Bulala*! Kill him! Kill

him!' They wanted to see the blood, for that was the fitting end and they hungered for it.

Zouga stabbed the blade deeply into the trunk of the acacia tree and then wrenched against the steel. It snapped at the hilt with a crack like a pistol shot, and he dropped the hilt contemptuously.

'Sergeant Cheroot,' he said, 'get him out of this camp.'

'I should shoot him,' the little Hottentot told him as he came up, and thrust the muzzle of the Enfield rifle into the fallen man's belly.

'If he tries to enter the camp again, you can shoot him. But now just get him out.'

'Big mistake,' Sergeant Cheroot's pug face took on a theatrically mournful expression. 'Always stamp on the scorpion – before he stings.'

'You are hurt.' Robyn was running towards him.

'It's a scratch.' Zouga unwound the bandanna from around his throat and pressed it to the wound in his hip as he strode away towards his tent, forcing himself not to favour it with a limp. He had to get away quickly for the reaction was on him, he felt dizzy and nauseated, the wound stung abominably, and he did not want anybody to see that his hands were trembling.

'I reset the shoulder,' Robyn told Zouga as she bound up his hip wound. 'I don't think there is anything broken, and it went in again very neatly – but you,' she shook her head, 'you won't be able to march with that. Every step will pull against the stitches.'

She was right, it was four days before the march could begin, and Camacho Pereira put that time to good use. He had left an hour after Robyn reset his dislocated shoulder, four paddlers taking a dugout canoe down the Zambezi with the current. When they would have pulled into the bank to make camp, Camacho snarled at them from the bows where he crouched, hugging the injured arm, that even after being set and strapped into a sling, still ached so fiercely that it lit little white sparks of agony behind his closed eyelids every time he tried to doze.

He also would have liked to rest, but his hatred drove him onwards, and the dugout canoe arrowed down current under a fat yellow moon that paled slowly at the coming of the new day.

Camacho went ashore on the south bank of the Zambezi at noon at the small native village at Chamba, a hundred miles below Tete.

He paid off the crew of the dugout and he hired two bearers to carry his rifle and blanket roll. Then he set off again immediately along the network of narrow foot paths that crisscross the entire African continent like the blood vessels of a living body, laid down by wandering men and migrating animals over the centuries.

Two days later he reached the Hyena Road that runs from the mountains of Dismay, Inyangaza, to the sea. The Hyena Road was a secret track. Although it paralleled the old road from the coast to Vila Manica, it kept forty miles north of it, following the course of the Pungwe river so that there would be water for the multitudes who unwillingly used the road on their long, last journey from their homeland to other lands, other continents.

Vila Manica was the last outpost of the Portuguese administration in East Africa. A decree by the Governor in council forbade any man, black or white, Portuguese or foreigner, to journey beyond that clay-walled fort towards the haunting range of mountains with the chilling name. It was for this reason that the Hyena Road had been secretly opened by enterprising men, and pushed up through the dense forests of the lower slopes to the bleak and open grasslands atop the mountains.

The march from Chamba to the Pungwe river was a

hundred and fifty miles. To make it in three days with the agony of a healing shoulder was good going, and once they reached it, the temptation to rest was almost irresistible. But Camacho kicked his two bearers to their feet and drove them with stinging words and lash along the deserted road towards the mountains.

The road was twice as wide as any of the other footpaths they had followed to reach it, wide enough for a double column rather than the Indian file that was the usual order of African travel. Although the surface had been beaten hard by the passage of thousands of bare feet, it was a source of satisfaction to Camacho that the road had clearly not been used for many months, except by the occasional herd of antelope, and once, perhaps a week before, by an old bull elephant, whose huge piles of dung had long dried out.

'The caravan has not passed yet,' Camacho muttered, as he scanned the trees ahead for the shapes of the vultures and searched without success for the sly skulking shapes of hyena in the undergrowth beside the road.

True there were human bones scattered along the route, here and there the thick knuckle of the thigh bone that had defied even the iron jaws of the scavengers, or other splintered fragments that they had overlooked, but even these were dried out and bleached white. They were the debris of the previous caravan that had passed this way three months before.

He had reached the road in time, and now he hurried along it, pausing now and then to listen or to send one of the bearers up a tree to search ahead.

However, it was two days later that they heard the first faint sound of many voices, and this time Camacho himself climbed to the highest fork of one of the umsisa trees beside the track, and peering ahead he saw the vultures circling, a wide slow wheel of tiny black specks turning against the silver and blue ranges of cloud, as though caught in a hidden vortex of the high heavens.

He sat in the fork thirty feet above the ground, while the sound of voices grew stronger, became the sound of singing. This was no sound of joy, but a terrible mourning dirge, slow and heart-breaking, rising and fading as flukes of the breeze and folds of the ground blanketed the sound, but each time it came back a little stronger, until Camacho could make out far away the head of the column, like the head of a maimed serpent writhing out of the forest into an open glade a mile ahead.

He slid down the trunk of the umsisa, and hurried forward. There was an armed party ahead of the main column, five blacks dressed in the tatters of cast-off European-style clothing and carrying muskets, but at their head was a white man, a little man with a face like a vicious gnome, wrinkled and burned darkly by the sun. The thick drooping black moustache was laced with grey, but he stepped out with a bouncing elastic stride and he recognized Camacho from two hundred paces and snatched his hat off his head and waved it.

He shouted 'Camacho!' and the two men ran to embrace, and then hold each other at arm's length, laughing with pleasure. It was Camacho who sobered first, the laughter changing to a scowl as he said,

'Alphonse, my beloved brother, I have bad tidings – the worst possible.'

'The Englishman?' Alphonse was still smiling, he had a tooth missing from the front of his upper jaw, which made the cold humourless smile seem less dangerous than it really was.

'Yes, the Englishman,' Camacho nodded. 'You know of him?'

'My father sent a message. I know.' Alphonse was the Governor of Quelimane's eldest surviving son, full-blooded Portuguese by the lawfully wedded bride who had come out forty years previously from Lisbon, a pale sickly mail-order bride, who had borne three sons in swift succession, the first two of which had succumbed to malaria and infantile dysentery even before the appearance of the little wizened yellowed mite whom they had named Alphonse Jose Vila y Pereira, and expected to bury with his brothers before the end of the rains. However, it was the mother they had buried in the end, and the child had flourished at the breast of a black wet nurse.

'He did not go north, then?' Alphonse demanded, and Camacho dropped his eyes guiltily, for he was speaking to the eldest, full-blooded and legitimate son.

Camacho himself was a bastard and a half-breed, son of one of the Governor's once beautiful mulatto concubines, now fat and faded and forgotten in one of the back rooms of the seraglio. He was not even recognized as a son, but had to bear the ignominious title of nephew. This in itself was enough for him to show respect for the other, but added to this Alphonse was as determined as their father had been at the same age, though even crueller and harder. Camacho had seen him sing a plaintive fado as he flogged a man to death, accompanying the traditional love song with the flute and percussion of the lash.

'He did not go north,' Camacho agreed uneasily.

'You were told to see that he did.'

'I could not stop him. He is English,' Camacho's voice croaked a little, 'he is stubborn.'

'We will speak again of that,' Alphonse promised coldly. 'Now, swiftly tell me where he is and what he plans to do.'

Camacho recited the explanation he had prepared, skirting delicately around the most offensive parts of the story, and dwelling on subjects such as the wealth that the Ballantyne expedition carried with it rather than his own brutal beating at the Englishman's hands.

Alphonse had thrown himself down in the shade of a tree beside the track and listened broodingly, chewing at the straggling ends of his moustache, filling in for himself the conspicuous gaps in his half-brother's recital, and speaking again only at the end.

'When will he leave the valley of the Zambezi?'

'Soon,' Camacho hedged, for the unpredictable Englishman might already be half-way to the escarpment. 'Although I cut him deeply, it may be he has had himself carried in a *mushila* (litter).'

'He must not be allowed to enter the Monomatapa,' Alphonse said flatly, and came to his feet with a single lithe movement. 'The best place to do the business would be in the bad ground below the rim of the valley.'

He glanced back along the winding road. The head of the column was a mile away still, across a glade of open golden grass. The shuffling double rank of bowed creatures, yoked at the neck, did not seem human, though the singing was sad and beautiful.

'I can spare fifteen men.'

'It will not be enough,' Camacho cut in swiftly.

'It will be,' said his brother coldly, 'if you do the business in the night.'

'Twenty men,' Camacho pleaded. 'He has soldiers with him, trained soldiers and he is a soldier himself.'

Alphonse was silent, weighing risk and advantage – but the worst part of the Hyena Road already lay behind the column, and each mile nearer the coast, the land was tamer, the risk diminished and the need for guards less pressing.

'Twenty!' he agreed abruptly, and turned to Camacho. 'But not one of the foreigners must escape.' Looking into his brother's cold black eyes, Camacho felt his skin crawl. 'Leave no sign, bury them deep, so the jackal and hyena do not dig them out. Use the porters to carry the expedition's equipment to the place in the hills, and when they have done so, kill them also. We will bring it down with the next caravan to the coast.'

'Si. Si. I understand.'

'Do not fail us again, my beloved cousin-brother.' Alphonse made the endearment a threat, and Camacho swallowed with a nervous little gulp.

'I will leave as soon as I have rested.'

'No,' Alphonse shook his head. 'You will leave immediately. Once that Englishman enters the land beyond the mountains, there will soon be no more slaves. It is bad enough that there has been no gold for twenty years and more, but if the river of slaves were to dry up, both my father and I would be displeased – very displeased.'

A t Zouga's order the long mournful blast of the kudu-horn trumpet shattered the silence of the utterly dark hour before the dawn.

The indunas took up the cry 'Safari! We march!' and they prodded the sleeping porters off their reed sleeping-mats. The camp fires had burned down to dim red mounds of coals smothered in the soft grey powder of their own ash. As fresh logs were thrown upon them they flared up in a false dawn that lit the underside of the umbrella-shaped acacia trees with wavering yellow light.

The smell of roasting *ropoko* cakes rose on tendrils of pale smoke straight into the windless dark sky. The muted voices became louder, more cheerful as the flames drove away the chills and the nightmares.

'Safari!' The cry was taken up, and the divisions assembled, ghostly figures in the gloom, emerging more clearly as the growing dawn light paled the sky and snuffed out the stars.

'Safari!' And the mass of men and equipment resolved itself and order grew out of chaos.

Like those long columns of big shiny black serowe ants

that endlessly cross and re-cross the African earth, the stream of porters moved steadily away into the still gloomy forest.

As each of them passed Zouga and Robyn standing together at the gate of the thorn *scherm*, they shouted a greeting and executed a few prancing steps to demonstrate their loyalty and enthusiasm – while Robyn laughed with them and Zouga called encouragement.

'We no longer have a guide, and we don't know where we are going.' She took Zouga's arm. 'What is to become of us?'

'If we knew, it would take all the fun out of it.'

'At least a guide.'

'While you thought I was hunting I went out as far as the escarpment which is further than that swaggering Portuguese ever went, further than any white man, except of course Pater, has ever been. Follow me, Sissy, I am your guide.'

She looked up at him now in the strengthening light of coming dawn.

'I knew you were not hunting,' she told him.

'The escarpment is rugged and very broken, but I have examined two passes through the telescope that I think will go—'

'And beyond that?'

He laughed, 'We will find out.' Then he squeezed her around the waist. 'That is the whole fun of the thing.'

She studied his face with full attention for a few moments. The new full beard emphasized the strong, almost stubborn, lines of his jaw. There was a piratical devil-may-care lift to the corner of his lips, and Robyn realized that no man of conventional mind would have proposed and engineered this expedition. She knew he possessed courage, his exploits in India had proved that beyond doubt, and yet when she looked at his sketches and water colours and read the rough notes he was making for the book, she discovered a sensitivity and an imagination she had never before suspected. He was a difficult person to know and understand.

Perhaps she could have told him about Sarah and the child or even about Mungo St John, and that night in the main cabin of *Huron*, for when he laughed like this, the stern features softened with humour and humanity, and green lights sparkled in his eyes.

'That's what we are here for, Sissy, the fun of it all.'

'And the gold,' she teased, 'and the ivory.'

'Yes, by God, the gold and ivory as well. Come on, Sissy, this is where it truly begins,' and he limped after the column as its tail disappeared into the acacia forest, favouring his injured leg and using a freshly cut staff to move across the sandy earth. For a moment Robyn hesitated and then she shrugged aside her doubts and ran to catch up with her brother.

That first day the porters were rested and eager, the valley floor flat and the going easy, so Zouga ordered *tirikeza*, the double march, so that even at their slow pace the column left many miles of dusty grey earth behind them that day.

They marched until the heat came up in the middle of the morning and the merciless sun dried the sweat the moment it burst through their pores and left tiny salt crystals on the skin, that sparkled like diamond chips. Then they found shade and lay like dead men through the heat of noon, stirring again only when the lowering sun gave the illusion of cooling the air and the blast of the kudu-horn trumpet forced them to their feet again.

The second stage of the *tirikeza* lasted until sunset when it became too dark to see the ground under their feet.

The fires were dying and the voices of the porters in their thorn bush *scherms* had slowly descended through the occasional mutter and soft murmur to ultimate silence before Zouga left his tent and limped silently as a night creature out of the camp.

He carried the Sharps rifle slung over his shoulder, the staff in one hand and a bull's eye lantern in the other, while the Colt revolver hung in its holster upon his belt. Once clear of the camp, he stepped out as briskly as his leg would allow two miles along the freshly beaten footpath that the column had made that afternoon until he reached the fallen tree trunk that was the agreed rendezvous.

He stopped and whistled softly, and a smaller figure stepped out from the undergrowth into the moonlight, carrying a rifle at high-port. The jaunty step and alert set of head on narrow shoulders was unmistakable.

'All is well, Sergeant.'

'We are ready, Major.'

Zouga inspected the ambush positions that Sergeant Cheroot had chosen for his men astride the path. The little Hottentot had a good eye for ground and Zouga found his trust and liking for him increasing with every such display of competence.

'A puff?' Jan Cheroot asked now, with the clay pipe already in his mouth.

'No smoking,' Zouga shook his head. 'They will smell it.' And Jan Cheroot reluctantly buttoned the pipe into his hip pocket.

Zouga had chosen a position in the centre of the line, where he could make himself comfortable against the trunk of the fallen tree. He settled down with a sigh, his leg thrust out stiffly ahead of him – after the *tirikeza* it was going to be a long wearying night.

The moon was a few days short of full, and it was almost light enough to read the headlines of a newspaper. The bush was alive with the scurry and rustle of small animals, and it kept their nerves tightened and their ears strained to catch the other sounds for which they waited.

Zouga was the first to hear the click of a pebble striking against another. He whistled softly and Jan Cheroot snapped his fingers, imitating the sound of a black scarab beetle to show that he was alert. The moon had dropped low upon the hills, and its light through the forest trees laid silver and black tiger stripes upon the earth and played tricks with the eye.

Something moved in the forest, and then was gone, but Zouga picked up the whisper of bare feet scuffling the sandy disturbed earth of the path, and then suddenly they were there, and very close, man shapes in file, hurrying, silent, furtive. Zouga counted them, eight – no nine. Each of them straight-backed under the bulky burden he carried balanced upon his head. Zouga's anger simmered to the surface and yet at the same time he felt a grim sense of satisfaction that he had not wasted the night.

As the leading figure in the file came level with the fallen tree trunk, Zouga pointed the muzzle of the Sharps rifle straight into the air and pressed the trigger. The crash of the shot broke the night into a hundred echoes that bounced and rebounded through the forest, and the silence magnified it until it seemed like the thunder of all the heaven.

The echoes had not dispersed, and the nine dark figures were still frozen with shock when Jan Cheroot's Hottentots fell upon them from every direction in a shrieking pack.

The sound of their cries was so shrill, so inhuman, that it even startled Zouga, while the effect on the victims was miraculous. They let fall the burdens they carried, and dropped to earth in a paralysis of superstitious awe, adding their wails and screams to the pandemonium. Then the thud and clatter of cudgels against skull and cringing flesh mingled with it all, and the screams and howls rose to a new pitch.

Jan Cheroot's men had spent much time and care on selecting and cutting their clubs and now they wielded them with a lusty glee, making up for a night of discomfort and boredom. Sergeant Cheroot himself was in the thick of it, and in his excitement he had almost lost his voice. He was yipping squeakily like a demented fox terrier with a cat up a tree.

Zouga knew he would have to stop it soon, before they killed or seriously maimed somebody, but the punishment was richly earned, and he gave it a minute more. He even joined in himself when one of the prostrate figures scrambled to its feet and tried to dart away into the undergrowth. Zouga swung his staff and brought him down again with a blow to the back of his knees, and when he sprang up again as though he were on springs, Zouga dropped him in the dust with a short right-handed punch to the side of the head.

Then, stepping back out of the fray, Zouga took one of his few remaining cheroots from his top pocket, and lit it from the chimney of the lantern, inhaling with deep satisfaction, while around him the enthusiasm of his Hottentots flagged a little as they tired and Jan Cheroot regained his voice and became coherent for the first time.

'*Slat hulle, kerels*! Hit them, boys!'

It was time to stop it, Zouga decided and opened the shutter of the lantern.

'That's enough, Sergeant,' he ordered, and the thuds of blows became intermittent and then ceased while the Hottentots rested on the cudgels, panting and streaming with the honest sweat of their exertions.

The deserting porters lay moaning and whimpering in pitiful heaps, with their loot scattered about them. Some of the packs had burst open, and trade cloth and beads, flasks of gunpowder, knives, mirrors and glass jewellery were strewn about and trodden into the dirt. Zouga's fury returned at full strength when he recognized the tin box which contained his dress uniform and hat. He delivered a last kick at the nearest figure and then growled at Sergeant Cheroot,

'Get them on their feet and clean it up.'

The nine deserters were marched into camp, roped together and bearing not only the heavy burdens which they had stolen – but also an impressive set of contusions, cuts and bruises. Lips were swollen and split, some teeth were missing, a good many eyes were puffed closed and most of their heads were as lumpy as newly-picked Jerusalem artichokes.

More painful than their injuries, however, was the ridicule of the entire camp which turned out to a man to jeer and mock them with laughter.

Zouga lined up the captives, with their booty piled in front of them, and in the presence of their peers made a speech in limping but expressive Swahili in which he likened them to sneaking jackals and lurking hyena and fined them each a month's wages.

The audience was delighted with the show, and hooted at every insult while the culprits tried physically to shrink themselves into insignificance. There was not one of the watching porters who would not have done the same thing. In fact, had the escape succeeded, most of them would have followed the next night, but now that it had been foiled, they could enjoy the vicarious pleasure of having escaped punishment, and the discomfort of their companions who had committed the sin of being caught.

During the noon rest that divided the two stages of the next day's *tirikeza*, the clusters of porters chatting in the shade of the mopani groves agreed that they had found a strong master to follow, one whom it would not be easy to cheat, and it gave them all confidence for the future of the safari. Coming directly after his defeat of the Portuguese, the recapture of the deserting porters added immeasurably to Zouga's standing.

The four indunas of the divisions agreed that it was fitting that such a man have a praise-name. They conferred at length, and after considering many suggestions, finally decided on 'Bakela'.

'Bakela' means 'the one-who-strikes-with-the-fist', for this was still the one of Zouga's many accomplishments which impressed them most.

Where Bakela led, they were now prepared to follow, and though Zouga spread a dragnet of his faithful Hottentots behind the column each of the following nights, no more fish swam into it.

'How many?' Zouga whispered, and Jan Cheroot rocked on his heels, sucking softly on the empty clay pipe and squinting his oriental eyes thoughtfully, before he shrugged, 'Too many to count. Two hundred, three hundred, perhaps even four.'

The ground had been ploughed up into soft fluffy dust by the multitude of huge cloven hooves, and the dark pats of dung were round and shaped in little concentric circles, completely indistinguishable from those of domestic cattle, and the rank smell of cattle was heavy on the heated air of the Zambezi valley.

For an hour they had followed a small herd through the open mopani forest, stooping under the low branches with the thick shiny double leaves, each of them shaped like the cloven spoor that they followed, and now where the spoor emerged from the forest it had been joined by another much larger herd.

'How close?' Zouga asked again, and Jan Cheroot slapped his own neck where one of the buffalo flies had settled. It was the size of a honey bee, but dull black and the long needle of its proboscis stung as though it was white hot.

'We are so close that the flies that follow the herd still

linger,' and he pushed his forefinger into the nearest pat of wet dung, 'and the body heat is still in the dung, but,' Jan Cheroot went on as he wiped his finger on a handful of dry grass, 'but they have gone into bad ground—' and he pointed ahead with his chin.

A week before they had reached the escarpment of the valley, but each of the possible passes that Zouga had examined through the telescope had proved on closer inspection to be dead ends, the gorges pinching out into abrupt rock faces, or falling off into some terrifying abyss.

They had turned westwards, following the edge of the escarpment, Zouga ranging ahead with his small scouting party. Yet day after day those impassable heights loomed at their left hand, rising sheer into the unknown. Even below the main escarpment, the ground was tortured and riven by deep gorges and ravines, by cliffs of dark rock and hills of enormous tumbled boulders. The ravines were choked with the drab grey stands of thorn, so densely interwoven that a man would have to crawl in on hands and knees, and his vision would be limited to a few feet ahead, yet the herd of many hundreds of buffalo that they were following had disappeared into one of these narrow gorges, their thick hides impervious to the cruel red-tipped thorn.

Zouga took the telescope from his haversack carried by his bearer, and carefully scanned the ground ahead. It had a wild and menacing beauty and for the hundredth time in the last few days he wondered if there was a way through this maze to the empire of Monomatapa.

'Did you hear that?' Zouga demanded, lowering the glass suddenly. It had sounded like the distant lowing of the milk herd as it returned to the farmyard.

'Ja!' Sergeant Cheroot nodded, as again the mournful sound echoed against the black ironstone cliffs, and was answered by the bleat of a calf. 'They are lying up in the jessie bush. They won't move again until sunset.'

Zouga glanced up at the sun. It was four hours or so from

its zenith. He had over a hundred mouths to feed, and they had rationed out the last of the dried fish two days before.

'We will have to go in after them,' he said, and Jan Cheroot removed the stem of the pipe from between yellow teeth and spat reflectively in the dust.

'I am a very happy man,' he said. 'Why would I want to die now?'

Zouga lifted the glass again, and while he scanned the ridges of higher ground about the choked valley, he imagined what it would be like in there. When the first shot was fired, the jessie bush would be filled with huge, furiously charging black animals.

The fluky breeze coming down the steep narrow valley brought with it another powerful whiff of the herd smell before it faded.

'The wind is down the valley,' he said.

'They have not smelled us,' Cheroot agreed, but that was not what Zouga had meant. Again he examined the nearest ridge of high ground. A man could work his way along the edge of it, up towards the head of the narrow valley.

'Sergeant, we are going to flush them out,' he smiled, 'like spring pheasant.'

Zouga had found the native names of his personal bearers hard to pronounce, and tiresome to remember. There were four of them. He had selected them with care, rejecting half a dozen others in the process, and he had rechristened them Matthew, Mark, Luke and John. They had earned enormous prestige by being so honoured, and had proved keen and willing to learn their duties. In a few days they were already proficient at reloading, though not yet of the same standard as Camacho Pereira's gunbearers – but that would come.

Zouga carried the Sharps rifle, but each of the four bearers had one of the heavy four-to-the-pound elephant guns that Harkness had recommended to Zouga. At any time he had only to reach back over his shoulder and a loaded and primed weapon would be thrust into his hand.

Apart from the elephant guns, his bearers carried his blanket roll, water bottle, canvas food bag, spare ball and powder, and the little clay fire-pot from which a smouldering ball of moss and wood pulp could be blown into flame in a few seconds. It was wise to conserve the amenities of civilization, such as Swan Vestas, for the months and years ahead.

Zouga relieved Luke, the quickest and most wiry of the four, of all his equipment except the fire-pot, pointed out the path along the cliff, and explained carefully to him what he was to do.

All of them listened with approval, even Sergeant Cheroot nodded sagely at the end. 'My old mother tells me, before she throws me out, "Jan", she says, "remember it's brains what counts." '

In the mouth of the valley, where it debouched out into the mopani forest, was a low outcrop of rock, the black ironstone boulders had been split into strange shapes by sun and erosion, and they formed a natural redoubt, with chest-high walls behind which a man could crouch. A hundred paces directly ahead, the dense palisade of iron-grey thorn blocked the valley, but the ground between was fairly open, with a few stunted second-growth mopani bushes and clumps of coarse dried razor grass as high as man's shoulder.

Zouga moved his party into the lee of the rocks, and himself scrambled on to the highest point to follow through his glass the progress of the almost naked bearer as he picked his way cautiously along the rim of the cliff. Within half an hour he had worked so far up the escarpment that he had disappeared from Zouga's view.

It was another hour before, from the head of the valley, a thin tendril of pure white smoke rose gently into the heated air, and then bent into the elegant shape of an ostrich plume before the gentle breath of the breeze.

With miraculous suddenness the rising column of white smoke was surrounded by another living cloud, hundreds of tiny black specks that weaved and darted about and around it. The faint but excited bird cries carried down to where Zouga waited, and through the glass he could make out the rainbow, turquoise and sapphire plumage of the blue jays as they rolled and dived for the insects put to flight by the flames. Competing with them for the feast, were the iridescent black drongas with their long, forked tails catching the sunlight with metallic glitter as they swirled above the spreading smoke clouds.

Luke was doing his job well. Zouga grunted with satisfaction, as new columns of smoke rose at intervals, sealing off the valley from side to side as they spread to meet each other. Now there was a solid wall of smoke from one cliff to the other, and as the smoke turned dirty black, billowing upwards, spinning upon itself, carrying flaming fragments of leaves and twigs within it, it began to roll ponderously down the valley.

It reminded Zouga of a snow avalanche he had watched in the high Himalayas, the slow majestic progress gathering weight and momentum, building up its own wind storm as it sucked the valley of air.

He could see the tops of the flames now, leaping above the thorn, and hear the sound of them, like the whispering waters of a distant river. The alarm bellow of a bull buffalo rang like the blast of a war trumpet from the ironstone cliffs, and the whisper of flames rose swiftly to a dull crackling roar.

The smoke clouds rose across the sun, plunging them into an unnatural gloom, and Zouga felt a sharp drop in his spirits at the extinction of the bright morning sun, that infernal swirling pall of dun smoke seemed to hold a world of menace.

From the edge of the jessie bush broke a herd of kudu, led by a magnificent bull with his corkscrew horns laid flat along his back. He saw Zouga standing on the pinnacle of rock, and snorted with alarm, swinging away out of easy shot with his cows flying big-eared and scared behind him, their fluffy white tails flickered away amongst the mopani groves.

Zouga scrambled down from his too obvious position, and propped himself comfortably against the rock, checked the nipple on the cap of the Sharps and then cocked the big hammer.

Ahead of the flames, a pale white dust cloud was rising over the tops of the jessie bush, and another sound was added to the roar of flames. It was a low thunder that made the earth tremble under their feet.

'They are coming,' Jan Cheroot muttered to himself, and his little eyes sparkled.

A single buffalo burst from the palisade of thorn. He was an old bull, almost bald across the shoulders and rump, the dusty grey skin criss-crossed by a thousand ancient scars and scabby with the bites of bush ticks. The big bell-shaped ears were torn and tattered, and one thickly curved horn was broken off at the tip. He came out at a crabbing gallop, dust exploding at each hoof beat like miniature mortar bursts.

He was on a line to pass the rocky redoubt at twenty paces, and Zouga let him come on to twice that distance before he threw up the Sharps rifle.

He aimed for the fold of thick skin under the throat that marked the frontal aiming point for the heart and its complex of arteries and blood vessels. He hardly noticed the recoil nor the blurt of the shot as he watched for the strike of the lead bullet. There was a little spurt of dust off the grey hide precisely on his aiming point, and the sound of the hit was exactly like his headmaster swinging the malacca cane against his own schoolboy backside, sharp and meaty.

The bull took the bullet without a stumble or lurch, instead it swung towards them, and seemed immediately to double in size as it lifted its nose into the high attitude of the charge.

Zouga reached for his second gun, but he groped in vain.

Mark, his number two, showed the whites of his eyes in a flash of terror, let out a squawk, hurled the elephant gun aside, and went bounding away towards the mopani grove.

The bull saw him and swerved again, thundering ten feet past Zouga as it went after the fleeing bearer. Waving the empty Sharps, Zouga shouted desperately for another rifle, but the bull was past him in a grey blur and it caught Mark as he reached the tree line.

The great bossed head dropped until the snout almost touched the earth, and then flew up again in a powerful tossing motion that bunched the muscle in the thick black neck. Mark was looking back over his shoulder, his eyes wide and glaring white in the black face, rivulets of sweat pouring down his naked back, his mouth a pink gape as he screamed.

Then he was in the air. Legs and arms tumbling wildly, he went up like a rag doll thrown by a petulant child and disappeared into the thick green canopy of mopani foliage overhead. Without missing a stride, the bull drove on into the forest, but that was all that Zouga saw, for a cry from Sergeant Cheroot made him turn again.

'Hier kom hulle! Here they come!'

Across their whole front, the earth seemed to move, as though racked by the convulsions of an earthquake. Shoulder to shoulder, nose to rump, the main herd broke from cover, flattening the thorn bush under the great wave of bodies, filling the valley from side to side.

They lifted behind them a dense curtain of pale dust, from which the front ranks seemed endlessly to emerge, their great bossed heads nodding in unison as they pounded on, long silver strings of saliva dangling from open jaws as they bellowed in alarm and anger, and the roar of their hooves drowned the sound of the flames.

Matthew and John, Zouga's two remaining bearers, had stood their ground, and one of them snatched away the empty Sharps and thrust the thick stock of an elephant gun into Zouga's hand.

The weapon seemed heavy and unbalanced after the Sharps, and the sights were crudely fashioned, a blunt cone for the foresight, and a deep vee for the backsight.

The solid wall of bodies was bearing down upon them with frightening speed. The cows were a dark chocolate colour, and their horns were more delicately curved. The calves that raced at their flanks were sleek russet with crowns of reddish curls between the rudimentary little horn spikes. The herd was so tightly packed that it seemed impossible that they could split open to pass the rock. There was a tall rangy cow in the leading rank, coming straight on to Zouga.

He held half a beat aiming into the centre of her chest, and squeezed off the shot. The firing cap popped with a tiny puff of white smoke, and a heartbeat later the elephant gun vomited a deafening gust of powder smoke and bright flame, the burning patches went spinning away over the heads of the charging buffalo, and Zouga felt as though one of them had kicked in his shoulder. He staggered backwards, the barrel thrown high by the recoil, but the big red cow seemed to run into an invisible barrier. A quarter of a pound of mercury-hardened lead drove into her chest, and brought her down in a rolling sliding tangle of hooves and horns.

'Tom Harkness! That one was yours!' Zouga shouted, offering the kill to the memory of the old white bearded hunter, and he grabbed the next loaded rifle.

There was a prime bull, big and black, a ton of enraged bovine flesh. It had seen Zouga, and was coming in over the rocks in a long scrambling leap – hunting him out, so close that Zouga seemed to touch it with the gaping muzzle of the four-to-the-pound. Again the great clanging burst of sound and flame and smoke, and half the bull's head flew away in a gust of bone chips and bloody fragments. It reared up on its hind legs, striking out with fore hooves, and then crashed over in a cloud of dust.

Impossibly, the herd split, galloping down each side of their rocky hide, a heaving, grunting, forked river of striving muscle and bone. Jan Cheroot was yipping shrilly with the fever of the chase, ducking down behind the rock to reload, biting open the paper cartridge with powder dribbling down his chin, spitting the ball into the muzzle and then plying the ramrod in a frenzy, before bobbing up again to fire into the solid heaving press of gigantic bodies.

It lasted for two minutes, which seemed to take a round of eternity, and then they were left choking and gasping in the swirling clouds of dust, surrounded by half a dozen huge black carcasses, with the drum beats of the herd fading away into the mopani forest, and a louder more urgent din roaring down on them from in front.

The first tongue of heat licked across them, and Zouga heard the lock of sunbleached hair that hung on his forehead frizzle sharply and smelt the stink of it. At the same instant, the dust cloud fell abruptly aside, and for seconds they stared at a spectacle which deprived them of power of movement.

The jessie bush was not burning, it was exploding into sheets of flame.

'Run!' shouted Zouga. 'Get out of here!'

The sleeve of his shirt charred, and the air he breathed scorched his lungs painfully. As they reached the edge of the mopani forest, the shiny green leaves about their heads shrivelled and yellowed, curling their edges in the heat, and Zouga felt his eyeballs drying out as the dark smoke clouds rolled over them. He knew that they were experiencing only the heat and smoke carried on the wind, but if the flames were able to jump the gap, then they were all doomed. Ahead of him, the Hottentots and the other bearers were shadowy wraiths, staggering forward but weakening and losing direction.

Then, as suddenly as they had been engulfed by them,

the billowing smoke clouds lifted. The flames had not been able to jump the open ground, and the heat came only in gusts. A ray of sunshine pierced the thick gloom overhead, and a puff of sweet fresh air came through. They sucked at it gratefully, and huddled in awed silence, beating at their clothing which still smouldered in patches. Zouga's face was blackened and blistered, and his lungs still convulsed in spasms of coughing. As he caught his breath, he grunted hoarsely,

'Well the meat is cooked already,' and he pointed back at the buffalo carcasses.

At that moment something fell limply out of the dense top branches of a mopani tree, and then picked itself up and limped painfully towards them. Zouga let out a husky growl of laughter.

'Oh, thou swift of foot,' he greeted Mark, the bearer, and the others took up the mockery.

'When you fly, the eagles are put to shame,' Jan Cheroot hooted.

'Your true home is in the treetops,' Matthew added with relish, 'with your hairy brethren.'

By evening they had hacked the buffalo carcasses into wet red chunks, and spread these on the smoking racks. The racks were waist-high cross-poles set in forked branches, with a slow smoking fire of wet mopani wood smouldering under it.

Here was meat for the caravan that would last them many weeks.

Camacho Pereira had no doubts that by simply following the line of the escarpment, keeping just below the bad broken ground, he must at last cut the spoor of the caravan. A hundred men, in column, would blaze a track that even a blind man would trip over.

His certainty dwindled with each day's march through the quivering, breathless heat that seemed to rebound from black kopjes and the ironstone cliffs which glittered in the aching sunlight like the scales of some monstrous reptile.

Of the men that his half-brother, Alphonse, had given him, he had already lost two. One had stepped on something that looked like a pile of dried leaves, but which had transformed itself instantly into six feet of infuriated gaboon adder, thick as a man's calf, with a repulsively beautiful diamond-patterned back, and a head the size of a man's fist. The gaping mouth was a lovely shade of salmon pink, and the curved fangs three inches long. It had plunged them into the man's thigh and squirted half a cupful of the most toxic venom in Africa into his bloodstream.

After blowing the serpent to shreds with volleys of rifle fire, Camacho and his companions had wagered all their expectations of loot on exactly how long it would take the victim to die. Camacho, the only one who owned a watch, was elected timekeeper, and they gathered around where the dying man lay, either urging him to give up the useless struggle or pleading with him raucously to hang on a little longer.

When he went into back-arching convulsions, with his eyes rolled up into his skull, his jaws locked into a grinning rictus and he lost control of his sphincter muscle, Camacho knelt beside him, holding a bunch of smouldering tambooti leaves under his nostrils to shock him out of it, and crooning, 'Ten minutes more – hang on for just ten minutes more for your old friend Machito!'

The last convulsion ended with a dreadful gargling

expulsion of breath, and when the heart beat faded completely, Camacho stood and kicked the corpse with disgust.

'He always was a dung-eating jackal.'

When they began to strip the corpse of all items of any possible value, five coins, heavy golden mohurs of the East India Company, fell out of the folds of his turban.

There was not one man in all that company who would not have willingly sold his mother into slavery for a single gold mohur, let alone five.

At the first gleam of gold, all their knives came out with a sardonic metallic snickering, and the first man to snatch for the treasure reeled away, trying to push his intestines back into the long clean slice through his stomach wall.

'Leave them lie,' Camacho shouted. 'Don't touch them until the lots are drawn!'

Not one of them trusted another, and the knives stayed out while the lots were cast, and grudgingly the winners were allowed one at a time to claim their prize.

The man with the belly wound could not march without his stomach falling out, and because he could not march, he was as good as dead. The dead, as everybody knows, have no need of personal possessions. The logic was apparent to all. They left him his shirt and breeches, both torn and badly stained anyway, but stripped him of all else as they had stripped the first corpse. Then, with a few ribald pleasantries, they propped him against the base of a marula tree, with the naked corpse of the snake-bite victim beside him for company, and they marched away along the line of the escarpment.

They had gone a hundred yards when Camacho was overcome by a rush of compassion. He and the dying man had fought and marched and whored together for many years. He turned back.

The man gave him a haggard grin, his dry crusted lips cracking with the effort. Camacho answered him with that marvellous flashing smile as he dropped the man's loaded pistol in his lap.

'It would be better to use it *before* the hyena find you tonight,' he told him.

'The thirst is terrible,' the man croaked, a tiny bead of blood appeared on his deeply cracked lower lip, bright as an emperor's ruby in the sunlight. He eyed the two-gallon water bottle on Camacho's hip.

Camacho resettled the water bottle on its strap so it was out of sight behind his back. The contents sloshed seductively.

'Try not to think about it,' he counselled.

There was a point where compassion ended and stupidity began. Who knew where and when they would find the next water? In this God-blasted desolation, water was an item not to be wasted on a man who was already as good as dead.

He patted the man's shoulder comfortingly, gave him a last lovely smile and then swaggered away amongst the grey ironthorn scrub, whistling softly under his breath with the plumed beaver cocked over one eye.

'Camachito went back to make sure we had forgotten nothing.' The one-eyed Abyssinian greeted him as he caught up with the column, and they shouted with laughter. Their spirits were still high, the water bottles more than half filled and the prospects of immense loot danced like a will-o'-the-wisp down the valley ahead of them.

That had been ten days ago, the last three of which without water, for you could not count the cupful of mud and elephant piss they had from the last puddled waterhole. Apart from the lack of water, the going had become appalling. Camacho had never marched through such broken and harsh terrain, toiling up one rocky slope and then battling down through tearing thorn to the next dry river course, and then up again.

Also, it now seemed highly probable that either the

Englishman had changed his mind and gone north of the Zambezi river after all, in which case they had lost him, or else, and Camacho's skin crawled at the thought, or else they had crossed the spoor of the caravan in the early dawn or late evening when the light was too bad to make it out clearly. It was an easy mistake to make, they had crossed hundreds of game tracks each day, and the spoor could have been wiped by a herd of game, or one of the fierce short-lived little whirlwinds, the dust devils which ravaged the valley at this season of the year.

To cap all Camacho's tribulations, his band of noble warriors was on the point of mutiny. They were talking quite openly about turning back. There never had been an Englishman and a caravan of riches, even if there had, he was now far from here and getting further every day. They were exhausted by these switchback ridges and valleys and the water bottles were nearly all of them dry, which made it hard to maintain enthusiasm for the venture. The ring-leaders were reminding the others that in their absence, their share of the profits of the slave caravan were blowing in the wind. Fifty slaves, for certain, were worth a hundred mythical Englishmen. They had many excellent reasons for turning back.

Camacho, on the other hand, had nothing to return for, apart from his half-brother's ire. He also had a score to settle, two scores. He still hoped that they might manage to take the Englishman and his sister alive, especially the woman. Even in the thirst and the heat, his groin swelled at the memory of her in men's breeches. He jerked himself back to reality, and he glanced over his shoulder at the straggling line of ruffians who followed him.

Soon it would be necessary to kill one of his men, he had decided hours ago. Dung-eaters all of them – it was the only language they truly understood. He must make an example to stiffen their backbones, and keep them slogging onwards.

He had already decided which one it would be. The one-eyed Abyssinian was the biggest talker, the most eloquent apostle of the return to the coast, and what made his choice even more attractive was that his left side was blind. The problem was that the job must be done properly. The others would be impressed by the knife but not the gun. However, the Abyssinian allowed no man into that blind spot. Without making it too obvious, Camacho had twice sidled up on his left, but each time the Abyssinian had swung his head with its frizzed-up halo of dense curly hair towards him, and given him a grin with a slow trickle of a tear running down his cheek from that obscenely empty eye-socket.

However, Camacho was a persistent man, and an inventive one, for he noticed that whenever he moved out of the Abyssinian's blind spot, the man relaxed, and immediately became more verbose and arrogant. Twice more Camacho tried an approach from the left, and twice more was met with a single cold beady stare. He was establishing a pattern, teaching the victim that threat came only from the left, and when they halted in the middle of the morning, he ostentatiously squatted on the right. The Abyssinian grinned at him as he wiped the spout of his almost empty water bottle on his sleeve.

'This is the place. I go no further.' The one-eyed man announced in fluent Portuguese. 'I make the oath on Christ's sacred wounds.' And he touched the Coptic gold cross that hung around his neck. 'Not another step forward. I am going back.'

Fanning himself with the beaver hat, Camacho shrugged, and answered the cold grin with his own sunny smile. 'Let's drink to your going then.' With his free hand, he lifted his own water bottle, and shook it slightly. There was a cupful, no more. All their eyes went instinctively to the bottle. Here water was life, even the Abyssinian's single eye fastened on it.

Then Camacho let it slip from his fingers. It looked like

an accident, the bottle rolled to the Abyssinian's feet with clear water glugging out on to the baked earth, and, with an exclamation, the man stooped for it with his right hand, his knife hand.

Nobody really saw Camacho move. He had been holding the knife in the lining of the beaver hat. Suddenly it seemed to reappear behind the Abyssinian's right ear, just the carved bone handle protruding, the blade completely buried. The Abyssinian lifted his hand with a mystified expression and touched the hilt of the knife, blinked his single eye rapidly, opened his mouth and then closed it firmly and fell forward on top of the water bottle.

Camacho was standing over them, a cocked pistol in each hand.

'Who else wishes to make an oath on Christ's sacred wounds?' he smiled at them, his teeth very big and white and square. 'Nobody? Very well then, I will make an oath. I make it on the long-lost maidenheads of your sisters which they sold a hundred times for an escudo the bunch.'

Even they were shocked by such blasphemy.

'I make it on your flaccid and puny manhoods which it will be my great pleasure to shoot off,' he was interrupted, and broke off in mid-sentence.

There was a faint popping sound on the hushed heated morning, so distant, so indistinct, that for a moment none of them recognized it as the sound of gunfire. Camacho recovered first, he thrust the pistols into his waistband. They were no longer needed – and he ran to the crest of the rocky kopje on which they sat.

There was a column of dun-coloured smoke rising into the washed-out blue sky far ahead. In this terrain, it lay a day's march away against the steep and fearsome edge of the escarpment.

Around him his men, once again loyal and filled with fire, were laughing and hugging each other with delight. It was a pity about the Abyssinian, Camacho conceded, like all his people he had been a good fighter, and now they had found the Englishman, he would be missed.

C amacho cupped the cheroot in both hands and inhaled deeply. Holding the smoke in his lungs, he squinted his eyes against the flat glare of rising noon. The sun had bleached all colour from the landscape, only the shadows were cut clearly, very sharp-edged and black below each tree and rock.

Across the narrow valley the long column of men he was watching moved at the pace of the slowest. Camacho doubted if they were making a single mile in an hour.

He removed the beaver hat from his head and blew the smoke into it gently, so that it dispersed into nothingness instead of standing in the still air to draw the eye of a watcher.

Camacho saw nothing odd in that the leading Hottentot musketeer in the column carried an English flag. Even in this barren, forsaken corner of the earth, although its gaudy folds were already dulled with dust and the ends shredded by grasping thorn, it was the promise of protection, a warning to those who might check their passage. All caravans in Africa went under a banner.

Camacho dragged on the black cheroot and considered once again how infallible was the advice of his brother Alphonse. The night was the only time to do the business. Here the column was spread out for more than a mile, there were wide gaps between each of the four divisions – and with himself he had eighteen men left. If he attacked in daylight, he would be forced to concentrate them on the detachments of Hottentot musketeers at the head and tail of the caravan. He imagined vividly what might happen at the first shot. One hundred porters would drop their packs and scatter away into the bush, and when the fighting was over he would have nobody to carry the loot.

It was necessary, furthermore, for him to wait until the Englishman rejoined the caravan. He guessed that Zouga Ballantyne was scouting or hunting, but that he would rejoin the column before nightfall.

The woman was there. He had another glimpse of her at that moment as she stepped up on to a log that lay across the path; for a moment she balanced there, long-legged, in those maddening breeches, and then jumped down off the log. Camacho passed a pleasant few minutes in erotic imaginings. He had been twenty days without a woman and it had put a razor edge to his appetite which even the hot hard trek had not dulled.

He sighed happily, and then squinted his eyes again as he concentrated on the immediate problems. Alphonse was right, they would have to wait until night. And tonight would be a good night for it, three days after the full moon, and yet the moon would rise very late, an hour or so after midnight.

He would let the Englishman come in, let the camp settle down, let the fires die, and let the Hottentot sentries drowse off. Then when the moon rose and the vitality of the camp was at its lowest ebb, he and his men would go in.

Every one of his men was good with the knife, the fact that they had lived so long showed that, Camacho smiled to himself, and they would have an opportunity to prove it again tonight. He personally would mark the position of the sentries while it was still light. They would start with them, Camacho did not expect more than three or four. After the sentries, then the Hottentot musketeers that were sleeping. They were the most dangerous, and after them there would be time for indulgence.

He would go to the woman's tent himself, he squirmed slightly at the thought and adjusted his dressing. It was just a terrible pity that he could not take care of the Englishman as well. He would send two of his best men to do that work. He had dreamed of driving a long wooden spike up between the Englishman's buttocks, and then taking wagers on how long he would take to die, amusing himself and the company while regaining some of his previous losses at the same time.

Then reluctantly, and prudently, he had decided not to chance such pleasures, not with a man like that. Best to cut his throat while he still slept. They could have their fun with the woman instead, Camacho decided firmly.

His single regret was that he would have only a short time with her, before the others demanded their turns, though a few minutes would probably be enough for him. Strange how months of craving could be satiated so swiftly, and after that brief out-pouring become indifference and even distaste. That was a very philosophical thought, Camacho realized. Once again he had astonished himself with his own wisdom and sensitivity to things of the mind. He had often thought that if he had ever learned to read and write he could have been a great man, like his father the Governor. After all, the blood in his veins was that of aristocrats and Dons, only slightly diluted.

He sighed, yes, five minutes would suffice and then the others could have her, and when they also had finished they could play the betting game with the long wooden stake – and, of course, there was a more amusing place to put it. He chuckled aloud at the thought and took a last draw on the cheroot, and the stub was so short that the ember scorched his fingers. He dropped it and crushed it under his heel. He moved like a panther, slipping quietly over the skyline and circling out stealthily to get ahead of the creeping caravan.

Zouga had left his bearers tending the smoking-racks, feeding the fire with chips of wet wood, and turning the hunks of red meat so they cured evenly. It was wearisome work, for the racks had to be guarded at all times from the hyena and jackals, the crows and kites which hung about the camp, while there was always wood to cut and the baskets of mopani bark to weave in which the smoked meat would be carried.

Jan Cheroot was happy to escape with Zouga as he marched back to find the main caravan and guide it to the buffalo camp. Even the tsetse fly could not repress his high spirits. They had been in 'fly-country' for a week or more now. Tom Harkness had called these importunate little insects 'the guardians of Africa' and certainly they made it impossible for man to move his domestic animals through vast tracts of country.

Here they were one of the many reasons why the Portuguese colonization had been confined to the low coastal littoral. Their cavalry had never been able to penetrate this deadly screen, nor had their draught animals been able to drag their war trains up from the coast.

They were the reason, also, why Zouga had not attempted to use waggons or beasts of burden to carry his stores. There was not even a dog with the caravan, for only man and wild game were immune to the dreadful consequences of the bite.

In some areas of the 'fly-belt' there were so few of the insects as to cause little annoyance, but in others they swarmed like hiving bees, plaguing and persecuting even during the moonlight periods of the night.

That day on the march back to rejoin the column, Zouga and Jan Cheroot suffered the worst infestation of fly they had yet met with in the valley of the Zambezi. They rose from the ground to sit thickly upon their legs, and clustered on the backs of their necks and between their shoulder blades, so that he and Jan Cheroot took turns to walk behind each other and brush them off with a freshly cut buffalo tail.

As suddenly as they had entered it, they found themselves out of the 'fly-belt' and with blessed relief from the torment they settled down to rest in the shade. Within half an hour they heard distant singing, and as they waited for the caravan to come up, they smoked and chatted in the desultory fashion of the good companions they had become.

During one of the long pauses in their talk Zouga thought he saw vague movement on the far side of the shallow valley in front of them. Probably a herd of kudu or a troop of baboon, both of which were plentiful in the valley, but apart from the buffalo herd the only game they had encountered since leaving Tete.

The approaching caravan would have alarmed whatever it was, Zouga thought, removing his cap and using it to drive away the persistent cloud of tiny black mopani bees which hovered around his head, attracted by the moisture of his eyes and lips and nostrils. The movement in the forest opposite was not repeated. Whatever it was had probably crossed the ridge. Zouga turned back to listen to Jan Cheroot.

'The rains only stopped six weeks ago,' Jan Cheroot was musing aloud. 'The water-holes and rivers up on the high ground will still be full, not here, of course – this ground drains too steeply.' He indicated the dry and rocky water course below them. 'So the herds spread out, and follow their old roads.' He was explaining the complete lack of elephant, or even recent evidence of elephant herds in the valley, and Zouga listened with attention, for here was an expert speaking. 'The old elephant roads, they used to run from the flat mountain of Good Hope to the swamps of the far Sud,' he pointed north. 'But each year they shrink as we, the *jagters* (hunters), and men like us, follow the herds and drive them deeper and deeper into the interior.'

Jan Cheroot was silent again, and his pipe gurgled noisily as he sucked on it.

'My father told me that he killed the last elephants south of the Olifants river – the river of elephants – when he was still a young man. He boasted that he killed twelve elephants that day, he alone with his old *roer* (muzzle-loader) that was too heavy to hold to the shoulder. He had to rest it on the crutch of a forked stick he carried for it. Twelve elephants in one day – by one man. That is a feat.' He gurgled his pipe again, and then spat a little yellow tobacco juice. 'But then my father was even more famous as a liar than he was as a hunter,' Jan Cheroot chuckled and shook his head fondly.

Zouga smiled, and then the smile vanished and his head jerked up. He narrowed his eyes, for a tiny dart of reflected sunlight had struck his eye coming from the same place across the valley. Whatever he had seen was still there, and it was neither kudu nor baboon. It had to be a man, for only metal or glass could have shot that reflection. Jan Cheroot had noticed nothing and he was musing on.

'When I rode with Cornwallis Harris, we found the first elephant on the Cashan mountains, that's a thousand miles further north of where my father killed his herd. There was nothing in between – it had all been hunted bare. Now there are no elephant in the Cashan mountains. My brother Stephan was there two years ago. He tells me that there are no elephant south of the Limpopo river. The Boers graze their herds where we hunted ivory – perhaps we will find no elephant even up there on the high ground, perhaps there are no elephant left in all the world.'

Zouga was not really listening any more. He was thinking about the man on the opposite side of the valley. It was probably somebody from the caravan, a party sent ahead to cut firewood for the night's bivouac, yet it was still early to think of making camp.

The singing of the bearers was louder now. There was a single voice carrying the marching song. Zouga recognized it. The man was a tall Angoni, with a fine tenor voice, and a poet who improvised his own verses, adding to them and altering them on the march. Zouga, cocking his head, could make out the words.

'Have you heard the Fish Eagle cry above Marawi?
Have you seen the setting sun turn the snows of Kiliman-
    jaro to blood?'

And then the chorus coming in after him, those haunting African voices, so beautiful, so moving,

'Who will lead us to these wonders, my brother?
We will leave the women to weep,
We will let their sleeping mats grow cold,
If a strong man leads, we will follow him, my brother.'

Zouga smiled at the next verse, as he recognized his name.

'Bakela will lead us like a father leads his children,
Bakela will give you a *khete* of *sam-sam* beads—
Bakela will feed you upon the fat of the hippopotamus
    and meat of the buffalo—'

Zouga closed his mind against the distraction, concentrating on the man across the valley. Here, a hundred miles from the nearest habitations of men, it must have been somebody from the caravan – woodcutter, honey-hunter, deserter, who knew?

Zouga stood up and stretched, and Jan Cheroot knocked out his pipe and stood with him. The head of the column appeared amongst the trees lower down the slope, the red, white and blue banner flapped lazily open and then drooped again dejectedly.

Zouga glanced once more at the opposite slope, it seemed deserted again. He was tired, the soles of his feet felt as though he had marched across burning embers, for they had been going hard since dawn, and the barely healed knife wound in his hip ached dully.

He should really go up that slope to check what he had seen, but it was steep and rocky. It would take another half an hour of scrambling to reach the crest and return. They went down to meet the caravan, and when Zouga saw Robyn striding along with that easy coltish grace of hers behind the standard bearer, he lifted his cap and waved it above his head.

She ran to meet him, laughing like a child with delight. He had been gone three days.

Below a polished face of smooth black rock, that when the river was in spate would be a roaring cascade, there was a bend in the dry waterbed filled with pure white sand.

On the banks above it the wild mahogany stood tall and vigorous, its roots in water and the baboon had been scratching in the sand below the bank.

Robyn and Zouga sat together on the edge of the dry waterfall, watching the men that Zouga had set to digging for water.

'I pray there will be enough.' Robyn watched them with interest. 'I have not used my bath since we left Tete.' Her enamelled hip bath was the expedition's single bulkiest item of equipment.

'I'll be satisfied with enough for a pot of tea,' Zouga replied vaguely, but he was clearly distracted.

'Something is worrying you?' she asked.

'I was thinking of a valley in Kashmir.'

'Was it like this?'

'Not really – it is just that,' he shrugged. In that far away valley, when he had been a young ensign leading a patrol ahead of the battalion, he had seen something as he had today. Something of no account, a stray movement, a glint of light that might have been off a gunbarrel, or the horn of a wild goat. Then, as now, it had been too much trouble to check it. That night he had lost three men, killed while they fought their way out of the valley. The fight had earned praise from his Colonel, but the men remained dead.

He glanced up at the slanting sun, there was an hour of light left. He knew he should have climbed that slope. While Robyn watched him with a puzzled expression, he wavered a few seconds longer and then with an exclamation of exasperation stood up wearily. His feet still ached abominably and he rubbed his hip where the knife-wound throbbed. It was going to be a long walk back down the valley.

Zouga used a deep and narrow ravine to get out of the camp unobtrusively, and once clear he scrambled out and kept to the thicker bush just above the river-bed until he reached a tangled barrier of driftwood carried down during the recent rains which blocked the dry watercourse from bank to bank.

He used this to cover his crossing, and started up the far slope. He went very carefully, dodging from one tree to another and watching and listening before each move.

On the crest there was a faint movement of cooler air, the evening breeze coming down the escarpment that cooled the sweat on his neck and almost made the hard climb worthwhile. It seemed that was all the reward he would get. The stony ground was too hard to carry sign, and it was deserted of life, animal or human. However, Zouga was determined to make up for his previous sloth. He stayed too long. It would be fully dark before he reached the camp again, moon-rise was late, and he risked a broken leg moving over this sort of terrain in the pitch dark.

He turned to go back, and he smelt it before he saw it, and the hair prickled along his forearms and he felt his belly muscles contract, yet it was such a commonplace smell. He stooped and picked up the small squashed brown object. He had smoked the last of his own cheroots two days before, perhaps that was why his nose was so sensitive to the smell of tobacco.

The cigar had been smoked down to a thin rind, and crushed out so it resembled a scrap of dried bark. Without the smell to guide him, he would never have found it. Zouga shredded it between his fingers, and there was still a little residual dampness of saliva in the chewed end. He lifted his fingers and sniffed them. He knew where he had smelled that particular scented type of Portuguese tobacco before.

Camacho left fifteen of his men well back off the crest, in a tumble of rock that looked like a ruined castle, and whose caves and overhangs gave shade and concealment. They would sleep, he knew, and he grudged it to them. His own eyelids were drooping as he lay belly down in cover, on the other side of the ridge watching the caravan making camp.

He had only two of his men with him to help him mark the sentries and *scherms*, the watch fires and the tent sites. They would be able to lead the others in, even in the complete darkness before the moon, if that should become necessary. Camacho hoped not. In the dark mistakes could be made, and it needed only a shot or a single shout. No, they would wait for the moon, he decided.

The Englishman had come into camp earlier, just before the caravan halted. He had the Hottentot with him, and they both hobbled stiffly like men who had made long, hard marches. Good, he would sleep soundly, was probably doing so already, for Camacho had not seen him in the last hour. He must be in the tent beside the woman's. He had seen a servant carrying a steaming bucket of water to her.

They had watched the Hottentot Sergeant set only two sentries. The Englishman must be feeling very secure, two sentries merely to watch against lions. They would probably both be fast asleep by midnight. They would never wake again. He, personally, would cut one of them. He smiled in anticipation, and he would send a good knife to cut the other.

The remaining Hottentots had built their usual lean-to shelter and thatched it in a rudimentary fashion. There was no chance of rain, not at this season and not with that unblemished eggshell-blue sky. It was almost two hundred paces from the tents, a groan or a whimper would not carry that far. Good, Camacho nodded again. It was working out better than he had hoped.

As always, the two tents were set close together, almost touching. The gallant Englishman guarding the woman, Camacho smiled again, and felt his drowsiness lifting miraculously as his groin charged once more. He wished the night away, for he had already waited so long.

Night came with the dramatic suddenness of Africa, within minutes the valley was filled with shadows, the sunset made its last theatrical flare of apricot and old gold light and then it was dark.

For an hour more Camacho could see the occasional dark figure silhouetted by the flames of the camp fires. Once the sound of singing carried softly and sweetly to the ridge, and the other camp sounds, the clank of a bucket, the thud of a log thrown on the fire, the drowsy murmur of voices, showed that the routine was unaltered, and the camp completely unaware.

The noises faded and the fires died. The silence and the darkness was disturbed only by the piping lament of a jackal across the valley.

The star patterns turned slowly across the sky marking the passage of the hours, and then gradually paled out before the greater brilliance of the rising moon.

'Fetch the others,' Camacho told the man nearest him, and rose stiffly to his feet, stretching like a cat to relieve numbed muscles. They came silently, and gathered close about Camacho, to listen to his final whispered instructions.

When he finished whispering, he looked from one to the other in turn. Their faces in the bright moonlight had the pale greenish hues of freshly exhumed corpses, but they nodded their agreement to his words, and then followed him down the slope, dark silent shapes like a troop of wolves; they reached the dry watercourse in the gut of the valley, and split into their prearranged groups.

Camacho moved up the newly beaten path towards the camp. He carried his knife in his right hand and his musket in the other, and his feet made a barely audible brushing sound through the short dry grass. Ahead of him, beneath the outspread branches of a mukusi tree, he could make out the shape of the sentry, where he had been placed six hours before. The man was asleep, curled like a dog on the hard earth. Camacho nodded with satisfaction and crept closer. He saw the man had pulled a dark blanket over his head. The mosquitoes had bothered him also, Camacho grinned and knelt beside him.

With his free hand he felt softly for the man's head through the blanket, then the hand stilled. He gave a little grunt of surprise, and jerked the blanket aside. It had been arranged over the exposed roots of the mukusi tree to look like the shape of a sleeping man, and Camacho swore quietly but with great vehemence.

The sentry had chosen the wrong time to sneak away from his post. He was probably back in the lean-to shelter snoring happily on a mattress of dry grass. They would get him with the others, when they cleaned out the shelter. Camacho went on up the slope into the camp. In the moonlight the canvas of the tents shone ghostly silver, a beacon on which his lust could concentrate. Camacho slipped the sling of the musket over his shoulder as he hurried forward, towards the left-hand tent, and then checked as another dark figure emerged from the shadows, the knife in his right hand instinctively came up and then he recognized his own man, one of those whom he had sent to cut the Englishman's throat.

The man nodded jerkily, all was well so far and they went forward together, separating only as they approached the two tents. Camacho would not use the fly opening of the woman's tent – for he knew it would be laced closed, and if there were any surprises they would be at the entrance. He slipped around the side of the tent, and stooped to one of the hooded ventilator openings. He ran the point of the blade into it and then drew it upwards in a single stroke. Although it was heavy canvas, the blade had been whetted expertly, and the side wall split with only a whisper of sound.

Camacho stepped through the opening, and while he waited for his eyesight to adjust to the deeper darkness of the interior, he fumbled with the fastening of his breeches, grinning happily to himself as he made out the narrow collapsible cot and the little white tent of the muslin mosquito net. He shuffled towards it slowly, careful not to trip over the cases of medical stores that were piled between him and the cot.

Standing over the cot, he ripped the muslin netting aside violently, and lunged full length on to the cot, groping for the woman's head to smother her cries, the loose ends of his belt flapping at his waist, and his breeches sagging around his hips.

For a moment he was paralysed with shock at the fact that the cot was empty. Then he groped frantically over every inch of it, before coming to his feet again and hoisting his breeches with his free hand. He was confused, disconnected, and wild ideas flashed through his mind. Perhaps the woman had left the cot to answer a call of nature, but then why was the fly carefully laced closed. She had heard him and was hiding behind the cases, armed with a scalpel, and he swung round panicking to lash out with the knife, but the tent was empty.

Then the coincidence of the missing sentry and the empty cot struck him with force, and he felt deep and urgent concern. Something was happening that he did not understand. He charged for the rent in the canvas, tripped and sprawled over one of the cases and rolled on to his feet again, nimble as a cat. He ran out, clinching his belt and looking about him wildly, unslinging his musket and only just preventing himself from calling aloud to his men.

He ran to the Englishman's tent, just as his man came running out of the long dark tear in the canvas side, brandishing his knife, his face pale and fearful in the moonlight. He saw Camacho, screamed and struck out at him wildly with the long silver blade.

'Silence, you fool,' Camacho snarled at him.

'He's gone,' the man panted, craning to stare about into the deep shadows that the moon cast under the trees. 'They've gone. They've all gone.'

'Come!' snapped Camacho, and led him at a run down towards the lean-to that the Hottentot musketeers had built.

Before they reached it they met their companions running towards them in a disorderly bunch.

'Machito?' somebody called nervously.

'Shut your mouth,' Camacho growled at them, but the man blurted on.

'The *scherm* is empty, they have gone.'

'The Devil has taken them.'

'There is nobody.'

There was an almost superstitious frenzy of awe on them all, the darkness and the silent empty camp turned them all into cowards. Camacho found himself, for once, without an order to give, uncertain of what to do. His men crowded around him helplessly, seeming to take comfort from each other's physical presence, cocking and fiddling with their muskets and peering nervously into the shadows.

'What do we do now?' A voice asked the question that Camacho had feared, and somebody else threw a log from the pile on to the smouldering watch fire in the centre of the camp.

'Don't do that,' Camacho ordered uncertainly, but instinctively they were all drawn to the warmth and comfort of the orange tongue of flame that soared up brightly, blowing like a dragon's breath.

They turned their backs towards it, forming a half circle and faced outwards into the dark which in contrast to the flames was suddenly impenetrably black.

It was out of this darkness that it came. There was no warning, just the sudden thunderous burst of sound and flame, the long line of spurting muzzle flashes, blooming briefly and murderously, and then the sound of the striking balls in their midst, like a handful of children's marbles hurled into a mud puddle, as the musket balls slogged into human flesh.

Immediately men were hurled lightly about by the heavy lead slugs, and the little band about the watch fire was thrown into struggling, shouting confusion.

One of them flew backwards, at a run, doubled in the middle where a ball had taken him low in the belly. He tripped over the burning log and fell full length into the blazing watch fire. His hair and beard flared like a torch of pine needles and his scream rang wildly through the tree tops.

Camacho himself threw up his musket, aiming blindly into the night from whence the Englishman's voice was chanting the ritual infantry orders for mass volley fire.

'Section One. Reload. Section Two. Three paces forward. One round volley fire.'

Camacho realized that the devastating blast of close range musketry that had just swept them would be repeated within seconds. With mild derision Camacho had watched on many a hot afternoon as the Englishman drilled his double line of red-jacketed puppets, the front rank levelling and firing on command, then the second rank taking three paces forward in unison, stepping through the gaps in the first rank and in their turn levelling and firing. The same evolutions which, when magnified ten thousand times, had broken the charge of French cavalry up the slope at Quatre Bras, now filled Camacho with unutterable terror, and he flung up his musket and fired it unsighted into the darkness, in the direction of the cool, precise English voice. He fired at the same instant as one of his own men who had been knocked down by the first volley scrambled to his feet only lightly wounded, directly in front of Camacho's musket muzzle. Camacho shot him cleanly between the shoulder blades at a range of two feet, so close that the powder burn scorched the man's shirt. It smouldered in little red sparks as he sprawled at full length on his face once more, until the glowing sparks were quenched by the quick flow of the man's heart blood.

'Fool!' Camacho howled at the corpse, and turned to run. Behind him the English voice called, 'Fire!'

Camacho threw himself down on to hard earth, howling again as his hands and knees sank into the hot ash of the watch fire and he felt his breeches char and his skin blister.

The second volley swept over his head, and around him more men were falling and screaming, and Camacho rebounded to his feet at a dead run. He had lost both his knife and musket.

'Section One, three paces forward, one round volley fire.'

The night was suddenly filled with running, shouting

figures, as the porters burst out of their encampment. There was no direction or purpose in their flight. They ran like Camacho, driven by gunfire and their own terrible panic, scattering away into the surrounding bush, singly and in small groups.

Before the command ordering the next volley of musket fire, Camacho ducked behind the hillock of porters' packs, which had been piled close to the Englishman's tent and covered with waterproof tarpaulin.

Camacho was sobbing with the agony of his scorched hands and knees, and with the humiliation of having walked so guilelessly into the Englishman's trap.

He found his terror giving way to bitter and spiteful hatred. As little groups of terrified porters stumbled towards him out of the darkness, he drew one of the pistols from his belt and shot the leader in the head and then leapt up howling like a demented ghost – they ran and he knew they would not stop until, miles away in the trackless wilderness, they dropped with exhaustion, easy prey for lion or hyena. It gave him a moment's sour satisfaction, and he looked around for some other damage he could wreak.

The pile of stores behind which he crouched and the dying fire in front of the Englishman's deserted tent caught his attention. He snatched a brand from the fire, blew it into flame and tossed it flaring brightly on to the high canvas-covered pile of stores and equipment, then flinched as another volley crashed out of the darkness, and he heard the Englishman's voice.

'As a line of skirmishers, take the bayonet to them now, men.'

Camacho jumped down into the dry river-bed, and blundered through the crunching sugary sand to the far bank, where he scrambled thankfully into the dense riverine bush.

At the re-assembly point on the ridge there were three of his men already waiting, two of them had lost their muskets and all three were as shaken and sweaty and breathless as Camacho himself.

Two more came in while they regained their breath and power of speech. One was badly hit, his shoulders shattered by a musket ball.

'There won't be any others,' he gasped, 'those little yellow bastards caught them with bayonets as they were crossing the river.'

'They'll be here any minute.' Camacho dragged himself to his feet again, looking back down into the valley. He saw with grim satisfaction that the pile of stores was blazing brightly, despite the efforts of half a dozen tiny dark figures to beat out the flames. He had only moments to enjoy the spectacle, for lower down the slope came the thin, but warlike cries of the Hottentot musketeers and the thud and flash of their musket fire.

'Help me,' cried the man with the shattered shoulder. 'Don't leave me here, my friends, my comrades, give me an arm,' he pleaded, trying to struggle back on to his feet, but he was speaking to the empty night, and as the rush of footsteps down the far slope of the hill dwindled, his knees buckled under him and he sank back on to the rock earth sweating with pain and the terror which lasted only until one of the Hottentots drove the point of a bayonet into his chest and out between his shoulder blades.

Zouga strode angrily through the camp in the rising heat and bright sunlight of morning. His face and arms were blackened with soot, his eyes still red and smarting from the smoke of the fires, his beard was scorched and his eyelashes burned half away from fighting the flames. They had lost most of their stores and equipment for the fire had run away through the tents and thatched shelters. Zouga paused to glance at the charred and trodden scraps of canvas that were all that were left of the tents; they would miss them when the rains broke, but that was the very least of their losses.

He tried to make a mental list of the most grievous damage they had suffered. There were firstly only forty-six porters left out of more than one hundred. Of course, he could expect Jan Cheroot and his Hottentots to bring in a few more. They were at this moment scouring the valleys and hills around the camp for scattered survivors. He could still hear the kudu-horn trumpets calling in the stragglers. However, many would have risked the long and dangerous journey back to Tete rather than a recurrence of the night's attack, and they would have seized the opportunity to desert. Others would be lost after their midnight flight and panic. They would fall prey to wild animals or succumb to thirst. Half a dozen had been killed by random musketry fire and by the retreating brigands who had deliberately fired into the masses of unarmed porters. Four others were so badly wounded that they would die before nightfall.

That was the most serious loss, for without porters they were helpless. Without porters to carry them, what remained of the carefully selected equipment and trade goods were as useless to them as if they had left them in London or dumped them overboard from *Huron*'s deck.

Of the equipment itself, it would take them hours to count their losses, to find what had burned and what they could salvage from the stinking, smouldering mass of cloth and canvas, what they could pick out of the trodden and dusty mess scattered down the rock hillside. The scene reminded Zouga forcefully of so many other battlefields, the terrible destruction and waste affronted him, as it had done before.

The few remaining porters were already at work, picking over the field like a line of harvesters, retrieving anything of value from the ash and the dust. Little Juba was with them concentrating on the search for Robyn's medical stores, and books and instruments.

Robyn herself had set up an emergency clinic under the wide green branches of the mukusi tree in the centre of the camp, and when Zouga paused to look at the wounded men who still awaited her attention, and at the dead bodies laid out in a neat row and covered with a blanket or a scrap of charred and dirty canvas, he was angry with himself all over again.

Though what alternative had there been for him, he wondered.

If he had turned the camp into an armed fortress, it would have meant enduring a long-drawn-out siege with Camacho's wolves skulking around the perimeter, sniping and harrying them until their opportunity came.

No, he had been right to set the trap and end it at a single stroke. At least now he could be certain that the Portuguese were still in full flight for the coast, but the price had been too high, and Zouga was still angry.

The expedition, so well conceived and lavishly equipped, had ended in disaster before it had achieved a single one of its objectives. The loss of equipment and life had been heavy, but that was not what burned so acidly in Zouga's stomach as he paused at the perimeter of the devastated camp and lifted his eyes longingly towards the high broken ground of the southern escarpment. It was the idea of having to give up, before he had begun, and when he was so close, so very close. Twenty, fifty, not more than a hundred miles ahead of him lay the frontier of the empire of Monomatapa. Behind him, one hundred miles to the north was the dirty little village of Tete, and the wide river which was the beginning of the long ignoble road back to England, back to obscurity, back to a commission in a third-rate regiment, back to conformity and the wearying discipline of the cantonments of the Indian army. Only now that he was doomed to return to that life did he realize how deeply he had hated and resented it, just how much the desire to escape it had brought him here to this wild untouched land. Like a long-term prisoner who has tasted one day of sweet freedom, so the prospect of return to his cage was that much more painful, now. The pain of it cramped his chest, and he had to breathe deeply to control it.

He turned away from the southern vista of ragged peaks and savage black rock cliffs, and he walked slowly to where his sister worked quietly in the shade of the mukusi. She was pale, with dark smudges of fatigue and strain under her green eyes. Her blouse was speckled with spots of her patients' blood, and her forehead appeared blistered with tiny beads of perspiration.

She had started work in the darkness, by the light of a bull's eye lantern and now it was mid-morning.

She looked up wearily as Zouga stood over her.

'We won't be able to go on,' he said quietly. She stared at him a moment without change of expression, and then dropped her eyes and went on smearing salve on the badly burned leg of one of the porters. She had treated the worst cases first, and was now finishing with the burns and abrasions.

'We've lost too much vital equipment,' Zouga explained. 'Stores that we need to survive.' Robyn did not look up this time, 'And we've not enough porters to carry what is left.'

Robyn began bandaging the leg with her full attention.

'Papa made the Transversa with four porters,' she observed mildly.

'Papa was a man,' Zouga pointed out reasonably, and Robyn's hands stilled ominously and her eyes narrowed, but Zouga had not noticed. 'A woman cannot travel or survive without the necessities of civilization,' he went on seriously. 'That is why I am sending you back to Tete. I'm sending Sergeant Cheroot and five of his Hottentots to escort you. You'll have no difficulty, once you reach Tete. I will send

346

with you what remains in cash, a hundred pounds for the launch down river to Quelimane and a passage to Cape Town on a trader. There you can draw on the money I deposited in Cape Town to pay for a passage on the mailship.'

She looked up at him. 'And you?' she asked.

Until then he had not made the decision. 'The important thing is what happens to you,' he told her gravely, and then he knew what he was going to do. 'You will have to go back and I am going on alone.'

'It'll take more than Jan Cheroot and five of his damned Hottentots to carry me,' she told him, and the oath was a measure of her determination.

'Be reasonable, Sissy.'

'Why should I start now?' she asked sweetly.

Zouga opened his mouth to reply angrily, then closed it slowly and stared at her. There was a hard uncompromising line to her lips, and the prominent, almost masculine, jaw was clenched stubbornly.

'I don't want to argue,' he said.

'Good,' she nodded. 'That way you won't waste any more of your precious time.'

'Do you know what you are letting yourself in for?' he asked quietly.

'As well as you do,' she replied.

'We won't have trade goods to buy our way through the tribes.' She nodded. 'That means we'll have to fight our way through if anyone tries to stop us.'

He saw the shadow in her eyes at that, but there was no wavering of her determination.

'No tents for shelter, no canned food, no sugar or tea.' He knew what that meant to her. 'We will live straight off the land, and what we can't scavenge or kill or carry, we go without. We'll have nothing but powder and shot.'

'You'd be a fool to leave the quinine,' she told him quietly, and he hesitated.

'The bare minimum of medicines,' he agreed, 'and remember, it won't be for just a week or a month.'

'We'll probably go a great deal faster than we have so far,' she answered quietly, as she stood up and brushed off the seat of her breeches.

The choice of what to take and what to leave had been nicely balanced, Zouga thought, as he listed and weighed the new loads.

He had chosen paper and writing equipment in place of sugar and most of the tea. His navigational instruments in place of spare boots, for the boots they wore could be resoled with raw buffalo hide. Quinine and other medicines together with Robyn's instruments in place of the extra clothing and blankets. Powder and shot in place of trade beads and cloth.

The pile of abandoned equipment grew steadily, cases of potted jams, bags of sugar, canned foods, insect nets, folding camp chairs and cots, cooking pots, Robyn's enamel hip bath and her flowered chamber pot, trade goods, *merkani* cloth and beads, hand mirrors and cheap knives. When the pile was complete, Zouga put fire into it, a token of finality and of determination. Yet they watched it burn with trepidation.

There were two small concessions Zouga had made: a single case of Ceylon tea for, as Robyn pointed out, no Englishman could be expected to explore undiscovered territories without that sovereign brew, and the sealed tin which contained Zouga's dress uniform, for their very lives might depend on impressing a savage African potentate. Otherwise they had divested themselves of all but the very essentials.

Chief of these essentials was the ammunition, the sacks of first-grade Curtis and May black powder and the ingots of soft lead, the bullet moulds, the flask of quicksilver to harden the balls and the boxes of copper caps. Out of the remaining forty-six porters, thirty of them carried this powder and shot.

Jan Cheroot's musketeers were horrified when they were informed that their field packs would in future hold two hundred, and not fifty, rounds of Enfield ammunition.

'We are warriors, not porters,' his Corporal told him loftily. Jan Cheroot used the metal scabbard of his long bayonet to reason with him, and Robyn dressed the superficial wounds in the Corporal's scalp.

'They now understand the need for carrying extra ammunition, Major,' Jan Cheroot reported to Zouga cheerfully.

It truly was interesting to realize how much fat they could shed, Zouga mused, as he watched the shorter, more manageable column start out. It was less than a hundred and fifty yards from head to tail, and the pace of its march was almost doubled. The main body nearly matched the speed of Zouga's advance party – falling only a mile or so behind during the first day's march.

That first day they reached the scene of the buffalo hunt before noon, and found more than bark baskets of cured black buffalo meat awaiting there. Zouga's head gunbearer, Matthew, came running to meet him through the forest and he was so excited as to be almost incoherent.

'The father of all elephant,' he gibbered, shaking like a man in fever, 'the grandfather of the father of all elephants!'

J an Cheroot squatted beside the spoor and grinned like a gnome in a successful piece of sorcery, his slant eyes almost disappearing in the web of wrinkles and folds of yellow skin.

'Our luck has come at last,' he exulted. 'This is indeed an elephant to sing about.'

He took a roll of twine out of the bulging pocket of his tunic and used it to measure the circumference of one of the huge pad marks. It was well over five feet, close to six feet around.

'Double that is how high he stands at the shoulder,' Jan Cheroot explained. 'What an elephant!'

Matthew had at last controlled his excitement enough to explain how he had awoken that dawn, when the light was grey and uncertain, and seen the herd passing close to the camp in deathly silence, three great grey ghostly shapes, moving out of the forest and entering the blackened and barren valley through which the fire had swept. They were gone so swiftly, that it had seemed that they had never existed, but their spoor was impressed so clearly into the soft layer of fire ash that every irregularity in the immense footprints, the whorls and wavy creases of the horny pads, were clearly visible.

'There was one of them, bigger and taller than the others, his teeth were long as a throwing spear and so heavy that he held his head low and moved like an old man, a very old man.'

Now Zouga also shivered with excitement, even in the stultifying heat of the burned-out valley, where it seemed the blackened earth had retained the heat of the flames. Jan Cheroot, mistaking the small movement, grinned wickedly around the stem of his clay pipe.

'My old father used to say that even a brave man is frightened three times when he hunts the elephant, once when he sees its spoor, twice when he hears its voice and the third time when he see the beast – big and black as an ironstone kopje.'

Zouga did not trouble to deny the accusation, he was following the run of the spoor with his eyes. The three huge animals had moved up the centre of the valley, heading directly into the bad ground of the escarpment rim.

'We will follow them,' he said quietly.

'Of course,' Jan Cheroot nodded, 'that is what we came for.'

The spoor led them over the cold grey ash, amongst the blackened and bared branches of the burned-out jessie bush and up the rising funnel of the narrow valley.

Jan Cheroot led. He had discarded his faded tunic for a sleeveless leather jerkin with loops for the Enfield cartridges across the chest. Zouga followed him closely, carrying the Sharps and fifty extra rounds, together with his two-gallon water bottle. His gunbearers, in strict order of seniority backed him, each with his burden of blankets and water bottles, food bag, powder flask and ball pouch, and of course the big smooth-bored elephant guns.

Zouga was anxious to see Jan Cheroot work. The man talked a very good elephant hunt, but Zouga wanted to know if he was as good on the spoor as he was at telling about it around the camp fire. The first test came swiftly when the valley pinched out against another low cliff of impassable rock, and it seemed as though the great beasts they followed had taken wing and soared away above the earth.

'Wait,' said Jan Cheroot and cast swiftly along the base of the cliff. A minute later he whistled softly and Zouga went forward.

There was a smudge of dark ash on a block of ironstone, and another above it, seeming to lead directly into sheer rock face.

Jan Cheroot scrambled up over the loose scree at the foot of the cliff, and disappeared abruptly. Zouga slung his rifle and followed him. The blocks of ironstone had fractured in the shape of a giant's staircase – each step as high as his waist so that he had to use a hand to climb.

Even the elephants would have extended themselves to make each step, rising on their back legs as he had seen them do at the circus, for an elephant is incapable of jumping. They must keep two feet on the earth before they can heave their ungainly bulks upwards.

Zouga reached the spot where Jan Cheroot had disappeared and stopped short in amazement at the threshold of the stone portals, invisible from below, which marked the beginning of the ancient elephant road.

The portals were symmetrically formed in fractured rock, and had eroded through the softer layers, leaving straight joints so they seemed to have been worked by a mason. The opening was so narrow that it seemed impossible that such a large animal could pass through, and looking above the level of his own head, Zouga saw how over the centuries their rough skin had worn the stone smooth as thousands upon thousands of elephant had squeezed through the gap. He reached up and plucked a coarse black bristle, almost as thick as a Swan Vesta, from a crack in the face. Beyond the natural gateway, the gap in the cliffs widened and rose at a more gentle pitch. Already Jan Cheroot was four hundred yards up the pathway.

'Come!' he called, and they followed him up.

The elephant road might have been surveyed and constructed by the corps of engineers, for never was the gradient steeper than thirty degrees and when there were natural steps they were never higher than a man or an elephant could comfortably negotiate, although it seemed there were always accidents, for within a quarter of a mile they found where one of the big animals had missed his footing and struck the tip of one tusk against the ironstone edge.

The tusk had snapped above the point, and twenty pounds of ivory lay in the path. The fragment was worn and stained, so thick around that Zouga could not span it with both hands, but where it had sheared the fresh ivory was a lovely finely grained porcelain white.

Jan Cheroot whistled again when he saw the girth of the fragment of tusk. 'I have never seen an elephant so big,' he whispered, and instinctively checked the priming of his musket.

They followed the road out of the rocky pass on to forested slopes, where the trees were different, more widely spaced, and here the three old bulls had paused to strip off long slabs of bark from the msasa trunks before moving on. A mile further along the road, they found the chewed balls of bark still wet and smelling of elephant saliva, a rank gamey smell. Zouga held one of the big stringy balls to his nostrils and inhaled the elephant smell. It was the most exciting odour he had ever known.

When they rounded a shoulder of the mountain, there was a terrifying drop of open blue space before them in which the tiny shapes of vultures soared. Zouga was sure that it was the end of the road.

'Come!' Jan Cheroot whistled and they stepped out on to the narrow secret ledge, with the deep drop close at hand and followed the road that had been smoothed for them by tens of thousands of roughly padded feet over the centuries.

Zouga at last was allowing himself to hope, for the road climbed always, and there seemed to be a definite purpose and direction to it – unlike the meandering game trails of the valley. This road was going somewhere, climbing and bearing determinedly southwards.

Zouga paused a moment and looked back. Far behind and below the sunken plain of the Zambezi shimmered in the blue and smoky haze of heat. The gigantic baobab trees seemed like children's miniatures, the terrible ground over which they had laboured for weeks seemed smooth and inviting, and far away just visible through the haze the serpentine belt of darker, denser vegetation marked the course of the great wide river itself.

Zouga turned his back upon it and followed Jan Cheroot up around the shoulder of the mountain, and a new and majestic vista opened around him as dramatically as though a theatre curtain had been drawn aside.

Another steep slope stretched away above the cliffs, and there were luxuriant forests of marvellously shaped trees, the colours of their foliage was pink and flaming scarlet and iridescent green. These lovely forests reached up to a rampart of rocky peaks that Zouga was certain must mark the highest point of the escarpment.

There was something different with this new scene, and it took Zouga some moments to realize what it was. Then suddenly he drew a deep breath of pleasure. The air fanning down from the crest was sweet and cool as that on a summer's evening on the South Downs, but carrying with it the scent of strange shrubs and flowers and the soft exotic perfume of that beautiful pink and red forest of msasa trees.

That was not all, Zouga realized suddenly. There was something more important still. They had left the tsetse 'fly-belt' behind them. It was many hours and a thousand feet of climbing since he had noticed the last of the deadly little insects. The land ahead of them now was clean, they were entering a land where man could live and rear his animals. They were leaving the killing heat of that harsh and forbidding valley for something softer and cleaner, something good, Zouga was certain of it, at last.

He stared about him with delight and wonder. Below him in the void a pair of vultures came planing in on wide-stretched pinions, so close that he could see every individual feather in their wingtips, and above the shaggy mound of their nest that clung precariously to a fissure in the sheer rock cliff they beat at the air, hovering, before settling upon the nest site. Clearly Zouga heard the impatient cries of the hungry nestlings.

On the rock shelf high above his head a family of hyrax, the plump rabbit-like rock dassie, sat in a row and stared down at him with patent astonishment, fluffy as children's toys, until at last they took fright and disappeared into their rocky warrens with the speed of a conjurer's illusion.

The sun, sinking through the last quadrant of its course, lit it all with richer and mellower light, and turned the cloud ranges to splendour, tall mushroom-shaped thunderheads of silver and brightest gold, touched with fleshy pink tongues and the tinge of wild roses.

Now at last Zouga could not doubt where the elephant road was leading him and he felt his spirits soaring, his body wearied from the climb was suddenly recharged with vigour, with vaulting expectations. For he knew that those rugged peaks above him were the threshold, the very frontiers of the fabled kingdom of Monomatapa.

He wanted to push past Jan Cheroot on the narrow track, and run ahead up the broad beautiful forested slope that led to the crest, but the little yellow Hottentot stopped him with a hand upon his shoulder.

'Look!' he hissed softly. 'There they are!'

Zouga followed his outflung arm and saw instantly. Far ahead amongst those magical red and pink groves something moved, vast and grey and slow, ethereal and insubstantial as a shadow. He stared at it, feeling his heartbeat quicken at this, his first glimpse of an African elephant in its savage habitat, but the slow grey movement was immediately screened by the dense foliage.

'My telescope.' He snapped his fingers urgently behind his back, never taking his eyes from the spot where the huge beast had disappeared, and Matthew thrust the thick cylinder of cool brass into his hand.

With fingers that shook slightly he pulled out the sections of the spy-glass, but before he could lift it to his eye the tree that had screened the elephant's body began to tremble and shake as though it had been struck by a whirlwind, and faintly they heard the rending crackle of tearing timber, the squealing protest of living wood, and slowly the tall tree leaned outwards and then toppled with a crack that echoed off the cliffs like a cannon shot.

Zouga lifted the telescope and rested it on Matthew's patient shoulder. He focused the eye-piece. Abruptly, in the rounded field of the glass, the elephant was very close. His head was framed by the foliage of the tree he had uprooted. The vast ears seemed wide as a clipper's mainsail as they flapped lazily and he could see the puffs of dust that their breeze raised from the massive grey withers.

Clearly he could see the dark wet tear-path from the little eye down the sered and wrinkled cheek, and when the animal raised its head, Zouga drew breath sharply at the unbelievable size of the double arches of stained yellow ivories. One tip was foreshortened, broken off cleanly, and the ivory was brilliant snowy white in the fresh break.

As Zouga watched, the bull used his trunk, with its rubbery fingers of flesh at the tip, to pluck a handful of the delicate new leaves from the fallen tree. He used his trunk with the finesse of an expert surgeon, and then drooped open his triangular lower lip and thrust the leaves far down his own throat, and his rheumy old eyes wept with slow contentment.

An urgent tap on his shoulder disturbed Zouga and irritably he lifted his eye from the telescope. Jan Cheroot was pointing further up the slope.

Two other elephants had drifted into view from out of the forest. Zouga refocused upon them and then gasped. He had thought the first elephant massive, and here was another bull just as big – yet it was the third elephant which daunted his belief.

Beside him Jan Cheroot was whispering with suppressed tension making his voice hoarse and his slanted black eyes shine.

'The younger bulls are his two askari, his indunas. They are his ears and his eyes. For in his great age he is probably almost deaf and more than half blind, but look at him. Is he not still a king?'

The oldest bull was tall and gaunt, taller by almost a head than his protégés, but the flesh seemed to have wasted off the ancient frame. His skin hung in baggy folds and creases from the massive framework of bones. He was thin in the way that some old men are thin; time had eroded him, seeming to leave only skin and stringy sinew and brittle bone. Matthew had been right in his description, the bull moved the way an old man moves, as though each joint protested with rheumatic pangs, and the weight of ivory he had carried for a hundred years was at last too much for him.

The ivory had once been the symbol of his majesty, and it was still perfect, flaring out from the lip and then turning in again so the points almost met. The gracious curves seemed perfectly matched, and the ivory was a lovely, butter yellow, unblemished despite the dominance battles he had fought with them, despite the forest trees that the bull had toppled with them and stripped of their bark or the desert roots he had dug from stony soil with them.

But now, at last, these great tusks were a burden to him, they wearied him and he carried his head low as they ached in his old jaws. It had been many years now since he had used them as fearsome weapons to keep control of the breeding herds. It had been as many years since he had sought the company of the young cows and their noisy squealing calves.

Now the long yellow ivories were a mortal danger to him, as well as a source of discomfort and pain. They made him attractive to man, his only enemy in nature. Always it seemed that the hunters were camped upon his spoor, and the man-smell was associated with the flash and thudding discharge of muzzle-loading firearms, or the rude stinging intrusion of sharpened steel into his tired old flesh.

There were pieces of beaten iron pot-leg and of round hardened lead ball deep in his body, the shot lying against the bone castle of his skull had become encysted with gristle and formed lumps as big as ripe apples beneath the skin, while the scars from arrows and stabbing spears, from the fire-hardened wood spikes of the dead-fall-trap had thickened into shiny grey scars and become part of the rough, folded and creased mantle of his bald grey hide.

Without his two askari he would long ago have fallen to the hunters. It was a strangely intimate relationship that knit the little herd of old bulls, and it had lasted for twenty years or more. Together they had trekked tens of thousands of miles, from the Cashan mountains in the far south, across the burning, waterless wastes of the Kalahari desert, along the dry river-beds where they had knelt and with their tusks dug for water in the sand. They had wallowed together in the shallow lake of Ngami while the wings of the water fowl darkened the sky above them, and they had stripped the bark from the forests along Linyati and Chobe, and crossed those wide rivers, walking on the bottom with just the tips of their trunks raised above the surface to give them breath.

Over the seasons they had swung in a great circle through the wild land that lay north of the Zambezi, feasting on the fruits of different forests scattered over a thousand miles, timing their arrival as each crop of berries came into full ripening.

They had crossed lakes and rivers, had stayed long in the hot swamps of the Sud where the midday heat, reaching 120°, soothed the aches in the old bull's bones. But then the wanderlust had driven them on to complete the circle of their migration, south again over mountain ranges and across the low alluvial plains of the great rivers, following secret trails and ancient passes that their ancestors had forged and which they had first trodden as calves at their mother's flank.

In these last dozen seasons, however, there were men

where there had never been men before. There were white-robed Arabs in the north around the lakes, with their long-barrelled jezails. There were big bearded men in the south, dressed in dark rough homespun and hunting from tough shaggy little ponies, while everywhere they met the tiny little bushmen with their wicked poisoned arrows, or the Nguni regiments hunting a thousand strong, driving the game into set positions where the plumed spearmen waited.

With each round of the seasons, the elephant ranges were shrinking, new terrors and new dangers waited in the ancient ancestral feeding grounds and the old bull was tired and his bones ached and the ivory in his jaws weighed him down. Still he moved on up the slope to the head of the pass with slow determination and dignity, driven on by his instincts, by the need for space about him, by the memory of the taste of the fruits he knew were already ripening in a distant forest on the shores of a far-away lake.

'We must hurry.' Jan Cheroot's voice roused Zouga, for he had been mesmerized by the sight of the regal old animal, filled with a strange feeling of déjà vu, as though he had lived this moment before, as though this meeting was part of his destiny. The old bull filled him with awe, with a sense of timelessness and grandeur, so he was reluctant to return to the reality of the moment.

'The day dies fast,' Jan Cheroot insisted, and Zouga glanced over his shoulder to where the sun was setting like a mortally wounded warrior bleeding upon the clouds.

'Yes,' he acknowledged, and then frowned as he realized that Jan Cheroot was stripping off his puttees and breeches, folding them and stuffing them together with his blanket and food bag into a crevice in the rock face beside him.

'I run faster like this,' he answered Zouga's silent enquiry with a twinkling grin.

Zouga followed his example, leaving his own pack and pulling off the webbing belt from which hung knife and compass, stripping down to good running order, but he stopped short of removing his breeches. Jan Cheroot's skinny naked yellow buttocks were totally devoid of dignity and his dangling penis played hide-and-seek from under his shirt tails. There were some conventions that an officer of the Queen must observe, Zouga decided firmly, and one was to keep his breeches on in public. He followed Jan Cheroot along the narrow ledge, until they stepped off it on to the forested slope and immediately their forward vision was limited to a dozen yards by the lichen-covered tree trunks. From higher up the slope, however, they could hear the crackle and the ripping sound as the bull with the broken tusk fed on the uprooted tree.

Jan Cheroot worked out swiftly across the slope to avoid the askari, to circle around him and come at the lead bull. Twice he paused to check the wind. It held steadily down the slope into their faces and the colourful leaves above their heads quivered and sighed at its passing.

They had gone a hundred yards when the sounds of the feeding bull ceased abruptly; again Jan Cheroot paused and the little group of hunters froze with him, every man instinctively holding his breath as they listened, but there was only the sound of the wind and the singing whine of a cicada in the branches above.

'He has moved on to join the others,' Jan Cheroot whispered at last. Zouga was also certain that the bull could not yet have suspected their presence. The wind was steady, he could not have scented them. Zouga knew that the eyesight of the elephant was as weak as his hearing and sense of smell were acute, but they had made no sound.

Yet this was a clear demonstration of the benefits that the three old bulls derived from their association. It was always difficult for the hunter to place each of them accurately, especially in thick forest such as this, and the two askaris seemed always to take station on the lead bull to cover and protect him. To come at him, the hunter must penetrate the screen they threw around him.

Standing now, listening and waiting, Zouga wondered if a genuine affection existed between the three animals, whether they derived the pleasure of companionship from each other, and whether the askaris would mourn or pine when the old bull fell at last with the musket ball in his brain or his heart.

'Come!' Silently Jan Cheroot made the open-handed signal to advance and they went on up the slope, stooping under the low trailing branches, Zouga keeping four paces out on the Hottentot's flank to open his field of vision and fire, concentrating his whole being in his eyes and his ears. Far up the slope there was the snap of a breaking twig and it stopped them dead once more, breathing shallowly with the tension, but the sound fastened all their attention ahead so none of them saw the askari.

The elephant waited with the stillness of granite, his sered hide grey and rough as the lichen-covered tree trunks, the shadows thrown by the low sun barred him and broke up the shape of his great body so he blended into the forest, grey and unearthly as mist, and they tiptoed past him at twenty paces without seeing him.

He let the hunters pass him and get up above the wind and when the acrid stench of carnivorous man was borne thickly down to him through the forest, he took it in his trunk and lifted it to his mouth and sprayed the tainted air over the little olfactory organs under his upper lip, and his smell buds flared open like soft wet pink rose buds and the askari bull squealed.

It was a sound that seemed to bounce against the sky, and ring from the peaks above them, it was an expression of all the hatred and pain, the terrible memories of that acrid man-smell from a hundred other encounters, and the askari bull squealed again and launched his huge body up the slope to destroy the source of that evil odour.

Zouga spun to the piercing din, his shocked eardrums still buzzing with the sound, and the forest shook with the bull's charge. The dense shiny vegetation burst open, like a storm surf running on to rock, and the bull came through.

Zouga was not conscious of his own movements, he was aware only that he was looking at the bull over the open sights of the Sharps rifle, and the blast of shot seemed muted and far off after the ringing squeal that initiated the charge. He saw dust fly from the bull's forehead in a brief little puff, saw the grey skin ripple like that of a stallion stung by a bee, and he reached back and found the stubby wooden stock of one of the big elephant guns in his hand. Again, there was no awareness of conscious movement, but over the crude vee of the sight the elephant appeared much closer. Zouga seemed to be leaning back to look up at the gigantic head and the long shafts of yellow ivory reached out over him, blotting out the sky. Clearly he could see the bright white porcelain break in one point of the left-hand tusk.

He heard Jan Cheroot beside him, and heard his excited shrieks.

'Skiet hom! Shoot him!'

Then the heavy weapon leapt against his own shoulder, driving him back a pace, and he saw the tiny fountain of bright blood squirting out of the elephant's throat like a lovely scarlet flamingo feather. He reached back for the next loaded gun, although he knew there would be no time to fire again.

He was surprised that he felt no fear, although he knew he was a dead man. The elephant was on him, his life was forfeit, there could be no question, yet he went on with the motions of living, hefting the new gun, thumbing back the clumsy hammer as he swung the barrel up.

The shape of the huge animal above him had altered, it was no longer so close, and he realized with a thrill that the bull was turning, it had been unable to endure the fearful punishment of the heavy-bored guns.

He was turning, passing by them with blood streaming down his head and chest. As he passed, he exposed his neck and flank, and Zouga shot him a hand's span behind the joint of his shoulder on the line of the lungs and the ball slogged into his rib cage.

The bull was going, crashing away up the slope, and with the fourth gun Zouga hit him high in the back, aiming for the bony knuckles of the spine where they showed through the scabby grey hide of the sloping back and the bull whipped his thick tufted tail at the agony of the strike and disappeared into the forest, gone like a wraith in the failing light of the sunset.

Zouga and Jan Cheroot stared at each other speechlessly, each of them holding a smoking weapon at high port across his chest, and they listened to the run of the bull up the slope ahead of them.

Zouga found his voice first, he turned to his gunbearers.

'Load!' he hissed at them, for they also were paralysed by the close passage of violent, thundering death, but his command liberated them and they each snatched a handful of black powder from the bag slung at their sides and poured it into the still-hot muzzles of the guns.

'The lead bull and the other askari will run,' lamented Jan Cheroot, himself frantically busy with the ramrod of his Enfield.

'We can still catch them before the summit,' Zouga told him, grabbing the first loaded gun. An elephant goes on a steep uphill at a very measured pace that a good runner can gain upon, but downhill he goes like a runaway locomotive, nothing can catch him, not even a good horse.

'We must catch them before the crest,' Zouga repeated, and launched himself at the slope. Weeks of hard going over bad terrain had toughened him, and the driving lust of the hunter was the spur. He flew at the slope.

Awareness of his own lack of experience that had resulted in such poor shooting made Zouga more fiercely determined to close with the lead bull and vindicate himself. He guessed that as a complete novice he had failed to find the vitals with any of the balls he had fired, he had missed brain and heart and lungs by inches, inflicting pain and mutilation, instead of the quick kill for which the true huntsman strives. He wanted desperately to have another chance to end it cleanly, and he pelted up the slope.

Before he had gone two hundred yards, he had evidence that his shooting had not been as wide as he had at first believed. He checked at a spot where it looked as though someone had thrown a bucket of blood across the stony earth. The blood was a peculiarly bright scarlet, and it frothed with tiny bubbles, lung blood. There was no question, that last ball fired into the bull's back as he went away up the slope must have raked the lungs. It was a killing shot, but a slow killer. The old bull was drowning in his own bright arterial blood, desperately trying to rid himself of it by squirting it out through his trunk as it bubbled up into his throat.

He was dying, but it would take time still, and Zouga raced on after him.

He had not expected the bull to stand again. He expected him to run until he dropped, or until the hunters caught up with him. Zouga knew the folly of attributing human motives and loyalties to wild animals, even so it seemed that the stricken bull had determined to sacrifice himself in order to allow the lead bull and the second askari to escape across the pass at the crest of the mountains.

Up the slope, he was waiting for Zouga, listening for him with those vast grey ears spread wide, his chest with the girth of a Cognac cask of Limousine oak racked and strained to force air into his torn lung.

He charged as soon as he heard the man. His huge ears cocked back and rolled at the tips, trunk curled at his chest, shrilling and squealing, blowing a fine red mist of blood from his trunk, pounding through the forest so the earth trembled with each massive footfall and the branches crackled and broke like a discharge of musketry.

Panting, Zouga stood to meet the charge, ducking his head and weaving for a clean shot through the thick forest. At the last moment the bull broke off the charge, and swerved up the slope again. Each time the hunters began to move forward, he launched another thunderous mock-attack, forcing them to stand to meet him, and then breaking off again.

Minutes passed between each charge, and the hunters were pinned down in the thick forest, fretting with the knowledge that already the lead bull and his surviving protégé must have reached the crest and gone away in a rush like an avalanche down the far side.

Zouga was learning two hard lessons: the first, as old Tom Harkness had tried to teach him, was that only a novice or a fool under-guns an elephant. The light ball of the Sharps might be highly effective on American bison, but the African elephant has ten times the body weight and resistance. Standing in the msasa forests, listening to the squealing and crashing of the wounded monster, Zouga determined never again to use the light American rifle on heavy game.

The second lesson he was learning was that if the first ball does not kill, then it seems to numb and anaesthetize heavy game to further punishment. Kill cleanly, or the subsequent shots into heart and lungs seem to be without effect. It was not only anger and provocation which made a wounded animal so dangerous, it was also this shock-induced immortality.

After standing down half a dozen mock charges, Zouga abandoned caution and patience, and he ran forward shouting to meet the next charge.

'Whoa there!' he called. 'Come on then, old fellow!'

This time he got in close, and crashed another ball through the bull's rib cage as he turned away. He had controlled his first wild excitement, and the ball struck precisely at the point of aim. He knew it was a heart shot, but the bull came again squealing, and Zouga fired a last time before the angry trumpeting and shrilling turned to a long sad bellow, that echoed off the peaks and rang out into the blue void of sky beyond the cliffs.

They heard him go down, the impact of the heavy body against the earth made it shiver under their feet. Cautiously, the little group of hunters went forward through the forest and they found him kneeling, his forelegs folded neatly under his chest, his long, yellow ivories propping up the dusty wrinkled old head, still facing down the slope as though defiant even in death.

'Leave him,' shouted Jan Cheroot. 'Follow the others.' And they ran on past him.

Night caught them before they reached the crest of the slope, the sudden impenetrably black night of central Africa, so they lost the spoor and missed the pass.

'We will have to let them go,' Jan Cheroot lamented in the darkness, his yellow face a pale blob at Zouga's shoulder.

'Yes,' Zouga agreed. 'This time we must let them go.' But somehow he knew there would be another time. The feeling that the old bull was part of his destiny was still strongly with Zouga. Yes, there would be another time, of that he was certain.

That night for the first time Zouga ate the hunter's greatest delicacy: slices of elephant heart threaded on to a green stick with cubes of white fat cut out from the chest cavity, salted and peppered, and roasted over the slow coals of the camp fire, eaten with cold cakes of stone-ground corn and washed down with a mug of tea, steaming hot, bitter, strong and unsweetened. He could not remember a finer meal, and afterwards Zouga lay down on the hard earth, covered by a single blanket and protected from the chill of the wind by the huge carcass of the old bull, and slept as though he also had been struck down, without dreams, without rolling over even once in his sleep.

In the morning, by the time they had chopped out one of the tusks and laid it under the msasa trees, they could already hear the singing of the porters as the main body of the caravan filed along the narrow ledge, rounding the shoulder of the mountain and then came up on to the slope.

Robyn was a hundred paces ahead of the standard-bearer, and when she reached the carcass of the bull she stopped.

'We heard the gunfire last evening,' she said.

'He's a fine old bull,' Zouga told her, indicating the freshly chopped tusk. It was the unbroken right-hand ivory, taller than Zouga, but a third of its length had been buried in the skull. This portion was smooth, unblemished white, whereas the rest of it was stained by vegetable juices.

'It will weigh almost a hundred pounds,' he went on, touching the tusks with the toe of his boot. 'Yes, he's a fine old bull.'

'Not any more, he isn't,' Robyn told him quietly, watching Jan Cheroot and the gunbearers hacking away at the enormous mutilated head. Little chips of white bone flew in the early sunlight, as they whittled away the heavy skull to free the second tusk. Robyn watched the butchery for a few seconds only before going on up the slope towards the crest.

Zouga was irritated and angry with her, for she had detracted from his own vaulting pleasure in his first elephant hunt. So, an hour later when he heard Robyn calling to him from higher up the slope, he ignored her cries. However, she was persistent, as always, and at last with an exclamation of exasperation, he followed her up through the forest. She came running down to meet him, with the unrestrained infectious joy of a child shining on her face.

'Oh Zouga.' She seized his hand, and began to drag him impetuously up the slope. 'Come and see, you must come and see.'

The old elephant road crossed the saddle, through a deep pass, guarded on each hand by grey buttresses of rough grey rock and as they took the last few paces over the highest point a new and beautiful world opened below and ahead of them. Zouga gasped involuntarily, for he had not anticipated anything like this.

Low foothills fell away from beneath their feet, regular as the swells of the ocean, covered with stately trees whose trunks were tall and grey as the oaks of Windsor Park, and then beyond the hills the undulating lightly forested grasslands, golden as fields of ripe wheat, spread to a tall blue horizon. There were streams of clear water meandering through the glades of pale grass, where herds of wild game drank or lazed upon the banks.

There were buffalo everywhere Zouga looked, black bovine shapes, standing shoulder to shoulder in dark masses under the umbrella branches of the acacia trees. Closer at hand a troop of sable antelope, that loveliest of all antelope, jet black above but with snowy bellies, their long symmetrical curve of horn extended backwards almost to touch the haunches, followed the herd bull in long file to the water, pausing unafraid to stare curiously at the interlopers, forming a frieze of stately, almost Grecian, design.

The endless stretch of land was dotted with hills like ruined natural castles of stone, seeming to have been built in past aeons by giants and ogres from mammoth blocks of stone, and tumbled now in fantastic shapes, some with fairy turrets and spires, others again flat topped, geometrically laid out as though by a meticulous architect with plumbline and theodolite.

This lovely scene was lit by a peculiar pearly luminosity of the morning light, so that even the furthest hills, probably more than a hundred miles distant, were sharply silhouetted through the sweet clear air.

'It's beautiful,' Robyn murmured, still holding Zouga's hand.

'The kingdom of Monomatapa,' Zouga answered her, his own voice husky with emotion.

'No,' Robyn answered softly. 'There is no sign of man here, this is the new Eden.'

Zouga was silent, letting his eyes rove across the scene, searching for, but not finding, any evidence of man. It was a land untouched, unspoiled.

'A new land, there for the taking!' he said, still holding Robyn's hand. They were as close, in that moment, as they had ever been or would ever be again, and the land awaited them, wide, limitless, empty and beautiful.

A t last, reluctantly, he left Robyn at the head of the pass and went back through the grey rock portals to bring on the caravan. He found the second tusk removed, and both of them bound up with bark rope on to the carrying-poles of newly cut msasa wood, but the porters had laid aside their packs and were indulging in an orgy of fresh meat, and that most sought-after of African spoils, thick white globs of elephant fat.

They had cut a trapdoor into the belly of the elephant carcass, and pulled out the entrails, and these glistened in the early sunlight, huge rubbery tubes of purple and yellow guts, already swelling and ballooning with the trapped gases they contained.

Half a dozen porters, stripped mother-naked, had crawled into the interior of the elephant's carcass, disappearing completely from view and wading almost waist-deep in the clotting, congealing bath of trapped blood. They crawled out, painted with it from head to foot, eyes and grinning teeth startlingly white in the grisly shining wet red visages, their arms filled with tidbits of liver and fat and spleen.

These delicacies were hacked into pieces with the blade of an assegai and thrown on the glowing coals of one of half a dozen fires, then snatched up again, black on the outside and more than half raw within to be wolfed down with every appearance of ecstatic pleasure.

There would be no moving them until they were satiated, Zouga realized. So he left instructions to Jan Cheroot, himself already pot-bellied with the meat he had gorged, to follow as soon as the carcass had been either eaten or packed up for carrying, and taking the Sharps rifle returned back up the slope to where he had left Robyn.

He called for her, fruitlessly, for almost half an hour, and was really becoming concerned for her safety when her reply echoed off the cliffs, and looking upwards he saw her standing on a ledge a hundred feet above him, waving him to come up to her.

Zouga climbed up swiftly to where she stood on the ledge, and checked the rebuke that he had ready for her when he saw her expression. She was pale, a sickly greyish colour, under the gilding of the sun, and her eyes were reddened and still swimming with tears.

'What is it, Sissy?' he asked with quick concern, but she seemed unable to reply, the words choking in her throat so she had to swallow thickly, and motion him to follow her.

The ledge on which they stood was narrow, but level – and was cut back under the cliff, forming a low long cavern. The cavern had been used before by other men, for the rocky roof was blackened with the sooty smoke of countless cooking-fires, and the back wall was decorated with the lyrically childlike paintings of the little yellow bushmen who over the centuries must have used this as a regular camp during their endless wanderings.

The paintings lacked both perspective and accurate form, but they captured the essential nature of all they recorded, from the graceful sweep of the giraffe's neck, to the bulky shoulders of the Cape buffalo with the mournful drooping horns framing the lifted nose.

The bushman artist saw himself and his tribe as frail,

370

sticklike figures, with drawn bows, dancing and prancing about the quarry, and again, out of all proportion to the rest of the painting, each little man sported a massively erect penis. Even in the heat of the chase, such was the universal conceit of all male kind, Zouga thought.

Zouga was enchanted by the frozen cavalcade of man and beast which covered the walls of the cave, and he had already determined to camp here so that he could have more time to study and record this treasure house of primitive art, when Robyn was calling him again.

He followed her along the ledge until they reached the point where it ended abruptly, forming a balcony over the dreaming land ahead of them. Zouga's attention was torn between this fresh vista of forest and glade and the cave art at his shoulder, but Robyn summoned him again impatiently.

There were strata of multi-coloured rock running horizontally through the rock face of the cliff. The different layers of rock varied in hardness, and the erosion of a softer layer had formed the long low cavern beyond the ledge.

This layer of rock was a soapy green colour, where it had not been painted over by the bushmen artists or discoloured with the smoke of their fires, and here at the point overlooking the empire of Monomatapa someone had used a metal tool to scrape a smooth square plaque into the green soapstone. The freshly cut surface stood out rawly as though it had been done that very day, but the words gave the lie to that impression.

There was a simple Christian cross chiselled deeply into the stone, and below it the name and the date, the lettering very carefully cut and designed by an expert penman.

'FULLER MORRIS BALLANTYNE'

Zouga exclaimed aloud at seeing his father's name, so clearly rendered by his father's own hand.

Despite the apparent freshness of the cut, the date was seven years previously – 20th July, 1853. After that single exclamation they were both speechless, staring at the inscription, each of them gripped by differing emotion – Robyn by a resurgence of filial love and duty, by a crushing desire to be with her father again after so many years, the vast empty place in her life aching more excruciatingly at the prospect of being soon filled. Her eyes refilled with tears, and they broke from her eyelids and ran down her cheeks.

'Please, God,' she prayed silently, 'lead me to my father. Dear God, grant me that I am not too late.'

Zouga's emotions were as strong, but different. He felt a corroding resentment that any other man, father or not, should have preceded him through these rocky gates into the kingdom of Monomatapa. This was his land, and he did not want to share it with another. Especially, he did not want to share it with that monster of cruelty and conceit that was his father.

He stared coldly at the inscription that followed the name and date, but inwardly he seethed with anger and resentment.

'In God's Holy name.' The words were carved below his father's name.

It was typical of Fuller Ballantyne that he should carve his own name here with cross and credentials as the Lord's ambassador, as he had on trees and rocks at a hundred other places across the continent which he regarded as a personal gift from his God.

'You were right, Zouga dear. You are leading us to him, as you promised. I should never have doubted you.'

If he had been alone Zouga realized that he might have defaced that inscription, scraped the rock bare with his hunting knife, but as he had the thought, he realized how futile it would be, for such an action would not wipe out the ghostly presence of the man himself.

Zouga turned away from the rock wall and its taunting plaque. He stared out over the new land – but somehow his heady pleasure in it had been dimmed by the knowledge that another man had passed this way ahead of him. He sat down with his feet dangling over the sheer drop to wait for Robyn to tire of staring at her father's name.

However, the caravan of porters came before she did that. Zouga heard the singing from the forested slope behind the pass long before the head of the line crossed the saddle. The porters had voluntarily doubled their own loads, and they struggled up the slope under the enormous weight of elephant meat and fat and marrow bones, bound up in baskets of green msasa leaves and bark rope.

If Zouga had asked them to carry that weight of trade cloth or beads, or even gunpowder, he would have had an immediate mutiny to deal with, he thought grimly, but at least they were carrying the tusks. He could see them near the head of the line. Each tusk slung on a long pole, a man at either end, but even here they had hung extra baskets of meat on the same pole as the tusk. The total weight must have been well over three hundred pounds, and they struggled up the slope uncomplainingly, even cheerfully.

Slowly, the caravan wound out of the forest and entered the gut of the pass, beginning to move directly under where Zouga sat, the figures of the porters and of Jan Cheroot's Hottentot musketeers foreshortened by the angle. Zouga rose to his feet, he wanted to order Jan Cheroot to camp just beyond the point where the pass debouched on to the foothills. From where Zouga stood, he could see a patch of green grass against the foot of the cliffs far below him, and a pair of pale grey herons hunting frogs in this verdant marshy area. There was certainly a spring, and with the meat upon which they had gorged, his servants would be burning with thirst by nightfall.

The spring would be a good place to camp, and it would allow him the following morning to copy and record the Bushman paintings in the cavern. He cupped his hands to his mouth to hail Jan Cheroot – when a crash like the broadside from a ship of the line filled the pass with thunder that echoed and bounced back and forth between the cliffs.

For many seconds Zouga could not understand what was happening for the thunderous bursts of sound were repeated, almost drowning the thin screams of his porters. They were throwing down their burdens and scattering like a flight of doves under the stoop of the falcon.

Then another movement caught his attention, a large round shape went bounding down the scree slope below the cliffs, charging straight at his panic-stricken caravan. For a moment Zouga believed it was some sort of living predator that was attacking his servants, and, running along the lip of the ledge, he unslung the Sharps rifle, ready to fire down into the pass as soon as he could get a sight on one of the dark bounding shapes.

Then he realized that at each leap the thing struck sparks and fine grey smoke from the scree slope, and he could smell, the faint smell, like burnt saltpetre that the sparks left in the air. He realized abruptly that they were giant rounded boulders rolling down upon his caravan, not one but a dozen or more, each weighing many tons, an onslaught which seemed to spring from the very air itself, and he looked wildly about for its source, goaded by the screams of his men and the sight of the rolling boulders smashing open packs of his precious irreplaceable provisions and scattering them across the rocky ground of the saddle.

Far below him, he heard the thudding report of an Enfield rifle, and glancing back he saw the tiny figure of Jan Cheroot aiming almost directly upwards at the sky, and following the direction of his rifle Zouga saw movement, just a flicker of movement on the edge of the cliff, outlined against the blue soaring vault of the heavens.

The deluge of huge boulders was coming from the very top of the cliff, and as Zouga stared, another and then another came raining down into the pass. Zouga squinted his eyes, head thrown back, as he studied the cliff rim. There was some animal up there. Zouga did not at first think of man, for he had already convinced himself that this new land was devoid of human presence.

He felt an almost superstitious chill of horror that some pack of giant apes was on top of the cliff, bombarding his men with huge rocks, then he shook himself free of the feeling, and looked quickly for some way to get higher up his side of the pass, to reach a position from where he could fire across the rocky gateway at the attackers on the opposite cliff and give some protection to his servants.

Almost immediately he discovered another ledge rising at a steep angle from the one on which he stood. Only a soldier's eye would have picked it out. The tiny feet of the little rock hyrax that used it had put a light sheen on the rock and it was this that had drawn Zouga's attention to the narrow pathway.

'Stay here!' he shouted at Robyn, but she stepped in front of him.

'Zouga, what are you going to do?' she demanded, and then before he could answer. 'Those are men up there! You cannot fire upon them!' Her cheeks were still smeared with tears, but her pale face was set and determined.

'Get out of the way,' he snapped at her.

'Zouga, it will be murder.'

'That's what they are trying to do to my men.'

'We must bargain with them.' Robyn caught at his arm as he pushed past her, but he shook her off and ran to the higher ledge.

'It will be murder!' Robyn's cry followed him, and as he climbed, Zouga was reminded of the words of old Tom Harkness. The accusation that his father would not stop at killing anybody who stood in his way. This was what he had meant, Zouga was suddenly sure of it. He wondered if his father had fought his way through this pass, just as Zouga himself was about to do.

'If the champion of the Almighty can do it, then what a fine example to follow,' he muttered to himself as he went up the steep ledge.

Below him the Enfield rifle thudded again, the sound muted by distance, almost lost in the roar of a new avalanche of murderous rock. Jan Cheroot could only hope to discourage his attackers with rifle fire from that angle; only if one of them actually leaned far out over the edge of the cliff would he be vulnerable to fire from the gut of the pass.

Zouga climbed in cold anger, stepping unhesitatingly over dangerous spots in the narrow ledge where small pieces of rock crumbled under his boots and went rattling down into the pass hundreds of feet below.

Abruptly he came out on to a broader ledge, formed by the strata of rock, which rose at a gentler pitch so that now Zouga could run along it without fear of losing his footing. He climbed swiftly, it was less than ten minutes since that first boulder had come crashing down into the pass; the attackers were continuing the bombardment, the hills reverberated with the crash and rumble of flying boulders.

Ahead of Zouga on the ledge, a pair of tiny grey klipspringer went bouncing up over broken rock, seeming to flit on the tips of their elongated hooves, terrified by the men and the rumble of falling rock. They reached the corner of the ledge and one after another they made what seemed suicidal leaps out into space, phenomenal leaps that carried them forty feet to the next sheer wall of rock, rock which seemed devoid of foothold, but they clung to it like flies and scrambled swiftly up out of sight over the top of the cliff.

Zouga envied them that birdlike agility as he toiled up the steep incline, sweat soaking his shirt and streaming stingingly into his eyes. He could not stop to rest, for far below him a thin wailing scream of agony told him that at least one of the flying boulders had struck down a porter.

He turned another steep corner in the goat track,

scrambled over the rim, and was suddenly out upon the flat table-like summit, dotted with little clumps of broom-bush and sparse stiff yellow grass, like hedgehog quills.

Zouga threw himself down on the edge of the plateau, heaving and straining for breath and he struck the sweat from his eyes, peering across the deep void of the pass at the cliffs on the opposite side. He found himself on a level with the heights opposite. It was three or four hundred yards across, easy range for the Sharps – though one of the smooth-bore four-to-the-pounds would have been hopelessly inaccurate at that distance.

While he primed the rifle, he studied the ground opposite him and saw almost immediately why the attackers had chosen that side of the pass in preference to the one on which Zouga lay.

They were on a flat-topped pinnacle of sheer rock, with no visible access to it from any direction, what path there was would be secret and highly defensible. The attackers had an inexhaustible supply of missiles for the rounded boulders were scattered everywhere upon the heights, varying in size from that of a man's head to that of an elephant carcass, and as Zouga watched them they were using heavy raw timber baulks to lever one of these over the edge of the cliff.

Zouga's hands were shaking and he fought to bring them under control, but the Sharps rifle wavered as he tried to take a sight on the little group of men across the open void of space. There were not more than two dozen of them, all of them naked except for a brief kilt of leather, their dark skins polished with a sheen of sweat in the sunlight.

He was regaining his breath swiftly, and now he wriggled forward on his belly and propped the stock of the Sharps rifle on a rock in front of him. It was a dead rest, and as he levelled his gaze over the open sights the group of men succeeded in working the huge boulder over the edge of the cliff.

It went with a brief grating that carried clearly to Zouga, and then it fell with the soft rushing of eagle's wings until it struck again in the pass two hundred feet below – and once again the hills rang and rumbled to the force and weight of it.

The little group of black men had drawn back from the edge of the cliff, resting a moment before they selected another missile. Only one of them wore a headdress. It looked like a cap made from the mane of a male lion, long tawny hair tipped with black. It made the man taller than his companions, and he seemed to be giving orders to them, gesticulating and pushing those nearest to him.

'You'll do, my beauty!' Zouga whispered. He had regained his breath now, and the sweat was cooling his back and his neck. He pushed up the leaf of his backsight to its 300-yard adjustment and then settled down on his elbows to peer over it. The rifle was rock-steady on its rest, while he took a fine bead on the man with the lion headdress.

He touched off the shot, and while the crack of it still stung his eardrums he saw a tiny chip of rock fly from the lip of the cliff across the valley. 'Low, but very nicely on line,' he told himself, opened the breech of the Sharps and forced the paper cartridge into it.

The shot had startled the little group of men. They peered about them, mystified, not certain of what it was or from where it had come. The tall figure in the headdress moved cautiously forward to the edge of the cliff and stooped to examine the fresh chip on the ironstone rim, touching it with one finger.

Zouga set the cap, and thumbed back the hammer. He gave it a full bead, and aimed at the waving yellow headdress, set the hair-trigger and then with a lover's touch stroked the curved trigger.

The bullet told loudly, a meaty slap like a housewife beating a carpet, and the man in the lion headdress spun round sharply, his arms jerking out wide, his legs shuffling in a grotesque little dance, they collapsed under him and he flopped on the very edge of the cliff like a harpooned catfish.

His companions stood frozen, making no effort to help him as he slid towards the edge of the cliff, and a final spasmodic jerk of his legs tumbled him into the void. He fell for a long time, his outspread limbs spinning like the spokes of a wagon wheel, and there was another meaty thump as he landed at last on the broken scree slope far below.

Zouga fired again, into the tight knot of men, not attempting to single out one of them, and he hit two of them with a single bullet, for even at that range the Sharps could drive the hardened lead ball through a man's body with hardly any loss of velocity, and the group was bunched up.

As the ball whacked into them, they split into separate racing figures, their yells of fright carrying clearly to where Zouga lay, and before he could fire again, they had disappeared into a narrow rocky gulley with the miraculous speed of a troop of little furry hyrax.

The silence was sudden and startling after the uproar of falling boulders and heavy gun fire and it lasted for many minutes, broken at last by the shrill voice of Jan Cheroot calling up from the deep gut of the pass. Zouga stood up, and hanging on to the branch of a monkey orange tree, leaned out over the edge of the cliff.

'Take the caravan through, Sergeant.' He pitched his voice to carry, and the echoes mocked him with his own words. 'Sergeant – Sergeant – Sergeant—'

'I will cover you – cover you – cover you,' taunted the echoes.

Zouga called the little Matabele maiden from the fireside where she knelt beside Robyn, helping her treat one of the porters who had been struck by a flying splinter of rock during the attack, and whose shoulder had been laid open to the bone.

'Juba,' he told her, 'I want you to come with me.' And the girl glanced back at Robyn, hesitating to obey him. Zouga's irritation flared again. He and Robyn had not spoken again since Zouga had broken up the attack with rifle fire, and Jan Cheroot had reassembled the caravan and led it out of the deadly trap of the gorge to where it was camped now in the foothills beyond the pass.

'Come,' Zouga repeated, and the child dropped her eyes at his tone and followed him submissively as he started back towards the pass.

Zouga moved back cautiously, pausing every fifty paces or so to survey the clifftops suspiciously, although he was fairly certain that there would be no renewal of the boulder rolling. However, he kept in close under the sheer rock face so that any boulder would fall beyond them.

The going was difficult over the loose scree, and through the thick scrub that covered it. It took them nearly an hour to get back to the spot where Zouga had marked the fall of the body of the warrior in the lion headdress, and then they had to search for nearly as long again before they found the corpse.

He had fallen into a deep crevice between two rocks, and he lay at the bottom of it on his back. The body appeared unmarked except for the small black bullet hole low on the left side of his naked chest.

The eyes were still wide open, but he had lost his headdress.

After a moment Zouga turned to Juba enquiringly. 'Who is he? Of what tribe is he?' he asked, and the child showed no concern at the presence of violent death. She had seen very much worse in her short life.

'Mashona!' she told Zouga scornfully. Juba was Matabele of Zanzi blood, and there was no more noble breeding outside the kraal of King Mzilikazi himself. She felt nothing but contempt for all the other tribes of Africa, especially for these people.

'Mashona!' she repeated. 'Eaters of dirt.'

It was the ultimate term of denigration that the Matabele applied to all the tribes that they have taken into slavery, or driven to the very point of extinction.

'It is always the way these baboons fight.' She indicated the dead man with a toss of her pretty head. 'They stand on the hilltops and throw down stones.'

Her dark eyes sparkled. 'It becomes more difficult each year for our young men to blood their spears, and until they do so the King will not give them leave to marry.' She broke off, and Zouga smiled ironically. The child's resentment was clearly not so much at the Mashona's alleged lack of sportsmanship, but at the havoc that this wrought on the Matabele marriage markets.

Zouga scrambled down into the rocky crevice, and stooped over the dead warrior. Despite Juba's contempt, the man was well formed, with straight strong limbs, and handsome intelligent features. For the first time Zouga felt a twinge of regret at having fired the ball which had caused that innocuous-looking little wound. It was a good thing that the warrior lay upon his back for the exit would have been ghastly. A soldier should never examine the bodies of the men he kills in the heat of battle. Zouga had made that discovery long ago in India, for afterwards there was always this moment of guilt and remorse. He shrugged it away, for he had not come either to gloat over the kill or to torture himself with it, but merely to identify his enemy.

Why had these warriors attacked the caravan without warning, he wondered? Were they the border guards of the fabled Monomatapa? It seemed the most likely explanation, though, of course, they could be ordinary bandits, like the dacoits of India with whom he had made bitter acquaintance.

Zouga stared moodily down at the corpse. Who had he been? And how much danger did his tribe still present to Zouga's caravan? But there was nothing more to learn and Zouga began to straighten up again when he noticed the necklace around the corpse's throat. He knelt again to examine it.

It was made of cheap trade beads on a thong of gut, a gaudy little trifle except for the pendant that had hung on to his chest but had slipped back under the armpit, half concealed so that Zouga noticed it now for the first time.

He pulled it free, examined it a moment and then slipped the entire necklace off over the man's head. When he put his hand around the back of the dead man's head to lift it he felt the pieces of skull grate together deep in his head like shards of broken pottery under his fingers. He lowered the shattered head and stood up with the necklace twined about his fingers, examining the pendant.

It had been carved from ivory that had yellowed with age, and tiny black cracks formed a fine web across its polished surfaces. Zouga held it to catch the sunlight, and turned it between his fingers to study it from every angle.

He had seen another figurine that was almost a twin to this one, a golden figurine that was now in a bank vault in Cape Town where he had deposited it before they had sailed on board the gunboat *Black Joke*.

This was the same stylized bird-like shape, perched on a round plinth. The plinth was decorated with the same triangular shark's-teeth pattern, and the bird had the same swelling chest and short, sharply pointed wings folded across its back. It could have been a pigeon or a dove, except for one detail. The beak was the curved and hooked weapon of a raptor.

It was a falcon, he knew it beyond any doubt, and it was certain that the heraldic bird had some deep significance.

The golden necklace that Tom Harkness had left him must have belonged to a king or a queen or a high priest. The choice of material, gold, was an indication of that. Now here was the same shape, worn by a man who seemed to be a chieftain, and once again the shape was faithfully copied in a precious material, ivory.

The falcon of Monomatapa, Zouga wondered, studying the ancient pendant, ancient it must be, for the ivory had acquired a deep patina and lustre.

Zouga looked up at the little Matabele girl, standing almost naked above him, and watching him with interest.

'Do you see this?' he asked.

'It is a bird.'

'Have you seen it before?'

Juba shook her head, and her fat little breasts joggled at the movement.

'It is a Mashona thing,' and she shrugged. What would a daughter of the sons of Senzangakhona and Chaka want with such a nonsense?

On an impulse Zouga lifted the necklace and dropped it over his own head, letting the ivory falcon fall inside the vee of his flannel shirt, where it nested against the darkly springing curls of his body hair.

'Come!' he told Juba. 'There is nothing more for us here,' and he led the way back towards the camp beyond the pass.

The land into which the old bull had led them by the secret road through the pass was an elephant kingdom. Perhaps the pressure of the hunters moving up from the far southern tip of the continent had driven them into this unpeopled world. The herds were everywhere. Each day Zouga and Jan Cheroot hunting far ahead of the main caravan came up with the huge grey pachyderms, and they shot them down.

They shot and killed forty-eight elephant the first month and almost sixty the second, and Zouga meticulously recorded each kill in his journal, the circumstances of the hunt, the weight of each tusk, and the exact location of the cache in which they buried them.

His small band of porters could not carry even a small part of such a mass of ivory, and the distance and direction which they must still travel was uncertain. Zouga buried his treasure, always near a readily recognizable landmark, a distinctive tree, or an unusual rock, a hilltop or a confluence of rivers, to enable him to find it again.

He would return one day for it, and when he did, it would have dried off its excess moisture and be easier for the porters to carry.

In the meantime, he spent all his daylight hours in pursuit of the quarry, walking and running great distances until his body was hard and fit as that of a highly trained athlete, and his arms and face a deep mahogany brown, even his full beard and moustache gilded to shades of gold by the bright sunlight.

Every day he learned from Jan Cheroot the tricks of bush-lore, until he could run a difficult spoor over rocky ground without a check. He learned to anticipate the turns and the twists that the driven herds made to get below the wind and take his scent. He learned to anticipate the pattern of their movements so that by cutting across the circuitous spoor he could save many hours of dogged pursuit. He learned to judge the sex and size and age and ivory of a beast by the mark of its rounded pads in the earth.

He found that if the herd was allowed to settle into that swinging gait between a trot and a canter, then they could keep it up for a day and a night without pause – whereas if they were taken in the full heat of noon, he could run them hard for the first five miles, winding them, bringing the tiny calves to a standstill so the cows stopped with them – flapping their huge ears to fan themselves, thrusting their trunk down their own throat to suck the water out of their belly and spray it over their head and neck.

He learned to find the heart and the brain, the lungs and the spine hidden in that amorphous mountain of grey hide and flesh. He learned to break the shoulder when the beast stood broadside, dropping him as though he had been struck by lightning, or to take the hip-shot as he ran choking in the dust cloud behind the herd, shattering the hip joint and pinning the beast for an unhurried *coup de grâce*.

He hunted the herds on the hilltops where they had climbed to catch the soothing evening breeze. At dawn he hunted in the thick forests and the open glades, and at noon he hunted them in the old overgrown gardens of a vanished people. For the land that he had first believed devoid of human presence was not, and had never been so, for unnumbered thousands of years.

Beyond the abandoned gardens where once man had planted his crops, but where now the elephant herds had returned to reclaim their heritage, Zouga found the remains of huge native cities, the long deserted keystones of a once-flourishing civilization, though all that remained were the circular outlines of the thatch-and-daub huts on the bare earth, the blackened hearth stones, and the charred fence poles of the cattle enclosures which must once have penned great herds. Judging by the height of the secondary growth in the ancient gardens, no man had farmed here for many decades.

It seemed strange to find these great grey herds of elephant ranging slowly through the ruined towns and fields. Zouga was reminded of a line from that strange exotic poetry that had been published in London the previous year, and which he had read just before sailing.

They say the Lion and the Lizard keep
The Courts where Jamshyd gloried and drank deep:
And Bahram, that great Hunter – the Wild Ass
Stamps o'er his Head, and he lies fast asleep.

Zouga scratched amongst foundations of the huts and found the deep ash, which must once have been the wooden walls and thatched roofs. In one ancient village Zouga counted a thousand such dwellings, before giving up the count. A numerous people, but where had they gone?

He found a partial answer in an ancient battlefield on the open ground beyond the thousand huts. The bones were white as daisies in the sunlight bleached and dried out, most of them half buried in the rich red soil or covered by the dense fields of waving fluffy-topped grass.

The human remains covered an area of many acres and they lay in clumps and chains like newly-cut wheat awaiting the reapers. Nearly every one of the skulls had been crushed in as if by a fierce blow with a club or mace.

Zouga realized that it was not so much a battlefield as a killing ground, for such slaughter could not be called warfare. If this toll had been repeated at each of the ruined towns he had stumbled upon, then the final total of the dead must have been in the tens or even hundreds of thousands. It was small wonder that what human presence remained was in tiny scattered groups, like the handful of warriors that had tried to prevent them crossing the pass of the elephant road. There were others. Occasionally, Zouga spied the smoke from a cooking fire rising from the highest point of one of the strangely-shaped rock hillocks that dotted the land in all directions. If these were the survivors of the vanished civilization, then they still lived in terror of the fate that had overtaken their forbears.

When Zouga and his hunting party approached any of these tiny elevated settlements, they found the crests were fortified with built up walls of rock, and they were greeted with a hail of boulders rolled down the slope upon them, that forced them to retreat hurriedly. Often there were small cultivated gardens on the level ground below the fortified hill tops.

In the gardens grew millet, ropoko and big sweet yams,

but most important for Zouga, dark green native tobacco. The soil was rich, the ropoko grew twice the height of a man and the corn sprays were loaded with red grain.

The tobacco leaves were thick-stemmed, the size of an elephant's ear. Zouga rolled the tip leaves into powerful cigars that had a rich distinction of taste and as he smoked them, he pondered how the plant had reached this distant land from its far origins. There must once have been an avenue of trade between these people and the coast. The trade beads on the necklace he had taken from the body at the pass, and now these exotic plants proved that, as did the tamarind trees, native of India, which grew amongst the ruins of the ancient villages.

Zouga wondered what a colony of British settlers with their industry and sophisticated agricultural technique, plough and crop rotation, seed selection and fertilization, might make of this lush rich soil, as he moved on slowly through the sparsely populated, well-timbered land that abounded with game and game birds, and was fed by strong clear streams of water.

Whenever he returned to the main body of the caravan, he made his meticulous observations of the sun and worked with chronometer and almanac to compute his exact positions, to add them, and his own succinct descriptions, to the map that old Tom Harkness had bequeathed him. The map increased in value, as new rivers were marked in, new boundaries set to the fly-belts and the 'fly-corridors' extended, as Zouga's observations of the terrain, of soil and vegetation types began to cover the blank portions of the old parchment.

If he was not immersed in his map, then he worked as long as the light was good enough on his journal and the manuscript which was an adjunct to it – and while he did so, Jan Cheroot and the porters brought in the latest harvest of ivory, only just beginning to stink, and buried it.

Totalling the harvest from the lists in his journal, Zouga

found that he had over twelve thousand pounds of tusks cached along his route. They were worth six shillings a pound in London, nearly four thousand pounds sterling. The trick was to get it to London. Zouga grinned to himself as he completed the calculation, a dozen wagons, or five hundred porters, and two thousand miles to carry it – that was all it required.

At each river crossing Zouga took the flat black iron pan, which doubled as laundry tub and cooking pot, and for miles in each direction along the river bed he worked the gravel. He would fill the pan from a likely spot under the bank in the bend of the river, and then set the contents swirling awash, dipping and turning the pan, spilling a little of the lighter gravel at each turn, refilling with water and spinning it again until at last he was left with a smear of the finest and heaviest material lying around the bottom of the pan in a 'tail'. Always the tail was dark and uninteresting, without the golden sparkle for which he longed so ardently.

When he detailed all these activities in his journal, only one thing gave Zouga a pause – and that was what to call this new and beautiful land. So far there was no evidence at all that it was the empire of Monomatapa, or even that Monomatapa existed. The timid, scattered and demoralized people he had so far encountered were certainly not the warriors of a powerful emperor. One other consideration decided him not to use that name. If he did so, it was tacit acknowledgement that the land had already been claimed, and each day that he travelled through the empty wilderness the dreams of himself claiming it for a queen and a nation seemed less far-fetched. Zouga began to use the name 'Zambezia' – the land below the Zambezi river – and that was how he wrote it in his journal and in the thick bundle of pages of his manuscript.

With all this work to impede progress, the pace of the caravan was leisurely, or, as Robyn told Zouga furiously, 'You would make a snail look like a Derby-winner.' For although Zouga might cover two hundred miles in the sweeping circular patterns of the hunt, the caravan camped and waited for his return, and then waited another four or five days as Jan Cheroot and the porters ferried in the loads of wet ivory.

'For all you know, Morris Zouga, your own father might be dying out there somewhere for want of a handful of medicine, while you . . .'

'If he has survived eight years already, the old devil is unlikely to turn up his toes for another few days.' Zouga's light tone covered the irritation he felt. Since he had killed the Mashona at the pass on the elephant road, the feeling between brother and sister had been strained to the point where each of them found it difficult to maintain a civil tone of voice on the few occasions when they spoke together.

Zouga's long and frequent absences from the main body were not entirely on account of his dedication to the chase and exploration of the surrounding countryside. He found it less wearying on the nerves to be away from his sister. That ecstatic mood when they had stood like two children on Christmas morning, hand in hand, upon the heights of the escarpment, was months in the past.

Brooding beside his solitary camp fire, with the hyena giggling and shrieking over the freshly killed elephant carcasses in the nearby forest, Zouga thought how it was really a miracle that two such definite personalities as his and Robyn's, whose objects were so widely divergent, should have come this far without serious disagreement. It had been too good to last indefinitely, and now he wondered how it might end. He should have followed his instincts and sent Robyn back to Tete and Cape Town when he had the excuse to do so, for the collision course on which they were so clearly set could only end in disaster for the entire expedition.

When he rejoined the main body the following day he

would have it out with her, one way or the other. She would have at last to accept that he was the leader of the expedition and as such, his decisions were final. If she did so, then he could make some concession to her wishes, though the hunt for Fuller Ballantyne was very low on Zouga's list of priorities. It would probably be best for all of them, Fuller Ballantyne not excluded, if he had long ago been laid by his faithful bearers in a hero's grave.

The thought gave Zouga a prick of guilt, and he knew he would never write it, not even in the most private pages of his journal, nor would he voice it to his sister. But the idea persisted, even while he rolled into his blanket, with a small fire at his feet and another at his head to break the thick crunching white frost which would cover the earth and grass at dawn, serenaded by Jan Cheroot's snores which took the basso profundo to the soprano of the hyena packs. Zouga fell asleep at last.

Having made the decision to assert his authority, Zouga rolled out of his blanket in the frosty dawn determined on a forced march that would take them back to where he had left Robyn and the caravan twelve days previously. He reckoned it was forty miles, perhaps a little less, to the main camp site and he set a cruel pace, not even taking the usual noon break, but pushing on remorselessly.

Zouga had deliberately left the main body encamped below a distinctively shaped kopje, whose rocky spires could be clearly seen from many miles around and which Zouga had named 'Mount Hampden' in memory of a childhood visit to that castle.

They were still far out when Zouga had his first misgivings. There was no smoke rising from the base of the hills, and there should have been. He had left almost a ton of elephant meat curing on the smoking-racks, and on the outward march he had been able to look back and see the rising column of smoke long after the crests of the hills had disappeared below the tops of the forest trees.

'There is no smoke!' he told Jan Cheroot, and the little Hottentot nodded.

'I did not want to be the first to say that.'

'Can Camacho have followed us this far?'

'There are other man-eating animals out here beside the Portuguese,' Jan Cheroot said, and cocked his head at an inquisitive bird-like angle as Zouga began swiftly to strip down to light running order. Then, still without another word, Jan Cheroot followed his example, and handed his breeches and other traps to the bearers.

'Follow as fast as you can!' Zouga told them, snatching a spare powder-bag from Matthew and turning away and breaking into a run.

Jan Cheroot fell in at his shoulder and they ran as they had run so often before, at the driving pace which could bring a breeding herd of elephant to a winded standstill within a few miles. All Zouga's feelings of antagonism towards his sister were swept away in the rush of deep concern for her safety. A series of horrific images flashed through his mind, of a sacked camp site, of mutilated bodies lying in the bloody trampled grass, shattered by the balls of the Portuguese muskets, or stabbed to death by the broad-bladed assegais of plumed and kilted warriors.

He found himself praying for her safety, repeating the formulas of his childhood which he had used so seldom since then, and unconsciously he increased the pace, until Jan Cheroot grunted a protest at his shoulder, and then slowly fell back, as Zouga forged powerfully ahead.

He reached the foot of the hill a mile ahead of Jan Cheroot, and turned to face the lowering red orb of the sun as he skirted the rocky base, crested a low rise and stopped there, hunting for breath, his chest swelling and subsiding and the sweat running down into his beard.

He stared down at the shaded dell, under the tall mukusi trees where he had left the caravan and his heart plunged. He felt physically nauseated with horror. The camp site was deserted, the fires were cold black ash and the thatched lean-to shelters had already acquired that dejected air of all deserted habitation. Still fighting for breath, Zouga plunged down the gentle slope into the abandoned camp, and looked about him wildly, for the bodies. There were none, and his first thought was of slavers. They would have taken them all, and he shuddered with horror at the thought of what Robyn must have suffered.

Zouga ran to her hut first. It was totally devoid of any trace of her. He ran to the next hut and the next – all of them were empty, but in the last hut he found a single body. It was curled on the sandy floor of the primitive shelter, and wrapped in a blanket. The blanket was drawn up over the head and wound tightly around the trunk.

Reluctant to discover his sister's horribly mutilated corpse, Zouga knelt beside it. Sweat half-blinded him as he reached out a hand that trembled with dread and exhaustion and gently drew back the fold of a grey blanket that covered the still head.

The corpse came to life with a howl of fright and leapt two feet into the air, gibbering wildly, trying to throw off the blanket and lashing out with feet and fists to defend itself.

'Hellhound!' Zouga had so christened the laziest of his porters, a skinny fellow with a vast appetite for meat and less enthusiasm for anything that involved physical exertion. 'What has happened? Where is Nomusa?'

Hellhound, once he had been quieted and had sufficiently recovered from shock, had a brief note to answer his questions. The single scrap of paper torn from Robyn's journal was double-folded and sealed with a splash of red wax, it read:

Dear Zouga,

I am of the opinion that further delay will seriously prejudice the interests of the sponsors of this expedition.

Accordingly, I have decided to move on at a speed better suited to achieving our objects before the onset of the rainy season.

I leave Hellhound to await your eventual return. Do follow at your own speed.

Your affectionate sister,
Robyn.

The note was dated ten days previously, and it was all she had left him. There was not even a pouch of salt or a bag of tea leaves, both of which commodities Zouga had been without for days.

Zouga's numbing shock lasted until Jan Cheroot reached the deserted camp, but by the time his exhausted bearers came, black rage had begun to replace all other emotion. He would have made a night march of it in pursuit of the caravan, but though he kicked their ribs as they lay and cursed them coldly, his bearers were so exhausted that they could not rise from where they had thrown themselves down.

R obyn had great difficulty getting the bearers to break camp and take up their packs. Her first efforts to do so were greeted with amusement and light laughter, for none of them believed she was serious. Even Juba could not understand that Nomusa, a woman, was taking command of the caravan.

When all argument failed, Robyn took the hippo-hide

kurbash whip to the little yellow Hottentot Corporal, whom Jan Cheroot had left in command of his musketeers. The Corporal shouted frantic orders to his men from the upper branches of the mukusi where Robyn had him treed.

They were on the march within the hour, but the light laughter had given way to scowls and sulks. They were all convinced that the safari was now doomed, for who had ever heard of a woman, a young woman – worst of all a young white woman – leading into the unknown. Within half a mile most of the porters were complaining of thorns in the feet or bad blood behind the eyes, afflictions common only to porters reluctant to advance.

Robyn got them to their feet again by firing a shot from the big naval Colt revolver over their heads, it almost sprained her wrist, but proved an amazingly effective cure for both feet and eyes, and finally they made a fair day's march of it, south and west, ten miles, as Robyn reckoned it as she marked her journal that night.

Despite the brave countenance she put on before the porters and musketeers, Robyn had her own doubts. She had watched closely as Zouga plotted a course with the prismatic compass, and she had mastered the technique of sighting a distant hill or other feature and marching upon it, before again sighting ahead on another feature. It was the only way to maintain a direct line of march in this rolling forested country.

She had studied, whenever she had the opportunity, the Harkness map, and had seen the wisdom of Zouga's choice of direction. He was aiming to traverse this wide unexplored land he had called Zambezia, and eventually to hit the road that their grandfather, Robert Moffat, had pioneered from his mission station at Kuruman to the town of the Matabele king, Mzilikazi, at Thabas Indunas.

However, Zouga's intention had been to cut south of the frontiers of the Matabele kingdom, avoiding the Burnt Land where, according to Tom Harkness, the rapacious border impis of Mzilikazi killed all travellers. Neither she nor Zouga could trust to their relationship to Moffat to protect them.

Once they gained the wagon road to Kuruman, they would be back in the known world, and the road would lead them to the series of waterholes that grandfather Moffat had marked. At Kuruman, after the family reunion, the road to Cape Town was long and wearying, but well travelled and in less than a year they could be back in London. The delicate part was to grope the way south of Matabele land, through the untold hazards still ahead and to reach the road.

Robyn did not truly anticipate that she would be called upon to perform this feat of navigation entirely on her own. It could only be a matter of days before Zouga reached Mount Hampden camp and then hurried to catch up with the main body. There would be an interesting clash of wills then. However, at the end of it, she was certain that Zouga would be convinced that their father's whereabouts and safety were more important than the slaughter of animals, whose teeth could probably never be retrieved from their burial mounds.

This was a gesture of defiance only, soon Zouga would be with her once more. In the meantime, however, she had an unpleasantly hollow and lonely sensation beneath her ribs as she strode along in her tight-fitting breeches, the Hottentot standard-bearer carrying the grubbied and tattered Union Jack a few paces behind her, and a sullen line of porters straggling along behind that again.

They camped the second night on the edge of a river that had been reduced to a string of still green pools in the sugary white sand of the river bed. On the steep bank stood a small grove of strangler figs, the smooth pale trunk and branches had crept up like thick and sinuous pythons to smother the host trees. The parasites were now taller and more robust than the rotting remains that supported them had ever been, and the clusters of ripening figs covered their branches. Fat green pigeons came to feed on the fruits, flicked their wings and whistled their shrill cry, unlike that of any other pigeon, that to Robyn sounded like 'Oh well, oh very well!' and they cocked their heads to peer down through the green leaves at the men camped below.

The porters had cut the *scherms* of thorn branches and lit the cooking fires when they all heard the lion roar. The sound checked the murmur of voices for only a few seconds, for it was very faint and distant, seeming to come from miles downstream, and it was a sound to which they had all long grown accustomed.

There had hardly been a night since they had crossed the pass on the elephant road that they had not heard the lions, and in the morning found the pug marks, sometimes almost the size of a soup plate, in the soft earth around the camp where the big inquisitive cats had circled them during the night.

However, Robyn had never seen one of them for they are almost entirely nocturnal, and her first tremors of trepidation had long turned to indifference. She felt quite secure behind the *scherm* of thornbush, and now at the distant muttering rumble she hardly bothered to look up from her journal in which she was exaggerating only slightly the competent manner in which she had directed the day's march.

'We are making as good a time as we ever did when Z was leading,' she wrote smugly, but did not go on to mention the mood of the porters.

The lion roared only once, and when it was not repeated, the sullen talk around the cooking fires was resumed and Robyn bowed her head once more over her journal.

A few hours after the setting of the sun the camp settled for the night, and lying under her hastily thatched shelter with Juba curled on the mattress of freshly cut grass beside her, Robyn listened to the melodious African voices of her porters fade gradually into silence. She sighed once, deeply, and fell instantly asleep – to wake with a confusion of sound and movement all around her.

She knew that it was late by the frostiness of the air, the utter darkness and her own sleep-drugged stupor. The night rang with the terrified shouts of men and the rush of their feet. Then there was the thudding report of a musket, the crash of heavy logs being thrown on the watch fires and then, heart-stoppingly, the screams of Juba close to her head.

'Nomusa! Nomusa!'

Still groggy with sleep, Robyn struggled up. She was not certain what she was dreaming and what was reality.

'What is it?'

'A devil!' shrieked Juba. 'Devils have come to kill us all.'

Robyn flung off her blanket, and ran out of the shelter bare-footed, dressed only in her flannel nightgown with ribbons in her hair.

At that moment the new logs blazed up on the watch fire, and she saw naked yellow and black bodies, and terrified faces, whites of rolling eyes and open shouting mouths.

The little Hottentot Corporal, stark naked, pranced beyond the fire, brandishing his musket, and as Robyn ran towards him he fired it blindly into the darkness.

Robyn caught his arm as he began to reload.

'What is it?' she shouted into his ear.

'*Leeuw*! Lion!' His eyes were glittering with fright and bubbles of spittle ran from the corners of his mouth.

'Where is it?'

'It has taken Sakkie! It pulled him out of his blankets.'

'Quiet!' Robyn shouted. 'Keep quiet all of you!'

Now, at last, they all turned instinctively to her for leadership.

'Quiet!' she repeated, and the gabble of fright and uncertainty died swiftly.

'Sakkie!' she called in the silence, and the missing Hottentot's voice answered her faintly from below the steep bank of the river-bed.

'*Die leeuw het my!* The lion has me! *Die duiwel gaan my dood maak*, the devil is going to kill me,' and he broke off with a shriek of agony.

Above the high-pitched shriek they all clearly heard the crunch of bone, and the muffled growl like that of a dog with food in its jaws. With a rush of horror that raised goose pimples along her arms, Robyn realized that she was listening to the sounds of a man being eaten alive not fifty yards from where she stood.

'*Hy vreet my bene,*' the voice out of the darkness rang with unbearable agony. 'He is eating my legs,' and the gruesome cracking, tearing sounds made Robyn's gorge rise to choke her. Without thinking, she snatched a burning brand from the fire and holding it aloft shouted at the Hottentot Corporal. 'Come on! We must save him!'

She ran forward to the lip of the bank before she realized she was alone, and unarmed.

She looked back. Not one of the men around the fire had followed her. They stood in a tight group, shoulder to shoulder, clutching muskets or axes or assegais, but rooted where they stood.

'He is finished.' The Corporal's voice shook with fright. 'Leave him. It's too late. Leave him.'

Robyn hurled the burning brand she held down into the river-bed below her feet, and before its flames faded and pinched out she thought she saw something big and dark and terrifying on the edge of the shadows.

Robyn ran back to the group and snatched a musket from one of the Hottentots. Thumbing back the hammer, she ran once more to the river bank and peered down into the dry bed. It was utterly dark, until suddenly there was somebody at her shoulder, holding high a burning branch from the fire.

'Juba! Go back!' Robyn snapped at the child. Juba was completely naked except for a single string of beads about her hips, and the firelight glinted on her sleek black body.

She could not answer Robyn, for tears were rolling down her plump cheeks and her throat was closed with terror, but she shook her head fiercely at the order to retreat.

Below them, outlined against the white sand of the dry river, was that grotesque dark shape. and the screams of the dying man blended with the grisly wet growls of the animal.

Robyn lifted the musket, but hesitated for fear of hitting the Hottentot. Disturbed by the light, the lion rose, becoming huge and black; swiftly it dragged the weakly wriggling body, dangling between its forelegs back into the darkness beyond the feeble circle of light from the flames.

Robyn drew a deep breath and the heavy musket shook in her hands, but she lifted her chin in a gesture of decision and holding up the skirts of her long nightdress in one hand, went down the path into the river-bed. Juba followed her like a faithful puppy, pressing so hard against her that they nearly lost their balance, but she kept the burning brand on high, though it shook, and the flames wavered smokily.

'Brave girl!' Robyn encouraged her. 'Good brave girl!'

They stumbled through the loose white sand that covered the ankles of their bare feet at each pace.

Ahead of them, at the extreme limit of their vision, moved the menacing black shadow, and the deep muttering growls seemed to fill the night around them.

'Leave it!' Robyn shouted, her voice quavering and breaking. 'Drop it – this instant!'

Unconsciously she was using the same commands as she had given her terrier as a child when it refused to deliver the rubber ball.

Ahead of her in the darkness Sakkie heard her, and bleated feebly. 'Help me – for God's love, help me.' But the lion pulled him away, leaving a long wet drag mark through the sand.

Robyn was tiring rapidly, her arms ached from the weight of the heavy weapon, each breath burned her throat as she panted, but she could not seem to get enough air, for an iron band of fear cramped her chest. She sensed that the lion would only retreat a certain distance before it lost patience with the shouting and harassment, and her instinct was right.

Suddenly she made out the full shape of the lion ahead of them. It had dropped the maimed body and stood over it now like a tomcat with its mouse, but it was as big as a Shetland pony, with the dark ruff of its mane fully erect, seeming to double its size.

In the light of the flames, its eyes glowed a bright ferocious gold, and it opened its jaws and roared. The very air dinned upon Robyn's eardrums, causing her actual physical pain as that great gust of sound struck her. It rose to an unbearable pitch, so that involuntarily she reeled backwards with Juba clinging to her. The child wailed with despair, losing control of her body so that in the light of the flames her urine shot in sharp little spurts down her legs, and as the lion launched into its charge she dropped the torch into the sand, plunging them all into complete darkness.

Robyn lifted the musket in front of her, as a defensive reflex rather than a planned act of aggression, and when the barrel was at waist-level she pulled the trigger with all her strength. The cap flashed, flaring brightly in the blackness and for an instant she saw the lion. It was so close that the long barrel of the musket seemed to touch its huge shaggy head. The mouth was wide open, still emitting those shattering gusts of sound, and the fangs that lined the deep, meat-red gape of jaws were long and white and cruel. The eyes burned yellow as living flames, and Robyn found that she was screaming, but the sound was lost completely in the roaring of the enraged animal.

Then an instant after the flash of the cap the musket

fired, bucking so savagely in her hands that it almost tore itself from her grip, and the butt, not anchored against her shoulder, was driven back into her stomach with a force that expelled the air from her lungs, and sent her reeling in the loose white sand. Juba, clinging to her legs and wailing with despair, tripped her and she went over backwards, sprawling full length at the same moment that the full weight of the lion lunged into her.

If Robyn had not fallen, the charge would have stove in her ribs, and snapped her neck, for the lion was over four hundred pounds weight of driving bone and muscle. As it was, it knocked her out of her senses for she never knew how long, but she became conscious again, with the strong cat-reek of the lion in her nostrils, and an immense weight crushing her into the sand. She wriggled weakly, but the weight was suffocating her, and gouts of hot blood, so hot it seemed to scald her, were spurting over her head and neck.

'Nomusa!' Juba's voice, high with anguish and very close, but those shattering roars were silent. There was just the unbearable weight and the rank smell of lion.

Robyn's strength came back to her with a rush, and she struggled and kicked, and the weight above her rolled loosely aside, slithering off her, and she dragged herself free of it. Immediately Juba clung to her again, throwing her arms about Robyn's neck.

Robyn comforted her as though she were an infant, patting her and kissing her cheek that was wet and hot with tears.

'It's over! There, now. It's all over,' she mumbled, aware that her hair was sodden with the lion's blood, and that a dozen men, led cautiously by the Hottentot Corporal, had lined the high river bank, each of them holding aloft a torch of burning grass.

In the dim yellow light the lion lay stretched out beside Robyn in the sand. The ball from the musket had struck him full in the nose, passed cleanly through the brain and lodged in the base of his neck, killing the great cat in mid-air, so the lifeless body had pinned Robyn to the sand.

'The lion is dead!' Robyn quavered as she called to the men, and they came down in a close bunch, timidly at first and then boldly when they saw the huge yellow carcass.

'It was the shot of a true huntress,' announced the Corporal grandly. 'An inch high and the ball would have bounced off the skull, an inch lower and it would have missed the brain.'

'Sakkie,' Robyn's voice still shook, 'where is Sakkie?'

He was still alive, and they carried him in his blanket up into the camp. His wounds were fearsome, and Robyn knew there was not the smallest chance of saving him. One arm from wrist to elbow had been chewed so that not a piece of bone bigger than the top joint of her finger remained. One foot was gone just above the ankle, bitten clean off and swallowed in one piece. He had been bitten through pelvis and spine, while through a tear in his diaphragm below the ribs the mottled pink of his lungs swelled out with each breath.

Robyn knew that to attempt to cut and sew that dreadfully torn flesh or to saw the splintered bone stumps would be inflicting futile agony on the little yellow man. She had him laid close to the fire, she plugged the worst holes gently, and then covered him with blankets and fur karosses. She administered a dose of laudanum so powerful as to be almost lethal in itself. Then she sat next to Sakkie and held his hand.

'A doctor must know when to let a man die with dignity,' her professor at St Matthew's had once told her. And a little before dawn Sakkie opened his eyes, the pupils dilated widely by the massive dose of the drug and smiled at her just once before he died.

His brother Hottentots buried him in a small cave in one of the granite kopjes and they blocked the opening with boulders that the hyena could not roll aside.

When the Corporal and his Hottentots came down from the hill, they indulged in a brief ritual of mourning which consisted mainly of emitting loud theatrical cries of anguish and firing their muskets in the air to speed Sakkie's soul on its journey, after which they ate a hearty breakfast of smoked elephant meat, and the Corporal came to Robyn, dry eyed and grinning broadly.

'We are now ready to march, Nomusa!' he told her, and with a stamp of his right foot, which began with the knee lifted under his chin, he gave her one of those widely extravagant salutes, a mark of deep respect that up to that date had been reserved exclusively for Major Zouga Ballantyne.

During that day's march, the porters sang again for the first time since leaving Zouga's camp at Mount Hampden.

> 'She is your mother and your father too,
> She will dress your wounds
> She will stand over you while you sleep
> We, your children, greet you, Nomusa,
> The girl child of mercy.'

It was not only the caravan's rate of advance under Zouga that had irritated and annoyed Robyn. It was also their complete failure to make contact with any of the indigenous tribes, with any of the inhabitants of the scattered and fortified villages.

To her it seemed completely logical that the only way that they would be able to trace Fuller Ballantyne through this wilderness was by asking questions of those who must have seen him pass, and almost certainly had spoken and traded with him.

Robyn could not believe that her father would have adopted the same high-handed actions to force a passage past anybody or anything that stood in the way of the caravan as Zouga had done.

When she closed her eyes she could still see clearly in her mind's eye the tiny falling body of the black man in the tall headdress, shot down ruthlessly by her brother. She had rehearsed in her mind how she or her father would have passed through the elephant road without gunfire and slaughter. The tactful withdrawal, the offering of small gifts, the cautious parley and eventual agreement.

'It was plain bloody murder!' she repeated to herself for the hundredth time. 'And what we have done since then has been blatant robbery.'

Zouga had helped himself to the standing crops of the villages they had passed, to tobacco, millet and yams, not even bothering to leave a handful of salt or a few sticks of dried elephant meat as token payment.

'We should try to communicate with these people, Zouga,' she had remonstrated.

'They are a sullen and dangerous people.'

'Because they expect you to rob and murder them – and, as God is my witness, you have not disappointed them, have you?'

The same argument had run its well-worn course many times, neither of them relenting, each holding stubbornly to their own view. Now at least she was free to attempt to establish contact with these people, Mashona, as Juba called them disparagingly, without her brother's impatience and arrogance to distract her and alarm the timid black people.

On the fourth day after leaving Zouga's camp, they came in sight of an extraordinary geological formation. It was as though a high dam wall had been constructed across the horizon, a great dyke of rock running almost exactly north and south to the very limit of the eye.

Almost directly in their line of march was the only breach in this rampart, and from the altered vegetation, the denser growth and deeper green, it was clear that a river flowed through the gap. Robyn ordered a small adjustment in their line of march and headed for the pass.

When they were still some miles distant Robyn was delighted to make out the first signs of human habitation that they had come across since leaving Mount Hampden.

There were fortified walls on the cliffs above the breach in the long low hill, high above the river bed, and as they drew closer, Robyn could see the gardens on the banks of the river, defended by high brush and thorn barriers with little thatched look-out huts standing high on stilts in the centre of the dark luscious green stands of young millet.

'We will fill our bellies tonight,' the Hottentot Corporal gloated. 'That corn is ripe enough to eat.'

'We will camp here, Corporal,' Robyn told him firmly.

'But we are still a mile . . .'

'Here!' Robyn repeated.

They were all puzzled and resentful when Robyn forbade entry to the tempting gardens, and confined them all to the perimeter of the camp, except for the water and wood parties. But resentment turned to genuine alarm when Robyn left the camp herself, accompanied only by Juba, and as far as they could see completely unarmed.

'These people are savages,' the Corporal tried to intercept her. 'They will kill you, and then Major Zouga will kill me.'

The two women entered the nearest garden, and carefully approached the look-out hut. On the earth below the rickety ladder that rose to the elevated platform a fire had burned down to ash, but flared again when Robyn knelt and fanned it. Robyn threw a few dry branches upon it and then sent Juba for an armful of green leaves. The column of smoke drew the attention of the watchers on the cliff above the gorge.

Robyn could see their distant figures on the skyline, standing very still and intent. It was an eerie feeling to know that so many eyes were upon them, but Robyn was not relying entirely on the fact that they were women, nor was she relying on their patently peaceful intentions, nor even upon the prayers which she had offered up so diligently to protect them. On the principle that the Almighty helps those who show willing, she had Zouga's big Colt pistol stuffed into the waistband of her breeches and covered by the tail of her flannel shirt.

Next to the smoking beacon fire, Robyn left a half pound of salt in a small calabash gourd, and a bundle of sticks of black smoked elephant meat which was the last of her stock.

Early the next morning Robyn and Juba again visited the garden and found the meat and salt had been taken, and that there were the fresh footprints of bare feet overlaying their own in the dust.

'Corporal,' Robyn told the Hottentot with a confidence she did not feel, 'we are going out to shoot meat.'

Corporal grinned beatifically. They had eaten the last of the smoked meat, weevils and all, the previous evening, and he flung her one of his more flamboyant salutes, his right arm quivering at the peak of his cap, his fingers spread stiffly, the stamp of his right foot raising dust, before he hurried away shouting orders to his men to prepare for the hunt.

Zouga had long ago declared the Sharps rifle to be too light for elephant, and left it in camp, favouring the big four-to-the-pound smooth bores to the more expensive breech-loading rifle. Robyn took it now and inspected it with trepidation. Previously she had only fired it at a target, and now in the privacy of her grass hut she rehearsed loading and cocking the weapon. She was not sure that she would be capable of cold-bloodedly aiming it at a living animal, and had to reassure herself of the absolute necessity of procuring food for the many mouths and stomachs that now depended upon her. The Corporal did not share her doubts, he had seen her shoot a charging lion between the eyes, and trusted her now implicitly. Within an hour's walk they found a herd of buffalo in the thick reed beds along the river. Robyn had listened to Zouga talking of the hunt with enough attention to know the necessity of keeping below the wind – and in the reeds with visibility down to a few feet and with the commotion created by two hundred cows and bleating calves they crept up to a range at which nobody could miss.

Her Hottentots blazed away with their muskets, while Robyn herself fired grimly into the galloping bellowing bodies that charged wildly past her after the first shot had startled them.

After the dust had settled and the thick bank of powder smoke had drifted away on the faint breeze, they found six of the big black animals lying dead in the reedbeds. Her entourage were delighted, hacking the bodies into manageable chunks which they slung on long poles and carried singing up to the camp. Their delight turned to amazement when Robyn ordered that an entire haunch of fresh buffalo be taken out and left next to the hut in the millet garden.

'These people are eaters of roots and dirt,' Juba explained patiently. 'Meat is too good for them.'

'To kill this meat we have risked our lives,' the Corporal began his protest, then caught the look in Robyn's eye, broke off and coughed and shuffled his feet. 'Nomusa, could we not give them a little less than a haunch? The hooves make a good stew, and these people are savages, they will eat anything,' he pleaded. 'A whole haunch . . .'

She sent him away muttering and shaking his head sorrowfully.

During the night, Juba woke her, and the two of them sat and listened to the faint throb of drums and the singing that carried down from the hilltop village, clearly the sounds of feasting and jubilation.

'They have probably not seen so much meat at one time in all their lives,' Juba murmured sulkily.

In the morning Robyn found, in place of the buffalo meat, a bark basket containing fifteen hen's eggs the size of pigeon's eggs, and two large earthen pots of millet beer. The look of the bubbling thin grey gruel almost turned Robyn's stomach. She gave the beer to the Corporal to distribute, and her followers drank it with such obvious relish, smacking their lips and nodding over it like connoisseurs over an ancient bottle of claret, that Robyn controlled her heaving stomach and tasted some; it was tart and refreshing and strong enough to set the Hottentots chattering and laughing raucously.

With Juba following her, each of them carrying a bundle of half-dried buffalo meat, Robyn returned to the gardens, certain that the exchange of gifts had proved it possible to establish friendly contact. They sat under the shelter and waited. The hours passed without a sign of the Mashona appearing. The hushed heat of noon gave way to long, cool shadows of evening – and then, for the first time, Robyn noticed a gentle stir amongst the millet plants that was neither wind nor bird.

'Do not move,' she cautioned Juba.

Slowly a human shape showed itself, a frail bent figure dressed in tatters of a skin kilt. Robyn could not tell whether it was man or woman, and she didn't dare stare openly at it, for fear of frightening it away.

The figure emerged from the stand of millet, crouched down on its haunches, and it hopped hesitantly towards them, with long pauses between each tentative hop. It was so thin and wrinkled and dried out that it looked like one of the unbandaged mummies that Robyn had seen in the Egyptian section of the British Museum.

It was definitely a man, she realized at last, sneaking a glance in his direction, for with each hop his shrivelled and stringy genitals flapped out from under the short kilt.

Closer still, Robyn saw that his cap of woolly hair was pure white with age, and in the seamed and pouched sockets his eyes wept slow tears of fright, as though these were the last drops of liquid that his desiccated old body contained.

Neither Juba nor Robyn moved or looked directly at him until he squatted a dozen paces from them, then slowly Robyn turned her head towards him. The old man whimpered with fear.

It was clear to Robyn that he had been selected as an emissary because he was the least valuable member of the tribe, and Robyn wondered what threats had been made to force him to come down from the hilltop.

Moving very slowly and calmly, as though she were dealing with a timid wild creature, Robyn held out a stick of half-cured buffalo meat. The old man stared at it, fascinated. As Juba had told her, these people probably existed almost entirely on their meagre crops and such roots and wild fruit as they could glean in the forest. Meat was a rare treat, and such an unproductive member of the tribe would be given only a very little of what there was.

The way he stared at the piece in Robyn's hand made her believe that the old man had not had so much as a taste of the buffalo haunch. He was more than half-starved. He rolled his tongue loosely around in his toothless mouth, gathering his courage, and then shuffled close enough to hold out the claws of his bony fingers, palms cupped upwards in the polite gesture of acceptance.

'There you are, dearie.' Robyn placed the stick in his hands and the man snatched it to his mouth, sucking noisily upon it, worrying it with his smooth gums, drooling silver strings of saliva as his mouth flooded, his eyes streaming again, this time with pleasure rather than fear.

Robyn laughed with delight, and the old man rapidly blinked his eyes and then cackled around the stick of meat, the sound so comical that Juba laughed also, the laughter of the two younger women rippling and tinkling without restraint. Almost immediately, the dense leaves of the millet garden stirred and rustled, as other dark, half-naked figures came slowly forward, their anxiety relieved by the sound of laughing women.

409

The hilltop settlement consisted of not more than a hundred individuals, men, women and children, and every one of them came out to stare and laugh and clap as Robyn and Juba climbed the steep twisting path. The old silver-headed man, almost unbearably proud of his achievement, led Robyn by the hand possessively, screeching out explanations to those around, pausing every now and then to perform a little shuffling dance of triumph.

Mothers held up their infants to look at this marvellous being, and the children ran forward to touch Robyn's legs and then squeal with their own courage, skipping away ahead of her up the path.

The pathway followed the contours of the hillside, and it passed between defensive gateways and under terraced walls. Above the path at every steep place were piled boulders, ready to be hurled down upon an enemy, but now Robyn's ascent was a triumphant progress, and she came out into the village surrounded by a welcoming throng of singing, dancing women.

The village was laid out in a circular pattern of thatched and windowless huts. The walls were of plastered clay with low doorways and beside each hut was a granary of the same materials but raised on poles to protect it from vermin. Apart from a few diminutive chickens, there were no domestic animals.

The space between the huts, and the central courtyard was swept, and the whole village had an air of order and cleanliness. The people themselves were handsome, though none of them carried any excess flesh or fat. Robyn was reminded by their slim, lithe bodies that they were almost exclusively vegetarian.

They had alert and intelligent faces, and the laughter and singing with which they welcomed her was easy and unaffected.

'These are the people whom Zouga shot down like animals,' she thought, looking around her with pleasure.

They had set a low carved stool in the shade for her and Juba squatted beside her. As soon as Robyn was seated, the old man squealed importantly and a giggling young girl brought her a pot of the millet beer. Only when she had drunk a mouthful of the beer did the crowd fall silent, and draw aside to let a commanding figure come through.

On his head was a tall headdress of animal fur, similar to the one worn by the chieftain on the elephant road pass. He wore a cloak of leopard skin over his shoulders, the skins so worn that they must have been very old, probably the hereditary symbol of his chieftainship. He sat down on another stool facing Robyn. He was a man of middle age, with a pleasant humorous face, and a lively imagination – for he followed Robyn's sign language with attention, and then acted out his own replies with facial expressions and gestures that Robyn understood readily.

This way he asked her from where she came and she showed him the north and made a circle of her hands towards the sun for each day's travel. He wanted to know who was her husband and how many children she had. That she was both unmarried and childless was a source of amazement to the whole village.

More beer was brought out in the clay pots and Robyn felt slightly light-headed, and her cheeks turned pink and her eyes shone. Juba was contemptuous of their hosts.

'They do not have even a goat!' she pointed out scornfully.

'Perhaps your brave young men have stolen them all,' Robyn answered tartly, and raised her beer pot in salute to the chief.

The chief clapped his hands, signalling his drummers to stoop to their instruments. The drums were hollowed tree trunks, beaten with a pair of short wooden clubs to a frantic rhythm that soon had the drummers running with rivulets of sweat and glassy-eyed with the mesmeric effects of the beat. The chieftain threw off his leopard-skin cloak and launched himself into the dance, swirling and leaping until his necklaces and bracelets jangled and rattled.

On his chest he wore a pendant of ivory, snowy-white polished ivory, and it caught the firelight, for by now the sun was long set. Robyn had not noticed the ornament before for it had been covered by the cloak, but now she felt her eye drawn repeatedly to the bouncing white disc.

The disc seemed too perfectly shaped, and as the chief came bounding up to her stool to perform a solo pass before her Robyn saw that it was decorated with a regular pattern around its border. The next moment her heart raced with excitement for the decoration was writing, she was not sure in what language, but it was Latin script, that was certain. Then the chief had gone, leaping away to prance in front of his drummers, exhorting them to greater efforts.

Robyn had to wait until the chief exhausted himself, and staggered panting back to his stool to quaff a pot of the thick grey beer, before she could lean forward and get her first close look at the ornament.

She had been mistaken. It was not ivory but porcelain, and its perfect shape and whiteness was immediately accounted for. It was an item of European manufacture, the lid of a small pot, the type in which tooth-powder or potted meats are sold. The writing was English, printed in neat capitals were the words:

'PATUM PEPERIUM – THE GENTLEMAN'S RELISH.'

She felt her skin go clammy with excitement. Clearly she remembered her father's rage when the pantry at King's Lynn had been bare of this delicacy. She remembered as a small girl running down the village street in the rain to the grocer to buy another pot.

'It is my one weakness, my only weakness,' she remembered her father's exact words while he spread the savoury paste on his toast, his anger mollified so that he came near to making a joke of it. 'Without my Gentleman's Relish, I doubt I would have had strength for the Transversa.'

When Robyn's mother left for Africa on that last ill-fated voyage, there had been a dozen cases of the relish in her luggage. There was only one possible way that the porcelain lid could have reached here.

Robyn stretched out a hand and touched it, but the chief's expression changed instantly and he jumped back, out of reach. The singing and drumming came to an abrupt halt, and the consternation of the entire village made Robyn realize that the porcelain lid was a charm of great personal magic, and that it was a disaster that another hand had touched it.

She made an attempt to mollify the chief, but swiftly he covered himself with his leopard cloak and stalked away to his hut at the end of the village. The festivities were clearly ended. The rest of the villagers were subdued and crept away after the chief, leaving only the silver-haired toothless ancient, possessive as ever, to lead Robyn to the hut which had earlier been set aside for her.

She lay awake most of the night on the plaited reed sleeping-mat, excited at the evidence that her father had passed this way, and worried that she had ruined her relationship with the Mashona chief and would learn no more of the ornament and, through it, of her father.

She did not have an early opportunity to meet the chief again, and make amends for her breach of manners. The villagers kept away from her, obviously hoping she would go, but she stayed on stubbornly in the hilltop village, attended only by the faithful old man. For Robyn was the most important thing that had ever happened in his long life, and he was not going to relinquish her for the chief or for anybody else.

In the end, there was nothing for it but to send the chief an extravagant gift. She used the last *khete* of *sam-sam* beads and one of the double-bladed axes.

The chief could not resist such princely bribes, and though his attitude was cooler and more reserved than at first, he listened attentively while Robyn asked her questions, acting out little charades, which the chief discussed seriously with his elders, before giving his answers.

The answer was southwards again, south for five circles of the sun, and the chief would send somebody to guide Robyn. He was obviously pleased to be rid of her at last, for, though her gifts were welcome, the chief was still deeply troubled by the ill-fortune that her sacrilegious action must bring upon the tribe.

For a guide the chief chose the silver-headed old man, ridding himself of a useless mouth and an unwelcome visitor at one stroke.

R obyn had doubted that her guide's thin legs could carry him either very fast or very far. However, the old man surprised her. He armed himself with a long throwing spear which looked as old and frail as he did himself and on his head he balanced a rolled sleeping-mat and a clay cooking pot – these clearly constituted his total worldly possessions. He girded up his tattered kilts and set off southwards at a pace that had Robyn's porters grumbling again. Robyn had to restrain him.

It took a little time to get the old man to understand that he was now her language tutor. As they marched she pointed at herself, and everything around them, naming them clearly in English, and then looking at him enquiringly. He returned the look with equal enquiry in his rheumy old eyes. However, she persevered, repeating her own name 'Nomusa' as she touched her chest, and suddenly he understood.

He slapped his own chest. 'Karanga,' he squeaked. 'Karanga!' Once again his enthusiasm for this new activity was such that she had to restrain him. Within a few days Robyn had dozens of verbs and hundreds of nouns which she could begin stringing together, much to old Karanga's delight.

However, it was four days before Robyn realized that there had been an initial misunderstanding. Karanga was not the old man's name, but the name of his tribe. It was too late to rectify, because by that time everybody in the caravan was calling him 'Karanga', and the old man answered to it readily. It was difficult to get him to leave Robyn's side. He followed her wherever she went, much to Juba's disgust and undisguised jealousy.

'He smells,' she told Robyn virtuously. 'He smells very bad.' Which was true, Robyn had to admit.

'But after a while you do not notice it, so much.' There was one thing, however, which could not be so readily overlooked, it appeared from under the old man's kilts whenever he squatted, which he did whenever at rest. Robyn solved this by giving old Karanga a pair of Zouga's woollen underwear and taking her chances with her brother's wrath later. The underwear filled old Karanga with almost unbearable pride. He preened and strutted like a peacock, as they flapped around his long thin legs.

Old Karanga led them cautiously wide of every occupied village along their route, although he assured Robyn they were also of the same tribe. There seemed to be no trade nor commerce between these settlements, each perched on its fortified hilltop in suspicious and hostile isolation.

By this time, Robyn could speak enough of the language to find out from Karanga more about the great wizard – from whom the chief had received the magical porcelain talisman, and the story filled her with excitement and anticipation.

Many rainy seasons ago, old Karanga was not sure how many, at his age every season blurred into the one before or the one after, anyway, at some not too far distant date an extraordinary man had come out of the forest, even as she had come, and like her he had been fair skinned. However, his hair and beard were the colour of flames (he showed her the camp fire), and he was without doubt a magician and prophet and rainmaker, for the day he arrived the long drought had broken with thunderous storms that filled the rivers for the first time in many years.

This pale wizard had performed other rare and wonderful feats, transforming himself first into a lion and then into an eagle, raising the dead from the grave, and directing the lightning with a wave of his hand. The tale had lost nothing in the retelling Robyn noted wryly.

'Did anybody speak to him?' Robyn asked.

'We were all too afraid,' Karanga admitted, shaking theatrically with terror, 'but I myself saw the wizard as an eagle fly over village and drop the talisman from the sky.' He flapped his skinny old arms in pantomime.

The strong anchovy smell would have attracted the bird to the discarded pot, Robyn reflected, but when it proved inedible the bird would have dropped it, by chance over Karanga's village.

'The wizard stayed a short time near our village and then went away to the south. We have heard that he travelled swiftly, obviously in his guise as a lion.

'We heard of his miracles, the word shouted from hill-top to hill-top or tapped out by the drums. How he cured others sick to death, how he challenged the ancestral spirits of the Karanga in their most sacred places shouting abuse at them so all who heard him trembled.

'We heard also how he slew the high priestess of the dead, an Umlimo of vast power in her own stronghold. This strange pale magician slew her and destroyed her sacred relics.'

In fact he had raged through the land like a man-eating

lion, which of course he was, until finally he came to rest upon a dark hilltop far to the south, the mountain of iron, Thaba Simbi, where he stayed to perform diverse curses and miracles so that the people came to him from far and wide to buy his services with corn and other offerings.

'Is he still there?' Robyn demanded.

Old Karanga rolled his watery eyes and shrugged. 'It is always difficult and dangerous to predict the comings and goings of wizards and magicians,' that eloquent gesture seemed to say.

The journey was not so straightforward as Robyn had hoped, for the further he went from his own village the less certain old Karanga became of his direction, or of the exact location of the Iron Mountain, which he had told her of.

At the beginning of each day's march, he informed Robyn confidently that they would reach their destination that day, and as they went into camp each evening he told her apologetically that it would be the next day for certain.

Twice he pointed out rocky kopjes, 'That is indeed the Iron Mountain,' but each time they were driven off with a hail of boulders and throwing spears from the heights.

'I was mistaken,' Karanga mumbled, 'there is a darkness in my eyes sometimes, even under the noonday sun.'

'Have you verily and truly seen this mountain?' Robyn demanded sternly, almost at the end of her patience, and Karanga hung his silver-wrinkled head and with great industry picked at his nostrils with a bony finger to hide his discomfiture.

'It is true that I have not personally seen this place, not with my own eyes, but I have been told by one who spoke with a man who himself . . .' he admitted, and Robyn was so angry that she shouted at him in English.

'You naughty old devil, why did you not say so before!'

Old Karanga understood the tone, if not the words, and his misery was so apparent that she could not sustain her anger for more than an hour, his gratitude when she once more allowed him to carry her water bottle and food bag was pathetic to see.

Robyn was now consumed with impatience. She had no way of knowing how far behind her was Zouga and his hunting party. He might have returned to the camp at Mount Hampden and found her note the very day after her departure, or he might still be killing elephant a hundred miles away, completely unaware that she had marched without him.

Her disapproval of her brother, and her anger at his recent actions, had gradually evoked a sense of competition in her. She had come so far and accomplished so much on her own – from the contact with the Karanga village to following the traces of her father so far and so doggedly – that she fiercely resented the idea of his arrival when she was at the very point of making the prayed-for and long-delayed reunion with Fuller Ballantyne. She guessed how the tale would be told in Zouga's journal, and in the book that would follow it. She knew who would get all the credit for the arduous search, and its brilliantly successful conclusion.

Once she had thought that fame and praise meant little to her, she believed she would be content to leave that to her brother. She had believed that her own reward would be her father's embrace, and the deep personal knowledge that she had brought some comfort or some surcease to the suffering black peoples of her Africa.

'I still do not know myself all that well,' she admitted, as she called the third successive *tirikeza*, double-marching her porters relentlessly to keep ahead of Zouga wherever he might be.

'I want to find Pater, I want to find him alone, and I want the world to know I found him.'

'Pride is sinful, but then I have always been a sinner. Forgive me, Sweet Jesus, I will make it up in a thousand other ways. Only forgive this one small unimportant sin,' she prayed in her rude grass shelter, and as she did so she listened with one ear for the shouts of Zouga's bearers coming into camp, and her heart tripped at each sudden noise. She was tempted to break camp and call a night march towards the next distant hill they had seen on the horizon at sunset and which old Karanga had once again confidently declared to be the Iron Mountain. The moon was full and would rise in an hour, that one march might be all that was necessary to keep ahead of her brother.

However, her porters were exhausted, even Juba was complaining of thorns in her feet. It seemed that only she and old Karanga were able to maintain this pace for day after day. She must let them rest.

The next morning she had them away when the grass was still bent under the weight of dew drops, and before she had gone a mile her breeches were soaked to the thighs. During the last few days' march the character of the land had altered. The elevated plateau of rolling grassland and open forest across which they had marched so long now seemed to be dipping southwards, and the single peak which she had seen the night before slowly evolved into a whole series of hills stretching across her horizon from west to east, and she felt her spirits sag.

What chance would she ever have of finding one man's camp, one single hilltop, amongst so many? But she slogged on doggedly, and she and Karanga reached the first foothills before noon well ahead of the column. She checked Zouga's barometer nestling in its velvet-lined wooden case, and found that the altitude was still well over 1,200 feet, though they had dropped two hundred feet in the last two days' march.

Then, followed closely by Karanga, and at a little distance by Juba, she climbed the rocky shoulder of one of the foothills and from its height had a clearer view ahead over the confused and broken ground. She could see that the hills descended sharply into the south. Perhaps they had crossed the highlands and before them lay the descent to one of the known rivers that Tom Harkness had marked upon his map. She tried to remember the names, Shashi and Tati and Macloutsi.

Suddenly she was starting to feel very lonely and uncertain again. The land was so vast, she felt like a tiny insect pinned to an endless plain beneath the high pitiless blue sky. She turned and looked back into the north, using the long, brass-bound telescope to search for any signs of Zouga's party. She was not certain if she was relieved or disappointed to find none.

'Karanga!' she called, and he scrambled to his feet readily and looked up at her on the pinnacle of rock on which she stood. His expression was trusting as that of a pet dog.

'Which way now?' she demanded, and he dropped his eyes and stood on one bird-thin leg, scratching his calf with the other foot as he pondered the question. Then with apologetic gesture he indicated the nearest half dozen promontories along the skyline ahead with a hesitant all-embracing gesture, and Robyn felt her heart sink further. She had to admit at last that she was lost.

She knew then that she would have to do one of two things. Either camp where she was until Zouga came up, or turn back along her own spoor until she met him. Neither alternative was attractive, and she put off the decision until the morrow.

There was water in the river-bed below her, the usual shallow warm green pools, foul with bird and animal droppings.

Suddenly she felt very tired. While the expectations of success had buoyed her up she had not noticed it, but now she felt deflated and the marrow of her bones ached with weariness.

'We will camp here,' she told the Corporal. 'Take two men and find meat.'

They had marched so hard and long since leaving Karanga's village that there had been no time to hunt. By this time the last of the dried buffalo meat smelt like badly cured hides and was full of bacon beetles. She could only eat it in a strong curry and the curry powder was almost finished. They needed fresh meat desperately, but she was too dispirited to lead the hunting party.

The porters had not finished thatching the low lean to roof that would be her home for the night when she heard a fusillade of musket fire close by in the forest, and an hour later the Corporal came into camp. They had found a large herd of the lovely sable antelope, Harris buck, as Zouga insisted on calling them, and had succeeded in bringing down five fat chocolate-coloured cows. The porters, chattering happily, left en masse to help bring in the meat, and Robyn wandered listlessly down the river-bed, accompanied only by Juba, until she found a secluded pool.

'I must smell as good as old Karanga,' she thought, scrubbing herself with handfuls of white sand for she had weeks previously used the last of her soap. She washed out her clothing and spread it to dry on the smooth, water-worn rocks around the pool. Then, still naked, she sat in the sunlight and Juba knelt behind her and combed out her hair so that it could dry.

Juba was obviously pleased to have Robyn to herself again, without old Karanga hovering nearby. Even though Robyn was silent and dejected, Juba loved to play with her hair and delighted in the reddish lights that flared in the sunlight as the comb stroked through it.

She chatted merrily as she worked and gurgled with laughter at her own sallies, so that neither of them heard the footsteps in the sand, and it was only when the shadow fell at Robyn's feet that she realized that they were not alone. With a cry of alarm she rose, snatching up her still wet breeches and holding them to her breast to cover her nudity.

The woman who stood before her was unarmed, nervous as she was and shy. Not a young woman, though her skin was smooth and unlined and she had all her teeth still. She was almost certainly Mashona with the finer, more Egyptian features than the Nguni, and she wore the short kilt that left her upper body bare. Her naked breasts were large, out of proportion to the slim upright body, the nipples were raised and drawn-out as though she had recently been nursing an infant.

'I heard the guns,' she whispered shyly, and Robyn felt a lift of relief when she understood the language. It was Karanga. 'I came when I heard the guns. I came to lead you to Manali.'

Robyn felt the quick rush of tears scald her eyes at the name, and the leap of her heart made her gasp aloud.

Manali, the man who wears a red shirt – her father had always stoutly maintained that the colour red discouraged tsetse fly and other stinging insects, and that good thick flannel staved off the ague of fever.

Robyn jumped to her feet, completely oblivious now of her nudity and rushed to the woman, seizing her arm and shaking it.

'Manali!' she cried, and then in English. 'Where is he? Oh, take me to him this instant.'

I t was more than mere chance, and old Karanga's faltering guidance, that had led her, Robyn decided exultantly as she followed her new guide along one of the narrow winding game paths. It was blood calling to blood. Instinctively, like a migrating swallow, she had flown straight to her father.

She felt like shouting aloud, singing her joy to the forest while the woman went swiftly ahead of Robyn, her narrow, smoothly muscled back and shoulders hardly moved above the gliding roll of her hips, that graceful walk of the African woman trained from childhood to carry a burden upon her head so smoothly that not a drop spills from a brimming pot.

She did not move swiftly enough for Robyn's expectations. Already Robyn could imagine the powerful figure of her father striding towards her, the great flaming bush of his beard, the deep compelling voice as he called her name, and swung her high as he had when she was a child, then the crushing embrace of his arms.

She imagined his joy matching hers, and after the first heady moments of reunion, then the serious hours of discussion, the recital of the long years between, the growing trust and intimacy between them that they had lacked before, so that finally they could march together to a common goal. In the long years ahead, he could hand the torch to her confident that his faith and work would go forward in loving and loyal hands.

What would be his first words when he saw and recognized her? How immense his surprise? She laughed aloud breathlessly, of course, he would be deeply touched and grateful that she had come so far, so determinedly to be with him, and she, Robyn, knew that she would not be able to hold back her own tears of joy, she could imagine her father tenderly wiping them away. The tone of his voice would betray the pent-up love of all the intervening years that had parted them, and which would be so sweet that she could hardly bear it.

Ahead of her, in the fading light of a dying day the Mashona woman led them on to a steep pathway, climbing at a traverse across the western slope of the highest hill. Robyn laughed again when she realized that it was the same hill that old Karanga had pointed out from twenty miles away. He had been right in the end, she must praise him lavishly for that. In her own happiness she wanted to give joy to all the world.

The path came out on a level shelf just below the crest, with a shallow cliff at the back of it and the slope of the hill falling away steeply towards the sunset, and a breathtaking view across the forest and savannah. The entire land turned pink and gold in the low sun and the stupendous flat-topped thunderheads of cloud rose along the dark blue horizon. The setting was right for this magical moment, but Robyn glanced at it only once and then her full attention fastened on what lay ahead of her.

In the face of the cliff was the mouth of a low cave, the slanting rays of the sun struck fully into it showing that it was not very deep, but had been occupied for a long time. The roof and walls were blackened with the soot of the cooking fire, the floor had been swept bare except for the fire at the entrance, with its circle of blackened hearth stones and a small clay pot standing upon them.

The clearing in front of the cave was bare also, trodden by feet over many years, and there was the offal of human occupation scattered about it, the bare bones of small animals, scraps of fur, chips of wood and shards of broken pottery. There was the odour of rotting food fragments, unwashed leather garments, wood smoke and human excrement that confirmed the other evidence that men had lived here for a long time.

There was a single human figure crouched over the smoky little fire, an old crone, bowed by age, a mere bundle of filthy fur blankets, moth-eaten and ragged, looking more like an ancient ape than a human being. It did not stir, and Robyn barely glanced at it for something else held her attention.

In the back of the cave, lit by the last fleeting sunlight, stood a bed. It was made of rudely cut poles, and tied together with bark rope, yet it stood on four legs in European style, not the African sleeping-mat, and it was piled with a stained fur kaross that might have contained a human shape.

On a ledge directly above the bed stood a brass telescope, a teak box similar to the one that held Zouga's sextant and chronometer, but scarred and battered with age, and a small cheap tin chest. The chest was much battered also, most of the original paint chipped away so the bare metal showed.

Robyn remembered that box so vividly, open in Uncle William's study at King's Lynn, the papers from it overflowing on to the desk top and her father bowed over them, steel-rimmed spectacles on the end of his beaked nose, tugging at his thick red beard as he worked.

Robyn gave a little choking cry, and ran forward passing the old crone sitting at the fire, crossing the cave, and flinging herself on her knees beside the crude bedstead.

'Pater!' Her voice husked over with emotion rasped her own throat. 'Pater! It's me – Robyn.'

There was no movement beneath the fur blanket and she put out a hand, then stopped before it touched.

'He is dead,' she thought miserably. 'I am too late!'

She forced her hand to move again, and touched the malodorous pile of old furs. They collapsed under her touch, and it took her seconds to realize that she had been mistaken. The bed was empty, the discarded blanket had fallen in the shape of a man, but the bed was empty.

Bewildered, Robyn rose to her feet and turned back towards the entrance of the cave. The Karanga woman stood by the fire, watching her expressionlessly, while little Juba hung back fearfully at the far side of the clearing.

'Where is he?' Robyn spread her hands to emphasize the question. 'Where is Manali?'

The Karanga woman dropped her eyes. For a moment Robyn did not understand, and then she too looked down at the grotesque figure that crouched by the fire at her feet.

She felt a cold steel band lock about her chest and squeeze her heart so that it took an effort of will to force her feet to move back across the swept floor of the cave.

The Mashona woman was watching Robyn expressionlessly. She had clearly not understood the English question, but she waited with the endless patience of Africa. Robyn was about to appeal to her again, when the skeletal figure across the smoky little fire started to rock from side to side agitatedly, and a querulous slurred old man's voice began to chant some strange litany, like a magical incantation.

It took Robyn some moments to realize that the accents were faintly Scots and the words, though blurred and jumbled, were a parody of the 23rd psalm.

'Yea! Though I walk through the valley of the shadow of death, I shall fear no evil.'

As suddenly as it had begun the chanting broke off, and the rocking ceased. The frail figure froze into stillness and silence again. Across the fire, the Mashona woman stooped and gently as a mother with a child, she drew back the kaross from the head and shoulder of the figure at her feet.

Fuller Ballantyne had shrivelled, his face lined and roughened like the bark of an old oak. It seemed as though the smoke of the fire had etched his skin, collecting in the creases, crusting it with soot.

His hair and beard had fallen out in clumps, as though from some disgusting disease, and what was left of it was pure white but stringed and darkened with dirt to a tobacco yellow at the corners of the mouth and at the nostrils.

Only his eyes still seemed to live, they rolled in their sockets, and it needed only one look at them for Robyn to realize that her father was mad. This was not Fuller Ballantyne, this was not the great explorer, the powerful evangelist and enemy of slavery. He had gone long ago, leaving a filthy shrivelled lunatic in his place.

'Pater,' she stared at him in disbelief, feeling the world spin and lurch beneath her. 'Pater,' she repeated, and across the fire the crouched figure gibbered with abrupt falsetto girlish laughter, and then began to rave incoherently, snatches of English giving way to a half-dozen dialects of Africa, the cries becoming more agitated, his thin pale arms thrown wildly in the air.

'I have sinned against thee, my God,' he screamed and clawed at his own beard, a tuft of the thin pale hair coming away in the hooked fingers. 'I am not worthy to be thy servant.' He tore at himself again, this time leaving a thin livid scratch down the pouched and wrinkled cheek, though it seemed that the wasted body had no blood left to shed.

The Mashona woman leaned over him and caught the bony wrist, restraining him. The action was so familiar that she must have performed it often. Then gently she stooped and lifted him. The body seemed to weigh no more than that of a child for she carried him without visible effort to the crude pole bed. One of his legs was bound up in a rudimentary splint and stuck straight out ahead of him.

Robyn stayed on beside the fire, hanging her head. She found that she was still shivering, until the woman came back to her, touching her arm.

'He is very sick.'

Only then could Robyn force back her revulsion and her horror. She stood, hesitated only a moment longer, and then went to her father. With Juba and the Mashona woman helping her, she began her examination, taking refuge behind her professional rituals and procedures while she regained control of her emotions. He was thinner than she had ever seen a living human body, thinner than the starved brats of gin-soaked slum sluts.

'There has been little food,' said the woman, 'and what there is, he will not eat. I have had to feed him like a small baby.' Robyn did not then understand what she meant, but she went on grimly with her examination.

The starved body was verminous, the bunches of little white nits hanging like grapes in the thin white pubic hair, and his whole body was crusted with filth and traces of his own incontinence.

Feeling under the staring rib cage her fingertips encountered the hard distended shape of liver and spleen, and Fuller Ballantyne screamed when she did so. The swelling and extreme tenderness were certain indications of massive malarial infection of long duration, and evidence of terrible neglect.

'Where is the medicine, the *umuthi*, of Manali?'

'It was long ago finished, together with the powder and shot for the gun. Everything was finished long ago,' the woman shook her head, 'long, long ago, and when it was finished, the people no longer came with gifts to feed us.' It was suicidal to remain in a malarial area without supplies of quinine. Fuller Ballantyne of all people knew that. The acknowledged world expert on malarial fever and its treatment – how could he have neglected his own often-repeated advice. She found the reasons almost immediately, as she opened his mouth, forcing open his lower jaw despite his feeble protests.

Most of his teeth had been rotted out by the disease, and his throat and palate were covered by the characteristic lesions. She released his jaw, allowing him to close the ruined mouth, and gently she touched the bridge of his nose, feeling the soggy collapsing bone and gristle. There could be no doubt at all, the disease was far advanced, had long ago begun its final assault upon the once magnificent brain. It was syphilis, in the terminal stages, the general paralysis of the insane. The disease of the lonely man that led inevitably to this lonely madman's death.

As Robyn worked, so her horror and revulsion gave way swiftly to the compassion of the healer, to the sympathy of one who had lived with human weakness and folly and had come far along the road of understanding. She knew now why her father had not turned back when his supplies of vital medicines ran perilously low, the half-destroyed brain had not recognized the dangers which he had previously described so clearly.

She found herself praying for him as she worked, praying silently but with the words coming more readily than they usually did.

'Judge him as he was, Oh Lord, judge him by his service in your name – not by his small sins, but by his great achievements. Look not upon this ruined pathetic thing, but on the strong and vital man who carried your work forward without flinching.'

As she prayed, she lifted the heavy kaross off his legs, and the smell of corruption made her blink and the frail figure began immediately to struggle with renewed strength, that needed both Juba and the Mashona woman to control.

Robyn stared at the legs, and realized the other reason why her father had never left this land. He had been physically unable to do so. The splints that held the leg had been whittled out of native timber. The leg had been fractured, probably at more than one place below the hip. Perhaps the hip joint itself had gone, that vulnerable neck of the femur. But what was certain was that the breaks had not mended cleanly. Perhaps the bindings of the splints had been too tight, for the deep suppurating ulcerations went down to the very bone, and the smell was a solid jarring thing.

Quickly she covered his lower body, there was nothing she could do until she had her medical chest and instruments, and now she was merely inflicting unnecessary pain and humiliation. Her father was still struggling and bleating like a petulant child, rolling his head from side to side, the toothless mouth darkly agape.

The Mashona woman leaned over him, and took one of her own dark tight breasts in her hand, squeezing out the nipple between her fingers, and then she paused and looked up shyly, imploringly at Robyn.

Only then did Robyn understand, and respecting the privacy of woman and the poor maimed thing that had been her father, she dropped her eyes and turned away towards the entrance of the cave.

'I must fetch my *umuthi*. I will return here later tonight.'

Behind her, the childlike bleats gave way abruptly to small snuffling sounds of comfort.

Robyn felt no shock or outrage as she went down the steep pathway in the moonlight. Instead she felt immense pity for Fuller Ballantyne who had made the full circle back to infancy. She felt also a deep gratitude to the woman, and a sense of wonder at her loyalty and dedication. How long had she stayed on with Fuller Ballantyne after all reason for staying was gone?

She remembered her own mother, and her devotion to the same man, she remembered Sarah and her child still waiting patiently beside a far-off river. And then there was Robyn herself, come so far and so determinedly. Fuller Ballantyne always had the power to attract as powerfully as he could repel.

Holding Juba's hand for her own comfort as well as that of the child, Robyn hurried along the moonlit path on the bank of the river, and with relief saw the glow of the camp fires in the forest ahead of her. On the return journey she would have bearers to carry her medical chest, and armed Hottentot musketeers as escort.

Her relief was short-lived, for as she answered the challenge of the Hottentot sentry and entered the circle of firelight, a familiar figure rose from beside the camp fire and came striding to meet her, tall and powerful, golden-bearded and handsome as a god from Greek mythology, and every bit as wrathful.

'Zouga!' she gasped. 'I didn't expect you.'

'No,' he agreed icily. 'I'm sure that you did not.'

'Why?' she thought desperately. 'Why must he come

now? Why not a day later, when I have had time to clean and treat my father? Oh God, why now? Zouga will never understand – Never! Never! Never!'

Robyn and her escort could not hope to keep pace with Zouga. They fell swiftly behind him as he climbed the pathway in the night, months of hard hunting had toned him to the peak of physical condition and he ran at the hill.

She had not been able to warn him. What words were there to describe the creature in the cave on the hilltop. She had told him simply:

'I have found Pater.'

It had deflected his anger instantly. The bitter accusations shrivelled on his tongue, and the realization dawned in his eyes.

They had found Fuller Ballantyne. They had accomplished one of the three major objectives of the expedition. She knew that Zouga was already seeing it in print, almost composing the paragraph that would describe the moment, imagining the newspaper urchins shouting the headlines in the streets of London.

For the first time in her life she came close to hating her brother, and her voice was crisp as hoar-frost as she told him, 'And don't you forget it was me. I was the one who made the march and broke trail, and I was the one who found him.'

She saw the shift in his green eyes in the firelight.

'Of course, Sissy.' He smiled at her thinly, an obvious effort. 'Who could ever forget that? Where is he?'

'First I must assemble what I need.'

He had stayed with her until they reached the foot of the hill, and then had been unable to restrain himself. He had gone at the slope at a pace that none of them had been able to equal. Robyn came out in the little clearing in front of the cave. Her heart was racing and her breathing ragged from the climb so that she had to pause and fight for breath, holding one hand to her breast.

The fire in front of the cave had been built up to a fair blaze, but it left the depths of the cave in discreet shadow. Zouga stood in front of the fire. His back to the cave.

As Robyn regained her breath, she went forward. She saw that Zouga's face was deathly pale, in the firelight his sun-bronzing had faded to a muddy tone. He stood erect, as though on the parade ground, and he stared directly ahead of him.

'Have you seen Pater?' Robyn asked. His distress and utter confusion gave her a sneaking and spiteful pleasure.

'There is a native woman with him,' Zouga whispered, 'in his bed.'

'Yes,' Robyn nodded. 'He is very sick. She is caring for him.'

'Why did you not warn me?'

'That he is sick?' she asked.

'That he had gone native.'

'He's dying, Zouga.'

'What are we going to tell the world?'

'The truth,' she suggested quietly. 'That he is sick and dying.'

'You must never mention the woman.' Zouga's voice, for the first time that she could remember, was uncertain, he seemed to be groping for words. 'We must protect the family.'

'Then what must we tell about his disease, the disease that is killing him?'

Zouga's eyes flickered to her face. 'Malaria?'

'The pox, Zouga. The French sickness, the Italian plague – or, if you prefer it, syphilis, Zouga. He is dying of syphilis.'

Zouga flinched, and then he whispered, 'It's not possible.'

'Why not, Zouga?' she asked. 'He was a man, a great man – but a man nevertheless.'

She stepped past him. 'And now I have work to do.'

An hour later when she looked for him again, Zouga had gone back down the hill to the camp beside the river pools. She remained to work over her father for the rest of that night and most of the following day.

By the time she had bathed and cleaned him, shaved off the infested body hair and trimmed the stringy beard and locks of yellowed hair, treated the ulcerations of his leg, she was exhausted both physically and emotionally. She had seen approaching death too many times not to recognize it now. She knew that all she could hope for was to give comfort and to smooth the lonely road that her father must travel.

When she had done all that was possible, she covered him with a clean blanket and then tenderly caressed the short soft hair which she had so lovingly trimmed. Fuller opened his eyes. They were a pale empty shade of blue, like an African summer sky. The last sunlight of the day was washing the cave, and as Robyn leaned over him, it sparkled in her hair in chips of ruby light.

She saw something move in the empty eyes, a shadow of the man who had once been there, and Fuller's lips parted. Twice he tried to speak and then he said one word, so husky and light that she missed it. Robyn leaned closer to him.

'What is it?' she asked.

'Helen!' This time clearer.

Robyn felt the tears choke up her throat at the sound of her mother's name.

'Helen.' Fuller said it for the last time, and then the flicker of comprehension in his eyes was gone.

She stayed on beside him, but there was nothing more. That name had been the last link with reality and now the link was broken.

As the last light of the day faded, Robyn lifted her eyes from her father's face and for the first time realized that the tin chest was missing from the ledge at the back of the cave.

Using the lid of his own writing-case as a desk, screened from the camp by the thin wall of thatch, Zouga worked swiftly through the contents of the chest.

His horror at the discovery of his father had long ago been submerged by the fascination of the treasures which the chest contained. The disgust, the shame, would return again when he had time to think about it, he knew that. He knew also that there would be hard decisions to make then, and that he would have to use all his force of personality and of brotherly superiority to control Robyn, and make her agree to a common version of the discovery of Fuller Ballantyne and a tactful description of the circumstances to which he had been reduced.

The tin chest contained four leather and canvas-bound journals, each of five hundred pages, and the pages were covered on both sides either with writing or with hand-drawn maps. There was also a bundle of loose sheets, two or three hundred of them tied together with plaited bark string, and a cheap wooden pen case with a partition for spare nibs, and cut-outs for two ink bottles. One bottle was dry, and the pen nibs had obviously been sharpened many times, for they were almost worn away. Zouga sniffed the ink in the remaining bottle. It seemed to be an evil-smelling mixture, of fat and soot and vegetable dyes that Fuller had concocted when his supplies of the manufactured item were exhausted.

The last journal and most of the loose pages were written with this mixture, and they had faded and smeared, making the handwriting that much more difficult to decipher, for by this stage Fuller Ballantyne's hand had deteriorated almost as much as his mind. Whereas the first two journals were written in the small, precise and familiar script, this slowly turned into a loose sloping scrawl as uncontrolled as some of the ideas expressed by it. The history of his father's madness was plotted therein with sickening fascination.

The pages of the leatherbound journals were not numbered, and there were many gaps between the dates of one entry and the next, which made Zouga's work easier. He read swiftly, an art he had developed when acting as regimental intelligence officer with huge amounts of reading, reports, orders and departmental manuals to get through each day.

The first books of writing were ground that had been travelled before, meticulous observations of celestial position, of climate and altitude, backed up by shrewdly observed descriptions of terrain and population. Sandwiched between these were accusations and complaints about authority, whether it was the directors of the London Missionary Society, or 'The Imperial Factor' as Fuller Ballantyne referred to the Foreign Secretary and his department in Whitehall.

There were detailed explanations of his reasons for leaving Tete and travelling south with a minimally equipped expedition, and then, quite suddenly, two pages devoted to an account of a sexual liaison with an ex-slave girl, an Angoni girl whom Fuller had christened 'Sarah' and who he suspected was about to bear his child. His reasons for abandoning her at Tete were direct and without pretence. 'I know that a woman, even a hardy native, carrying a child would delay me. As I am on God's work, I can brook no such check.'

Although what Zouga had seen on the hilltop should have conditioned him for this sort of revelation, still he could not bring himself to terms with it. Using his hunting-knife, whetted to a razor edge, he slit the offending pages from the journal, and as he crumpled them and threw them

435

in the camp fire he muttered, 'The old devil had no right to write this filth.'

Twice more he found sexual references which he removed from the journal, and by then the handwriting was showing the first deterioration, and passages of great lucidity were followed by wild ravings and the dreams and imaginings of a diseased mind.

More often Fuller referred to himself as the instrument of God's wrath, his blazing sword against the heathen and the ungodly. The weirdest and patently lunatic passages Zouga cut from the journal and burned. He knew he must work swiftly, before Robyn came down from the hilltop. He knew that what he was doing was right, for his father's memory and place in posterity and also for those who would have to live on after him, Robyn and Zouga himself, and their children and children's children.

It was a chilling experience to see his father's great love and compassion for the African people, and the very land itself changing and becoming a bitter unreasoning hatred. Against the Matabele people, whom he referred to as the Ndebele – or the Amandebele – he railed: 'These leonine peoples who acknowledge no God at all, whose diet consists of the devil's brew and half-cooked meat, both in vast quantities, and whose greatest delight is spearing to death defenceless women and children, are ruled over by the most merciless despot since Caligula, the most grossly blood-besotted monster since Attila himself.'

Of the other tribes he was at least as scornful. 'The Rozwis are a sly and secret people, the timid and treacherous descendants of the slave-trading and gold-mad kings they called the Mambos. Their dynasties destroyed by the marauding Ndebele and their monstrous Nguni brothers the Shangaans of Gungundha and the blood-smattered Angoni.'

The Karangas were 'cowards and devil-worshippers, lurking in their caves and hilltop fortresses, committing unspeakable sacrilege and offending in the face of the

Almighty by their blasphemous ceremonies in the ruined cities where once their Monomatapa held sway'.

The reference to Monomatapa and ruined cities checked Zouga's eyes in the middle of the page. Then he read on eagerly, hoping for elaboration of the mention of ruined cities, but Fuller's mind had flown on to other ideas, the theme of suffering and sacrifice which has always been the spine of Christian belief.

'I thank God, my Almighty Father, that he has chosen me as his sword, and that as the mark of his love and condescension he has made his mark upon me. This dawn when I awoke there were the stigmata in my feet and hands, the wound in my side, and the bleeding scratches from the crown of thorns upon my forehead. I have felt the same sweet pain as Christ himself.'

The disease had reached that part of his brain that affected his eyes, and sense of feeling. His faith had become religious mania. Zouga cut out this and the following pages and consigned them to the flames of the camp fire.

Ranting madness was followed by cool sanity, as though the disease had tides which ebbed and flowed within his brain. The next entry in the journal was dated five days after that claiming the stigmata. It began with a celestial observation that placed him not far from where Zouga sat reading the words, always making allowances for the inaccuracy of a chronometer that had not been checked for almost two years. There was no further reference to the stigmata. They had healed as miraculously as they had bloomed. Instead there was a brisk and businesslike entry in the old neat script.

'The Karanga people practise a form of ancestor worship, which calls for sacrifice. It is extremely difficult to make any one of them discuss either this ceremony or even the basic precepts of this abominable religion. However, my command of the Karanga language is now sufficient to have earned the respect and trust of those members of the tribe with whom I have succeeded in making contact. The spiritual centre of their religion is in a place which they refer to as "the burial place of the kings", or in their language "Zimbabwe" or "Simbabwee" where the idols which represent their ancestors stand.

'It would seem that this place is situated to the south and east of my present position.

'The head of this foul belief is a priestess, referred to as the "Umlimo" who once dwelt in the "burial place of the kings" but fled from there at the coming of the Angoni marauders. She lives in another sacred place, and commands such sway that even the godless Ndebele, that sanguine tyrant Mzilikazi, send gifts for her oracle.

'So deeply ingrained is the power of this evil belief into the minds of all these people, that they are strongly resistant to the word of Christ which I bear.

'It has come to me in a revelation which can only be the voice of God Almighty himself that he has chosen me to march upon this citadel of evil, this "Simbabwee", and to throw down the images of the ungodly – even as Moses threw down and destroyed the golden calf on his descent from the mountain.

'God Almighty has further revealed to me that I have been chosen to seek out the High Priestess of evil in her secret place and to destroy her, and by so doing break the hold that she has over the minds of these people, that they might become receptive to the Holy Word of Christ which I bring them.'

Zouga raced on through the pages, it seemed as though they were being written by two different men, the rational being with the neat script and the raving religious maniac with the wild looping hand. In some passages the change came from one line to the next, in others the one character was maintained for page after page. Zouga could not afford to miss a word of it.

It was well after noon, and his eyes felt grainy and tender

from continually scanning the cramped sheets, blurred and faded from the improvised ink which Fuller was now using.

'November 3rd. Position. 20° 05 S. 30° 50 E. Temperature 103° in the shade. Heat unbearable. Rain threatens each day and never comes. Have reached the lair of the Umlimo.'

The single laconic entry electrified Zouga. He almost missed it, for it was squeezed into the bottom of a page. He turned that page and on the next the madman had taken over again, writing in vaunting hyperbole and thunderous religious ecstasy.

'I praise God, my Maker. The one true and Almighty Saviour for whom all things are possible. Thy will be done!

'The Umlimo knew me as the instrument of God's wrath when I confronted her in that reeking charnel house, for she spoke in the voices of Belal and Beelzebub, the hideous voices of Azazel and Beliar, all Satan's myriad alter-egos.

'But I stood before her strong and proud in God's word, and when she saw she could not move me, she fell silent.

'So I slew her, and cut off her head and carried it out into the light. And God spoke to me in the night and said in his small still voice "Go on, my faithful and well-beloved servant. You cannot rest until the graven images of the godless are cast down."

'So I rose, and God's hand held me up and carried me onwards.'

How much of this was fact, and how much was the ranting of madness, fantasy of a diseased brain, Zouga could not know, but he read on furiously.

'And the Almighty guided me until I came at last alone to the foul city where the devil-worshippers commit their sacrilege. My bearers would not follow me, terrified of the devils. Even old Joseph who was always at my side could not force his legs to carry him through the gateway in the high stone wall. I left him cringing in the forest, and went in alone to walk between the high towers of stone.

'As God had revealed to me, I found the graven images of the heathen all decked with flowers and gold, the blood of the sacrifice not yet dried, and I broke them and cast them down and no man could oppose me for I was the sword of Zion, the finger of God's own hand.'

The entry broke off abruptly, as though the writer had been overwhelmed by the strength of his own religious fervour, and Zouga flipped through the next one hundred pages of the journal searching for further reference to the city and its gold-decked images, but there was none.

Like the miraculous blooming of the stigmata upon Fuller's hands and feet and body and brow, perhaps this was also the imaginings of a lunatic.

Zouga returned to the original entry, describing Fuller's meeting with the Umlimo, the sorceress whom he had slain. He wrote the latitude and longitude into his own journal, copied the rough sketch map and made cryptic notes of the text, pondering it for clues that might lead or guide him. Then, quite deliberately, he cut out the pages from Fuller's own journal and held them one at a time over the fire, letting them crinkle and brown, then catch and flare before he dropped them and watched them curl and blacken. He stirred the ashes to dust with a stick before he was satisfied.

The last of the four journals was only partially filled, and contained a detailed description of a caravan route running from 'the blood-soaked lands where Mzilikazi's evil impis hold sway' eastwards five hundred miles or more 'to where the reeking ships of the traders surely wait to welcome the poor souls who survive the hazards of this infamous road'.

'I have followed the road as far as the eastern rampart of mountains, and always the evidence of the passing of the caravans is there for all the world to see. That grisly evidence which I have come to know so well, the bleaching bones and the circling vultures. Is there not a corner of this savage continent which is free from the ravages of the traders?'

These revelations would interest Robyn more than they did him. Zouga glanced through them swiftly and then marked them for her attention. There was a great deal on slavery and the traders, a hundred pages or more – and then the penultimate entry.

'We have today come up with a caravan of slaves, winding through the hilly country towards the east. I have counted the miserable victims from afar, using the telescope and there are almost a hundred of them, mostly half-grown children and young women. They are yoked together in pairs with forked tree trunks about their necks in the usual manner.

'The slave-masters are black men, I have been unable to descry either Arabs or men of European extraction amongst them. Although they wear no tribal insignia, no plumes nor regalia, I have no doubt they are Amandebele, for their physique is distinctive, and they come from the direction of that tyrant Mzilikazi's kingdom. They are furthermore armed with the broad-bladed stabbing spear and long ox-hide shields of that people, while two or three of them carry trade muskets.

'At this moment they are encamped not more than a league from where I lie, and in the dawn they will continue their fateful journey towards the east where the Arab and Portuguese slave-masters no doubt wait to purchase the miserable human cattle and load them like cargo for the cruel voyage half across the world.

'God has spoken to me, clearly I have heard his voice as he enjoined me to go down, and, like his sword, cut down the ungodly, free the slaves and minister to the meek and the innocent.

'Joseph is with me, that true and trusted companion of the years, and he will be well able to serve my second gun. His marksmanship is not of the best, but he has courage and God will be with us.'

The next entry was the last. Zouga had come to the end of the four journals.

'God's ways are wonderful and mysterious, passing all understanding. He lifteth up and he casteth down. With Joseph beside me I went down, as God had commanded, to the camp of the slave-masters. We fell upon them, even as the Israelites fell upon the Philistines. At first it seemed that we must prevail for the ungodly fled before us. Then God in all his knowing wisdom deserted us. One of the ungodly leapt upon Joseph while he was reloading, and though I put a ball through his chest he impaled poor Joseph from breast bone to spine with that terrible spear, before himself falling dead.

'Alone I carried on the fight, God's fight, and the slave-masters scattered into the forest before my wrath. Then one of them turned and at extreme range fired his musket in my direction. The ball struck me in the hip.

'I managed, I do not know how, to drag myself away before the slave-masters returned to slay me. They did not attempt to follow me, and I have regained the shelter which I left to make the attempt. However, I am sorely smitten and reduced to dire straits. I have managed to remove the musket ball from my own hip, but I fear the bone is cracked through and I am crippled.

'In addition I have lost both my firearms, Joseph's musket lies with him where he fell, and I was so badly hurt as to be unable to carry my weapon off the field. I sent the woman back to find them, but they have been carried away by the slave-traders.

'My remaining porters, seeing the state to which I have been reduced and knowing that I could not prevent it, have all deserted, but not before they had looted the camp and carried away almost everything of value, not excluding my medicine chest. Only the woman remains. I was angry at first when she attached herself to my party, but now I see God's hand in this – for although she is a heathen, yet she

442

is loyal and true beyond all others, now that Joseph is dead.

'What is a man in this cruel land without a musket or quinine? Is there a lesson in this for me and posterity, a lesson that God has chosen me to teach? Can a white man live here? Will he not always be the alien, and will Africa tolerate him once he has lost his weapons and expended his medicines?'

Then a single poignant cry of agony.

'Oh God, has this all been in vain? I came to bring your Word and nobody has listened to my voice. I came to change the ways of the wicked and nothing is changed. I came to open a way for Christianity, and no Christian has followed me. Please, my God, give me a sign that I have not followed the wrong road to a false destination.'

Zouga leaned back and rubbed his eyes with the heels of both hands. He found himself deeply moved, his eyes stinging not only from fatigue.

Fuller Ballantyne was an easy man to hate, but a hard man to despise.

R obyn chose the place with care. The secluded pools on the river, away from the main camp, where nobody could overlook or overhear them. She chose the time, in the heat of midday – when most of the Hottentots and all the porters would be asleep in the shade. She had given Fuller five drops of the precious laudanum to quiet him, and left him with the Mashona girl and Juba to care for him while she went down the hill to Zouga.

They had barely exchanged a dozen words during the ten days since he had caught up with her. In all that time he had not returned to the cave on the hilltop, and she had seen him only once when she had gone down to the river camp for supplies.

When she had sent Juba down with a tersely worded note, demanding the return of Fuller's tin chest of papers, he had sent a porter with it immediately. In fact, with such alacrity that Robyn was immediately suspicious.

This distrust was a symptom of their rapidly deteriorating relationship. She knew that she and Zouga must talk, must discuss the future, before the opportunity to talk was past.

He was waiting for her beside the green river pools, as she had asked him to be, sitting in the mottled shade beneath a wild fig tree, quietly smoking a hand-rolled cheroot of native tobacco. He stood up courteously as soon as he saw her, but his expression was reserved and his eyes guarded.

'I do not have much time, Zouga dear.' Robyn tried to lessen the tension between them by using the small endearment, and Zouga nodded gravely. 'I must get back to Pater.' She hesitated. 'I did not want to ask you to come up the hill since you find it distasteful.' She saw the green sparks in his eyes kindle immediately, and went on quickly. 'We must decide what we should do now. Obviously we cannot stay on here indefinitely.'

'What do you suggest?'

'Pater is much stronger. I have subdued the malaria with quinine and the other disease,' she worded it tactfully, 'has responded to the mercury. It is only the leg that truly worries me now.'

'You told me he was dying.' Zouga reminded her levelly, and she could not help herself, despite her good intentions she flared at him.

'Well, I am sorry to disappoint you, then.'

Zouga's face stiffened into a handsome, bronzed mask. She could see the effort it took him to control his own temper, and his voice was thick with it as he answered.

'That's not worthy of you.'

'I'm sorry,' she agreed, and drew a deep breath. 'Zouga, he has rallied strongly. Food and medicine, care and his

444

own natural strength have made an immense difference. I am even convinced that if we could get him to civilization to a skilled surgeon – we could cure the ulceration of his leg, and possibly even induce the bone to mend.'

Zouga was silent for a long time, and though his face was expressionless, she could see the play of emotion in his eyes.

He spoke at last. 'Father is mad.' She did not answer. 'Can you cure his mind?'

'No.' She shook her head. 'That will get worse, but with care and skilled attention in a good hospital, we can improve his body and he could live for many years still.'

'To what purpose?' Zouga insisted.

'He would be comfortable and perhaps happy.'

'And all the world would know that he was a syphilitic madman,' Zouga went on quietly for her. 'Would it not be kinder to let the legend stand untarnished? No, more than that, to add to it by our own account, rather than drag back this poor diseased and demented thing for all his enemies, his numerous enemies, to mock?'

'Is that why you tampered with his journals?' Robyn's voice was shrill, even in her own ears.

'That's a dangerous accusation.' He was losing his control also. 'Can you prove it?'

'I don't have to prove it; we both know it's so.'

'You cannot move him.' Zouga changed direction. 'He is crippled.'

'He could be carried on a litter. We have more than enough porters.'

'Which way would you take him?' Zouga demanded. 'He would never survive the route over which we have travelled – and the route southwards is uncharted.'

'Pater himself has charted the slave road in his journal. We will follow that. It will lead us directly to the coast.'

'With the major objectives of the expedition unfulfilled?' Zouga asked quickly.

'The major objectives were to find Fuller Ballantyne, and

report on the slave trade, both of which we can accomplish if we march down the slave road to the sea.' Robyn broke off, and then made a show of dawning comprehension. 'Oh dear, how silly of me, you mean the gold and the ivory. Those were the major objectives all along, were they not, my dear brother?'

'We have a duty to our sponsors.'

'And none to that poor sick old man up there?' Robyn flung out one hand dramatically, and then spoiled the effect by stamping her foot. Angry with herself as well as with him, she yelled at him. 'I am taking Pater down to the coast, and quickly as I can.'

'I say you are not.'

'And I say the hell with you, Morris Zouga Ballantyne!' The oath gave her a dark pleasure, and she turned and strode away from him, long-legged in her tight-fitting breeches.

Two days later Robyn was ready to march. All exchanges between Zouga and herself since their final meeting at the river had been in the form of written notes, and Robyn realized that her brother would be keeping copies of all this correspondence to justify his actions later.

She had briefly repudiated his written command not to attempt a march with the sick man. Zouga had listed half a dozen reasons, each neatly enumerated, why she should remain. Once he had her written defiance, his next note sent up the hill in Juba's sweating little hand was magnanimous, written for future readers other than herself, Robyn decided sourly.

'If you insist on this folly,' he began and he went on to offer her the protection of the entire force of Hottentot musketeers – with the exception of Sergeant Cheroot who had expressed a desire to remain with Zouga. Under the Corporal they would form an escort capable, as Zouga worded it 'of bringing you and your charge safely to the coast, and protecting you from any hazards upon the way'.

He insisted that she take most of the remaining porters. He would keep five porters to carry his essential stores, together with his four gunbearers, Matthew, Mark, Luke and John.

He also ordered her to take the Sharps rifle and all the remaining stores, 'leaving me only sufficient powder and shot, and the bare minimum of medicines to enable me to complete the further objectives of this expedition which I deem to be of prime importance'.

His final note reiterated all his reasons for keeping Fuller Ballantyne on the hilltop, and asked her once more to reconsider her decision. Robyn saved him the trouble of making fair copies by simply turning over the note and scribbling on it. 'My mind is made up. I will march at first light tomorrow for the coast.' Then she dated and signed it.

The next morning, before sunup, Zouga sent a team of porters up the hill, carrying a litter of mopani poles. The poles had been peeled of their rough bark and bound together with rawhide strips from the green hide of a freshly killed roan antelope, and the body of the litter was made from the same interwoven hide strips. Fuller Ballantyne had to be strapped into the litter to prevent him throwing himself out of it.

When Robyn brought them down again, walking beside the litter to try and calm the crazy old man on it, the Hottentot escort and the porters were ready to join her on the march. Zouga was waiting also, standing a little aside as though he had already dissociated himself, but Robyn went to him directly.

'At least we know each other now,' she said huskily. 'We may not be able to rub along together, Zouga. I doubt we ever could have, or ever will be able to, but that does not mean I do not respect you, and love you even more than I respect you.'

Zouga flushed and looked away. As she should have known, such a declaration could only embarrass him.

'I have made sure that you have a hundred pounds of gunpowder, that is more than you could ever need,' he said.

'Do you not wish to say farewell to Pater?'

Zouga nodded stiffly, and followed her to the litter, avoiding looking at the Mashona woman who stood beside it and spoke formally to Fuller Ballantyne.

'Goodbye, sir. I wish you a swift safe journey and a speedy return to good health.'

The wizened toothless face, revolved towards him upon its scrawny neck. The shaven head had a pale porcelain gloss in the grey dawn light, and the eyes were bird-bright, glittering with madness.

'God is my shepherd, I shall fear no evil,' Fuller cawed, mouthing the words so they were barely understandable.

'Quite right, sir,' Zouga nodded seriously. 'No doubt at all about that.' He touched his cap in a military salute and stepped back. He nodded to the porters and they lifted the litter and moved away towards the pale orange and yellow sunrise.

Brother and sister stood side by side for the last time watching the column of escort and porters file past, and when the last of them had gone and only little Juba remained beside her, Robyn reached up impulsively and threw her arms around Zouga's neck, embraced him almost fiercely.

'I try to understand you, won't you do the same for me?'

For a moment she thought he might unbend, she felt the hard erect body sway and soften, and then Zouga straightened again.

'This is not goodbye,' he said. 'Once I have done what is necessary, I shall follow you. We'll meet again.'

Robyn dropped her arms to her sides, and stood back.

'Until then,' she agreed wistfully, sad that he had not been able to make even a show of affection.

'Until then,' she repeated, and turned away. Juba followed her away into the forest, after the departing column.

Zouga waited until the singing of the porters dwindled, and the only sound was the sweet wild bird chorus that greets each dawn in Africa, and the distant melancholy whooping of a hyena slinking away to its earth.

There were many emotions warring in him. Guilt that he had let a woman, however well supported, attempt the journey to the coast; worry that once she reached it, her accounts would be the first to reach London; doubt as to the authenticity of the clues which Fuller Ballantyne had left for him to follow, but overlying it all a sense of relief and excitement that he was at last answerable only to himself, free to range as fast and far as hard legs and harder determination would take him.

He shook himself, a physical purging of guilt and doubt, leaving only the excitement and soaring sense of anticipation, and he turned to where Sergeant Cheroot waited at the perimeter of the forlorn and deserted camp.

'When you smile, your face makes the children cry,' Zouga told him, 'but when you frown . . . What troubles you now, oh mighty hunter of elephants?'

The little Hottentot lugubriously indicated the bulky tin box that contained Zouga's dress uniform and hat.

'Say not another word, Sergeant,' Zouga warned him.

'But the porters complain, they have carried it so far.'

'And they will carry it to the gates of hell itself, if I say so. Safari!' Zouga raised his voice, elated with the sense of excitement still strongly upon him. 'We march!'

449

Zouga was prepared for wide discrepancies in the positions that his father had fixed by celestial observation, and his own. A few seconds of error in the chronometers would put them many miles out.

So he treated with suspicion the terrain features which he saw ahead and which seemed to match with uncanny accuracy the sketch maps he had copied from Fuller's journals.

Yet as each day's march that he made opened up country that fitted his father's descriptions, he became more confident, more certain that the Umlimo and the ruined city were real and that they lay not many days' march ahead.

It was beautiful country they passed through, though the air was more sultry as they descended the sloping plateau towards the south and west. The long dry season, now drawing to its close, had seared the grasslands to the colour of fields of ripening wheat, and turned the foliage of the forests to a hundred shades of plum reds and soft apricots. Many of the trees were bare of all leaves, lifting arthritically contorted limbs to the sky as if beseeching it for the relief of rain.

Each day the thunderheads built up, tall silver ranges of cloud turning purple and sullen leaden blue, threatening rain, but never making good that threat, though the thunder muttered, and in the evenings the lightning flickered low on the horizon as though great armies were locked in battle far to the east.

The big game was concentrated on the remaining water, the deeper river pools and the strongest waterholes, so that each day's march was through a wonderland of wild animals.

In one herd Zouga counted thirty-two giraffe, from the old stink-bull almost black with age, his long neck taller than the trees on which he fed, to the pale beige splotched calves on their disproportionately long legs, galloping away in that slow rocking gait with their long tufted tails twisted up over their backs.

Every clearing had its family of rhinoceros, the cows with the distinctively long slender nose horn, running their calves ahead of them, guiding them with a touch of the horn on the flank. There were herds of Cape buffalo, a thousand strong, flowing in a black dense mass across the open glades, steaming with pale dust like the lava from an active volcano.

Then there were elephant. There was not one day they did not cut fresh spoor, veritable roads through the forest, tall trees pushed down or still standing but stripped of their bark so the trunks were naked and weeping with fresh sap, the earth beneath them strewn with chewed twigs and bunches of picked leaves only just beginning to wither, the huge piles of fibrous dung standing like monuments to the passing of the great grey beasts, and the baboon and the plump brown pheasant scratching and foraging enthusiastically in them for the half digested wild nuts and other tidbits.

Zouga could seldom resist him when Jan Cheroot looked up from his examination of the pad marks and said, 'A big bull, this one, walking heavy in the front quarters. Good teeth, I'd stake my sister's virtue on it.'

'A commodity which was staked and lost many years ago,' Zouga observed drily. 'But we will follow, none the less.'

Most evenings they could cut teeth and, having buried them, carry the bleeding heart to where they had left the porters, two men to carry the forty-pound hunk of raw flesh slung on a pole between them, a feast for the whole party. Because of the hunt, progress was slow and not always direct, but steadily Zouga identified and passed the landmarks that his father had described.

Then at last, knowing he was close, Zouga withstood the temptation to hunt, for the first time refusing to follow the fresh spoor of three fine bulls, and disappointing Jan Cheroot most grievously by doing so.

'You should never leave a good elephant, or a warm and willing lady,' he advised dolefully, 'because you never know where or when you are going to meet the next one.'

Jan Cheroot did not yet know the new object of their quest and Zouga's behaviour puzzled him. Zouga often caught him watching him with a quizzical sparkle in his bright little slitty eyes, but he avoided the direct question diplomatically and accepted Zouga's orders to abandon the spoor with only a little further grumbling, and they went on.

It was the porters who first baulked. Zouga never knew how they guessed, perhaps old Karanga had spoken of the Umlimo around the camp fire, or perhaps it was part of their tribal lore, although the gunbearers and most of the porters were from the Zambezi many hundreds of miles to the north. Yet Zouga had learned enough of Africa by now to recognize the strange, almost telepathic knowledge of far events and places. Whatever it was, and however they had acquired forewarning, there were thorns in the porters' feet for the first time in months.

At first Zouga was angry, and would have lived up to his nickname of 'Bakela', the Fist, but then he realized that their reluctance to continue towards the range of bald hills that showed above the horizon was confirmation that he was on a hot scent and close to his goal.

In camp that night, he drew Jan Cheroot aside and, speaking in English, explained what he was seeking and where. He was unprepared for the sickly expression that slowly spread over Jan Cheroot's wizened yellow features.

'Nie wat! Ik lol nie met daai goed nie!' The little Hottentot was driven in his superstitious terror to fall back on the bastard Cape Dutch. 'No what! I don't mess around with that sort of thing,' he repeated in English, and Zouga smiled tauntingly across the camp fire at him.

'Sergeant Cheroot, I have seen you run, with a bare backside, right up to a wounded bull elephant, and wave your hat to turn him when he charged.'

'Elephants,' said Jan Cheroot without returning the smile, 'is one thing. Witches is another thing.' Then he perked up and twinkled like a mischievous gnome.

'Somebody must stay with the porters or they'll steal our traps and run for home.'

Zouga left them camped near a muddy little water-hole, within an hour's march of the northernmost granite kopje. At the water-hole he filled the big enamelled water bottle, and wet its thick covering of felt to keep the contents cool, slung a freshly charged powder sack on one hip and a food bag on the other, and, with the heavy smooth-bore elephant gun over his shoulder set out alone while the shadows were still long on the earth, and the grass wet with the dew.

The hills ahead of him were rounded domes of pearly grey granite, smooth as a bald man's pate and completely free of vegetation. As he trudged towards them across the lightly forested plain, his spirit quailed at the task ahead of him.

With each step the hills seemed to rise higher and steeper, the valleys between them deeper and more sheer, the thorny bush that choked the gorges and ravines more dense. It could take months to search all of this broken wilderness, and he did not have a guide as his father had had. Yet, in the end, it was so easy that he was irritated with his own lack of foresight.

His father had written in the journals 'Even Mzilikazi, that sanguine tyrant, sends gifts for her oracle.'

He struck the well-defined road, leading out of the west, broad enough for two men to walk along it, and aimed directly into the maze of smooth, granite hills. It could only be the road used by the emissaries of the Matabele king.

It led Zouga up the first gentle slope of ground, and then turned abruptly into one of the gorges between peaks. The path narrowed, and jinked between huge round granite boulders, the bush so thick on each side that he had to duck below the thorny branches that interlinked to form a gloomy tunnel overhead.

The valley was so deep that the sunlight did not penetrate to the floor, but the heat was thrown out by the granite as though it had been baked in an open fire, and the sweat soaked Zouga's shirt and slid in cool, tickling drops down his flanks. The bush thinned and the valley narrowed, and then pinched out into a narrow neck between the converging rock walls. It was a natural gateway where a few good spearsmen could have held a regiment. On a ledge high above was a small thatched watch hut, and beside it an idle blue tendril of smoke rising from a watchfire into the still hot air. But if there had been a guard he had deserted his post at Zouga's approach.

Zouga grounded the butt of his elephant gun, and leaned upon it to rest from the climb and at the same time surreptitiously to search the cliffs above him for a hidden enemy, or for the spot from which one could send the familiar boulders bounding and clanging down upon him.

The gorge was silent, hot and deserted. There was not even the chittering of birds or the murmur of insects in the undergrowth. The silence was more oppressive than the heat, and Zouga threw back his head and hallooed up at the deserted watch hut.

The echoes, boomed grotesquely back and forth across the gorge, and then descended through confused whispers to the same foreboding silence. The last white man to pass this way was the Sword of God, in person, intent on decapitating the oracle, Zouga thought bitterly. He could not expect a hero's welcome.

He shouldered the gun again and went into the natural granite gateway, his instinct telling him that the bold approach was the only one open to him. The narrow passageway was floored with crunching grey sand, full of mica chips that sparkled like diamonds even in the subdued light. The passage curved gently until he could see neither the entrance behind him nor the end of it. He wanted to hurry, for this was like a cage, or a trap, but he controlled his feet and showed neither fear nor indecision in his tread.

Around the curve the passageway fanned open, and from one wall a small freshet rippled down the granite cliff, spilled with a tiny liquid gurgle into a natural basin of rock and then overflowed to run down into the hidden valley beyond. Zouga came out of the natural gateway and paused again to stare about him. It was a pleasant valley, probably a mile long and half as wide. The rivulet watered it, and the grass was cool green.

In the centre of the valley was a huddle of neatly thatched huts, around which scratched a few scrawny fowls. He went down to them. The huts were all deserted, although there was everywhere evidence of very recent occupation, even the porridge in the cooking pot was still hot.

Three of the largest huts were crammed with treasures, leather bags of salt, tools and weapons of iron, ingots of smelted red copper, a pile of small ivory tusks, and Zouga guessed that these were the tributes and gifts sent by petitioners and supplicants to the oracle. Payments made for her intercession with the rain gods, fees for a spell cast upon an enemy, or to soften a coquette's heart.

The fact that these treasures were left unguarded was proof of the Umlimo's power, and her own belief in that power. However, if Fuller Ballantyne's journal was the truth 'the foul and midnight hag' as he referred to her was long ago dead, and her severed skull crunched by the hyenas or bleaching somewhere in the hot African sun.

Zouga stooped out through the low entrance of the last

hut, into the sunlight once more. He called again, but again there was no answer. There were people here, many of them, but to make contact with them, and then to learn from them the location of the 'burial place of the kings' was going to be more difficult than he expected.

Leaning on the long gun he turned his attention to the steep side of the valley, and again it was the pathway that caught his eye and led it to the entrance of the cave. For the path continued beyond the village, running down the centre of the valley and climbed the far slope of the valley, then came to an abrupt end against the granite cliff. The mouth of the cave was low and wide, a narrow horizontal gash in the base of the cliff like a toad's mouth and the path led directly into it.

Zouga climbed the gentle slope to the cave entrance. He had left his food bag and water bottle at the village, and, lightly burdened, he strode upwards, tall and lithe, his beard sparkling like gold thread in the sunlight, so that any hidden watcher could not have doubted that this was a chieftain and warrior to treat with respect.

He reached the cave entrance and checked, not from fatigue, for the climb had not taxed him, but merely to take his bearings. The cave entrance was a hundred paces or so wide, and the roof so low that he could reach up and touch the rough rock.

There was a guard wall built to close off the opening, a wall of dressed granite blocks, fitted so closely that it would have been impossible to drive a knife blade between the joints, clearly the work of skilled masons, but done long ago for in places the wall had tumbled, the blocks piled upon each other in disarray.

The path led into one of these gaps, and disappeared into the gloom beyond. It was a most unwelcoming entrance. Going in he would have the light at his back, and his eyes would be unaccustomed to the poor light, there would be many places where a man could wait with spear or axe.

456

Zouga felt his first ardour cooling as he peered into that forbidding opening, and he called again in the Matabele language.

'I come in peace!'

He was answered almost immediately, in a childish piping voice speaking the same language, close behind his shoulder, so close that his heart tripped, and he whirled.

'White is the colour of mourning and death,' piped the voice, and Zouga looked about him in confusion. There was no child, no human, no animal even, the valley behind him was deserted, silent. The voice had emanated from the very air.

Zouga felt his mouth drying, and the skin on his forearms and at the base of his skull crawled with the loathsome little insects of fear, and while he stared another voice screeched from the cliff above him.

'White is the colour of war.'

It was the voice of an old woman, a very old woman, quavering and shrill. Zouga's heart jumped again, and then raced as he looked up. The cliff face was bare and smooth. His heart was fluttering against his ribs like a trapped bird, and his breath rasped and sawed in his throat.

'White is the colour of slavery,' sang a young girl's voice, filling the air about his head, having no direction and no substance, sweet and liquid as the burble of running water.

'She spoke in the voices of Belial and Beelzebub, the hideous voices of Azazel and Beliar, all Satan's myriad alter-egos,' his father had written, and Zouga felt the slow leaden spread of superstitious terror weighing down his legs.

Another voice, roaring like a bull, boomed from the mouth of the cave. 'The white eagle has cast down the stone falcons.'

He took a long slow breath, to bring his mutinous body under control, and he cast his mind back deliberately to a childhood memory. Brighton pier on August bank holiday, the small boy clutching his Uncle William's hand and staring up in wonder at the magician on the stage who made a doll come to life and speak in a quaint piping voice, answering the voice from the box too small to contain even a rabbit. The memory steadied him, and he laughed. It was a clear firm laugh, surprising himself even.

'Keep your tricks for the children, Umlimo. I come in peace to speak with you as a man.'

There was no reply, though he thought he caught the silken whisper of bare feet on stone, coming from the darkness beyond the turned stone wall.

'See me, Umlimo! I lay aside my weapons.'

He unslung the powder bag and dropped it at his feet, and then he laid the elephant gun across it and holding his empty hands before him advanced slowly into the cave.

As he reached the step in the wall, there was the crackling spitting snarl of an angry leopard from the shadows just ahead of him. It was a terrible sound, fierce and real, but Zouga had himself in hand now. He did not miss a pace but stooped under the sill through the gap, and straightened on the far side.

He waited for a minute while his eyes adjusted, and he could make out shapes and planes in the gloom. There were no other voices or animal sounds. There was a faint source of light somewhere ahead of him in the depths of the cave, and he could now make out a way amongst the scree and fallen rock that choked the cave, in some places as high as the low roof.

Zouga began to pick his way carefully forward. The light grew stronger, and Zouga realized that it was a single beam of sunlight, shining through a narrow crack in the roof.

Looking up, he stumbled and put out a hand to save himself. It was not rock that he touched but something sticklike that moved beneath his touch. There was a rattle, and a loose dry rush of debris. Zouga caught his balance and glanced down. A disembodied human skull gaped up at him from empty black eye sockets, the cheekbones still covered with a parchment of dried skin.

With a jolting little shock Zouga realized that what he had taken for loose scree and rock was instead piles of human remains, dried and desiccated corpses, lying in mounds and heaps, choking the passages and deepest recesses of the cave, here and there a single body, crouched or sprawled alone, bone shining dully from gaps in its covering of dark dried skin or through the rotting leather garments.

'That reeking charnel house,' Fuller Ballantyne had called it.

Instinctively Zouga wiped the hand that had touched the long-dead skeleton, and then went on towards the light. There was the smell of smoke now, and of human presence, and another sweet mousy odour that was hauntingly familiar but which Zouga could not place at that moment. The floor of the cave sloped downwards under his feet, and he turned a rocky shoulder and looked down into a small natural amphitheatre, with a floor of smooth granite.

In the centre of the floor burned a low fire of some aromatic wood, the smoke spiced the air and rose in a slow spiral towards the crack in the rocky roof, clouding the beam of sunlight with milky blue. There seemed to be other fingers of the cave leading further into the hillside, like the adits of a mine shaft, but Zouga's attention was focused on the figure that sat across the fire from him.

Zouga went slowly down on to the floor of the stone amphitheatre without taking his eyes from the figure.

'That foul and midnight hag,' his father had called the Umlimo, but this was no hag. She was young, in full physical prime, and as she knelt facing him Zouga realized that he had seldom seen such a fine-looking woman, certainly not in India or Africa, and very seldom, if ever, in the northern lands.

She had a long regal neck on which her head balanced like a black lily on its stem. Her features were Egyptian, with a straight fine nose and huge dark eyes above high moulded cheekbones. Her teeth were small and perfect, her lips chiselled like the flutes of a pink seashell.

She was naked, her body slim, her limbs long and fine, her hands and feet narrow and delicately shaped with tapered fingers and pale pink palms. Her small breasts rode high and were perfectly round, her waist narrow but flaring into tight hard buttocks and hips like the curve of a Venetian vase. Her sex was a wide triangle, deeply cleft, the inner lips bursting out unashamedly, like the wings of a dark exotic butterfly emerging from its woolly chrysalis.

She was watching him with those huge dark eyes, and when he stopped across the fire from her, she made a slow gracious gesture with those delicately long fingers. Obediently Zouga squatted down opposite her and waited.

The woman took up one of the calabash gourds from the array beside her, holding it between the palms of her hands and poured from it into a shallow earthen bowl. It was milk. She set the calabash aside, and Zouga expected her to offer him the bowl, but she did not. She continued to watch him inscrutably.

'I come from the north,' Zouga said at last. 'Men call me Bakela.'

'Your sire slew the one before me,' said the woman. Her voice was compelling, for although the fluted lips barely moved, it was thrown with the power and timbre of the skilled ventriloquist. The sound of it seemed to quiver in the air long after she finished speaking and he knew now with certainty who had spoken in the voice of child and maid, of warrior and wild animal.

'He was a sick man,' Zouga replied, not questioning the source of her knowledge. Not querying how she had known that he was the son of the father.

Her words explained much to Zouga, and it was logical

that the Umlimo was a hereditary duty, the office of high priestess being passed on down the years. This magnificent woman was the latest bearer of the title.

'My father was driven mad by the sickness in his blood. He did not know what he did,' Zouga explained.

'It was part of the prophecy.' The Umlimo's statement shimmered against the cave walls, but she did not stir while the silence spun out over many minutes.

'These,' Zouga spoke at last, indicating the dusty, crumbling piles of dead, 'who were they, and how did they die?'

'They are the people of the Rozwi,' said the woman, 'and they died by fire and smoke.'

'Who laid the fire?' Zouga persisted.

'The black bull from the south. The Angoni.'

Zouga was silent again imagining the tribe fleeing here, to their holy place, their refuge, the women carrying the children, running like driven game ahead of the beaters, looking back over their shoulders for a glimpse of the waving tufts on the crowns of the warshields and the plumed headdresses of the Angoni *amadoda*.

He imagined them lying here in the darkness listening to the ring of axes and shouts of the besieging warriors as they cut the timber and piled it in the mouth of the cave, and then the crackle of the flames as they put fire to it and the first choking acrid clouds of smoke boiled into the cave.

He could hear again in his imagination the screams and cries of the choking, dying victims, and the shouts and the laughter of the men beyond the flaming, smoking barricade of timber.

'That too was part of the prophecy,' said the Umlimo and was silent. In the silence there was a soft rustling sound like a leaf blown by the breeze across the tiles at midnight, and Zouga's eyes turned towards the sound.

A dark thing flowed out of the shadows at the back of the cave, like a trickle of spilt blood, black in the gloom but catching the firelight in pinpricks of reflected light. It rustled softly across the stone floor, and Zouga felt his skin crawl and his nostrils flare at the sweetish mousy odour which he had noticed before, but which he recognized now.

It was the smell of snake.

Zouga stared at it, frozen with fascinated horror, for the reptile was as thick as his wrist and the full length of it was lost in the recesses of the cave. The head slid into the circle of orange firelight. The scales glittered with a marble lustre, the lidless eyes fixed Zouga with an unblinking stare and from the slyly grinning lipless mouth the silken black tongue vibrated as it tasted his scent upon the air.

'Sweet Christ!' Zouga whispered hoarsely and dropped his hand to the hilt of his hunting-knife on his belt, but the Umlimo did not move.

The snake lifted its head from the stone and dipped it over the bowl of milk. It began to drink.

It was a mamba, a black mamba, the most venomous of all the reptiles, the death it could inflict was swift but agonizing beyond all nightmares of pain. Zouga had not believed that a mamba could grow to such dimensions, for as he watched it drink half its length was still lost in the shadows.

After a minute the monstrous reptile lifted its head from the bowl and turned towards the Umlimo, it began to slide forward, the muscles under the glittering scales convulsing in little waves that came running down its length towards the broad spatulate head.

It touched the woman's bare knee with the flickering black tongue, seeming to use the tongue like a blind man uses his cane to grope its way along her thigh, licking briefly at the lips of her bulging sex, and then lifting itself up over her belly, over her breasts, still licking at her smoothly oiled skin, it rose up around her neck and then slid down over the other shoulder until it came to rest at last suspended around her neck, the head reaching out the length of a man's arm ahead of her at the level of her breasts; swaying slightly, it fixed Zouga once more with that cold ophidian stare.

Zouga licked his lips, and relaxed his grip on the hilt of the knife.

'I have come to seek wisdom,' he said hoarsely.

'I know what you seek,' replied the Umlimo. 'But you will find more than you seek.'

'Who will lead me?'

'Follow the little seeker of sweetness in the treetops.'

'I do not understand.' Zouga frowned, still watching the huge snake, and the Umlimo did not reply. Her silence was clearly an invitation to ponder the reply, and Zouga did so in silence, but found no explanation. He memorized the words, and would have asked another question but there was a silken rush out of the darkness near him, and he started half to his feet as a second snake slid swiftly past him.

It was another mamba, but a smaller snake, not much thicker than his thumb, and as long as twice the stretch of his arms. Half its length was raised arrow-straight in the air, and it sailed on its tail up to the kneeling woman with her grotesque living necklace.

The woman did not move, and the smaller snake stood before her, swaying gently from side to side, lowering itself gradually until it touched tongue to darting black tongue with the thick reptile around her neck.

Then it slid forward and began to roll itself around the body of the other snake, throwing turn after turn like a sailor lashing a sheet about a mast, and each time it rolled, it showed its soft white pulsating underbelly with the narrow scales reaching from flank to flank.

Neither the woman nor the bigger snake moved, nor removed their steadfast gaze from Zouga's fascinated pale face. The thinner lighter-coloured body of the second snake began a slow rhythmical and sensual movement, expanding and contracting about the thicker and darker body of the other and Zouga realized that they were a mating pair.

Two-thirds of the way down the underbelly of the male were the elongated scales that guarded the genital sac. As the male's excitement mounted so the scales gaped apart and the penis began to extrude. It was the colour and shape of the bloom of a night-flowering cactus, a pale lilac belled flower that gleamed like wet satin.

Insistently the male caressed the thick, dark body, and gradually his ardour was rewarded. The female rolled a portion of her own length, the white belly throbbing softly in acquiescence, exposing the scaled genital purse.

With a long shuddering movement the male slid his length down hers, white belly pressed to belly, and the swollen lilac flower prised open the female sexual purse, distorting the lips. The female mamba opened her mouth wide, and her throat was a lovely buttercup yellow. In her top jaw the small bony needles of her fangs were erect, each tipped with a pearly drop of venom, and she emitted a low sibilant hiss of ecstasy or pain as the male locked his penis deeply into her.

Zouga found he was sweating. A droplet slid down from his temple into his beard. The bizarre courtship and copulation had taken only minutes during which neither he nor the Umlimo had moved, but now she spoke.

'The white eagle has stooped on the stone falcons, and cast them to earth.' She paused. 'Now the eagle shall lift them up again and they will fly afar.'

Zouga leaned forward, listening intently,

'There shall be no peace in the kingdoms of the Mambos or the Monomatapa until they return. For the white eagle will war with the black bull until the stone falcons return to roost.'

While she spoke, the slow convulsing copulation of those

interlocked bodies continued, giving to her words an obscene and evil weight.

'Generation will war with generation, the eaglet will strive against the bull calf – white against black, and black against black, until the falcons return. Until the falcons return.'

The Umlimo raised her narrow pink-palmed hands, and lifted the garland of intertwined serpents from off her own neck. She laid them gently on the stone floor of the cave and with a single flowing movement, she stood erect, and the firelight glinted upon her oiled satin body.

'When the falcons return,' she spread her arms, and the round breasts changed shape at the movement, 'when the falcons return, then once again the Mambos of Rozwi and the Monomatapa of Karanga will hold sway in the land.' She lowered her arms and her breasts sagged weightily. 'That is the prophecy. That is the whole prophecy,' she said, and turned away from the fire and, with a gliding walk, moved across the irregular stone floor, her back straight and her naked buttocks swaying in stately rhythm.

She disappeared into the dark shadows that shrouded one of the fingers of the cave beyond the amphitheatre.

'Wait!' Zouga called after her, scrambling to his feet and starting after her.

The huge female mamba hissed sharply, like the steam of a boiling kettle, and rose up as high as Zouga's head. The butter yellow mouth gaped again, and a crest of dark glittering scales came angrily erect down the length of her neck.

Zouga froze, and the snake hissed again and flared a little higher, the raised body arching gently into a taut 'S' shape. Zouga backed off, one pace and then another. The crest of scales subsided a little. He took another step backwards and the tense bow of the serpentine body relaxed, the head lowered a few inches. He moved steadily backwards towards the entrance of the cave, and before the amphitheatre was obscured by the shoulder of rock he saw the huge snake coiled in a knee-high pile of glinting scales still locked in sexual congress with her deadly consort.

The prophecy of the Umlimo, cryptic and unrevealed, stayed with Zouga during the long march back to where Jan Cheroot waited with his porters.

That night, by the firelight, Zouga copied it word for word into his journal, and afterwards the sweet smell of snakes haunted his nightmares – and lingered in his nostrils for long days after.

Now the wind turned fickle, sometimes completely still in the enervating hush and heat of noon, at other times dancing in the tall swaying vortex of the 'dust devils' across the plains, lifting leaves and dried blades of grass hundreds of feet in the yellow columns of dust, then again it gushed in turn from every point of the compass, one minute firm out of the north and the next as firmly from out of the south.

It was impossible to come up with elephant while the wind played so loosely. Often when the spoor was hot and true, and they had already laid aside their heavy traps and stripped down to light running order, Zouga would feel the cool touch of the breeze upon the back of his sweating neck and almost immediately afterwards hear the alarm squeal of an elephant ahead of them in the forest, and after that first alert, it was impossible to close with the herd, for they went into that long sloping gait that they could keep up for mile upon mile, hour upon hour, day upon night, that would kill a man who tried to match it for more than a few miles.

Thus they killed no elephant in the days following Zouga's meeting with the Umlimo, and once, when they had a good spoor which would have led them back into the north, away from the direction in which Zouga was convinced lay his quest, it was Zouga himself who called off the hunt. Jan Cheroot muttered bitterly for the rest of that day and the next while they made those seemingly aimless casts, eastward and then westward again, through the unmarked and uncharted wilderness.

Each day now the heat became fiercer, for the suicide month that ushers in the rains was upon them. Not even Zouga could march during the hours before and after noon. They would find the best shade and throw themselves down under it, sweating out the worst of the heat, trying to sleep when the buffalo flies would let them, but it was an effort to speak, an effort to wipe away the sweat that welled up on their bodies, and dried in white crystals on their skin and clothing. The salt rotted the fabric of Zouga's shirt and breeches so they tore like paper at the first touch of thorn or rock, and Zouga was gradually reduced to a beggar's rags, patched and stitched until little of the original material remained.

His boots had been resoled more than once with the rawhide peeled from the inner surface of an elephant's ear, and his belt and webbing, the sling of his elephant gun renewed with uncured buffalo hide.

He made a strange gaunt figure, for hard hunting had burned all the fat and loose flesh from his body and limbs. His height was enhanced by his leanness and the breadth of his bony shoulders tapering sharply to his waist. The sun had burned his skin dark, yet bleached his hair and beard to white gold. His pale hair hung to his shoulders, and he tied it at the back of his neck with a leather thong. He kept his beard and whiskers trimmed neatly, using scissors to cut and the heated blade of his hunting knife to singe them.

The sense of well-being from his superb physical condition, together with the vaunting anticipation of a successful outcome to his search drove him onwards so that the days seemed too short for him and yet when the night fell he lay

467

down on the hard earth and slept the deep dreamless refreshing sleep of a child, to awake again long before the dawn's first glimmering, impatient for what the new day would bring forth.

However, the time was passing. After each hunt, the powder bags were lighter, and though he dug the spent musket balls from the carcasses of his victims and recast them, there was always wastage.

The precious little store of quinine dwindled as swiftly, and the rains were coming. No white man could survive the rains without ammunition and quinine. Soon he would have to abandon the search for the ruined city with its idols decked in gold. He would have to march to beat the rains, south and west, five hundred miles or more, if his observations were still accurate, to cut his grandfather's road and follow it down to the mission station at Kuruman, which was the nearest outpost of European civilization.

The later he left it, the harder would be the march when he made it. Hard and fast, stopping for nothing, neither elephant nor gold, until he was out into the drier, safer land to the south.

The thought of leaving depressed him, for he knew with a certain deep gut feeling that there was something here, very close to where he now lay, and it irked him terribly that the oncoming rains would frustrate his search. But then he consoled himself that there would be another season, and he knew with the same deep gut certainty that he would return. There was something about this land – an insistent, irritating sound interrupted his thoughts. He pushed the cap up from his eyes and looked into the dense branches of the marula tree under which he lay. The harsh cries were repeated, and the drab little bird that uttered them hopped agitatedly from branch to branch, flirting its wings and tail with a sharp whirring. The bird was the size of a starling, its back dull brown and its chest and belly a muddy yellow.

Zouga rolled his head and saw that Jan Cheroot was awake also.

'Well?' he asked.

'I haven't tasted honey since we left Mount Hampden,' Jan Cheroot answered. 'But it's hot, and perhaps the bird is a liar, perhaps he will lead us to a snake or a lion.'

'He only leads you to a snake if you cheat him of his share of the comb,' Zouga told him.

'So they say,' Jan Cheroot nodded and then they were silent, considering the effort involved in following the honey guide, and weighing it against the possible rewards. The bird will very often lead a badger or a human being to a wild hive, and wait for its share of the wax and honey and bee grubs. The legend is that if his payment is not made, then next time he will lead the man who cheated him to a venomous snake or a man-eating lion.

Jan Cheroot's sweet tooth got the better of him. He sat up and the bird's cries were immediately shriller and more excited. It flashed away across the clearing to the next tree, flitting wings and tail noisily, calling them impatiently, and when they did not follow it darted back to the branches above them and continued its display.

'All right then, old fellow,' Zouga agreed resignedly, and stood up. Jan Cheroot took an axe from Matthew and the clay fire-pot in its plaited bark carrying net.

'Make camp here,' Jan Cheroot told the bearers. 'We will bring you honey to eat tonight.'

Salt and honey and meat, the three greatest delicacies of the African bush. Zouga felt a twinge of guilt at wasting so much of the time that remained to him on this frivolous side-journey, but his men had worked hard and travelled fast and honey would revive their flagging spirits.

The little brown and yellow bird danced ahead of them, burring and rattling with a sound like a shaken box of Swan Vestas, darting from tree to bush, turning immediately it settled to make sure they were following.

For almost an hour it led them along a dry river course, and then it turned and crossed a rocky ridge of ground. At the saddle, they looked down into a heavily wooded valley bounded by the familiar rocky kopjes and hillocks.

'The bird is teasing us,' Jan Cheroot grumbled. 'How much further will he make us dance?'

Zouga shifted his elephant gun to the other shoulder. 'I think you are right,' he agreed, the valley ahead of him was forbidding, the floor choked with tall stands of the razor-edged elephant grass, higher than a man's head. It would be even hotter down there, and the dried seeds of the grass had arrowheads to them that could work themselves into the skin and cause festering little wounds.

'It comes to me that I am not so fond of honey as I thought I was,' Jan Cheroot cocked his head at Zouga.

'We will turn back,' he agreed. 'Let the bird find another dupe. We, will look for a fat kudu cow on the way back, meat instead of honey.'

They started back down the ridge and instantly the bird flashed back, and renewed its entreaties above their heads.

'Go and find your friend the *rattel* (honeybadger)!' Jan Cheroot shouted at it, and the bird's contortions became frantic. It dropped lower and lower, until it was in the branches barely an arm's length above their heads, and its cries were irritating and distracting.

'*Voetsak!*' Jan Cheroot yelled at it. The bird's cries would alert all wild game for miles about to the presence of man, and thwart any chance of killing an animal for the night's meal. '*Voetsak!*' He stooped and picked up a stone to shy at the bird. 'Go away and leave us, little sugar mouth.'

The name stopped Zouga in his tracks. Jan Cheroot had used the bastard Dutch, *'klein Suiker bekkie'*, and now he drew back his right arm to hurl the stone.

Zouga caught his wrist. 'Little sugar mouth,' he repeated, and the Umlimo's voice rang in his ears, that strange shimmering tone that he had memorized, 'the seeker of sweetness in the treetops.'

'Wait,' he told Jan Cheroot, 'Do not throw.'

It was ridiculous, of course it was. He would not make himself ludicrous by repeating the Umlimo's words to Jan Cheroot. He hesitated a moment. 'We have come so far, already,' he told the Hottentot reasonably, 'and the bird is so excited that the hive cannot be far.'

'It could be another two hours,' Jan Cheroot growled, but lowered his throwing arm. 'That makes six hours back to camp.'

'You don't want to grow fat and idle,' Zouga said. Jan Cheroot was lean as a whippet that had been coursing rabbit all season, and he had walked and run a hundred miles in the last two days. He looked pained at the accusation, but Zouga went on remorselessly, shaking his head in mock sympathy. 'But when a man grows older, he cannot march as far or as fast, and he is slower with the women too.'

Jan Cheroot dropped the stone, and went back up the ridge at a furious pace with the bird flitting and screeching ecstatically ahead of him.

Zouga followed him, smiling at the little man and at his own folly in placing any store on the words of that naked witch. Still the honey would be welcome, he consoled himself.

An hour later, Zouga was convinced that Jan Cheroot had been correct. The bird was a liar, and they were wasting what remained of the day, but there would be no turning Jan Cheroot now, he had been deeply insulted by Zouga's gibes.

They had crossed the valley, blundering through the stands of elephant grass, for the bird did not pick one of the game trails to follow. It moved on a direct line, and as they followed, the grass seeds showered upon them, working their way down the backs of their shirts and the sweat of their bodies activated them, as the first rains should have done, so that they began to worm like living things, trying to pierce the skin.

The view ahead was cut off by the tall grass, and they reached the far side of the valley with little prior warning. Suddenly there was a smooth rock cliff looming over them, almost obscured by the tall leafy trees of the forest, and by its own covering of lianas and dense climbing plants. It was not a very high prominence, perhaps forty feet, but it was sheer. They stopped below it and peered upwards.

The wild beehive was almost at the top of the cliff. The honey guide fluttered triumphantly above it, twisting its head to look down at them with a single bright beadlike eye.

The rock below the opening to the hive was stained with a dark dribble of old melted wax and the detritus of the hive, but it was almost entirely masked by a lovely creeping plant. The stem of the plant climbed the cliff, branching and twisting and doubling on itself, its pale leaves a cool green, but its flowers a lovely shade of cornflower blue.

The bees leaving and returning to the hive caught the sunlight like golden dust-motes, but their trajectory was swift and straight through the hot still air.

'Well, Sergeant, there is your hive,' Zouga said. 'The bird did not lie.' He felt a deep sense of disappointment. Although he had told himself to place no store on the Umlimo's words, yet there had been a sneaking anticipation, the forlorn hope persisting against common sense, and now that sense had prevailed, there was this regret.

Zouga leaned his gun against the bole of a tree, shed his traps, and threw himself down to rest and watch Jan Cheroot making his preparation to rob the hive. Jan Cheroot cut a square of bark from a mukusi tree, and rolled it into a smoking tube. He filled it with wood pulp scraped from a dead tree trunk.

Then he swung the clay fire-pot on its carrying sling of bark rope, fanning it until the smouldering moss and wood chips that it contained burst into flame. He transferred fire to the smoking tube, and when it had fairly taken he hooked his axe over one shoulder and began climbing the interwoven branches of the flowering creeper up the sheer cliff.

The first of the defender bees buzzed fiercely about his head when he was a few feet from the opening to the hive, and pausing, Jan Cheroot lifted the bark tube to his mouth and blew a gentle stream of blue smoke towards his attacker. The smoke drove the insect away, and he resumed the climb.

Lying under the mukusi tree, idly swatting at the buffalo flies, picking at the grass under his shirt, and harbouring his disappointment, Zouga watched the Hottentot work.

Jan Cheroot reached the hive, and blew another stream of smoke over the hole in the cliff, stupefying the bees which had now formed a swirling defensive cloud about him. One of them darted in, despite the smoke, and stung him in the soft of the neck. Jan Cheroot swore bitterly, but he did not make the mistake of slapping at the attacker, or of trying to scratch away the barbed sting lodged in his flesh. He worked on calmly, unhurriedly, with the smoke tube.

Minutes later, the hive was sufficiently smoke-drugged to allow him to begin cutting away the screen of flowering branches that hid the entrance, and he balanced on a fork of the creeper, using both hands to swing the axe, perched like a little yellow monkey forty feet above where Zouga lay.

'What the thunder ...' He stopped after a dozen axe blows, and stared at the cliff face that he had exposed. 'Master, there is devil's work here.'

The tone of his voice alerted Zouga, and he scrambled to his feet.

'What is it?'

Jan Cheroot's body obscured the object of his amazement,

and impatiently Zouga crossed to the foot of the cliff and climbed hand over hand up the serpentine stem of the creeper.

He reached Jan Cheroot's side and clung to a handhold.

'Look!' Jan Cheroot exhorted him. 'Look at that!' He pointed at the stone face that he had exposed with the axe.

It took Zouga a few seconds to realize that the entrance to the hive was through a geometrically perfect, sculptured aperture, one of a series which pierced the cliff face in a broad horizontal band, that seemed to extend unbroken in both directions along the cliff face. The decorated band was in chiselled stone blocks, arranged in a chevron pattern, a lattice work that was without question the work of a skilled stonemason.

The realization jolted Zouga so that he almost missed his hand-hold, and immediately afterwards he saw something else that up to that moment had been hidden by the dense covering of climbing plants and the thick coating of old outpouring wax from the beehive.

The entire cliff face was formed of perfect blocks of dressed stone, small blocks, so neatly fitted that they had appeared on casual inspection to be a solid sheet of rock. Zouga and Jan Cheroot were suspended near the top of the massive stone wall, forty feet high, so thick and long that it had seemed to be a granite hillock.

It was a monumental work of stonemasonry, to compare with the outer wall of Solomon's temple, a vast fortification which could only be the periphery of a city, a city forgotten and overgrown with trees and creepers, undisturbed over the ages.

'Nie wat!' whispered Jan Cheroot. 'This is a devil's place – this is the place of Satan himself. Let us go, master,' he pleaded. 'Let us go far away, and very fast.'

The circuit of the walls took Zouga almost an hour, for the growth was thicker along the northern curve of the stone rampart. The wall seemed to be laid out in an almost perfect circle, and without openings. At two or three likely points, Zouga hacked away the growth and searched the foot of the wall, looking for a postern or a gateway. He found none.

The decorative chevron pattern did not seem to extend around the entire circumference, but covered the eastern quadrant. Zouga wondered at the significance of that. The immediate explanation seemed to be that the decoration would face the rising sun. It seemed likely that the peoples who had built this massive edifice had been sun-worshippers.

Jan Cheroot trailed him reluctantly, prophesying the wrath of the devils and hobgoblins who guarded this evil place, while Zouga hacked and chopped his way around the walls, completely oblivious of his advice.

'There must be a gateway,' he grumbled. 'How did they get in and out?'

'Devils got wings,' Jan Cheroot pointed out broodingly. 'They fly. Me, I wish I had wings also, to fly the hell away from here.'

They came again to the point in the wall where they had discovered the beehive, and by then it was almost dark, the sun had disappeared below the treetops.

'We'll search for the gate in the morning,' Zouga decided.

'We are not going to sleep here?' Jan Cheroot demanded, horrified.

Zouga ignored the protest. 'Honey for supper,' he suggested mildly, and for once he did not sleep that deep hunter's sleep, but lay under his single blanket, his imagination filled with golden idols and treasure houses built out of massive worked stone blocks.

Zouga resumed the search again when it was light enough to see the top of the wall silhouetted against the misty pearl sky of dawn. The previous day he had been blinded by his own eagerness and careless in his haste. He had missed the area, only a few yards from where they were camped, where the creepers covering the wall had once been hacked away and then had regrown, even more thickly than before. Now, however, a lopped branch beckoned him like a finger, the clean cut unmistakably made by an axe.

'Jan Cheroot,' Zouga called him from the cooking fire. 'Clean out this rubbish.' Zouga showed him the dense secondary growth, and Jan Cheroot sauntered away to fetch his axe.

While he waited for him to return, Zouga decided that only one person could possibly be responsible for these old overgrown and healed cut marks on the stems of the vine. Once again Fuller Ballantyne had been his guide, yet this time he did not resent it so fiercely; the experience of treading squarely in his father's footsteps was no longer new, and he had his excitement and anticipation to lessen the sting of it.

'Hurry,' he called to Jan Cheroot.

'It's been here a thousand years. It's not going to fall down now,' Jan Cheroot answered saucily, and spat on his palms before hefting the axe.

The little Hottentot was a great deal happier this morning. He had survived a night under the wall, without being assailed by even a single hobgoblin, and Zouga had passed the unsleeping hours in describing to him the treasures that might lie beyond that wall. Jan Cheroot's temporarily paralysed avarice had revived sufficiently to imagine himself with his pants filled with gold coin seated in his favourite tavern on the Cape Town dockside with a dozen butter-yellow Hottentot beauties crowding close to him to hear the story, while the bartender prised the wax seal off another bottle of Cape Smoke. Now his enthusiasm almost matched that of Zouga himself.

He worked swiftly and when Zouga stooped and peered into the passage that he had cut through the dense secondary growth that had sprung up behind Fuller Ballantyne's axe, he saw the outline of the curved portals of the gateway, and the chiselled granite steps that led up to the narrow opening.

The steps had been worn into a smooth dish shape by the passage of thousands of feet over the centuries, but the gateway had been deliberately blocked with stone and rubble, not the neat stonework of the main wall, but a careless and hurried attempt to seal the entrance, probably in the face of an advancing enemy, Zouga thought.

Somebody, probably Fuller Ballantyne, had pulled down this barrier sufficiently to enter. Zouga followed his tracks, the loose stone rolling under foot, he squeezed his way through the gateway, and found that it turned abruptly to the left, into a narrow vegetation-choked passage, between high walls open to the sky.

His disappointment was intense. He had hoped that once he forced the gateway, all the city and its wonders and treasures would lie before him. Instead there were many hours of hard sweating labour facing him. It had been years, four years at least, since Fuller Ballantyne had entered through this gate and passageway, and it was as though he had never been.

Where the stonework had collapsed, Zouga clambered over it gingerly, the thought and dread of snakes was very much with him since the cave of the Umlimo. The long narrow passage, obviously constructed as a defensive measure against intruders, followed the curve of the main wall until abruptly it opened into a clear space, again choked with dense green thorny growth, and dominated by a tall cylindrical tower of lichen-covered granite blocks. The tower was immense, seeming in Zouga's excitement to reach to the very clouds.

Zouga started across the intervening courtyard, hacking impatiently at the bush and creepers, and halfway across he saw that there was a second tower, an identical twin to the first that had until that moment obscured it. Now his heart was pounding fiercely against his ribs, not from the exertion of swinging the axe – but with an intuitive belief that the towers were the centre of this strange ancient city, and that they held the key to the mystery.

He stumbled in his haste, and went down on his knees, tearing another long rent in his breeches and abrading a strip of skin from his shin, so he swore in his impatience and his pain. He had lost the axe but when he groped for it in the tangled roots and interwoven branches, he found it almost immediately, and at the same time, uncovered the stone that had tripped him.

It was not the granite of which the walls and tower had been constructed. That fact caught his attention, and still on his knees he used the axe to clear the bush around the stone. He felt his nerves thrill as he realized that it was a work of sculpture.

Jan Cheroot had come up behind him, and now he knelt also and tore at the plants with his bare hands – then the two of them rocked back on their heels, and still squatting examined the statue that they had uncovered. It was not large, probably weighed less than a hundredweight. It was carved in satiny, greenish soapstone, sitting on that familiar ornamental plinth, the simple pattern of triangles, like a row of shark's teeth.

The head was smashed off the statue, seemingly by a blow from a sledge hammer, but more likely from a rock used as a hammer. The body of the statue was still intact, the body of the raptor, with the folded pointed wings of a bird of prey, crouching on the point of flight.

Zouga slipped his hand into the opening of his shirt and drew out the little ivory charm on its leather thong, that he had taken from the body of the Mashona chieftain he had killed at the pass of the elephant road.

He let it nestle in the palm of his hand, comparing it with the statue. Beside him Jan Cheroot murmured, 'It is the same bird!'

'Yes,' Zouga agreed softly. 'But what does it mean?' He dropped the ivory charm back inside his shirt.

'It is from long ago.' Jan Cheroot shrugged. 'We will never know.' And dismissing it thus, he would have risen to his feet again, but something else caught his beady bright eye and he darted forward, his hand pecking at the loose earth beside the statue like a greedy hen, and held it up between thumb and forefinger to catch the slanted morning light.

It was a perfectly round bead of metal, pierced for the string of a necklace, a tiny bead only slightly bigger than the head of a wax Vesta, and it was irregular in shape, as though beaten out under the hammer of a primitive smithy, but the colour was red-yellow, and its surface was undulled by either tarnish or corrosion; there is only one metal that has that peculiar lustre and sheen.

Zouga held out his hand for it almost reverently, and it had the weight and the warmth of a living thing.

'Gold!' said Zouga, and beside him Jan Cheroot giggled ecstatically, like a young bride at her first kiss.

'Gold,' he agreed. 'Good yellow gold.'

Zouga was always aware of the very limited time left to him, and every hour or so as they worked he would lift his head to the sky with the sweat streaming down his face and neck, and greasing the flat hard muscles of his naked upper body, and always the clouds were taller and blacker, the heat more punishing, and the wind sullen as a captive tribe on the point of rebellion.

In the night he would start awake, breaking up through the drugged surface of exhausted sleep, to lie and listen to the thunder growling below the horizon like a man-killing monster.

Each dawn he shook his men from their blankets and drove them in a suppressed frenzy of impatience to their labour, and when Matthew, the gunbearer, refused to rise again after the short rest which Zouga allowed them in the hottest hour of the day, Zouga dragged him to his feet and hit him once, a short, perfectly timed chopping blow that sent him spinning backwards full length into his own excavation. Matthew crawled out again with blood dribbling from his chin, and picked up the crude sieve of plaited split bamboos with which he was sifting the earth from the digging, and began again working over the piles of loose earth and rubble.

Zouga drove himself harder than he drove his small band of temple plunderers. He worked shoulder to shoulder with them as they cut out all the undergrowth from the courtyard below the twin stone towers, exposing the broken cobbles, and piles of loose rubble amongst which lay the fallen statues.

He found six more of the bird carvings virtually undamaged, except for minor chips and the attrition of the ages, but there were the fragments of others that had been broken with a savagery which could only have been deliberate, so that he was uncertain of the original number of statues. Zouga spent little time puzzling over them. The loose earth and rubble on which they lay was rich ground for his band to pick over, though they were handicapped by the lack of tools. Zouga would have paid a hundred guineas for a set of good picks and spades and buckets. However, they had to make do with sharpened wooden stakes, the tips hardened in the fire, to dig out the loose stuff, and Jan Cheroot wove baskets of split bamboo, like the flat baskets used by the African women to winnow the stamped corn meal, and with these they sieved the fine earth after picking over it by hand.

It was tedious, back-breaking work, and the heat was murderous, but the harvest was rich. The gold was in small pieces, mostly in the pierced round beads, from which the string had long ago rotted away, but there were flakes and flecks of thinly beaten foil, which might once have been used to decorate a votive wooden carving, there were coils of fine gold wire, and more rarely small ingots of the metal the size and thickness of a child's finger.

Once the green stone birds must have stood in a circle, facing inwards like the granite columns of Stonehenge, and the gold had probably formed some part of the offerings and sacrifices made to them. Whoever had thrown down the statues had scattered and trampled the sacrifice, and time had corroded all except that lovely yellow incorruptible metal.

Within ten days of first hacking away the undergrowth that choked the inner courtyard, the temple yard, as Zouga called it, they had gleaned over fifty pounds weight of native gold, and the interior of the stone courtyard had been gutted, the earth rutted and harrowed as though a troop of wild bush pig had rooted it out.

Then Zouga turned his full attention to the twin towers. He measured them around the base, over a hundred paces, and inspected each joint in the masonry for a secret opening. There was none, so he built a rickety ladder of raw timber and bark rope and risking neck and limb reached the top of the tallest tower. From this vantage point he could look down into the roofless passageways and courtyards of the city. It was a maze, all of it choked with growth – but there was no other part as promising as the temple courtyard of the bird statues.

He turned his attention back to the tower on which he stood. There was again no sign of a secret opening, although he searched diligently for one. It puzzled him that the ancient architect would have built such a solid structure with no apparent use or motive, and the possibility occurred to him that it might be a sealed treasure house, built around an inner chamber.

The work of trying to penetrate the massive stonework daunted even Zouga, and Jan Cheroot declared the attempt to be madness. But Zouga had exhausted the digging below the tower, and this seemed to be the only fruitful area left to him.

Complaining bitterly, a small team led by Matthew climbed the rickety ladder, and under Zouga's supervision began prising loose the small blocks from the summit of the tower. However, such was the skill and dedication of the original masons that progress in the demolition was painfully slow, and there was a long pause between each crash of one loose block into the courtyard below and the next. It needed three days' unremitting toil to break a jagged aperture through the first layer of dressed blocks and to discover that the interior of the tower consisted merely of a fill of the same grey granite.

Standing beside him on the summit of the tower, Jan Cheroot voiced Zouga's own disappointment.

'We are wasting our time. It's stone and more stone.' He spat over the side of the tower and watched the speck of phlegm float down into the ransacked courtyard. 'What we should look for is the place the gold came from.'

Zouga had been so obsessed by his search and plunder of the ruined and deserted city that he had paid no thought to the mines which must lie somewhere beyond the walls. Now he nodded his head thoughtfully.

'No wonder your mother loved you,' he said. 'Not only are you beautiful, but you are clever too.'

'Ja,' Jan Cheroot nodded smugly. 'Everybody tells me that.'

At that moment a fat weighty drop of rain struck Zouga's forehead, and ran down into his left eyeball so that his vision blurred. The drop was warm as blood, warm as the blood of a man racked by malarial fever.

B eyond the high walls there were other ruins, none of such proportions or importance as the inner city, however, and all of them so scattered, so overgrown and thrown down as to make detailed exploration of them out of the question in the time still available to Zouga.

The kopjes around the city had been fortified, but were deserted, the caves empty as the eye-sockets of a long-dead skull, smelling of the leopards and rock rabbits who were the latest tenants, but Zouga concentrated his search on the ancient mine workings which he had convinced himself had formed the backbone of this vanished civilization. He imagined deep shafts driven into a hillside, and dumps of loose rock like the ancient Cornish tin mines, and he scoured the densely wooded country for miles in each direction, eagerly checking each irregularity of ground, each eminence that could possibly be an abandoned mine dump.

He left Jan Cheroot to oversee the scratching and sweeping up of the last tiny scraps of gold in the temple yard, and all the men profited by the new relaxed supervision. They shared views with Jan Cheroot on the role of menial work in the life of a warrior and hunter.

The first spattering of rain had been only a warning of

the fury to come, and it had barely wet Zouga's shirt through to the skin before passing, but it was a warning that Zouga realized he was ignoring at his own peril. Yet still the hope of the ancient mine workings with their fat golden seams of the precious metal tantalized him, and he spun out the days until even Jan Cheroot began to worry.

'If the rivers spate, we will be trapped here,' he brooded beside the camp fire. 'Besides we have taken all the gold. Let us now live to spend it.'

'One more day,' Zouga promised him as he settled into his single blanket, and composed himself to sleep. 'There is a valley just beyond the southern ridge, it will take me only another day to search it – the day after tomorrow—' he promised sleepily.

Z ouga smelled the snake first, the sweet sickening stench of it filled his nostrils, so he drew each breath with difficulty, yet trying not to gasp or choke lest he called the snake's attention. He could not move, he was pinned under an immense dark weight which threatened to crush his ribs and the smell of snake suffocated him.

He could barely turn his head towards the place from which he knew the snake would come, and it came flowing sinuously, coil upon thick undulating coil. Its head was lifted, its eyes were unblinking and glassily fixed in the cold and deadly reptilian stare, the ribbon of its tongue flickered in a soft black blur through the icy smile of its thin curved lips. Its scales scratched softly across the earth, and they glittered with a soft metallic sheen, the colour of the polished gold foil that Zouga had gleaned from the temple courtyard.

Zouga could not move nor cry out, his tongue had swollen with terror to fill his mouth and choke him, but the snake slid past him, close enough to touch if he had had command of his arms to reach out. It slid on into the circle of soft wavering light, and the shadows drew back so the birds emerged from the darkness, each on its elevated perch.

Their eyes were golden and fierce, the cruel yellow curve of their beaks echoed by the proud pout of their russet-flecked breasts and the long pinions folded across their backs like crossed blades.

Though Zouga knew they were hunting falcons with belled tresses on their legs, yet they were the size of golden eagles. They were decked with garlands of flowers, crimson blossoms of King Chaka fire, and the snowy virginal white of arum lilies. They wore necklaces and chains of brightest gold about their arrogant necks, and as the serpent slid into the midst of the circle they stirred upon their perches.

Then as the serpent raised its glittering head with the crest of scales erect upon the back of its neck, the falcons burst into thunderous flight and the darkness was filled with the roar of their wings and the plaintive lament of their wild hunting cries.

Zouga lifted his hands to shield his face, and great wings beat all about him, as the flock of falcons took flight and the presence of the snake was no longer of significance – what was important was the departure of the birds. Zouga felt a tremendous sense of doom, of personal loss, and he opened his mouth, able to utter again. He shouted at the birds to call them back to roost.

He shouted into the darkness, after the soaring, buffeting thunder of the birds' wings and his own shouts and the cries of his servants woke him from the coils of the nightmare.

He woke to find the night was thunderous with the wind of the storm that swept down upon the camp. The trees tossed and thrashed their branches overhead, showering them with leaves and small twigs, and the rush of air was glacially cold. It stripped the thatch from their crudely built huts and it scattered ash and live coals from the fire. The coals, fanned into new life, were the only source of light, for overhead, the stars were obscured by the thick rolling banks of cloud that pressed close upon the earth.

Shouting to each other above the wind, they scrambled to collect their scattered equipment.

'Make sure the powder bags are kept dry,' Zouga bellowed, naked except for his tattered breeches, and barefooted as he groped for his boots. 'Sergeant Cheroot, where are you?'

The Hottentot's reply was lost in the cannon's roar of thunder that drove in their eardrums, and the flash of lightning that followed immediately seared their eyeballs, and imprinted on Zouga's vision the unforgettable picture of Jan Cheroot dancing stark naked on one foot, a red hot coal from the scattered fire stuck to the sole of the other, his wild curses lost in the drawn-out roll of the thunder and his face contorted like that of a gargoyle on the parapet of Notre Dame cathedral. Then the darkness fell on them again, like an avalanche, and out of it came the rain.

It came in cutting horizontal sheets like the blade of a harvester's scythe, so thick that it filled the air with water so they coughed and gasped like drowning men, it came with such hissing force that it stung their naked skins as though coarse salt had been fired at them from a shotgun barrel, and the cold chilled them to the bone, so that they crawled into a forlorn huddle, crowding together for comfort and warmth with the sodden fur blankets pulled over their heads, and stinking like a pack of wet dogs.

The cold gloomy dawn found them still huddled from the silver streams of falling rain, under the swollen bruised sky that pressed down upon them like the belly of a pregnant sow. Scattered and sodden equipment floated or was submerged in the ankle-deep flood of water that poured through the wind-shattered camp. The lean-to shelters had been wrecked, the camp fire was a muddy black puddle of ash, there was no prospect of rebuilding it, and with that went any chance of hot food or comfort for their stiff cold bodies.

Zouga had wrapped the powder bags in strips of oilskin,

and he and Jan Cheroot had held them in their laps, like ailing infants, during the night. However, it was impossible to open the bags and check the contents for damp, for the rain still teemed down out of the low grey sky in long thin silver lances.

Slipping and sloshing in the muddy footing, Zouga drove his men to make up their loads for the outward march, while he made his own final preparations. In the middle of the morning they ate a miserable and hasty meal of cold millet cakes and the last scraps of smoked buffalo meat. Then Zouga stood, with a cloak of half-cured kudu skin draped over his head and shoulders, the rain dripping from his beard and plastering his patched clothing to his body.

'Safari!' he shouted. 'We march!'

'And not too damned soon either,' muttered Jan Cheroot, reversing his musket on its sling so the barrel pointed at the ground and the rain could not run into the muzzle.

It was then that the porters discovered the extra burden that Zouga had for them. It was lashed to carrying poles of mopani wood with bark rope, and protected by a plaited covering of elephant grass.

'They are not going to carry it,' Jan Cheroot told him, squeezing the rain from his woolly eyebrows with his thumb. 'I told you they would refuse.'

'They'll carry it.' Zouga's eyes were cold and green as cut emeralds, and his expression was fierce. 'They carry it – or they'll stay here with it, dead!'

He had carefully selected the best specimen of the carved stone birds, the only one that was completely undamaged and the one on which the carving was the most artistically executed, and he had packed and prepared it for porterage himself.

For Zouga the carving was physical proof of the existence of the ancient abandoned city, proof that could not be denied when even the most cynical critic read his account in far-away London. Zouga guessed that the intrinsic value of this relic probably surpassed the equivalent weight in pure gold. The value of the artefact was not the most important consideration in Zouga's determination to carry the carving out to civilization. The stone birds had come to have a special superstitious significance to him. They had come to symbolize for him the success of his endeavours, and by possessing one of them he had in some strange manner taken possession of this entire savage and beautiful land. He would return for the others, but he must have this single perfect specimen. It was his talisman.

'You and you.' He picked two of his strongest and his usually most willing porters, and when they still hesitated, he unslung the heavy elephant gun from his shoulder. They saw his expression and knew that his intention was serious, deadly serious. Sulkily they began breaking down their own loads and distributing them amongst their comrades.

'At least let us leave this other thundering piece of rubbish.' The rain and the cold had affected Jan Cheroot as much as the others, and he indicated the tin box that contained Zouga's dress uniform with a hatred and contempt usually reserved for animate objects. Zouga did not bother to reply, but gestured to Matthew to take it up.

It was noon before the bedraggled little column struggled through the long sodden grass that choked the valley floor and began to climb the far side, slipping and cursing in the mud.

It rained for five days and five nights, sometimes in thick drumming bursts that seemed to fall in solid sheets of water from the sky, at other times it was a cold drizzling mist, that swirled gently about them as they trudged on in the soft treacherous footing, a fine silver mist that blanketed and muted all sound except the eternal dripping of the forest and the soft sighing passage of the wind in the upper branches.

The fever vapours seemed to rise from the very ground, entering their lungs with each breath, and in the icy cold mornings they writhed and twisted like the wraiths of tormented souls down in the hollows of the valleys. The porters were the first to show the symptoms of the disease, for the fever was in their bones and the cold rain brought it out so they shivered in uncontrollable spasms and their teeth chattered in their jaws until it seemed they might crack like porcelain. However, they were seasoned to the rigours of the disease and they were still able to march.

The bulky statue in its ungainly packing of grass and bark was borne painfully up the rocky ridges and down the other side by half-naked men staggering like drunks from the fever boiling in their veins, and when they reached the bank of another watercourse they dropped it gratefully and fell in the mud to rest without the will to cover themselves from the relentless rain.

Where there had been dried river-beds, with drifts of white sand shining like alpine snowfields in the sun, with quiet pools of still green water, and with steep high banks in which the brilliant kingfisher and little jewelled bee-eaters burrowed to nest, there were now maddened torrents of racing brown water, which brimmed over the high banks and carved out the roots of tall trees, toppling them into the flood and whisking them away as though they were mere twigs.

There was no possible means by which a man could cross these racing, foaming deluges; the corpse of a drowned buffalo with bloated pink belly and its limbs sticking stiffly into the air was borne downstream at the speed of a galloping horse, while Zouga stood morosely on the bank, and knew that he had left it too late. They were trapped by the spate.

'We will have to follow the river,' he grunted, and wiped his streaming face on the sleeve of his sodden hunting jacket.

'It goes towards the west,' Jan Cheroot pointed out with morbid relish, and it was not necessary for him to expand on the thought.

To the west lay the kingdom of Mzilikazi, King of the Matabele, and already they must be close to that vaguely defined area that old Tom Harkness had marked on his map.

'The Burnt Land – here Mzilikazi's impis kill all travellers.'

'What do you suggest, my ray of golden Hottentot sunshine?' Zouga demanded bitterly. 'Have you got wings to fly this?' He indicated the broad expanse of wild water, where the curled waves stood as stationary as carved sculptures as they marked the position of submerged rocks and hidden snags. 'Or what about gills and fins?' Zouga went on. 'Let me see you swim, or if you have neither wings nor fins, then surely you have good advice for me?'

'Yes,' Jan Cheroot answered as bitterly. 'My advice is that you listen to good advice when first it is given, and second that you drop those in the river.' He indicated the bundled statue and the sealed uniform box. Zouga did not wait for the rest of it, but turned his back and shouted.

'Safari! On your feet, all of you! We march!'

They worked slowly west and a little south, but too much westward for even Zouga's peace of mind, though his route was dictated by the network of rivers and flooded valleys.

On the sixth day the rain relented, and the clouds broke open, revealing a sky of deep aquamarine blue and a fierce swollen sun that made their clothing steam, and stilled the fever shakes of the porters.

Even with the accuracy of his chronometer in doubt, Zouga was able to observe the meridian passage that local noon and establish his latitude. He was not as far south as he had calculated by dead reckoning, so he was probably even further west than his suspect calculations of longitude suggested.

'The land of the Mzilikazi is drier,' Zouga consoled himself, as he wrapped his navigational instruments in their oilskin covers, 'and I am an Englishman, and the grandson of Tshedi. Not even a Matabele would dare deny me passage, despite what old Tom writes.' And he had his talisman, the stone bird, to add its protection.

Resolutely he faced west again, and drove his caravan onwards. There was one other misery to add to their sufferings. There was no meat, and there had been none since the day they left the abandoned city.

With the first onslaught of the rains, the great herds of game that had been concentrated upon the last few pools and waterholes had been freed to scatter widely across the vast land where every ditch and irregularity was now at last brimming with fresh sweet water and where the baked and sun-seared plains were already blooming green with the tender first shoots of new growth.

In five days' march in the rain, Zouga had seen only one small herd of waterbuck, the least palatable of all African game with its rank turpentine-scented musk which permeates the flesh. The heavily built bull, in his shaggy plum-brown coat led his small troop of hornless females at a frantic gallop across Zouga's front with his wide lyre-shaped horns cocked high and the perfect white circle over his buttocks flashing with each bound. He tore through the drizzling rain and dense wild ebony bush not twenty paces from Zouga. Zouga threw up the heavy gun and led on his driving shoulder.

Behind him his hungry, exhausted porters yipped like a pack of hunting dogs with anticipation, and Zouga held his aim for an instant to make deadly certain and then squeezed off the trigger.

With a sharp crack, the cap exploded under the falling hammer, but there was not the long spurt of flame from the

491

muzzle and the great clangour of the shot, followed by the heavy thumping impact of the lead ball into flesh. The gun had misfired and the handsome antelope led his harem away at full gallop, disappearing almost immediately into the bush and rain while the dwindling clatter of their hooves mocked Zouga. He swore with frustration as he laboriously drew the ball and charge with the corkscrew tool fitted to his ramrod, and found that the insidious rain had somehow entered the barrel, probably through the nipple and that the powder was as wet as though he had dipped it in the raging brown flood waters.

Those few hours of fierce sunlight on the sixth day gave Zouga and Jan Cheroot an opportunity to spread the coarse grey contents of their powder bags on a flat rock and dry it out so there would be less chance of another disastrous misfire, and while they did so the porters let drop their packs and limped off to find a dry spot to stretch their aching limbs.

Then too swiftly the sunlight was blotted out once more, and hurriedly they scooped the powder back into the pouches and as the fat raindrops began to hiss and splat about them they wrapped them in the worn oilskins, tucked them under their voluminous leather capes and resumed the westward trudge, heads bowed, silent and hungry and cold and miserable, Zouga's ears singing with the quinine-buzz, the first apparent side-effect of massive doses of the drug taken over long periods. The quinine-buzz that can lead to eventual, irreversible deafness.

Despite the heavy daily doses of the bitter powder, the morning arrived at last when Zouga woke with the deep ache in the marrow of his bones, the dull weight like a heavy stone lying behind his eyes and by midday he was shaking and shivering with the alternate flood of fierce heat and deathly sepulchre cold through his veins.

'The seasoning fever,' Jan Cheroot told him philosophically, 'it kills you, or hardens you.'

'Some individuals would appear to have a natural resistance to the ravages of this disease,' his father had written in his treatise, *The Malarial Fevers of Tropical Africa: Their Causes, Symptoms and Treatment*, 'and there is evidence to suggest that this resistance is hereditary.'

'We'll see now if the old devil knew what he was talking about,' Zouga mumbled through chattering teeth, hugging the stinking wet leather coat around his shoulders. It never occurred to him even briefly to halt for his affliction; he had not accorded that courtesy to any of his men, and he did not expect it himself.

He trudged on grimly, with his knees giving a rubbery little bounce at each pace, his vision blurring and starring into little pinwheels of light, then emerging again though phantom worms and gnats still wriggled across his sight. Every now and then a touch upon his shoulder from the little Hottentot who marched behind him directed Zouga's stumbling feet back on to the path from which they had strayed.

The nights were hideous with the nightmares of his fever-inflamed brain, they were filled with the buffeting thunder of dark wings and the sickly stench of snakes so that he would wake panting and screaming, often to find Jan Cheroot holding him with a comforting arm around his shaking shoulders.

The lifting of that first bout of his seasoning fever coincided with the next brief break in the rains. It seemed that the bright sunlight, magnified by the lingering moisture in the air, burned away the mists from his mind and the poisonous miasma from his blood, leaving him clear-headed, with a fragile sense of well-being but a weakness in his legs and arms and a dull ache up under the right-hand side of his rib cage where his liver was still swollen and hard as a rock, the typical after-effect of the fever.

'You will be all right,' Jan Cheroot prophesied. 'You threw it off as quickly as I've seen any man with his first hit of the fever. Ja! You are a man of Africa – she will let you live here, my friend.'

It was while he still walked on wavering legs, light-headed, so that it felt as though his feet did not touch the muddy earth but danced inches above it, that they cut the spoor.

The weight of the great bull had driven the spoor a foot deep into the sticky red mud, so that it was a series of deep potholes, strung across the earth like beads on a necklace. The exact impression of the huge pads had been cast in the holes as though in plaster of Paris, each crack and fissure in the skin of the sole, each irregularity, even the outline of the blunt toenails were there in precise detail, and at one place where the soft earth had been unable to bear his weight and the elephant had sunk almost belly deep, he had left the impression of his long thick ivories in the earth when he had used them to push himself free.

'It is him!' breathed Jan Cheroot, without looking up from the enormous spoor. 'I would know that spoor any-where.' He did not need to say more to identify the great old bull that they had last seen so many months before on the high pass of the elephant road on the escarpment of the Zambezi river.

'Not an hour ahead of us,' Jan Cheroot went on in a reverential whisper, like a man at prayer.

'And the wind stands fair.' Zouga found he was whisper-ing also. He remembered his premonition that he would encounter this animal again. Almost fearfully he looked up at the sky. In the east the storm clouds were rolling ponderously towards them once more, the brief respite was almost over. The next onslaught of the storm promised to be as fierce, and even those deep and perfect prints would soon dissolve into liquid mud and be washed into oblivion.

'They are feeding into the wind,' he went on, trying to put the threat of rain out of his mind and concentrating his still fever-dulled wits to the problem of the hunt.

The old bull and his remaining consort were feeding and moving forward with the wind into their faces. That way they would not walk into trouble. Yet these two old veterans, with their decades of accumulated experience, would not hold steadily into the wind for long, they would turn at intervals to get below the wind of a possible tracker.

Every minute now was of vital consideration, if the hunt was to succeed – for despite the weakness in his legs and the silliness in his head, Zouga had not for a single moment considered letting the spoor go. They might be a hundred miles within the borders of Mzilikazi's country with a Matabele impi of border guards closing swiftly, and the hours lost in following the two old bulls might make all the difference between escaping from these fever-haunted forests or leaving their bones here for the hyena to crunch, but neither Zouga nor Jan Cheroot hesitated. They began to shed their unneeded equipment, they would not need water bottles for the land was overflowing, the food bags were empty anyway, and the blankets sodden. They would shelter tonight against the old bull's massive carcass.

'Follow at your best pace,' Zouga shouted to his heavily burdened porters, dropping his unwanted equipment into the mud for them to pick up. 'You can fill your bellies with meat and fat tonight, if you put your feet to it now.'

They had to gamble all Zouga's remaining strength on the opening play, using speed to beat the rain and to reach the bulls before they made a turn into the wind and took the scent. They ran at the spoor, going hard from the first, knowing that even a healthy man could not hold that pace beyond an hour or two at the most before his heart broke.

In the first mile Zouga's vision was starring and fragmenting again, sweat drenched his lean body and he reeled like a drunkard as his legs threatened to give under him.

'Run through it,' Jan Cheroot counselled him grimly.

He did it, by a sheer effort of will. He drove himself through to that place beyond the pain. Quite suddenly his vision cleared and though there was no feeling at all in his legs they drove on steadily under him so he seemed to float over the ground without effort.

Running at his side Jan Cheroot recognized the moment when Zouga broke the shackles and went clear of his own weakness. He said nothing but glanced sideways at him, eyes bright with admiration and he nodded once. Zouga did not see that nod, for his head was up and his dreaming gaze was fixed far ahead.

They ran the sun to its zenith, Jan Cheroot not daring to break the rhythm for he knew that Zouga would drop like a man shot through the heart if they stopped to rest. They ran on as the sun began to drop, pursued by the ponderous cohorts of the oncoming storm that threatened it with extinction, and their own shadows danced ahead of them along the deeply driven elephant spoor. In a tight bunch behind Zouga, his four gunbearers matched him pace for pace, ready to hand him a weapon at the instant he required it.

The hunter's instinct warned Jan Cheroot. He twisted his head every few minutes to look back along the trail they had already run. That was how he picked them up.

They were two grey shadows, merging softly into the darker acacia shadows below the dripping trees, but they were moving with steadfast purpose, circling to strike their own spoor again, throwing a loop about their pursuers and taking the wind from them.

The bulls were half a mile away, moving with that swinging deceptively leisurely gait that would bring them, within minutes, full on to the hot trail with which the small band of hunters had overlaid their own huge pug marks; the trail would be reeking with the rank odour of man, the air thick with it.

Jan Cheroot touched Zouga's arm, turning him back upon their own run without checking him nor breaking the driving rhythm of his numbed legs.

'We have to catch them before they cut our spoor,' he called softly. He saw Zouga's eyes come back into focus and the colour flare in his waxen pale cheeks as Zouga turned and saw for the first time the two huge shapes cruising serenely through the open forest, under the tall umbrella-shaped acacias, moving with a stately deliberation down towards the string of reeking man-prints in the red mud.

The big bull was leading, his gaunt frame too tall and bony for the wasted flesh over which the skin hung in deep folds and bags. The huge yellow tusks were too long and heavy for the ancient head, and his ears were ripped and torn into scarred tatters that dangled on to his creased cheeks. He had been wallowing in a mud hole and his body was slick and shining with wet red mud.

He stepped out on his long, heavily boned legs around which the thick loose skin bagged and drooped like a badly tailored pair of breeches, and close behind him strode his askari, a big heavily toothed elephant, but dwarfed by his leader.

Zouga and Jan Cheroot ran together, stride for stride, their breath hissing and gasping in their throats, as they spent their last reserves to get in gunshot range before the bulls took the scent.

They traded all stealth, any attempt at concealment for speed, trusting that the weak eyesight of the old bulls would betray them. This time, the vagaries of the weather favoured them, for as they ran, the storm burst about them again.

It had held off just long enough to allow them to come up, and now the thick streamers of pale grey rain were drawn across the forest like lace curtains and the bad light beneath the thick cloud banks gave them cover to cross the last few hundred yards unseen, and the tapping rain and the rush of the wind in the branches of the acacias muffled their racing footfalls.

A hundred and fifty yards ahead of Zouga the old bull hit the man spoor, and it stopped him as though he had run into the side of an invisible cliff of glass. He flared back on his hind quarters, his back humping and his wrinkled ivoried head flying up high, the ragged banners of his ears filling like the mainsail of a tall ship, and clapping thunderously as he flapped them against his shoulders.

He froze like that for a long moment, groping at the tainted earth with the tip of his trunk, then he lifted it to his mouth and sprayed the scent into the open pink buds of his olfactory organs. The dread and hated odour struck him like a physical force and he went back another pace, then his trunk lifted straight into the air above him, and he wheeled and like a well-trained pair of coach horses his tall askari wheeled with him, shoulder to shoulder, and flank to flank, they began to run – and Zouga was still a hundred yards behind them.

Jan Cheroot dropped on to one knee into the mud, and flung up his musket. At the same instant, the askari checked slightly and swung left, crossing his leader's rear. Perhaps it was unintentional, but neither Zouga nor Jan Cheroot believed that. They knew that the younger bull was drawing fire, protecting the other with his own body.

'You want it? Take it then, you thunder!' Jan Cheroot shouted angrily, he knew he had lost too much ground by pausing to fire.

Aiming for the younger bull, Jan Cheroot took the hip shot, and the bull staggered to it, flecks of red mud flying from his skin where the ball struck, and he broke his stride, favouring the damaged joint, swinging out of the line of his run, broadside to the hunters, while the great bull ran on alone.

Zouga could have killed the crippled bull with a heart shot, for the animal was down to a dragging, humpbacked trot and the range was less than thirty paces, but Zouga ran straight past him, never checking, hardly glancing at him, knowing he could leave Jan Cheroot to finish that business.

He ran after the big bull, but despite his utmost endeavour, losing ground to him steadily.

Ahead of them, the ground dipped into a shallow open saucer, and beyond that it rose to another ridge on which the wild teak trees stood like sentinels in the grey rain. The bull went down into the saucer, still in his initial burst of speed, stretching out so that his padfalls sounded like the steady beat of a bass drum, opening the gap between himself and his hunter – until he reached the bottom of the dip, and then he was almost halted.

His weight broke the surface of the swampy ground, and he sank through almost to his shoulders, and had to lunge for each step, with the glutinous mud sucking and squelching obscenely at each of his frantic movements.

Swiftly Zouga closed the range, and his spirits took wing, his exultation driving back his weakness and fatigue. He felt intoxicated with battle lust. He reached the swampy ground, and leapt from tussock to tussock of coarse swamp grass, while the bull struggled on ahead of him.

Closer and closer still Zouga came up to him, almost point-blank range, less than twenty yards and at last he stopped and balanced on one of the little islands of grass roots.

Just ahead of him the bull had reached the far side of the swamp, and was heaving himself out on to the firm ground at the foot of the slope. The bull's front legs were higher than his still-bogged hindquarters, the whole slope of his back was exposed to Zouga, the knuckles of his massive spine stood out clearly through the mudpainted skin and the arched staves of his rib cage were like the frames of a Viking longboat. Zouga fancied he could actually see the pounding rhythm of the great heart beating against them.

There could be no mistake this time. In the months since their first encounter, Zouga had become a skilled huntsman, he knew the soft and vital places in the mountainous bulk of an elephant's body. At this range and from this angle, the heavy ball would shatter the spine between the shoulder blades without losing any of its velocity, and it would go on deep, to the heart, to those thick serpentine arteries that fed the lungs.

He touched the hair trigger, and with a pop like a child's toy the gun misfired. The great grey beast heaved himself clear of the mud, and went away up the slope, now at last settling into the swinging ground-eating gait which would carry him fifty miles before nightfall.

Behind him Zouga reached the firm ground and flung down the useless weapon, dancing with impatience as he screamed at his bearers to bring up the second gun.

Matthew was fifty paces behind him, slipping and staggering in the swampy ground. Mark, Luke and John were strung out behind him.

'Come on! Come on!' screamed Zouga, and snatched the second gun from Matthew, and dashed away up the slope. He had to catch the bull before he reached the crest of the slope, for down the other side he would go like an eagle on the wind.

Zouga ran now with all his heart, with all his will and the very last dregs of his strength, while behind him Matthew stopped, snatched up the misfired weapon from where Zouga had thrown it down, and, acting instinctively in the heat and excitement of the chase, he reloaded it.

He poured another handful of black powder on top of the charge and ball already in the barrel and tamped down a second quarter-pound ball of lead on top of it all. In so doing he changed the gun from a formidable weapon to a lethal bomb that could maim or even kill the man who attempted to fire it. Then Matthew slipped another percussion cap over the nipple and ran on up the slope after Zouga.

The bull was nearing the crest of the ridge, and Zouga was coming up on him, but slowly, the difference in their speed just barely discernible. At last Zouga's strength was failing, he could keep this pace for minutes more and he knew when he finally stopped he would be on the verge of total physical collapse.

His vision was swimming and wavering, and his feet stumbled and slipped on the wet lichen-covered rocks of the slope, and the rain beat into his face, almost blinding him. Sixty yards ahead of him the bull reached the crest, and there he did something that Zouga had never seen a hunted elephant do before – he turned broadside, flaring his ears, to look at his pursuers.

Perhaps he had been pushed too hard, perhaps he had been hunted too often and the hatred had accumulated like weed below the waterline of an old ship, perhaps this was his last defiance.

For a moment he stood tall, and glistening black with mud and rain, silhouetted against the low grey sky, and Zouga hit him in the shoulder, the gun ringing like a bronze cathedral bell, and the long lick of red flame blooming briefly in the gloom.

Both man and beast reeled to the shot, Zouga driven off-balance by the recoil and the bull taking the hardened lead ball through the ribs and going back on his haunches, the rheumy old eyes clenching tightly to the shock of it.

The bull kept his feet, though he was hard hit, and he opened his eyes and saw the man, that hated, evil-smelling and persistent animal that had persecuted him so relentlessly down the years.

He launched himself back down the slope, like an avalanche of dark grey granite, and his repeated blood squeals rang against the low sky, and Zouga turned and ran from the charging bull, while the earth trembled beneath his feet at the weight and nearness of the stricken beast.

Matthew stood his ground, even in the terrible press of the moment. Zouga loved him for that. He stood to do his duty, to deliver the second gun to his master.

Zouga reached him, just ahead of the charging bull, dropped his smoking weapon, snatched the second gun from Matthew, never suspecting that it had been double-shotted, and as he turned he thumbed the hammer back and swung up the long thick barrel.

The bull was on top of him, blotting out the rain-sodden sky, the long yellow ivories raised like roof-beams over his head, and the trunk already uncoiling to reach down and snatch Zouga up.

Zouga pressed the hair trigger, and this time the gun fired. With a shattering roar the barrel burst, the metal opening like the petals of a flower, and burning powder flew back into Zouga's face, singeing his beard and blistering the skin of his face. The hammer was blown clean off the barrel and it hit Zouga in the cheek, just under the right eye, cutting a jagged wound clean down to the bone. The shattered weapon flew out of his grasp, and slammed back into his shoulder with such force that he felt the ligaments and tendons tearing. Zouga was hurled into a backward somersault that carried him just beyond the grasp of the bull's questing trunk.

He fell heavily behind a pile of loose stone chips, and for a moment the elephant checked, going back on its hind legs to avoid the flashing flame and smoke of the explosion, blinded and unsighted for a moment – and then it saw the gunbearer still standing.

Matthew started to run, poor, loyal, brave Matthew, but the bull had him before he had gone a dozen paces. It took him about the waist with a single coil of its long trunk, and it threw him into the air as though he were light as a child's rubber ball. Matthew went up forty feet, with his arms and legs windmilling, his scream of terror unheard in the deafening squeals of the elephant. It sounded like the whistle of a steam engine blown by a crazed engineer, and Matthew seemed to rise very slowly into the air, hang for a long moment and then drop just as slowly downwards.

The bull caught him in mid-air and threw him again, this time even higher.

Zouga dragged himself into a sitting position. His right arm hung limply on its torn muscles and tendons, blood streamed from his ripped cheek into his beard and his eardrums were so tortured by the explosion that the elephant's squeals seemed muted and far off. He looked up groggily and saw Matthew high in the air, beginning to fall, saw him hit the ground, and the elephant begin to kill him.

Zouga dragged himself to his knees, and began to creep over the mound of loose stone towards the empty gun, the gun from which he had fired the first shot and which he had dropped when he snatched the double-shotted gun from Matthew, the gun which now lay five paces from him, five paces which seemed an infinite distance to drag his maimed body.

The elephant placed one foot on Matthew's chest and his ribs crackled like dry sticks in a fierce fire. It took his head in its trunk and plucked it from Matthew's shoulders, as easily as a farmer kills a chicken.

The elephant tossed Matthew's head aside and as it trundled down the slope close to where Zouga sat, he saw that Matthew's eyelids were blinking rapidly over his bulging eyeballs and that the nerves flickered under the skin of his cheeks.

Tearing his eyes from the gruesome object, Zouga lifted the empty gun into his lap and began to reload it. He had no use or feeling in his right arm, which still hung limply at his side.

Twenty paces away the elephant knelt over Matthew's decapitated body and drove one of the long yellow ivories through his belly.

Painfully Zouga poured a handful of powder into the gun muzzle, trying not to be distracted from his task.

Matthew hung impaled through the middle from the

bloody tusk like a wet shirt on a laundry line, the elephant's trunk came up and coiled python-like about his battered body.

Zouga dropped a ball from his pouch into the barrel of the gun, and one-handed tamped it home with the ramrod.

The elephant tore an arm from the body, and Matthew slid from the point of the tusk and dropped once more to earth.

Moaning softly with the pain of each movement, Zouga primed the gun and hauled back the hammer against the powerful tension of the spring.

The elephant was kneeling with both front legs on what remained of Matthew, grinding him into a red mush against the rocky earth.

Dragging the gun with him, Zouga crawled back to the mound of rock chips behind which he had fallen. Using only his left hand, he balanced the stock of the big elephant gun over the top of the mound.

The elephant was still squealing in unabated fury as it crushed Matthew's corpse.

Grovelling flat on his belly, Zouga sighted over the thick barrel, but with only one hand it was almost impossible to hold the clumsy weapon true, and his vision swam and wavered with pain and exhaustion.

For an instant the shaking foresight aligned with the crude vee of the backsight, and he let the shot fly, in flame and a billowing cloud of burnt powder smoke.

The elephant's squeals stopped abruptly. As the powder smoke was blown aside on the cold breeze, Zouga saw that the bull had hoisted himself wearily upright and was swaying slowly from foot to foot. The massive head dropped under the weight of its blood-smeared tusks and its trunk hung as limply as Zouga's own damaged right arm.

The elephant was making a strange mournful humming sound deep in his chest, and from the second bullet wound just behind his bony shoulder joint, his heart blood spurted in short regular jets, to the beat of the huge heart, and poured down his body in a honey-thick stream.

The bull turned towards where Zouga lay, and shuffling like an old and very weary man came towards him, the tip of his trunk twitching with his last fading warlike instincts.

Zouga tried to drag himself away, but the bull came on faster than he could crawl, and the trunk reached out, touched Zouga's ankle, the vast bulk of the elephant filled the sky above Zouga, and he kicked out frantically but the trunk tightened its grip on his ankle with agonizing unbearable strength, and Zouga knew that it would tear his leg out of its socket at the hip.

Then the elephant groaned, a shuddering exhalation of air from the torn lungs, the grip on Zouga's ankle relaxed and the old bull died on his feet, his legs collapsed under him and he went down.

He fell with a weight and force that made the earth bounce and tremble under Zouga's prostrate body and with a thud that Jan Cheroot, who was crossing the swamp, heard clearly from a mile away.

Zouga dropped his head against the earth and closed his eyes, and the darkness overwhelmed him.

J an Cheroot made no attempt to move Zouga from where he lay beside the old bull's carcass. He built a rough shelter of saplings and wet grass over him and then coaxed smoky flames from a fire at his head, and from another at his feet. This was all he could do to warm him until the porters came up with the blankets at sunset.

Then he helped Zouga into a sitting position and between them they strapped the damaged arm.

'God Almighty,' croaked Zouga, as he prepared the needle and thread from his sewing kit, 'I'd give both the tusks for a dram of good malt whisky.'

Jan Cheroot held the hand mirror, and using one hand only Zouga stitched the flap of his torn cheek back into place, and as he pulled the last knot tight and snipped the thread, he collapsed back into his damp and stinking fur kaross.

'I will die rather than march again,' he whispered.

'That is your choice exactly,' Jan Cheroot agreed, without looking up from the chunks of elephant liver and heart that he had begun wrapping in yellow fat before stringing them on a green twig. 'You can march or die here in the mud.'

Outside the hut the porters were wailing and chanting the mourning dirge for Matthew. They had gathered the fragments of his horribly mutilated body and made a package of them, wrapping them in Matthew's own blanket and binding it up with bark rope.

They would bury him the following morning, but until then they would keep up the haunting cries of mourning.

Jan Cheroot scraped coals from the fire and began to grill the kebab of liver and fat and heart over them, 'They will be useless until they have buried him, and we must still cut the tusks.'

'I owe Matthew a night of mourning at least,' Zouga whispered. 'He stood the bull down. If he had run with the second gun . . .' Zouga broke off, and groaned as a fresh stab of pain transfixed his shoulder. Using his good left hand, he scratched under the skin blanket on which he lay, moving the lumps of stone which had caused the discomfort.

'He was good – stupid, but good,' Jan Cheroot agreed. 'A wiser man would have run.' He turned the kebab slowly over the coals. 'It will take all day tomorrow to bury him and then cut the tusks from both elephants. But we must march the day after that.'

Jan Cheroot had killed his bull down on the plain, under the outspread branches of a giant acacia. Looking through the low opening of the hut Zouga could see the carcass of his own bull lying on its side not twenty feet away. Already it was swelling with trapped gas and the upper legs thrust out stiffly above the grey balloon of the belly. The tusks were unbelievable. Even as he stared at them Zouga thought they must be a fantasy of his exhaustion and agony. They were as thick as a girl's waist, and the spread of them must have been twelve feet from tip to tip.

'How much will they weigh?' he asked Jan Cheroot, and the Hottentot looked up and shrugged.

'I have never seen a bigger elephant,' he admitted. 'We will need three men to carry each of them.'

'Two hundred pounds?' Zouga asked, the conversation distracted him from the agony in his shoulder.

'More,' Jan Cheroot decided. 'You will never see another like him.'

'No,' Zouga agreed. 'That is true. There will never be another like him.' Deep regret blended with his pain, making it more intense. Regret for the magnificent beast and sorrow for the brave man who had died with him.

The pain and the sorrow would not let him sleep that night, and in the dawn when they gathered in the rain to bury Matthew, Zouga strapped his damaged arm into a bark sling and had two men help him to his feet, then he walked unaided but slowly and stiffly up the slope to the grave, using a staff to balance himself.

They had wrapped Matthew's body in his fur blanket and placed his possessions with him, his axe and spear, his food bowl and beer calabash, to serve him on the long journey ahead.

Singing the slow mournful song of the dead, they packed the rock over and around him so that the hyena would not dig him out. When they had finished, Zouga felt drained of strength and emotion. He staggered back to his hut and crawled under the dank blanket. He had only that day to gather his strength for the march that must be resumed in

507

the dawn. He closed his eyes, but could not sleep for the thud of the axes into bone as Jan Cheroot supervised the chopping of the tusks from the casket of the old bull's skull.

Zouga rolled on to his back, and once again a loose rock chip dug into his aching body. He reached back and pulled it from under the blanket, was about to throw it aside when something caught his attention and he arrested the movement.

The rock was as white and as crystalline as the candied sugar that Zouga had loved so as a boy, a pretty little fragment, but that was not what had stopped his hand.

Even in the subdued light of the hut, the thin irregular seam of metal that wavered uncertainly between quartz crystals flicked a pin-prick of bright gold into his eyes. Zouga stared at it numbly, twisting the lump of quartz to catch the light and make it twinkle. There was a sense of unreality about the moment, as there always is when something sought for and longed for is at last held in the hand.

He found his voice at last, a hoarse croak through his swollen, blistered, powder-scorched lips, and Jan Cheroot came almost immediately.

'The grave,' he whispered urgently, 'Matthew's grave, it was dug so swiftly in such rocky soil.'

'No,' Jan Cheroot shook his head. 'It was there. There are other holes like it along the ridge.'

Zouga stared at him for a long moment, his face lopsided with the scabbed and stitched wound, his one eye a mere slit in the puffed and bruised flesh. He had let himself sink low from the wound. It had been under his nose and he had almost missed it. He started to drag himself out of his blanket.

'Help me!' he ordered. 'I must see them. Show me these holes.'

Leaning on Jan Cheroot's shoulder, stooped to favour his shoulder, he dragged himself along the ridge in the rain, and when at last he was satisfied, he limped back to the hut and used the last feeble light of that day to scrawl in his journal, holding it in his lap and bowing over it to protect the pages from the drip of rain through the rough thatching, using his left hand so the writing was barely decipherable.

'I have named it the Harkness Mine, for this must be very similar to the ancient workings that old Tom described. The reef is white sugar quartz and runs along the back of the ridge. It would appear to be very narrow but rich, for there is visible gold in many of the samples. My injury prevents me crushing and panning these, but I would estimate values well in excess of two ounces of fine gold to the ton of quartz.

'The ancient miners have driven four shafts into the hillside. There may be more that I overlooked, for they are heavily overgrown and an attempt has been made to refill the shafts, possibly to conceal them.

'The shafts are large enough to admit a small man crawling on hands and knees. Probably they used child-slaves in the diggings and the conditions of labour in these rabbit warrens must have been infernal. In any event they were only able to go down as far as the water table, and without sophisticated machinery to pump the flooded working, they would have been abandoned. This is probably what happened here at the Harkness Mine and there is almost certainly a great amount of gold-bearing ore to be recovered by modern methods.

'The rock dump on which stands my rude hut is composed almost entirely of the gold matrix, awaiting crushing and refining, and the miners were probably driven away by an enemy before they could complete their labours.

'I am couched upon a mattress of gold, and like King Midas all around me is the precious metal. Like that unfortunate King, there seems to be little profit in it for me that I can perceive at this moment—'

Zouga paused, and laid his pen aside, warming his icy hands at the smoky fire. He should have felt wildly elated.

He picked up his pen once again. He sighed and then wrote tortuously,

'I have a huge store of ivory, but it is spread across this land, buried in small caches. I have fifty pounds and more of native gold in ingot and nugget, and I have discovered the mother lode of untold fortune, but it cannot buy me a pound flask of gunpowder nor an unguent for my grievous injuries.

'I will not know until tomorrow if I have the strength remaining to me to continue the march to the south, or if I am destined to remain here with Matthew and the great elephant as my only companions.'

Jan Cheroot shook him awake. It took a long time. Zouga seemed to be swimming up from great depths through cold and murky water, and when at last he surfaced, he knew immediately that his gloomy prophecy written in the journal the previous evening had become reality. There was no feeling or strength in his legs. His shoulder and arm were bound rock hard with spasmed muscle.

'Leave me here,' he said to Jan Cheroot, and the Hottentot heaved him to a sitting position, snarling at him when Zouga cried out at the agony of each movement, and forced him to drink the steaming hot soup made from elephant marrowbones.

'Leave one gun with me,' Zouga whispered.

'Here.' Jan Cheroot ignored the order and instead made him take the bitter white powder. Zouga gagged on the quinine.

It took two porters to get him on his feet.

'I am leaving that stone.' Jan Cheroot pointed to the packaged statute. 'We cannot carry both of you.'

510

'No!' Zouga whispered fiercely. 'If I go, the bird must go with me.'

'How?'

Zouga shrugged off their hands.

'I will walk,' he said. 'Carry the bird.'

They made less than five miles that day, but the following day the sun emerged again to cheer them on. Once it warmed Zouga's abused muscles, he could increase the pace.

That night he logged ten miles in his journal when they camped in open grassland. In the dawn Zouga was able to crawl from his blankets and gain his feet unaided. His injuries still stiff, he used the staff to leave the thorn *scherm* by its single gate and limp to the periphery of the camp. When he urinated his water was a dark amber colour from the fever and the quinine, but he knew now that he was going to be able to continue the march.

He looked up at the sky. It would rain again soon. They should start at once. He was about to turn back to the camp and rouse the porters when movement in the tall grass caught his attention.

For a moment he thought it might be a troop of wild ostrich passing the camp, then suddenly he realized that the whole plain was alive with swift but stealthy movement, the fluffy grass tops rustling and nodding with the passage of many bodies, only now and again there was a brief glimpse of ruffled plumes above the grass. The movement spread swiftly around both sides of the small camp, where Zouga's men still slept.

Zouga stared uncomprehendingly, leaning on his staff, still muzzy with sleep and fever and anchored by his injuries, he did not move until the swift encircling movement had been completed, and then the stillness and silence descended again so for a moment he believed he had been imagining phantoms.

Then there was a soft fluting whistle, like a blast on Pan's pipe, sweet and hauntingly melodious in the dawn, and immediately there was movement again, an encroaching movement, like a strangler's hand upon the throat. Zouga saw the ostrich plumes clearly now, snowy white and dead black they swayed and danced above the grass tops, and immediately afterwards he saw the war shields, long oval shields of dappled black and white cowhide. The long shields – the Matabele.

Dread was a cold, heavy lump under his ribs, yet instinct warned him that to show it would mean death, just when he had once again believed in life.

There were a hundred, he calculated swiftly as he glanced around the closing ring of warriors. No, there were more than that, at least two hundred Matabele *amadoda* in full war plumage, only the plumes and their eyes showing above the tops of the long dappled shields. The grey dawn light glinted on the broad-bladed stabbing spears, held underhand so the points protruded beyond the ring of shields. The ring was unbroken, shield overlapped shield, the encircling horns of the bull, the classic tactics of the Matabele, the finest and most ruthless warriors that the continent of Africa had ever spawned.

'Here Mzilikazi's border impis kill all travellers,' Tom Harkness had written.

Zouga drew himself up and stepped forward, holding up his one good arm with the palm extended towards the ring of shields.

'I am an Englishman. A commander of the great white Queen, Victoria. My name is Bakela, son of Manali, son of Tshedi – and I come in peace.'

From the ranks stepped a man. He was taller than Zouga and his tossing ostrich plumes turned him into a giant. He

512

swept aside his shield, and he was lean and muscled like a gladiator. On his upper arms he wore the tassels of cow tails, each one awarded him by his King for an act of valour. The cow tails were thick bunches, layer upon layer. His short kilt was of spotted civet cat tails, and there were more cow tails bound around his calves just below the knees. He had the handsome smooth moon-face of the true Nguni, with a broad nose and full sculptured lips. His bearing was noble, the carriage of his head proud.

He looked at Zouga slowly and with grave attention. He looked at his tattered rags, at the untidy bindings that held his damaged arm, the staff on which he leaned like an old man.

He studied Zouga's singed beard, and powder-burned cheeks, the blisters on his lips and the black scabs that clung obscenely on his swollen discoloured cheek.

Then the Matabele laughed. It was a deep musical laugh, and then he spoke.

'And I,' he said, 'am Matabele. An Induna of two thousand. My name is Gandang, son of Mzilikazi, son of the high heavens, son of Zulu, and I come with a bright spear and a red heart.'

R obyn Ballantyne realized within the first day's march that she had seriously miscalculated her father's strength and resilience when she made the decision to try for the coast. Perhaps Zouga had divined instinctively what she a trained physician should have known. That thought made her angry with herself. She found that since parting with Zouga her hostility and sense of rivalry towards him had, if anything, increased. It made her angry that he should have given the correct advice.

By noon of that first day Robyn had been forced to call a halt and to go into camp. Fuller Ballantyne was very weak, weaker than he had been when first she found him. His skin was burning hot and dry to the touch. The movement of the litter, the jolting and bumping over uneven ground had aggravated Fuller's leg. It was grotesquely swollen, and so tender that he screamed and fought at the lightest touch upon the discoloured skin.

Robyn had one of the bearers begin work on a cradle of green twigs and bark to place over the leg and keep the fur blanket off it, and then she sat by the litter applying a damp, cool cloth to her father's forehead and speaking to little Juba and the Mashona woman, not expecting nor receiving advice from them, but taking comfort from the human contact.

'Perhaps we should have stayed at the cave,' she fretted. 'At least he would have been more comfortable there, but then for how long could we have stayed?' She spoke her thoughts aloud. 'The rains will be on us soon. We could not have stayed, and even if we march as slowly as this, we will still be trapped here by them. We simply must increase the pace – and yet I do not know if he can survive it.'

However, on the following day Fuller seemed stronger again, the fever had cooled, and they made a full day's march, but in the evening when they went into camp he had once more sunk very low.

When Robyn removed the dressing from the leg, it seemed to be less sensitive, and she was relieved – until she saw the colour of the skin around the ulceration. When she lifted the soiled dressing to her nose and sniffed it, she caught the taint that her professor of medicine at St Matthew's had taught her to watch for. It was not the usual taint of benign pus, but a more pervasive odour, the smell of a decomposing corpse. Her alarm flared, and she threw the dressing on the fire and with dread returned to her examination of the leg.

From the inside of the groin, down the wasted thigh

muscles, there were the unmistakable scarlet lines beneath the thin pale skin, and the extreme sensitivity of the area seemed to have passed. It was almost as though Fuller had no further feeling in the leg.

Robyn tried to console herself that the change and mortification in the leg was unconnected with having carried him two days in a litter over rough ground. But what other reasons were there? She could find no answer. Before the move, the ulceration had been stabilized, for it was almost eighteen months since the slaver's ball had shattered the bone.

The movement in the litter must have precipitated some serious change in the limb, and this was the result.

Robyn felt herself culpable. She should have listened to Zouga. She had brought this on her own father. Gasgangrene. She could only hope that she was wrong, but she knew she was not. The symptoms were unmistakable. She could only continue the march and hope they would reach the coast and civilization before the disease swept to its inevitable climax – but she knew that hope was futile.

She wished that she had been able to develop the same philosophical acceptance that most of her fellow physicians cultivated in the face of disease or injury which were beyond their training and ability to alleviate. But she knew she never could, always she would be victim of this helpless sense of frustration, and this time the patient was her own father.

She bound up the leg in a hot compress and knew that it was a pathetic gesture, like trying to hold back the tide with a child's sand-castle. In the morning the leg felt cooler to the touch, and the flesh seemed to have lost resilience so that her fingers left depressions as though she had touched unleavened bread. The smell was stronger.

They made a full day's march, and Fuller was silent and comatose in the litter as Robyn walked beside it. He no

515

longer chanted psalms and wild exhortations to the Almighty, and she thanked God that at least there was no pain.

In the late afternoon they met a broad pathway, well travelled and running east and west as far as Robyn could see it. It fitted exactly the description and the location that her father had written of in his journal. Little Juba burst into tears and was rendered almost helpless with terror when she saw the road.

They found an encampment of deserted and dilapidated huts, that might have been those used by the slave-traders, and Robyn ordered camp there. She left the Mashona woman and the still snivelling and shivering Juba to tend Fuller Ballantyne, and she took only old Karanga with her. He armed himself with his long spear, and strutted like an ancient peacock to be so honoured. Within two miles, the pathway climbed steeply to pass through a saddle in a line of low hills.

Robyn was seeking evidence that this was indeed the slave road, the Hyena Road, as Juba called it tearfully.

She found her evidence on the saddle, lying in the grass a few paces from the edge of the track. It was a double yoke, hewn from a forked tree trunk and roughly dressed with an axe.

Robyn had studied the sketches in her father's journal, and she recognized it immediately. When the slave-masters did not have chains and cuffs, they used these yokes to bind their captives around their necks; two slaves linked together and forced to do everything in concert, march, eat, sleep and defecate – everything except escape.

Now all that remained of the slaves who had once worn the yoke were a few fragments of bone that the vultures and hyena had overlooked. There was something terribly sad and chilling about that roughly carved fork of wood, and Robyn could not bring herself to touch it. She said a short prayer for the unfortunate slaves who had died on that spot and then, with her knowledge that she was on the slave road confirmed, she turned back towards the camp.

That night she held counsel with the Hottentot Corporal, old Karanga and Juba.

'This camp and road have not been used for this many days,' Karanga showed Robyn both hands twice with fingers spread, 'twenty days.'

'Which way did they go?' Robyn asked. She had come to have confidence in the old man's tracking ability.

'They were moving along the road towards the sunrise, and they have not yet returned,' quavered Karanga.

'It is even as he says,' Juba agreed, and it must have been an effort to agree with somebody for whom she had such disdain and jealousy. 'The slavers will make this the last caravan before the rains. There will be no trading after the rivers are full, and the Hyena Road will grow grass until the dry season comes again.'

'So there is a caravan of slavers ahead of us,' Robyn mused. 'If we follow the road we may overtake them.'

The Hottentot Corporal interrupted. 'That will not be possible, madam. They are weeks ahead of us.'

'Then we will meet them returning, after having sold their slaves.'

The Corporal nodded and Robyn asked him, 'Will you be able to defend us, if the slavers decide to attack our column?'

'Me and my men,' the Corporal drew himself to his full height, 'are a match for a hundred dirty slavers,' he paused and then went on, 'and you shoot like a man, madam!'

Robyn smiled. 'All right,' she nodded. 'We will follow the road down to the sea.'

And the Corporal grinned at her. 'I am sick of this country and its savages, I long to see the cloud on Table Mountain and wash the taste of dust from my throat with good Cape Smoke once again.'

The hyena was an old male. There were patches of fur missing from his thick shaggy coat, and his flat, almost snakelike head was covered with scars, his ears ripped away by thorns and a hundred snapping, snarling encounters over the decomposing carcasses of men and animals. His lip had been torn up into the soft of the nostril in one of these fights and had healed askew, so that the yellow teeth in one side of his upper jaw were exposed in a hideous grin.

His teeth were worn with age, so that he could no longer crush the heavy bones that made up the bulk of hyena diet, and unable to compete, he had been driven from the hunting pack.

There had been no human corpses along the trail since the slave column had passed weeks before, and game was scarce in this dry country. Since then the hyena had subsisted on scrapings, the fresh dung of jackal and baboon, a nest of field mice, the long-abandoned and addled egg of an ostrich which had burst in a sulphurous geyser of gas and putrefied liquid when he pawed at it. However, even though the hyena was starving, he still stood almost three foot high at the shoulder and weighed 140 pounds.

His belly under the matted and scruffy coat was concave as that of a greyhound. From high ungainly shoulders, his spine slanted back in a bony ridge to his scraggy hindquarters.

He carried his head low, snuffling at the earth for offal and scourings, but when the scent came down on the wind to him, he lifted his head on high and flared his deformed nostrils.

There was the smell of wood smoke, of human presence, which he had grown to associate with a source of food, but sharper, clearer than all the others was a smell that made the saliva run from his twisted, scarred jaws in drooling silver ropes. He went lolloping into a swaying, uneven trot up into the wind following the tantalizing drift of that odour. The scent that had attracted the old dog hyena was the cloyingly sweet taint of a gangrenous leg.

The hyena lay on the outskirts of the camp. It lay like a dog, with its chin on its front paws and its hindlegs and its bushy tail drawn up under its belly, flat behind a clump of coarse elephant grass. It watched the activity about the smoking watch fires.

Only its eyes rolled in their sockets, and the ragged stumps of its ears twitched and cocked to the cadence of human voices, and the unexpected sounds of a bucket or an axe on a stump of firewood.

Once in a while a puff of wind would bring a whiff of the scent that had first attracted the hyena, and it would snuffle it, suppressing with difficulty the little anxious cries that rose in its throat.

As the evening shadows thickened, a human figure, a half-naked black woman, left the camp, and came towards its hiding place. The hyena gathered itself to fly but before she reached the place where it lay, Juba paused and looked about her carefully without seeing the animal, then she lifted the flap of her beaded apron, lowered herself and squatted. The hyena cringed and watched her. When she stood and returned to the camp, the creature, emboldened by the oncoming night, crept forward and wolfed down that which Juba had left.

Its appetite was piqued and as the night fell, it inflated its chest, curled its bushy tail up over its back and uttered its drawn-out haunting cry, ascending sharply in key, 'Oooo-auw! Oooo-auw!' – a cry so familiar to every man and woman in the camp that hardly one of them bothered to look up.

Gradually the activity about the camp fires subsided, the sound of human voices became drowsy and intermittent, the fires faded, the flames sinking and the darkness crept in upon the camp, and the hyena crept in with it.

Twice a sudden loud voice put it to trembling flight and it galloped away into the bush, only to gather its courage at the renewed silence and creep back. It was long after midnight when the beast found a weak place in the protective *scherm* of thorn branches about the camp and quietly, furtively pushed its way through the opening.

The smell led it directly to an open-sided, thatched shelter in the centre of the enclosure, and with its belly low to the earth the huge dog-like animal slunk closer and fearfully closer.

R obyn had fallen asleep beside her father's litter, still fully dressed and in a sitting position; she merely let her head fall forward on to her crossed arms and then overwhelmed by fatigue and worry and guilt she at last succumbed.

She awoke to the old man's shrill shrieks. There was complete darkness blanketing the camp, and Robyn thought for a moment that she was blinded by a nightmare. She scrambled wildly to her feet, not certain where she was and she stumbled over the litter. Her outflung arms brushed against something big and hairy, something that stank of death and excrement, a smell that blended sickeningly with the stench of her father's leg.

She screamed also, and the animal growled, a muffled sound through clenched jaws like a wolfhound with a bone. Fuller's shrieks and her screams had roused the camp, and somebody plunged a torch of dried grass into the ashes of the watch fire. It burst into flames, and after the utter blackness the orange light seemed bright as noonday.

The huge humpbacked animal had dragged Fuller from

his litter in a welter of blankets and clothing. It had a grip on his lower body, and Robyn heard the sharp crack of bone splintering in those terrible jaws. The sound maddened her and she snatched up an axe that lay beside the pile of firewood and struck out at the dark misshapen body, feeling the axe strike solidly, and the hyena let out a choking howl.

The darkness and its own starvation had emboldened it. It had the taste in its mouth now, seeping through the blankets into its locked jaws and it would not relinquish its prey.

It turned and snapped at Robyn, its huge round eyes glowing yellow in the light of the flames and those terrible yellowed fangs clashing like the snap of a steel man-trap, closing on the axe handle inches from her fingers, jerking it out of her hand. Then it turned back to its prey, and once more locked its jaws on to the frail body. Fuller was so wasted that he was light as a child and the hyena dragged him swiftly towards the opening in the thorn *scherm*.

Still screaming for help, Robyn stumbled after them and seized her father's shoulders, while the hyena had him by the belly. The woman and the animal fought over him, the blunted yellow teeth ripping and tearing through the lining of Fuller's belly as the hyena strained back on its hind-quarters, the neck stretched out at the pull.

The Hottentot Corporal dressed only in unlaced breeches, but brandishing his musket, ran towards them in the firelight.

'Help me,' screamed Robyn. The hyena had reached the thorn fence, her feet were slipping in the loose dust, she was not able to hold Fuller.

'Don't shoot!' Robyn screamed. 'Don't shoot!' The danger from the musket was as great as from the animal.

The Corporal ran forward, reversing the musket and swung the butt at the hyena's head. It struck with a sharp crack of wood on bone, and the hyena released its grip.

Finally, its natural cowardice overcame its greed. It turned and shambled through the opening in the thorn hedge and disappeared into the night.

'Oh, sweet merciful God,' Robyn whispered as they carried Fuller back to the litter, 'has he not suffered enough?'

Fuller Ballantyne lived out that night, but an hour after dawn that tenacious and tough old man at last relinquished his grip on life without having regained consciousness. It was as though a legend had passed, and an age had died with him. It left Robyn feeling numbed and disbelieving and she washed and dressed the frail and rotting husk for burial.

She buried him at the foot of a tall mukusi tree and carved into the bark with her own hand:

FULLER MORRIS BALLANTYNE
3rd Nov. 1788   17th Oct. 1860
'In those days there were giants upon the earth.'

She wished that she had been able to cut the words in marble. She wished that she had been able to embalm his body and carry it back to rest where it belonged in the great Abbey of Westminster. She wished that he had recognized and known who she was just once before he died, she wished she had been able to allay his suffering, and she was consumed with grief and guilt.

For three days she maintained the camp astride the Hyena Road, and she spent those days sitting listlessly beside the mound of newly turned earth under the mukusi tree. She drove old Karanga and even little Juba away, for she needed to be alone.

On the third day she knelt beside the grave and she spoke aloud. 'I make an oath to your memory, my dear father. I swear that I will devote my entire life to this land and its people, just as you did before me.'

Then she rose to her feet and her jaw-line hardened. The

522

time for mourning was past. Now her duty lay plain before her – to follow this Hyena Road to the sea, and then to bear witness before all the world against the monsters who used it.

When the lions are hunting, the prey animals seem able to sense it. They are seized by a restlessness and will graze for only seconds at a time before throwing up their horned heads and freezing into that peculiar antelope stillness, only the wide trumpet-shaped ears moving incessantly; then, skittering like thrown dice, they rearrange themselves upon the grassy plains, snorting and nervous, aware of danger but uncertain of its exact source.

Old Karanga had the same instinct bred into him for he was Mashona, an eater of dirt, and as such he was natural prey. He was the first to become aware that there were Matabele somewhere close at hand. He became silent, nervous and watchful, and it infected the other bearers.

Robyn saw him pick a broken ostrich plume from the grass beside the path and study it gravely, puckering his lips and hissing quietly to himself. It had not fallen from the wing of a bird.

That night he voiced his fears to Robyn.

'They are here, the stabbers of women, the abductors of children—' He spat into the fire with a bravado that was hollow as a dead tree trunk.

'You are under my protection,' Robyn told him. 'You and all the people in this caravan.'

But when they met the Matabele war party, it was without further warning, in the dawn when the Matabele always attack.

Suddenly they were there, surrounding the camp, a solid phalanx of dappled shields and nodding plumes, the blades of the broad stabbing assegai catching the early light. Old Karanga had gone in the night, and with him had gone all the other porters and bearers. Except for the Hottentots the camp was deserted.

Karanga's warning had not been in vain, however, and behind the thorn *scherm* all the Hottentot guards were standing to their muskets, with their bayonets fixed.

The encircling Matabele stood silent, and still as statues carved from black marble. There seemed to be thousands upon thousands of them – though common sense told Robyn that it was merely a trick of her heated imagination and the poor light. A hundred, at the most two hundred, she decided.

Beside her Juba whispered, 'We are safe, Nomusa. We are beyond the Burnt Land, beyond the border of my people. They will not kill us.'

Robyn wished she was as confident, and she shivered briefly, not merely from the dawn chill.

'See, Nomusa,' Juba insisted. 'The baggage boys are with them, and many of the *amadoda* carry their *isibamu* (firearms). If they intended to fight, they would not so burden themselves.'

Robyn saw that the girl was right, some of the warriors had rusty trade muskets slung upon their shoulders, and she remembered from her grandfather's writings that whenever the Matabele intended serious fighting they handed their muskets, which they neither trusted nor used with any accuracy, to the baggage boys and relied entirely upon the weapon that their ancestors had forged and perfected, the assegai of Chaka Zulu.

'The baggage boys carry trade goods, they are a trading party,' Juba whispered. The baggage boys were the young apprentice warriors, and beyond the ranks of fighting men they were still in column. As soon as Robyn recognized the boxes and bundles that the baggage boys carried balanced on their heads, her last qualms faded to be replaced by anger.

They were traders, that she was sure of now, and returning along the road from the east there was little doubt in Robyn's mind as to what they had traded for these paltry wares.

'Slavers!' she snapped. 'In God's name and mercy, these are the slavers we seek, returning from their filthy business. Juba, go and hide immediately,' she ordered,

Then, with her Sharps rifle tucked under her arm, she stepped out through the opening in the wall of thorn bush, and the nearest warriors in the circle lowered their shields a little and stared at her curiously. This small change in attitude confirmed Juba's guess, their intentions were not warlike.

'Where is your Induna?' Robyn called, her voice sharp with her anger, and now their curiosity gave way to astonishment. Their ranks swayed and rustled, until a man came from amongst them, one of the most impressive men she had ever laid eyes upon.

There was no mistaking his nobility of bearing, the arrogance and pride of a warrior tried in battle and covered in honours. He stopped before her and when he spoke his voice was low and calm. He did not have to raise it to be heard.

'Where is your husband, white woman?' he asked. 'Or your father?'

'I speak for myself, and all my people.'

'But you are a woman,' the tall Induna contradicted her.

'And you are a slaver,' Robyn flared at him, 'a dealer in women and children.'

The warrior stared at her for a moment, then lifted his chin and laughed, it was a low clear musical sound.

'Not only a woman,' he laughed, 'but an insolent one also.'

He shifted his shield on to his shoulder and strode past her. He was so tall that Robyn had to lift her chin to look up at him. He moved with a sinuous balance and assurance of carriage. The muscles in his back shone as though they were covered in black velvet, the tall plumes of his headdress nodded and the war rattles on his ankles whispered with each pace.

Swiftly he moved through the gap in the thorn hedge and at Robyn's gesture the Hottentot Corporal lifted the point of his bayonet into the 'present' position and stepped back to let the Induna pass.

With a sweeping gaze the Induna took in the condition of the camp and laughed again.

'Your bearers have run,' he said. 'Those Mashona jackals can smell a real man a day's march away.'

Robyn had followed him into the camp and now she demanded with anger that was not feigned,

'By what right do you enter my kraal and terrify my people?'

The Induna turned back to her.

'I am the King's man,' he said. 'On the King's business.' As though that was all the explanation that was necessary.

Gandang, the Induna, was a son of Mzilikazi, the King and Paramount Chief of the Matabele and all the subservient tribes.

His mother was of pure Zanzi blood, the old pure blood of the south, but she was a junior wife and as such, Gandang would never aspire to his father's estate.

However, he was one of his father's favourites. Mzilikazi, who mistrusted nearly all of his sons, and most of his hundreds of wives, trusted this son, not only because he was beautiful and clever and a warrior without fear, but because he lived in strict accordance with the law and custom of his people, and because of his unquestioned and oft-proven loyalty to his father and his King.

For this and for his deeds, he was covered in honours to which the ox-tail tassels on his arms and his legs bore witness. At four and twenty summers, he was the youngest *indoda* ever to be granted the head-ring of the Induna and a place on the high council of the nation, where his voice was listened to with serious attention even by the old grey pates.

The ageing King, crippled with gout, turned more and more towards this tall and straight young man when there was a difficult task, or a bitter battle in the offing.

So when Mzilikazi learned of the treachery of one of his Indunas, a man who commanded the border guards of the south and eastern strip of the Burnt Land, he had not hesitated before summoning Gandang, the trusted son.

'Bopa, son of Bakweg, is a traitor.'

It was a mark of Gandang's favour that his father condescended to explain his orders as he issued them.

'At first, as he was ordered, he slew those who trespassed in the Burnt Land, then he grew greedy. Instead of killing, he took them as cattle and sold them in the east to the *Putukezi* (Portuguese) and the *Sulumani* (Arabs) and sent word to me that they were dead.' The old King shifted his swollen and painful joints and took snuff, before going on,

'Then because Bopa was a greedy man, and the men with whom he deals are greedy also, he began to seek other cattle to trade. On his own account, and secretly, he began to raid the tribes beyond the Burnt Land.'

Gandang, kneeling before his father, had hissed with astonishment. It was contrary to law and custom, for the tribes of the Mashona beyond the Burnt Land were the King's 'cattle', to be raided only at the King's direction. For another to usurp the powers and gather the booty that belonged to the King was the worst form of treason.

'Yes, my son,' the King agreed with Gandang's horror. 'But his greed was without frontiers. He hungered for the

527

baubles and the trash which the *Sulumani* brought him, and when this supply of Mashona "cattle" was not enough, then he turned upon his own people.'

The King was silent and his expression one of deep regret, for though he was a despot with powers that were subject to neither check nor limitation, although his justice and his laws were savage, yet within those laws he was a just man.

'Bopa sent to me messengers accusing our own people, some of them nobles of Zanzi blood, one of treachery, another of witchcraft, another of stealing from the royal herds – and I sent the messengers back to Bopa ordering him to slay the offenders. But they were not slain. They, and all their people were taken along the road that Bopa had opened to the east. Now their bodies will not be buried in this land and their spirits will wander homelessly for all time.'

That was a terrible fate, and the King lowered his chin upon his chest, and brooded on it. Then he sighed and lifted his head. It was a small neat head and his voice was high-pitched, almost womanish, not that of a mighty conqueror and a warrior without fear.

'Take your spear to the traitor, my son, and when you have killed him, return to me.'

When Gandang would have crawled from his presence, the King halted him with one finger raised.

'When you have killed Bopa, you and those of your *amadoda* who are with you when the deed is done may go in to the women.'

It was the permission for which Gandang had waited for so many years, the highest privilege, the right to go in to the women and take wives.

Gandang shouted his father's praises as he crawled backwards from the royal presence.

Then Gandang, the loyal son, had done what his father commanded. He had carried his spear of retribution swiftly across all of Matabeleland, across the Burnt Land, and along the Hyena Road until he had met Bopa returning from the east laden with the spoils he so dearly coveted.

They had met at a pass through a line of granite hills, not a day's march from where Gandang now confronted Robyn Ballantyne.

Gandang's *Inyati* impi (buffalo) in their ostrich plumes and civet-tail skirts, carrying the dappled black and white ox-hide shields had surrounded the slave-guards formed from selected warriors of Bopa's *Inhlambene* impi (the Swimmers). The slavers wore white egret plumes and kilts of monkey tails, while their war shields were of chocolate-red ox-hide – but right was on the side of the Inyati, and after the swift *jikela* (encirclement) they raced in to crush the guilty and confused slave-guards in a few terrible unholy minutes of battle.

Gandang himself had engaged the grizzle-headed but powerfully built Bopa. He was a wily, scarred fighter and veteran of a thousand such conflicts. Their shields, the one dappled black and the other red, collided with a thud like charging bulls, and they wrestled for the advantage until Gandang, the younger and stronger, with a shift of weight and feet had hooked the point of his shield under the red shield of Bopa and prised it aside to open his enemy's flank.

'*Ngidla* – I have eaten!' Gandang sang out as he sent the broad blade cleaving between Bopa's ribs, and when it was withdrawn against the reluctant cling of flesh with a sucking noise like a man walking in thick ankle-deep mud, Bopa's heart blood burst out behind it and splattered against Gandang's shield, drenching the ox tail tassels on his arms and his legs.

Thus it was for good reason that Gandang had laughed when Robyn called him 'Slaver'.

'I am on the King's business,' he repeated. 'But what do you do here, white woman?' He knew very little of these strange people, for he had been a child when the impis of

Mzilikazi had fought them in the land to the south, and had been driven by them northwards into what was now Matabeleland.

Gandang had met only one or two of them. They had been visitors to his father's great kraal at Thabas Indunas, travellers and traders and missionaries who had been 'given the road' by the King and allowed to cross the strictly guarded frontiers.

Gandang was suspicious of them and their gaudy trade goods. He distrusted their habit of breaking pieces off the rocks along their path, he disliked their talk of a white man who lived in the sky and seemed to be in serious competition with the 'Nkulu-kulu', the great God of the Matabele.

Had he met this woman and her followers in the Burnt Land, he would have followed his orders without hesitation, and killed them all.

However, they were still ten days' march from the frontier and his interest in them was casual; he was impatient to return to his father and report to him the success of his expedition. He would not waste much further time.

'What is your business, woman?'

'I come to tell you that the Great Queen will no longer allow human beings to be sold like cattle for a few beads. I come to put an end to this evil business.'

'That is man's work,' Gandang smiled. 'And besides, it has already been seen to.'

The woman amused him, at another time he might have enjoyed bantering with her.

He would have turned and strode from the camp when suddenly a small movement seen through a gap in the thin thatch of one of the temporary shelters caught his attention. With uncanny speed for such a big man, he ducked into the hut and pulled the girl out by her wrist; holding her at arm's length he studied Juba gravely.

'You are of the people, you are Matabele,' he said flatly.

530

Juba hung her head and her face had a pale greyish sheen of terror. For a moment Robyn thought that Juba's legs would no longer bear her weight.

'Speak,' Gandang commanded in that low but imperious voice. 'You are Matabele!'

Juba looked up at him and her whisper was so soft that Robyn hardly caught it.

'Matabele,' she agreed, 'of Zanzi blood.'

The warrior and the maid considered each other carefully. Juba lifted her chin, and the greyness vanished from her face.

'Your father?' Gandang asked at last.

'I am Juba, daughter of Tembu Tebe.'

'He is dead, and all his children, at the King's orders.'

Juba shook her head. 'My father is dead – but his wives and his children are in the land of the *Sulumani* beyond the sea. I alone escaped.'

'Bopa!' Gandang said the name as though it were a curse. He considered a moment. 'It is possible that your father was wrongly sentenced, for Bopa sent false accusation to the King.'

Juba made no reply, but in the silence that followed, Robyn saw a subtle change coming over the girl, something altered in the carriage of her head, she shifted her weight, thrusting out one hip, a small but provocative movement.

Her eyes, when she looked up at the tall Induna, grew wider and softer, and her lips were held slightly apart so that the pink tip of her tongue just showed between them.

'What is this white woman to you?' Gandang asked, and there was just a trace of huskiness in his own voice. He held her wrist still, and she made no effort to pull away.

'She is as my mother was,' Juba replied, and as the Induna looked down from her face to her sweet young body the ostrich plumes fanned softly about his head, and Juba changed the angle of her shoulders slightly, offering up her breasts to his gaze.

'You are with her by your own will?' Gandang insisted, and Juba nodded.

'So be it.' It seemed to require an effort for the warrior to break his gaze, but he dropped Juba's wrist and turned back to Robyn. His smile was mocking once more.

'The slavers you seek are not far from here, white woman. You will find them at the next pass in the road.'

He went as swiftly and as silently as he had come, and his warriors followed him in a dense black column. Within minutes the last of them had disappeared along the winding narrow trail into the west.

Old Karanga was the first of the servants to return to the camp. He came in through the thorn *scherm* like a bashful stork on his thin legs.

'Where were you when I needed you?' Robyn demanded.

'Nomusa, I could not trust my temper with those Matabele dogs,' old Karanga quavered, but he could not meet her eyes.

Within the hour the other porters and bearers had crept down from the hills and out of the forest, all of them now endowed with amazing enthusiasm to continue the march in the opposite direction to that of the *Inyati* impi.

Robyn found the slavers where Gandang had promised her she would. They were scattered over the neck of the pass, they lay in knots and windrows, like leaves after the first storm of autumn. Nearly all of them had their death wounds in the chest or throat, proof that at the end they had fought like Matabele.

The victors had slit open the dead men's bellies to allow their spirits to escape, a last courtesy to men who had fought gallantly, but the vultures had used the openings to enter the belly pouches.

The birds hopped and flapped and squabbled raucously

over the cadavers, tugging and dragging at them so that their dead limbs kicked and twitched as though they were still alive, and dust and loose feathers flew around them. The croaking and squawking of the birds was deafening.

In the trees and on the cliffs above the pass, the birds that had already gorged crouched somnolently, puffing out their feathers and hunching their naked scaly heads and necks upon their shoulders, digesting the contents of their bulging crops before returning to the feast.

The little caravan passed slowly, in fascinated horror at the carnage, speechless in the raucous chorus of the scavengers, stepping carefully over the ragged, dust-covered remains of brave men, reminded by them of their own mortality.

Once they had crossed the pass they hurried down the far slope with fearful backward glances. There was a stream at the bottom of the slope, a tiny trickle of clear water springing from the slope and threading its way from pool to small shaded pool. Robyn went into camp upon the bank, and immediately called Juba to follow her.

She had to bathe herself, she felt as though death had touched her with its putrid fingertips and she needed to wash away the taint of it. She sat under the trickle of clear water, waist deep in the pool below the waterfall and let the stream flow over her head, her eyes closed trying to blank out the horrors of the battlefield. Juba was not so affected, she was no stranger to death in its most malevolent forms, and she splashed and played in the green water, completely absorbed in the moment.

At last Robyn waded to the bank, and pulled her shirt and breeches over her still-wet body. In that heat, her clothes would dry upon her within minutes, and while she twisted her wet hair into a rope on top of her head she called to Juba to come out of the pool.

In a mischievous and rebellious mood the girl ignored her, and remained rapt in her own game, singing softly as she picked wild flowers from a creeper that hung over the pool and plaited them into a necklace over her shoulders. Robyn turned away and left her, climbing back along the bank towards the camp, and the first turn hid her from view.

Now Juba looked up and hesitated. She was not certain why she had refused to obey, and she felt a little chill of disquiet at being alone. She was not yet accustomed to this new mood of hers, this strange and formless excitement, this breathless expectancy for she knew not what. With a toss of her head she returned to her song and her play.

Standing above the bank, half screened by the trailing creeper and mottled like a leopard by the slanting dappled sunlight through the leaves of the forest, a tall figure leaned against the bole of a wild fig tree and watched the girl.

He had stood there, unseen and unmoving since he had been led to the pool by the sound of splashing and singing. He had watched the two women, comparing their nakedness – the bloodless white against the luscious dark skin, the skinny angular frame against sweet and abundant flesh, the small pointed breasts tipped in the obscene pink of raw meat against the full and perfect rounds with their raised bosses, dark and shiny as new-washed coal, the narrow hips of a boy against the proud wide basin which would cradle fine sons, the mean little buttocks against the fullness and glossiness that was unmistakably woman.

Gandang was aware that by returning along the trail he was for the first time in his life neglecting his duty. He should have been many hours' march away from this place, trotting at the head of his impi into the west, yet there was this madness in his blood, that he had not been able to deny. So he had halted his impi and returned alone along the Hyena Road.

'I am stealing the King's time, just as surely as Bopa stole his cattle,' he told himself. 'But it is only a small part of a single day, and after all the years I have given to my father, he would not grudge me that.' But Gandang knew that he would, favourite son or not, Mzilikazi had only one punishment for disobedience.

Gandang was risking his life to see the girl again, he was risking a traitor's death to speak a few words to a stranger, daughter of one who had himself died a traitor's death.

'How many men have dug their graves with their own *umthondo*,' he mused, as he waited for the white woman to leave the pool, and when she had covered her skinny boy's body with those stiff and ugly garments and called to the lovely child in the pool to follow her, Gandang tried to reach out with his own will to hold Juba there.

The white woman, clearly piqued, turned and disappeared amongst the trees and Gandang relaxed slightly, giving himself once more to the pleasure of watching the girl in the water. The wild flowers were a pale yellow against her skin, and the waterdrops clung to her breasts and shoulders like stars against the midnight sky. Juba was singing one of the children's songs that Gandang knew so well, and he found himself humming the chorus under his breath.

Below him the girl waded to the bank and standing in the sugar-white sand began to wipe the water from her body; still singing she bent forward to wipe her legs, encircling them with long slim pink-lined fingers and running her hands slowly down from thigh to ankle. Her back was to Gandang, and as she stooped he gasped aloud at what was revealed to him, and instantly the girl flew erect and spun to face him. She was trembling like a roused fawn, her eyes huge and dark with fright.

'I see you, Juba, daughter of Tembu Tebe,' he said, there was a husky catch to his voice as he came down the bank to her.

The expression in her eyes changed, they glowed with golden lights like sunshine in a bowl of honey.

'I am a messenger of the King, and I demand the right of the road,' he said, and touched her shoulder. She shivered under his fingers. He saw the little goose bumps rise upon her skin.

The 'right of the road' was a custom from the south, from the old country beside the sea. It was the same right which Senzangakhona had demanded of Nandi, 'the sweet one', but Senzangakhona had not respected the law, and he had penetrated the forbidden veil. From this transgression one had been born, the bastard 'Chaka', 'the worm in the belly', who had grown to become both the King and the scourge of Zululand, the same Chaka from whose tyranny Mzilikazi had flown with his tribe to the north.

'I am a loyal maiden of the King,' Juba answered him shyly, 'and I cannot refuse to comfort one who follows the road on the King's business.' Then she smiled up at him. It was neither bold nor provocative, but a smile so sweet, so trusting and filled with admiration, that Gandang felt his heart squeezed afresh.

He was gentle with her, very gentle and calm and patient, so that she found herself impatient to render the service he desired, found herself desiring it as strongly as he so evidently did. When he showed her how to make a nest for him between her crossed thighs, she responded instantly to his word and touch, and there was something wrong with her throat and her breathing, for she was unable to answer him aloud.

While she held him in this nest she felt herself gradually overwhelmed by a strange wildness of heart and body. She tried to alter the angle of her pelvis, she tried to unlock her tightly crossed thighs and spread them for him, she strove to engulf him for she could no longer bear that dry and tantalizing friction against the inside of her upper legs. She wanted to feel him breast the warm and welcoming flood that she sent down for him and she wanted to feel him gliding upon it deeply up inside her. But his resolve, his respect for custom and law, was as powerful as that muscular body that drove above her, and he held her captive until the moment when she felt his grip break and his seed spring strongly from him to waste itself in the white sand beneath them. At that moment she felt such a sense of loss that she could have wept aloud.

Gandang held her still, his chest heaving and the sweat forming little shiny runnels across that smooth dark back and down the corded neck. Juba clung to him with both arms wrapped tightly about him, her face pressed into the hollow between his shoulder and his neck, and for a long time neither of them spoke.

'You are as soft and as beautiful as the first night of the new moon,' Gandang whispered at last.

'And you are as black and as strong as the bull of the *Chawala* festival,' she instinctively chose the simile that would mean most to a Matabele, the bull as the symbol of wealth and virility and the *Chawala* bull the most perfect specimen of all the King's herds.

'You will be only one of many wives,' Robyn was horrified at the thought.

'Yes.' agreed Juba. 'First of all of them, and the others will honour me.'

'I would have taken you with me to teach you many things and show you great wonders.'

'I have already seen the greatest wonder.'

'You will do nothing but bear children.'

Juba nodded happily. 'If I am truly lucky, I will bear him a hundred sons.'

'I will miss you.'

'I would never leave you, Nomusa, my mother, not for any person nor reason in the world, except this one.'

'He wants to give me cattle.'

'Since the death of my family, you are my mother,' explained Juba, 'and it is the marriage price.'

'I cannot accept payment – as though you were a slave.'

'Then you demean me. I am of Zanzi blood and he tells me that I am the most beautiful woman in Matabeleland. You should set the *lobola* as one hundred head of cattle.'

So Robyn called the Induna to her.

'The marriage price is one hundred head of cattle,' Robyn told him sternly.

'You make a poor bargain,' Gandang answered loftily. 'She is worth many times that amount.'

'You will keep the cattle at your kraal, against my coming. You will tend them carefully and see that they multiply.'

'It will be as you say, *amekazi*, my mother.' And this time Robyn had to return his smile, for it was no longer mocking and his teeth were so white and he was, as Juba had said, truly beautiful.

'Look after her well, Gandang.'

Robyn embraced the young woman and their tears mingled and smeared both their cheeks. Yet when she left Juba did not look back once, but trotted behind Gandang's tall erect figure carrying her rolled sleeping-mat balanced upon her head, and her buttocks jiggled merrily under the short beaded apron.

Man and woman reached the saddle of the pass and disappeared abruptly from view.

The Hyena Road led Robyn and her little party into the mountains, into the mist and the strangely desolate valleys of heather and fantastically shaped grey stone. It led her to the slave stockades which Juba had described to her, the meeting place where the white man and black man made their trade for human life, where the slaves exchanged their carved yokes for cuff and chains. But now the stockades were deserted, the thatch already sagging and falling in untidy clumps, only the sour smell of captivity lingered, and the swarming vermin that infected the empty buildings. In a futile gesture, Robyn put fire to the buildings.

From the misty mountains the road led on, down through dark gorges and at last to the low littoral where once more the heat clamped down upon them from a sullen overcast sky and the grotesque baobab trees lifted their twisted arthritic branches to it like crippled worshippers at a healing shrine.

The rains caught them here upon the coastal plain. The flood swept three men away at a ford, four more, including one of the Hottentots, died of fever and Robyn herself was smitten with the first onslaught of the disease. Shivering, half demented by the phantoms of malaria, she toiled on along the rapidly overgrown trail, slipping and stumbling in the mud, and cursing the fever miasma that rose from the brimming swamps and hung like a silver wraith in the sickly green glades of fever trees through which they hurried.

Fever and the rigours of the last stage of the journey had tired and weakened them all. They knew that they were, at the most, only a few days' march from the coast, deep into Portuguese territory and therefore under the protection of a Christian king and a government of civilized men. It was for these reasons that the Hottentot sentries slept beside the smouldering watch-fire of damp wood, and it was there that they died, their throats slit with a blade sharp enough to cut off the least cry.

So Robyn woke to rough hands upon her, twisting her arms up between her shoulder blades and a bony knee in the small of her back, while steel cuffs clicked coldly about her wrists. Then the hands released her and she was wrenched cruelly to her feet, and dragged from the leaky hut beside the Hyena Road.

The previous evening she had been too tired and feverish to undress, so now she was still clad in a stained and rumpled flannel shirt and patched moleskin breeches. She had even kept the cloth cap on her head, covering her hair, thus in the darkness her captors did not realize that she was a woman.

She was bound with her own porters and Hottentots, forced to wear the light marching chains which were proof, if proof was needed, as to who her captors were. The dawn revealed them to be half-breeds and blacks, all of them dressed in the cast-off finery of European style, but carrying modern weapons.

These were the men she had crossed half a continent to meet, but now she cowered in the rags that were her only disguise. She shuddered to think on her fate if they should discover her sex, and she berated herself for having so blithely believed that she and her entourage would be safe from these predators merely because she was white and English. Their prey was human flesh, of whatever colour and condition and that was all she was now, human flesh on the hoof. A chained creature, of little real value, a few dollars on the auction block, and she knew that her captors would think nothing of taking their pleasure upon her, or of leaving her beside the road with a ball through the temple if she provoked them in any way. She kept silent and obeyed instantly the least word or gesture from her captors. Slipping and dragging in the mud they were marched on eastwards, forced to carry their remaining stores and equipment which had now become slavers' booty.

They were closer to the coast than Robyn had calculated, they smelt the iodine and salt of the sea from afar, and later as night began to fall, they caught the smell of woodsmoke and the unmistakable odour of captive humanity. Then at last they saw the firelight flickering in the darkness ahead, and the awful loom of the barracoons.

Their captors marched between the dark stockades of pole and daubed mud, from which the chilling dirge rose of men without hope singing of a land they would never see again.

At last they came into the central square around which

the barracoons were built. It was an open area of trampled mud where a raised platform of rough-sawn planks had been built. Its purpose was immediately clear, for the first of Robyn's servants was dragged up the steps and stood upon the platform, while the fires around the clearing were heaped with dry wood to light the scene. The platform was the auction block, and it seemed that the sale was to take place immediately.

The auctioneer was clearly of pure Portuguese stock, a little man with the wrinkled, sun-browned face of a vicious gnome. He had the bland smile and the unblinking eyes of a serpent. He was dressed in elegantly cut jacket and breeches, and his boots and belt were of the finest Iberian leather, ornately tooled and with solid silver buckles. He carried a pair of expensive pistols in his belt, and wore the wide-brimmed flat-crowned hat of a Portuguese gentleman upon his small wrinkled head.

Before climbing up on the block, he sent one of his personal slaves with a casual kick and cuff to the carved wooden drum at the edge of the clearing. On it the slave began pounding out a summons to the buyers. The slave went to it lustily, his bare torso gleaming with sweat and raindrops in the firelight.

And in answer to the urgent staccato rhythm of the drum, men came from out of the shadows of the grove and from the living huts between the barracoons. Some of them had been drinking, they came arm in arm brandishing their rum bottles and bellowing in drunken chorus, others came singly and silently – but they came from every direction and gathered in a circle about the auction block.

The men who formed the circle seemed to be of every hue that the human skin is capable of assuming, from purple black through all the shades of brown and yellow to dead shark's belly white, and their features were African and Arabic, Asian and European. Even their dress differed widely, from the flowing robes of Arabia to the faded finery

541

of embroidered jackets and high boots. They had only one thing in common, the hawk-fierce eyes and merciless mien of those who deal in human misery.

One at a time Robyn's servants were prodded up on to the block, and their ragged clothing was ripped away to expose their physique to the buyers. One of these might come forward to feel the muscle tone, or force open a slave's mouth to examine the teeth, like a gypsy horse-dealer at the fair.

Then, when the buyers had satisfied themselves as to the quality of the wares on offer, the small Portuguese would step lightly to the front of the block and begin the bidding.

The men in the circle below him called him Alphonse and though they exchanged coarse banter with him, yet they all treated him with wary respect, there could be no clearer proof of the man's reputation than fear and respect from these men.

Under his control the sale went swiftly. The Hottentots, small wiry men, butter yellow and with flat pug-like features attracted little interest from the circle of buyers, knocked down for a few silver rupees apiece, while the porters, taller men and well muscled from many months of hard marching and porterage, fetched better prices – until they came to old Karanga, ancient and toothless, hobbling on to the block on storklike legs, seeming barely able to support the weight of his chains.

The laughter was derisive, and the little Portuguese pleaded in vain for a single bid before dismissing the old man with disgust. It was only when he was hauled down off the block and dragged away into the darkness beyond the fires that Robyn realized what was about to happen to him and, forgetting her resolve not to draw attention to herself, 'No – let him go!'

Hardly one of them glanced in her direction, and the man who held her chain hit her a careless open-handed blow across the side of her face that blinded her for a moment. She dropped to her knees in the mud, and through the buzzing in her ears heard the thud of a pistol shot from the darkness.

She began quietly to weep, and still weeping she was hoisted to her feet and in her turn dragged forward into the circle of firelight and hoisted by her chain on to the block.

'A young skinny one,' said the Portuguese. 'But white enough to make a choice bum-boy for the harems of Omani, once he has had his knackers clipped. Who will give me ten rupees?'

'Let's have a look at him,' a voice shouted from the circle, and the Portuguese turned to Robyn, hooked a finger into the top button of her flannel shirt and ripped it down to the level of her belt buckle.

She doubled over, trying to conceal her upper body, but the man behind her twisted the chain and forced her upright. Her breasts pushed out pertly through the torn shirt, and the ring of watchers growled and moved restlessly, the mood changing instantly.

Alphonse touched the butt of one of the pistols in his belt significantly, and the growl of comment died, the ring of men drew back a little.

'Ten rupees?' Alphonse Pereira asked.

From across the circle a powerfully-built man swaggered into the firelight. Robyn recognized him instantly. He wore a tall beaver hat tilted back on his head, and from under the brim curled thick shiny black hair. His teeth, when he opened his mouth, sparkled in the light of the flames. His face was flushed with excitement, his voice was thick with it.

'Gold,' he shouted. 'I'll bid gold, a gold mohur of the East India Company, and a plague on any of you that goes above that.'

'A gold mohur,' called Alphonse, the slave-master. 'A

543

gold mohur bid by my brother Camacho Pereira, and good luck to him,' he chuckled. 'Come now, who wants to deny my brother Camacho a fair tup at the wench?'

One of his men slapped Camacho's back.

'Sweet Christ, you always were a hot one, at that price you can have my turn on top of her.'

And Camacho laughed delightedly and came to the front of the block to stare up at Robyn, tipping the beaver hat forward he whispered,

'I've had to wait a long time—'

Robyn felt the little insects of loathing crawling over her skin, and she backed away to the limit of her chain.

'Come now,' Alphonse called. 'Who will go beyond a gold mohur for a fine piece—'

'She's mine,' Camacho told his brother. 'Strike the bargain.'

His brother lifted his hand to knock down the sale, when another voice stopped him.

'A double eagle, sir. Twenty golden dollars American bid.' The voice was not raised, yet it carried clearly to every man there, as it could carry from quarterdeck to maintop in a force eight gale.

Robyn started, and swung her chains disbelievingly in that direction; she would have known that lazy drawling inflection if she had not heard it for a lifetime. He stood at the very edge of the firelight, but as every head in the circle turned to him, he stepped forward.

The smile had frozen on Alphonse's face, and he hesitated.

'Call the bid!' The advancing figure in plain white shirt and dark breeches made the men about him seem small and grubby, and after a moment's hesitation, Alphonse obeyed him.

'A double eagle bid,' he said harshly. 'Captain Mungo St John of the clipper *Huron* bids a double golden eagle.'

Robyn felt her legs start to sag under her with relief, but

the men behind her jerked her upright by her chain. Camacho Pereira had whirled to face the American, and to stare at him furiously. Mungo St John answered him with a smile, indulgent and patronizing. Robyn had never seen him look more handsome and dangerous, his dark wavy hair catching the firelight, and the gaze of his yellow-flecked eyes level and unflinching in the face of Camacho's fury.

'A thousand rupees, Camacho,' he said softly. 'Can you match it?'

Camacho hesitated, and then turned back quickly to his own brother, his voice low and urgent.

'Stake me?' he asked, and Alphonse laughed.

'I never lend money.'

'To a brother?' Camacho insisted.

'Especially not to a brother,' Alphonse answered. 'Let the wench go, you can buy a dozen better for fifty rupees each.'

'I must have her.' He whirled back to face Mungo St John. 'I must have her. It is a matter of honour. Do you understand?' He took the beaver from his head, and spun it away. One of his men caught it, and Camacho ran both hands through his thick black locks, and then stretched his arms down at his sides, flexing the fingers like a conjurer about to perform a sleight of hand.

'I will make one more bid,' he said ominously. 'I bid one mohur of gold – and ten inches of Toledo steel.' The knife seemed to appear in his hand from out of the air, he lifted the point to the level of Mungo St John's belt buckle.

'Walk away, Yankee, or I will take the woman *and* your gold double eagle.'

The watchers growled, a low blood-thirsty sound, and swiftly rearranged themselves into a ring about the two men, jostling for a better view.

'One hundred rupees says Machito slits the Yankee's guts.'

'Done!' And there was a rising hubbub as the wagers were called and accepted.

Mungo St John had not stopped smiling, but now he held out his right hand without once taking his eyes off the Portuguese's face.

Out of the ranks of the watchers emerged a large, toad-like figure with a head as round and bald as a cannon ball. Tippoo moved with reptilian speed to Mungo St John's right side. He placed a knife in the outstretched hand, and then unknotted the embroidered sash from his waist and handed that to his Captain. Mungo wrapped the sash around his left forearm, still smiling softly to himself.

He had not once looked up at Robyn, though she had not been able to tear her own eyes from his face.

He seemed godlike to her at that moment, everything about him, the darkly classical features, the wide shoulders under the white cloth of his shirt, the narrow waist clinched with a broad belt of polished leather, the strong straight legs in tightly fitted breeches and soft leather boots, seemed to have come down directly from Olympus. She would have gladly thrown herself at his feet and worshipped him.

Just below Robyn, Camacho was stripping off his own jacket and wrapping his guard arm with it. Then with the long knife in his right hand he made a low swift cut, forehand and then backhand, so the steel whispered as it dissolved into a silver blur like the wing of a dragonfly in flight. At each stroke he ducked his head slightly and flexed his knees, loosening and warming his muscles like an athlete before the contest.

Then he moved forward, stepping lightly in the treacherous mud and weaving the point of the knife to distract and intimidate his adversary.

The smile went from Mungo St John's lips, to be replaced by a grave and attentive expression, like a mathematician considering a complex problem. He kept his own knife low, advancing his wrapped forearm, and balancing easily, stood his full height and turned gently to face the Portuguese as he circled. It reminded Robyn of the night she had watched him on the dance floor at Admiralty House, so tall and graceful, so balanced and controlled in each movement.

Now at last the watchers were silent, straining eagerly for the first glimpse of blood, but when Camacho charged they roared the way the crowd roars when the bull first bursts into the ring. Mungo St John barely seemed to move, swaying his body at the hips so the knife slid past him, and then he was facing Camacho again.

Twice more the Portuguese attacked, and both times Mungo St John moved effortlessly aside, but each time he gave a little ground, until he was backed up to the first rank of watchers, they began to fall back to give the American room to fight, but Camacho saw his opportunity as Mungo was crowded like a prizefighter against the ropes and he swung back to attack. At the same moment, almost as though it had been rehearsed, a booted foot shot out from the crowd.

Nobody was sure whose foot it was, for the throng was closely packed and the light uncertain, but the kick to the back of Mungo St John's heel almost brought him down sprawling in the mud, he lunged to catch his balance, but before he could do so, Camacho hit him with the long bright blade. Robyn screamed and Mungo St John spun away from the sting of the steel with scarlet spreading wetly down his shirt-front like rich Burgundy wine spilled on a damask tablecloth, and his own knife flicked out of his hand and was lost in the red mud.

The crowd bellowed, and Camacho swarmed in eagerly, following the wounded man the way a good dog hunts a pheasant with a broken wing.

Mungo was forced to give him ground, falling back, clutching the wound, dodging and weaving, catching a forehand slash on his wrapped guard arm so the embroidered cloth split almost to the flesh beneath it.

Skilfully Camacho herded him towards the auction block, and when Mungo felt the poles catch in the small of his back, he froze for a moment as he realized that he was trapped. Camacho drove in at him, going for the belly, his lips drawn back baring his perfect white teeth.

Mungo St John caught the knife on his guard and then snatched a grip on the wrist with his right hand. The two men stood chest to chest, their arms entwined like vines on a trellis, swaying slightly as they strained together, and the effort brought a fresh flood of bright blood from Mungo's wound, but slowly he forced Camacho's knife hand upwards, bending it at the elbow, until the point was no longer aimed at Mungo's belly but at the night sky above them.

Mungo shifted his feet, gathering himself and then his face darkened, his jaw clenched and his breath sobbed with effort. Slowly Camacho's wrist gave to the pressure, and his eyes widened as the point of his own knife reversed towards him.

Now he also was wedged against the side of the auction block and could not break away, and infinitely slowly but inexorably, the long blade moved towards his own chest. Both men stared down at it, their hands and arms inter-locked, pitting their strength to hold each other, but the point touched Camacho's chest, a drop of blood welled up at the tiny prick.

On the block beside Robyn, Alphonse Pereira drew the pistol from his belt with a furtive movement – but before she could shout a warning there was a blur of movement and Tippoo the mate towered beside him, his own huge smooth-bore pistol pressed to the side of Alphonse's skull. The little Portuguese rolled his eyes sideways at Tippoo, and then hurriedly returned the weapon to his belt, and Robyn could watch again with fascinated horror the contest at her feet.

Mungo St John's face was congested with dark blood, every muscle in his shoulders and arms raised in knots under the thin shirt, his whole existence concentrated on the knife, and he slid his left foot back until it was anchored against the auction block, and then using it as a pivot hurled all his weight forward on to the knife, the final effort like the matador going over between the horns for the kill.

For a moment longer Camacho resisted him, and then the blade resumed its forward movement entering Camacho's chest as slowly as a python swallows a gazelle.

Camacho's mouth opened in a cawing burst of despair, and suddenly his fingers opened as all resistance and strength went out of them. His own blade with Mungo St John's full weight behind it shot its length into his chest with such force that the cross piece of the hilt struck against his ribs with a sharp thump.

Mungo St John released his grip and let him fall, face forward into the mud, while he himself caught at the edge of the auction block for support. Only then did he lift his chin to look up at Robyn.

'Your servant, ma'am,' he murmured, and Tippoo rushed forward to catch him before he fell.

*H*uron's seamen, all of them armed, formed a guard about them, and Tippoo led them holding aloft a bull's-eye lantern which he shone into the shadows as they hurried down the path.

Mungo St John was on his feet, but supported by Nathaniel, his bosun, and Robyn had bound up the wound roughly with a strip of linen torn from a seaman's shirt and had used the rest of the shirt to make a sling for Mungo's right arm.

Through a grove of mangroves they reached the bank of the creek on which the barracoons had been built and in the centre of the stream, her bare masts and yards silhouetted against the starry sky, lay the lovely clipper.

She had lanterns in her rigging and an alert anchor watch, for at Tippoo's first hail the whaler swung away from her side and was rowed swiftly to the bank where they stood.

Mungo climbed the ship's side unaided, but sank down with a grateful sigh on to his bunk in the stern cabin, the bunk that Robyn remembered so vividly.

She tried to force the memory from her mind, and keep her manner brisk and businesslike.

'They have taken my medical chest,' she said as she rinsed her hands in the porcelain basin at the head of the bunk.

'Tippoo.' Mungo looked up at his mate, and the bald, scarred head bobbed once and then Tippoo ducked out of the cabin. Mungo and Robyn were alone, and she tried to remain remote and professional as she made her first examination of the wound in good lantern light.

It was narrow, but very deep. She did not like the angle of the thrust, just below the collar bone but angled in towards the point of the shoulders.

'Can you move your fingers?' she asked. He lifted his hand towards her face and touched her cheek lightly.

'Yes,' he said, as he stroked her. 'Very easily.'

'Don't,' she said weakly.

'You are sick,' he said. 'So thin and pale.'

'It is nothing – lower your arm, please.'

She was terribly conscious of her matted hair and filthy mud-stained clothing, of the yellow tinges of fever on her skin and the dark smudges of fatigue and terror under her eyes.

'Fever?' he asked quietly, and she nodded as she went on working on the wound.

'Strange,' he murmured. 'It makes you seem so young, so fragile,' he paused, 'so lovely.'

'I forbid you to talk like that.' She felt flustered, uncertain of herself.

'I said I would not forget you,' he ignored the instruction, 'and I did not.'

'If you don't stop, I will leave immediately.'

'When I saw your face tonight in the light of the fires, I could not believe it was you, and at the same time it was as though all our lives we had a rendezvous to keep here tonight. As though it had been destined from the moment of our births.'

'Please,' she whispered, 'please stop.'

'That's better – please is better. Now I will stop.'

But he watched her face attentively as she worked. In the ship's medical chest which he kept in the locker below his bunk Robyn found most of what she needed.

He neither flinched nor grimaced as she laid the stitches in the wound, but went on watching her.

'You must rest now,' she said as she finished, and he lay back on the bunk. At last he looked tired and drained, and she felt a rush of gratitude, of pity, and of that other emotion which she had believed that she had long ago subdued.

'You saved me.' She dropped her eyes, no longer able to look at him and busied herself with repacking the ship's chest. 'I will always be grateful for that, just as I will always hate you for what you are doing here.'

'What am I doing here?' he challenged her lightly.

'Buying slaves,' she accused. 'Buying human lives, just as you bought me on the slaving block.'

'But for a much lower price,' he agreed as he closed his eyes. 'At twenty dollars gold a head there is not much profit in it, I assure you.'

551

S he awoke in the small cabin, the same cabin in which she had sailed the length of the Atlantic Ocean and in the same narrow uncomfortable bunk.

It was like homecoming, and the first thing she saw after her eyes had adjusted to the harsh beam of sunlight through the skylight were the chests of her medical instruments, her remaining medicines and her few personal possessions.

She remembered the unspoken command that Mungo had given to the mate the previous evening. Tippoo must have gone back ashore during the night, and she wondered what price he had paid or what threat he had made to get them back for her.

She rose swiftly from her bunk, ashamed of her sloth; whoever had left the chests, had also filled the enamel jug with fresh water. With relief she washed away the mud and filth and combed the tangles of her hair before finding worn but clean clothing in her chest. Then she hurried from her cabin down to the master's quarters. If Tippoo had been able to find her chests, then he might be able to find and free her people, the Hottentots and porters who had gone upon the auction block in the firelight.

Mungo's bunk was empty, the vest and bloodstained shirt bundled and thrown into a corner of the cabin, and the bedclothes in disarray. She turned swiftly for the deck, and as she came out into the sunlight she saw that it would be only a temporary respite from the monsoon, for already the thunderclouds were boiling up over the horizon.

She looked about her quickly. *Huron* lay in the centre of a broad estuary, with mangroves on each bank, and the bar and the open sea were not in sight, though the tide was ebbing, rustling down the ship's hull and leaving the mud flats half exposed.

There were other vessels in the roadstead, mostly big dhow-rigged buggaloos typical of the Arab coastal traders, but there was another fully rigged ship at anchor half a mile further downstream, flying the flag of Brazil at her peak.

Even as Robyn paused to watch her, there came the clank of her capstan, and men ran up the ratlines and spread out along her yards. She was getting under way. Then Robyn realized that there was unusual activity all about her. Small boats were plying from the shore to the anchored dhows, and even on *Huron*'s deck there was a huddle of men on the quarterdeck.

Robyn turned towards them, and realized that the tallest of them was Mungo St John. His arm was in a sling and he looked drawn and pale, but his expression was forbidding, the dark curved brows drawn together in a frown, and the mouth a thin cruel line as he listened attentively to one of his seamen. So absorbed was he that he did not notice Robyn until she was only a few paces away. Then he swung towards her, and all the questions and demands stayed behind her lips for his voice was harsh.

'Your coming was an act of God, Doctor Ballantyne,' he said.

'Why do you say that?'

'There is a plague in the barracoons,' he said. 'Most of the other buyers are cutting their losses, and leaving.' He glanced downstream to where the Brazilian schooner had set reefed main and jib and was running down towards the bar and the open sea, and there was activity aboard most of the other vessels.

'But I have over a thousand prime blacks afattening ashore – and I'll be damned if I'll run now. At least, not until I know what it is.'

Robyn stared at him. Her mind was a whirl of doubts and fears. 'Plague' was a layman's word, it covered everything from the Black Death to syphilis, the grand pox, as it was called.

'I will go ashore immediately,' she said, and Mungo St John nodded.

'I thought you would say that,' he said, 'I will go with you.'

'No.' Her tone brooked no argument. 'You will aggravate that wound, and in your weakened condition you will be easy prey to this plague, whatever it is.' She glanced at Tippoo, and his face split laterally into that broad toadlike grin and he stepped up beside her.

'By God, ma'am, I've had them all,' said Nathaniel, the little pockmarked bosun. 'And none of them killed me yet.' And he stepped up to her other hand.

Robyn sat in the stern while Tippoo and Nathaniel handled the oars, and as they pulled across the ebb towards the shore the bosun explained what they would find ashore.

'Each of the traders has his own barracoon built and guarded by his own men,' he told Robyn. 'He buys from the Portos as the blackbirds are brought in.'

As Robyn listened to Nathaniel, she realized the answers to questions that had worried her and Zouga. This was the reason why Pereira had tried so desperately to persuade them not to bring the expedition south of the Zambezi river, and why, when all else had failed, he had attacked it with his armed brigands and tried to destroy it. He had been protecting his brother's trade routes and selling area. It was not mere avarice and lust, but a logical attempt to preserve this lucrative enterprise from discovery.

She went on listening to Nathaniel.

'Each trader fattens his wares ashore, like pigs for the market. That way they are stronger for the crossing, and he makes sure that they are healthy and not going to bring sickness aboard with them.

'There are twenty-three barracoons here, some small ones with twenty blacks or so, belonging to the small traders, right on up to the big ones like *Huron*'s, with a thousand and more prime blackbirds in the cage.

'We have the slave-decks set up in *Huron*'s hold, and we would have begun taking them aboard any day, but now—'

Nathaniel shrugged, and spat on the horny calloused palms of each hand in turn, and then plied himself to the oars once more.

'Are you a Christian, Nathaniel?' Robyn asked softly.

'That I am, ma'am,' he said proudly. 'As good a Christian as ever sailed out of Martha's Vineyard.'

'Do you think God approves of what you are doing here to these poor people?'

'Hewers of wood, ma'am, and drawers of water, like the Bible says,' the weather-beaten sailor told her, so glibly that she knew that the reply had been put in his mouth, and she guessed by whom.

Once they were ashore, Tippoo led the small party with Robyn in the centre and Nathaniel carrying her chest in the rear.

Captain Mungo St John had chosen the best site available for his barracoon, on a rise of ground at a distance from the river. The sheds were well built, with floors of sawn timber raised above the mud and good roofs thatched with palmetto leaves.

Huron's guards had not deserted, proof of the discipline which Mungo St John maintained and the slaves in the barracks had evidently been carefully chosen. They were all well set-up men and women, and the copper cookers were filled with boiling farina so that their bellies bulged and their skins were glossy.

At Robyn's direction they were lined up and she passed swiftly down the ranks. There were some mild ailments, which she singled out for later treatment, but she found none of the symptoms which she so dreaded.

'There is no plague here,' she decided. 'Not yet.'

'Come!' said Tippoo.

He led her through the palm groves, and the next barracoon had been deserted by the traders who had built it and stocked it. Already the slaves were hungry and confused by their sudden liberation.

'You are free to go,' Robyn told them. 'Go back to your own land.'

She was not certain that they understood her. They squatted in the mud and stared at her blankly. It was as though they had lost all power of independent thought or action, and she knew that they would never be able to make their way back along the Hyena Road, even if they survived the coming epidemic.

With a flash of horror Robyn realized that without their slave-masters these poor creatures were doomed to a lingering death by starvation and disease. Their masters had cleared out the store rooms before they left, there was not a cupful of farina or corn meal left in any of the barracoons they visited that morning.

'We will have to feed them,' Robyn said.

'We have food for our own, that is all,' Tippoo told her impassively.

'He is right, ma'am,' Nathaniel confirmed. 'We feed them, then we'll starve our own blackbirds – besides, most of them are poor goods, not worth the price of a cup of farina.'

In the second barracoon Robyn thought that she had at last discovered the first plague victims, for the low thatched sheds were crammed with rows of prostrate naked figures, and their low moaning and whimpering was a heart-breaking sound, while the smell of corruption was thick and oily on the palate.

It was Tippoo who corrected her. 'China birds,' he grunted, and for a moment Robyn did not understand, and she stooped over the nearest body, then straightened immediately. Despite her training, cold blisters of sweat formed on her forehead.

By Imperial Decree from Peking, no black African slave could be landed on the shores of China unless he had been rendered incapable of reproducing his own kind. The Emperor was concerned that future generations would not be plagued by the growth of an alien population in their midst. The traders found it expedient to castrate their purchases in the barracoons, so that losses caused by the operation could be absorbed before the expense of the long voyage was incurred.

It was crudely done, a tourniquet applied to the root of the scrotum and then the entire scrotal sack removed at a single knife stroke and the wound cauterized immediately with a heated iron or a daub of hot pitch. About sixty per cent would survive the shock and subsequent mortification, but their price per capita was so enhanced that the trader could face forty per cent losses with equanimity.

There was nothing that Robyn could do for so many, she felt overwhelmed by the suffering and misery all around her, and she stumbled out on to the muddy pathway, blinded by her own tears. In the next barracoon, the one nearest to the central auction block, she found the first plague victims.

Once again the sheds had been deserted by the slave-masters, and the dimly lit thatched sheds were filled with naked figures, some squatting motionlessly, others lying on the damp earthen floor, knees drawn up, shaking with the cold of fever, and powerless to lift themselves out of their own bodily wastes. The sound of delirium and suffering was murmurous as of insects in an English orchard on a hot summer's day.

The first sufferer that Robyn touched was a young girl, just beyond puberty, and her skin was burning hot. She rolled her head from side to side, endlessly and senselessly, mouthing snatches of gibberish. Swiftly Robyn ran her fingertips down the girl's naked bulging stomach and immediately she felt the tiny lumps under the hot skin, like pellets of buckshot. There could be no doubt.

'Smallpox,' she said simply, and Tippoo drew back fearfully.

'Wait outside,' she told him, and he went swiftly and with obvious relief. She turned to Nathaniel. She had noticed the little pitted scars in his folded sun-toughened skin, and now there was no fear in his expression.

'When?' she asked.

'When I was a boy,' he said. 'It killed my old ma and my brothers.'

'We have work to do,' she told him.

In the gloomy stinking shed the dead were piled with the living, and on some of the wracked and furnace-heated black bodies the plague had already burst into full flower. They found it in all its stages. Papules beneath the skin had erupted into vesicles, bubbles of clear thin fluid that thickened into pustules, which in turn burst and released a custard-thick trickle of matter.

'These will live,' Robyn told Nathaniel. 'The plague is purging from their blood.' She found a man whose open pocks had already crusted over.

While Nathaniel held the man from moving, Robyn scraped away the crusty scabs with a spatula and gathered them in a wide-mouthed glass bottle that had once held quinine powder.

'This strain of the disease has been attenuated,' Robyn explained impatiently, and for the first time she saw fear in the flecked eyes of Mungo St John. 'The Turks first used this method two hundred years ago.'

'I would prefer to sail away from it,' Mungo St John said quietly, staring at the stoppered bottle which was half filled with damp yellow matter in which were small flecks of blood.

'It would be no use. The infection is already aboard.' Robyn shook her head firmly. 'In a week or less *Huron*

558

would be turned into a stinking plague-ship filled with dying men.'

Mungo turned away from her and went to the ship's rail. He stood there with one hand clasped into a fist behind his back, the other still in its sling, staring at the shore where the thatched roofs of the barracoons just showed above the mangroves.

'You cannot leave those poor wretches,' Robyn said. 'They will starve. I alone could never find food to feed that multitude. You are responsible for them.'

He did not answer her for a moment, then he turned back to study her curiously.

'If *Huron* sailed, with her holds empty, would you stay here on this fever and smallpox-ridden coast to tend this multitude of doomed savages?' he asked.

'Of course.' She was still impatient, and he inclined his head. His eyes no longer mocked, but were sober, perhaps even filled with respect.

'If you will not stay for common humanity, then stay for self-interest.' She scorned him with her tone. 'A million dollars' worth of human cattle, and I will save them for you.'

'You would save them to be sold into captivity?' he insisted.

'Even slavery is better than death,' she replied.

Again, he turned away from her, taking a slow turn of the quarterdeck, frowning thoughtfully, puffing on the long black cheroot so that wreathes of tobacco smoke drifted behind him, and Robyn and half *Huron*'s crew watched him, some fearfully, others with resignation.

'You say that you have yourself undergone this – this thing.' His eyes were drawn back, with loathing fascination, to the little bottle that stood in the centre of his chart table.

For answer Robyn lifted the sleeve of her shirt and showed him the distinctive deeply pocked scar on her forearm.

A minute longer he hesitated, and she went on persuasively, 'I will give you a strain of the disease that is "passant" that has been weakened and attenuated by passage through another man's body, rather than the virulent form of the plague which you will breathe on the very air and which will kill most of you.'

'There is no risk?'

She hesitated and then replied firmly, 'There is always risk, but one hundred, nay a thousand times less risk, than if you take the disease from the air.'

With an abrupt gesture Mungo St John ripped open the sleeve on his left arm with his teeth and offered it to her.

'Do it,' he said. 'But in God's name do it quickly, before my courage fails.'

She drew the point of her scalpel across the smooth deeply tanned skin of his forearm and the tiny crimson droplets rose behind it. He did not flinch, but when she dipped the scalpel into the bottle and scraped up a speck of the noisome yellow stuff, he blanched and made as if to jerk his arm away, then with an obvious effort controlled himself. She smeared the pus over the tiny wound, and he stepped back and turned from her.

'All of you.' His voice was rough with his horror and disgust. 'Every last one of you,' he told the gaping terrified seamen.

With Nathaniel, the bosun, there were three others who had survived the disease, and were speckled by the small dimpled scars which were its stigma.

Four men were not enough to help Robyn care for a thousand slaves, and her losses were much higher than she had expected. Perhaps this strain of the disease was more virulent, or perhaps the black men from the interior did not have the same resistance as the Europeans whose forbears had for generations been exposed to smallpox.

She introduced the crusted pus into the scratches on their limbs, working in the noon sunlight on into the gloom of dusk and then by the lantern's gleam, and they submitted with dumb resignation of the slave which she found pitiful and repugnant, but which none the less made her work much easier.

The reaction began within hours, the swelling and fever and the vomiting, and she went out into the other deserted barracoons to gather more of the loathsome pus from the bodies of those who had survived the smallpox and were now dying of starvation and neglect – resigning herself to the fact that she had only the strength and time to care for those in *Huron*'s barracoons, resigning herself to the fact that there was farina to feed them only, and closing her mind to the cries and entreaty, to the silent dying stare from wizened faces that streamed pus from open pocks.

Even in her own barracoon the four of them working hour after hour, night and day, could give only perfunctory attention to each of the slaves, a handful of the cold pasty farina and a mugful of water once a day during the period of the most violent reaction to the inoculation. Those who survived this were left to care for themselves, to crawl to the water bucket when they could or to wolf a lump of farina from the spadeful that Nathaniel left on a wooden platter at intervals between the rows of supine figures.

Then when they were strong enough to stand they were put to work at piling the rotting bodies of their less fortunate peers upon a gun carriage and dragging them out of the barracoon. There was not the remotest chance of either burying or burning the bodies, and there were too many for the bloated vultures. They piled the corpses in heaps in the coconut grove, well downwind of the barracoon and went back for more.

Twice a day Robyn went down to the edge of the creek

and hailed the anchor watch on *Huron*'s deck and had the whaler row her out to the ship, and she spent an hour in the stern cabin.

Mungo St John's reaction had been frighteningly severe, perhaps he had been weakened by the knife wound. His arm had swollen to almost twice normal size, and the scratch Robyn had inflicted turned into a hideous canker with a thick black crusty scab. His fever was intense, his skin almost painfully hot to the touch, and the flesh seemed to melt off his big-boned frame like candle wax under the flame.

Tippoo, himself suffering from a raging fever, his own arm swollen grotesquely, could not be made to leave the side of Mungo's bunk.

Robyn felt easier knowing that Tippoo was there, strangely gentle, almost like a mother with a child, to care for Mungo while she must go back ashore to the suffering multitude that choked the barracoons.

On the twelfth day when she went abroad *Huron*, Tippoo met her at the companionway with that wide toadlike grin which she had not seen for so long, and when she hurried into the cabin she saw why.

Mungo was propped up on the bolsters, thin and pale, his lips dry and dark purple bruises under his eyes, as though he had been beaten with a heavy club, but he was lucid and his skin cool.

'God's breath,' he croaked. 'You look awful!' And she felt like weeping with relief and chagrin.

When she had bathed and bandaged the shrinking canker on his forearm and was ready to leave, he took her wrist.

'You are killing yourself,' he whispered. 'When did you last sleep, and for how long?'

Only when he spoke did she realize the depth of her exhaustion – it had been two days previously and then only for a few hours that she had slept, and she felt *Huron*'s deck swing and lurch under her feet as though she rode a high sea and was not lying quietly at anchor in a placid creek.

Mungo drew her down gently beside him on to the bunk, and she did not have either the strength or will to resist. He made a cradle for her head on his shoulder, and almost immediately she was asleep, her last memory was the feel of his fingers smoothing back the dank ringlets from her temples.

She awoke with a guilty start, not sure how long she had slept, and still fuzzy with sleep struggled out of Mungo St John's arms brushing the hair out of her eyes with her fingers, trying ineffectually to straighten her rumpled and sweat-dampened clothing.

'I must go!' she blurted groggily, still exhausted and half-asleep. How many had died while she slept, she wondered. Before he could prevent it, she was stumbling up the companionway to the deck calling for Nathaniel to row her ashore.

The few hours of rest had refreshed her so that she looked about the estuary with a new and lively interest again. It was the first time that she became aware that one other vessel, besides *Huron*, was still lying at anchor in the river. It was one of the small dhow-rigged buggaloos, the coastal slavers similar to the one from which she had rescued Juba. On impulse, she had Nathaniel row her alongside, and when nobody answered her hail, she went aboard. The vessel had clearly been overwhelmed by the plague before she could flee, perhaps she was the vessel from which the original infection had been carried ashore.

Robyn found the same conditions prevailing aboard as there were ashore, the dead, the dying, and those who would recover. Although they were slavers she was still a physician, and she had taken the Hippocratic oath. What she could do was very little, but she did it and the Arab captain, stricken and weakened, thanked her from his sleeping-mat on the open deck.

'May Allah walk beside you,' he whispered, 'and may he give me the opportunity to return this kindness one day.'

'And may Allah show you the error of your ways,' Robyn told him tartly. 'I will send some fresh water, before nightfall, but now there are others more deserving.'

In the days that followed, the plague ran its inexorable course, the weaklings died, some of them consumed by the terrible thirsts of fever crawled from the abandoned barracoons on to the mud flats of the estuary to fill their bellies with the salt water. Their bodies were twisted into grotesque contortions by the cramps of the salt in their blood, and their insane rantings were like the cries of seabirds across the water, extinguished at last by the incoming tide. The surface of the river was troubled by the swirl and splash of the crocodile and the big sharks that had come upstream to gather this grisly harvest.

Others had crawled away into the forest and groves; they lay under every bush and even before they died they were covered with a red mantle of the fierce safari ants that overnight picked the skeleton to gleaming whiteness.

Some of those that survived, encouraged by Robyn, crept weakly away towards the west. Perhaps a few would complete the long hazardous journey back to their razed villages and devastated countryside, she hoped.

However, most of the survivors of the plague were too weak and confused and demoralized to move. They stayed on in the squalid stinking barracoons, pathetically dependent upon Robyn and her tiny band of helpers for every mouthful of water and farina, watching her with the eyes of dumb and suffering animals.

All this time the piles of corpses in the palm groves grew taller, and the stench more penetrating. Robyn knew too well what would happen next.

'The battlefield plagues,' she explained to Mungo St John. 'They always follow when the dead are left unburied, when the rivers and wells are choked with bodies. If they strike now, then none will survive. All of us are weakened, we would be unable to resist the typhoid and enteric plagues. Now is the time to leave, for we have saved all those who are for saving. We must fly before the new onslaught, for unlike the smallpox, there is no defence against them.'

'Most of my crew are still sick and weak.'

'They will recover swiftly out on the open ocean.'

Mungo St John turned to Nathaniel to ask, 'How many slaves have survived?'

'More than eight hundred, thanks to the missus here.'

'We will begin to take them aboard at dawn tomorrow,' he decided.

T hat night Robyn came back to his cabin after dark. She could not stay away, and he was waiting for her, she could tell by his expression and the quickness of his smile.

'I was beginning to fear that you preferred the company of eight hundred plague-ridden slaves,' he greeted her.

'Captain St John, I wish to make one more appeal to you. As a Christian gentleman, will you not release these poor creatures and have them escorted and fed on the journey back to where they came—'

He interrupted her, his tone light and that smile hovering on his lips.

'And will you not call me Mungo, rather than Captain St John?'

She ignored the interruption, and went on. 'After all they have suffered, that terrible march down from the highlands, the error and humiliation of slavery and now this plague. If you would consent to release them, I would lead them back to their homes.'

He rose from the canvas chair and came to stand over her. His leanness and pallor made him seem even taller.

'Mungo!' he insisted.

'God would forgive you, I am sure of that, he would forgive you the sins that you have already committed against humanity—'

'Mungo!' he whispered, and placed his hands upon her shoulders. She felt herself begin to tremble uncontrollably.

He drew her to his chest. She could feel his ribs, he was so thin, and her voice choked up in her throat when she tried to continue her appeal. Slowly he stooped over her, and she closed her eyes tightly; her arms stiff at her sides, her fists clenched.

'Say Mungo,' he commanded quietly, and his lips were cool and soft on hers. Her trembling became uncontrolled shaking.

Her lips opened under his, and her arms went up around his neck.

'Mungo,' she sobbed. 'Oh, Mungo, Mungo.'

She had been taught that her naked body was shameful, but it was the one lesson that she had learned imperfectly, and much of her shame had been mitigated, firstly in the lecture rooms and dissection rooms of St Matthew's, and secondly in the company of Juba, the little Matabele dove, whose unaffected delight in her nudity she had transmitted to Robyn. Those childlike romps together in the cool green pools of an African river had served to blow away most of the cobwebs of shame.

Now Mungo St John's delight in and admiration for her body gave her joy and, far from shame, brought her pride that she had never known before. Their love-making was no longer accompanied by pain, there were no longer barriers between them – so locked together they could ride the dips and swings of emotion from Himalayan heights where the great winds blew, down to the sweet languorous depths where they seemed to be drowning in honey, each movement slowed, every breath drawn out as though it would last for ever, their bodies damp and hot, pressed together and without form, like clay in the hands of a child.

The night was too short, while the wick of the lantern guttered and smoked, neglected and untrimmed. In the dawn, their loving seemed to have filled them with new strength, to have driven away the weakness of fever and the exhaustion of the past weeks.

It was only the sound of the first slaves coming aboard that roused Robyn and brought her back from that far frontier to the hot and cramped little cabin on a slave ship in a fever creek on a wild and brutal continent.

She heard the whisper of bare feet and the clank and drag of slave chains, the sound of men's voices raised, hectoring and impatient, on the deck above.

'Hurry them, or we load for a week.' Tippoo's voice.

Robyn raised herself on her elbow and looked down at Mungo. His eyes were closed but he was not asleep, she knew.

'Now,' she whispered. 'Now you cannot but release them. After last night, I know that something has changed in you.'

She felt a strange joy, the zeal of the prophet looking down upon a convert for whose soul she had wrestled with the devil and won.

'Call Tippoo,' she insisted, 'and give him the order to free the slaves.'

Mungo opened his eyes, even after the long night in which neither of them had slept, his eyes were clear. There was the shadow of new beard, dense and dark, carpeting his bony jawline. He was magnificent and she knew then that she loved him.

'Call Tippoo,' she repeated, and he shook his head – a little gesture of perplexity.

'You still do not understand,' he answered her. 'This is my life. I cannot alter that, not for you nor for anybody.'

'Eight hundred souls,' she pleaded, 'and you have their salvation in your hands.'

'No.' He shook his head again. 'You are wrong, not eight hundred souls but eight hundred thousand dollars – that is what I have in my hands.'

'Mungo.' His name still felt awkward on her lips. 'Jesus has said that it is easier for a camel to pass through the eye of a needle than that a rich man should enter the kingdom of Heaven. Let them go, you cannot judge human lives in gold.'

He laughed and sat up,

'With eight hundred thousand bucks, I can bribe my way into Heaven, if I want to, but between us, my dear, it sounds an awfully dull place. I think the devil and I might have more to talk about.'

The mocking gleam was back in his eye again, and he swung his legs out over the edge of the bunk, and naked crossed to where his breeches hung on a wooden peg from the bulkhead.

'We have lain too long abed,' he said briskly. 'I must see to the loading, and you had best begin your own preparations for the voyage.' He belted his breeches and stuffed his shirt into them. 'It will take us three days to load, I would be obliged if you would test the water barrels.' He came to sit on the edge of the bunk and began dragging on his boots, talking the while in crisp business-like fashion, detailing the preparations that she should make for the welfare of the slaves during the voyage. 'We will have less than a full cargo, which will make it easier to exercise them on deck and keep the holds clean.' He stood up and looked down at her.

With a rush like a roused fawn she threw off the blanket that covered her and knelt on the edge of the bunk, seizing him about the waist with both arms.

'Mungo,' she whispered urgently, 'you cannot torture us so.' She pressed the side of her face to his chest, feeling the harsh springing curls of his body hair even through the linen of his shirt. 'I cannot offend further against my God and my conscience – unless you free these poor damned souls, then I can never marry you.'

His expression changed swiftly, becoming tender and concerned. He lifted his hand to stroke the dense russet locks of her hair, still damp and tangled from the loving of the night.

'My poor darling,' his lips formed the words, soundlessly, but her face was still pressed to his chest and she could not see his lips. He drew a deep breath, and though his eyes were still marked with regret and his expression sober, his tone was light and casual.

'Then it is just as well that I have no intention of freeing a single one of them – for what would my wife say otherwise?'

The words took many seconds to make sense to Robyn, and then her whole body spasmed, her grip around his waist tightened and then slowly relaxed. She released him. Slowly she sat back on her heels, naked in the midst of his disordered bunk, and she stared at him with an expression of desolation and disbelief.

'You are married?' Robyn's voice echoed strangely in her own ears, as though from the end of a long bare corridor, and Mungo nodded.

'These ten years past,' he answered quietly. 'A French lady of aristocratic birth, a cousin of Louis Napoleon. A lady of great beauty who with the three fine sons she has borne me awaits my return to Bannerfield.' He paused and then went on with infinite regret. 'I am sorry, my dear, I never dreamed that you did not know.' And he reached out to touch her cheek, but she cringed away as though he held a poisonous serpent in his hand.

'Will you go away, please,' she whispered.

'Robyn—' he began, but she shook her head violently.

'No,' she said. 'Please don't say any more. Just go. Go away! Please go away.'

Robyn locked the door of her cabin and sat down at the sea chest which she used for a desk. There were no tears. Her eyes felt dry and burning as though blasted by a wind off the desert. She had very little paper left and had to tear the end sheets from her journal. They were speckled with mildew and distorted from the heat and dryness of the highlands and the humidity of the monsoon-ridden littoral.

She smoothed out the first sheet carefully on the lid of her writing case, dipped her pen in the remaining half-inch of India ink and headed the sheet with a hand that was calm and unshaken.

16th November, 1860.
Aboard the Slaver *Huron*.

And then in the same clear unhurried script began to write:

My Dear Captain Codrington,

My trust in an all-merciful Providence and my belief in the true and one God, and in his gentle son and our Saviour Jesus convinces me that this will come into your hands while there is still time for you to act.

Through a series of incredible adventures and misfortunes I now find myself devoid of friends or protectors, in the power of the notorious American Slave Master and trader Mungo St John. Against both my will and my conscience I am being forced to act as the physician for this infamous vessel which is at this moment preparing for the voyage around the Cape of Good Hope, across the Atlantic Ocean for a port in the Southern States of America.

As I write, I can hear the doleful sounds from the deck above me, and from the hold below where the poor creatures, eight hundred forsaken souls in all, wearing only their chains are being brought aboard and incarcerated for the voyage which many of them will not survive.

We are lying at anchor in a hidden creek, concealed from the open sea by a sweep of the channel and the mangrove swamps, a perfect hide-away for the nefarious business in hand.

However, I have been able to study the ship's chart and from the navigator's markings learn the name of the estuary and its exact position. The river is the Rio Save, and it lies 20° 58′ south latitude and 35° 03′ east longitude.

I will do all in my power to delay the sailing of this vessel, though at this moment I cannot think what that will be. If this letter reaches you in time, there will be no difficulty for an officer of your courage and experience to blockade the river mouth and seize this slave ship when she attempts to leave the river.

If we have sailed before your arrival, then I implore you to follow in the same course as the Captain of the *Huron* must set to round Good Hope, and I will pray for adverse winds and weather that will enable you to come up with us.

Robyn went on pouring out the tale of her capture, of the plague that had swept through the barracoons, of her fear and hatred of the slavers, the detailed accounts of their barbaric practices and cruelties, and suddenly she realized that she had filled many pages with her account, and she began her last paragraph.

> You were gracious enough to express your belief that our destinies were linked in some mysterious way. I know that you share with me the same hatred of this abominable trade, and for these reasons I have made bold enough to appeal to you, confident that you will hearken to my anguished cry.

Robyn paused again, and then searched swiftly in her pen case and found the pair of the earring which she had given to Clinton Codrington so many months ago.

> I enclose with this letter a token of my friendship and trust which I hope you will recognize, and I will search every day to see the topsails of your fine ship hurrying to give succour to myself and to the other unfortunates who are my ship-mates on this cursed and iniquitous voyage.

She signed it with her bold, rounded signature and stitched the folded pages and the single item of cheap jewellery into a square of duck canvas.

There was only one address that she had for it. Clinton had told her that he was under orders to call at Zanzibar Island, and she knew that Her Majesty's Consul on the island was a man of substance and integrity, a staunch adversary of the slave trade, one of the few men that her father, Fuller Ballantyne, had ever written about with respect and affection.

When she had finished, she tucked the small canvas package up under her skirts, and went up on deck. Mungo St John was on his quarterdeck, gaunt and lean and pale, and he took a step towards her, but she turned from him immediately.

'Nathaniel,' she called to the bosun. 'I wish to visit the

buggaloo.' She indicated the Arab dhow which was still anchored downstream from *Huron*.

'She's making ready to sail, ma'am.' Nathaniel knuckled his forehead. 'She'll be gone before we can get across—'

'She will if you continue talking,' Robyn told him briskly. 'I must see if there is aught they need, the poor devils, before they sail.'

Nathaniel glanced at his Captain, and after a moment's hesitation, Mungo nodded his assent and turned back to watch the stream of slaves coming aboard through the entry port.

The Arab Captain of the dhow, just strong enough to take his place at the tiller, greeted her respectfully and listened attentively while she spoke.

Nathaniel was waiting in the gig, out of sight below the level of the dhow's deck, and Robyn made sure that they were shielded from a casual watcher on *Huron*'s deck before she passed the canvas package to the Arab, and followed it with a gold English sovereign.

'There will be another sovereign for you, from the man you deliver it to,' she told him, and the Arab bit the coin, and smiled wanly as he tucked it into a fold of his turban.

'**A**nd I am Matabele. Induna of two thousand. My name is Gandang, son of Mzilikazi, son of Zulu, and I come with a bright spear and a red heart.'

Zouga understood the words with difficulty, for they were spoken rapidly, in accents that were strange to his ears, but there was no misunderstanding the Induna's intention. His tone was clear, the murderous determination in his voice evident, and around him the circle of long black shields was unbroken and steadfast.

Unconsciously Zouga had straightened, forcing his aching muscles erect, and he held the Induna's gaze without flinching. They stared at each other, and Zouga found himself exerting all his will, all the force of his personality, trying to stay the Induna's spear arm. He knew it needed only for the bright broad blade to drop and two hundred *amadoda* would sweep into the rudimentary camp. It would be over so swiftly, the resistance that Zouga and his tiny band could offer would be so puny that they would not even earn the compliment of disembowelment from the victors.

He knew that only his steady gaze and the completely fearless mien that he offered to the Matabele had so far stayed the spear arm, but the silence was drawing out. At any second the spell would snap. He must choose his next action and words as though his life depended upon them, as indeed they did.

Gandang watched the strange pale man before him with his features impassive, yet for possibly the first time in his life while on his father's service, he was uncertain.

The man who called himself Bakela had spoken familiar names. Tshedi and Manali, they were names that his own father revered, yet that in itself would not have been enough to stay his hand, for the King's orders were clear: all who entered the Burnt Land must die. It was more than that. He knew who this man was. The maiden who he would soon take as wife had spoken of him. This was the brother of the white woman who had delivered Juba to his care, and whom he had called *amekazi*, mother.

Juba had spoken of the man Bakela as she lay beside him on her sleeping-mat. She had spoken of him with admiration and awe, as a mighty hunter of elephant, as a warrior honoured by an all-powerful Queen who lived far beyond great waters. Juba had spoken of this man Bakela as a friend and a protector.

So Gandang paused before giving the order 'Bulala! Kill them!'

A Matabele Induna is never influenced by the words or whims of a woman, if he has fifty wives their voices are still as the chattering of the waters over the rocks in the shallow rapids of the Nyati river, and a man does not heed them, or rather it must never be apparent that he heeds them.

Juba had travelled to strange places and spoken of wonders and witchery, and Gandang while seeming not to listen had indeed listened and been impressed. The girl was not only comely and high bred, but sensible far beyond that mere simpering giggling sexuality to which he was accustomed in other girls of her age.

Gandang was learning that a Matabele Induna is never influenced by the whims and words of a woman, unless those words are spoken and the whims expressed on the privacy of the sleeping-mat, by a senior wife whose good sense has been proven.

Then it is folly not to hearken, for a senior wife can make a man's life unbearable, even if that man is an Induna of two thousand, and the favoured son of the most powerful monarch in Africa.

Behind the dark impassive mask of his handsome face, Gandang was thinking furiously. Instinct and Juba's words had warned him that it would be folly to slay this man, yet the warriors at his back knew his orders, and if he failed to carry them out, that failure would immediately be construed as weakness, and his treason reported to the King.

Before him the tattered figure took a pace forward, his whole being ludicrously arrogant. Gandang could see no trace of fear in the steady gaze of his strangely coloured eyes.

'I come as an emissary to the great King Mzilikazi, ruler of the Matabele people – and I bring greeting from the White Queen from across the waters.'

At the words Gandang felt a small warm flame of relief. The fact that the white man spoke the language of the people, albeit with a strange accent, made it more plausible that he was indeed an emissary. It was plausible, also, that this Queen of his would want to seek the protection and favour of a king as powerful as his father, and that she should be so ignorant as to send her emissary through the Burnt Land instead of along the open road from the south. Zouga saw the shift of mood in the Induna's eyes, that tiny crack in his determination.

'Wait,' he said. 'There is something that I have for you.'

In Zouga's writing-case still reposed the impressive hand-written letters, with seals of wax and scarlet ribbon that had been provided him by the Under-secretary at the Foreign Office, in the usual form.

In the name of Her Britannic Majesty, Ruler of Great Britain and Ireland, Defender of the Faith – To the representatives of all foreign governments or to whosoever it may concern,

We do, by these presents, request and require that our beloved Morris Zouga Ballantyne be allowed to pass freely without let or hindrance and that he be afforded that assistance of which he may stand in need.

Zouga turned his back on the silent menacing ranks of spearsmen, and walked back slowly through the gap in the *scherm* of cut thorn branches.

Jan Cheroot was waiting for him, his face the colour of the watch-fire ash. He and the gunbearers were crouched below the thorn barrier, staring through the chinks with expressions of such utter terror that Zouga felt emboldened in comparison.

'Lay down those guns,' he snapped, for all the weapons were cocked and primed and a nervous finger on a hair-trigger could let fly the shot that brought a solid black wave of Matabele sweeping through the camp.

Gandang suddenly found himself in a position of uncertainty. From being the merciless bearer of the King's justice, he found himself waiting like a timid suitor outside the gate in the thorn barrier, and every second detracted from his dignity.

Behind him he heard the stir of one of his men, the soft tap of assegai spear on hide shield. His men were growing restless already, sensing the passing of advantage to the little group of ragged starvelings they had surrounded. Gandang turned slowly, and his stony gaze passed over the ranks. They froze once more into utter stillness.

'Gandang, son of Mzilikazi, Induna of two thousand. Come forward.'

The hail from beyond the thorn barrier was unexpected and startlingly loud, but it came the moment before Gandang reached the limit of his patience and loosed his eager warriors. Gandang moved forward to the gateway. His plumes nodding about his head, his tread dignified, his carriage proud, so that no men might guess at his uncertainty. At the gateway he paused, and though his expression did not alter nor his gaze waver, he experienced a profound relief that his own wisdom and the words of his little dove had stayed his blade.

Before him stood a figure of almost incredible beauty. It took him many seconds to recognize the ragged individual of a few minutes previously. The figure wore cloth of that same peculiarly rich shade of red of the bushshrike's chest, brighter than the colour of freshly spilled blood. Though this was enough to stop a man's breath, it was not all. Bright metal ornaments on breast and shoulders sparkled in the morning sunlight, the belt buckle was of the same metal. The belt and cross-straps were of the same blinding whiteness as an egret's wing. The tall shako swept down to an elegant point between the eyes, and the helmet badge blazed like a sunrise upon the man's forehead.

There was no doubt now in Gandang's mind that here was indeed an important man, and a soldier of repute, as Juba had warned him, and he made a silent resolution to listen to her words in future with even greater attention. He felt a little shiver of dismay at the thought that he might have followed his first instincts and had this man cut down as though he was a worthless Mashona, a mere eater of dirt.

The magnificent figure took one pace towards him and lifted a hand to the peak of that beautiful helmet in a formal gesture that Gandang answered instinctively with a sweeping salute of his stabbing spear.

'I, Bakela, request that my token be conveyed to your father, the honoured and victorious Mzilikazi, and that he be informed that I request from him the right of the road,' said the man in his atrocious Sindebele and Gandang accepted the token from his hand, the small package with strange signs and marks upon it, bound up with strips of coloured cloth so beautiful that they would have delighted the heart of even the most vain and spoiled woman.

'It shall be done,' he agreed.

I n the moments of his confrontation with Gandang, Zouga had been thinking as furiously as the Matabele and making his own calculations of survival. Now that he had fallen in with a border impi he knew he must put aside any thought of escaping southwards. Apart from the fact that they were completely surrounded and heavily outnumbered, he knew that no unmounted man could run ahead of these warriors. They were like machines built for the pursuit and the annihilation of an enemy.

The meeting had not taken him completely unprepared. There had been many a night in the preceding weeks when he had woken in the darkest hours, and he had lain on the hard earth and dreaded a moment such as this.

He had mentally rehearsed his actions – from the concealment of any fear while he won time to don his dress uniform, to the demand to be taken to the King's kraal. When it had gone as he planned it, when the tall Induna had agreed, 'It shall be done,' it had taken another enormous extension of Zouga's will not to show relief. He had stood aloof, disinterested, while Gandang had picked and called out five of his swiftest runners and had recited a long message that they must memorize, and take to Mzilikazi.

It began with a long recitation of the King's praises which began,

'Great black elephant who shakes the earth with his tread – ' went on to list the deeds that Gandang had performed since leaving the great kraal at Thabas Indunas, the march eastward, the battle at the pass and the slaying of Bopa, the slave master, right up to this day's encounter with the white man. After a flowery description of the man's magnificent finery (which Gandang knew would intrigue his father), it ended with a repetition of Bakela's request to be 'given the road' to Thabas Indunas.

The chosen messengers, each in turn repeated the long message, and though he showed no change of expression Zouga was amazed that each of them had it word-perfect. It was an impressive demonstration of the developed memory of people who do not have the art of writing and reading.

Gandang handed them the sealed parchment envelope that contained Zouga's letter of credentials, and the messengers sprang to their feet from where they had squatted, saluted their Induna, formed file and trotted away towards the west.

Gandang turned back to Zouga. 'You will stay encamped here until the King sends word.'

'When will that be?' Zouga asked, and Gandang answered him sternly.

'Whenever the King is pleased to do so.'

Zouga's little party was left unmolested. Although there were a dozen Matabele *amadoda* positioned about the camp, guarding it day and night, not one of them attempted to enter the gateway in the thorn *scherm*. Until the moment when they might slaughter them, the persons and property of their prisoners were inviolate.

The main body of the impi camped a quarter of a mile downstream. Each evening the tall Induna visited Zouga and they sat for an hour or so across the fire from each other, speaking gravely and seriously.

As the days of waiting became weeks so the two men developed each for the other a deep respect, if not actual friendship. They were both warriors, and they had common ground when they spoke of old campaigns and of skirmishes and battles fought. They recognized in each other the strengths, the essential decencies of men who live by and respect the laws of their society, though those laws might diverge widely.

'I account him a gentleman,' Zouga wrote in his journal. 'One of nature's own.'

While Gandang, speaking to Juba on the sleeping-mat said merely, 'Bakela is a man.'

Gandang allowed Zouga's bearers to leave the enclosure, to cut and bring in thatch and timber to strengthen and improve the buildings, so that at last Zouga could sleep dry and warm. This, together with the rest and respite from endless marching brought an immediate improvement in Zouga's health. The deep wound in his cheek healed cleanly, leaving a pink shiny scar. The shoulder mended, the bruising abated, and he no longer needed a staff or a sling for his arm. Within a week he knew he was well enough to shoot the heavy four-to-the-pound.

One evening he proposed to Gandang that they hunt together, and the Induna who by this time was finding the waiting as tedious as Zouga was, agreed with alacrity. Gandang's *amadoda* surrounded a herd of Cape buffalo, and drove them down in a bellowing, stampeding black wave to where they waited. Zouga saw the tall Induna rise from cover and race in, bare-footed, shieldless, and kill a mature bull buffalo with a single thrust of the broad-bladed assegai through the ribs behind the heaving shoulder. Zouga knew that he lacked both the skill and courage to emulate that feat.

Gandang watched when Zouga stood to meet the squealing, thunderous charge of an enraged bull elephant, and when the beast went down to the crash of the shot, dropping on to its knees in a storm of dust, Gandang stepped past Zouga and touched the little black hole that punctured the elephant's thick grey hide an inch above the first crease at the top of the trunk.

Gandang inspected the smear of blood on the tip of his forefinger, and said 'Hau!' quietly but with force, which is an expression of deep amazement. For Gandang, himself, owned a musket, a Tower Musket manufactured in London in 1837. When he first acquired the weapon, Gandang had fired it at buffalo, elephant and Mashona, all of whom had fled headlong, but unscathed.

Gandang understood that when firing it was necessary to close the eyes and the mouth firmly, to hold the breath and at the moment of discharge to shout a rebuke to the devil who lived in the gunpowder smoke, otherwise the devil could enter through the eyes or mouth and take possession of the marksman. In order to throw the musket ball to any distance, it was also necessary to pull the trigger with sudden and brutal force, as in hurling a spear. Furthermore, to minimize the recoil of the weapon, the butt should not touch the shoulder, but be held a hand's span from it. Despite all these precautions, Gandang had never succeeded in hitting the target at which he aimed, and had long abandoned the weapon to rust away, while he kept his assegai polished brightly.

Thus Gandang appreciated to the full the magnitude of the feat that Zouga performed with such apparent ease. So their mutual respect deepened with each day spent in each other's company, and became almost friendship. Almost, but not quite, there were chasms of culture and training between them that could never be bridged, and always the knowledge that on any day a swift runner might come from the west with a message for Gandang from his father.

'*Bulala umbuna*! Kill the white man!' And both of them knew that Gandang would hesitate not an instant longer.

Zouga had much time alone in camp, and he spent it planning his audience with the King. The longer he dwelt upon that the more ambitious became his plans. The memory of the ancient disused mine-shafts returned to plague these idle hours, and at first merely to amuse himself, and then with truly serious intent, Zouga began to draw up a document which he headed:

'Exclusive Concession to mine gold and hunt ivory in the Sovereign territory of Matabeleland.'

He worked on it each evening polishing and reshaping it in the gibberish that the layman takes for legal jargon and which he fondly believes will dignify his creation. 'Whereas I, Mzilikazi, ruler of Matabeleland, hereinafter referred to as the party of the first part—'

Zouga had completed this document to his entire satisfaction when a fatal flaw in his plans became evident. Mzilikazi could not sign his name. Zouga pondered this for a day and then the solution occurred to him. Mzilikazi should by this time have in hand the sealed package. The crimson wax seals must surely impress him, and in his writing-case Zouga had two full sticks of sealing wax.

Zouga began to design a great seal for King Mzilikazi. He sketched the design on the back cover of his journal, and the inspiration came from the first of the King's praise names.

'Great Black Elephant who shakes the earth.'

In Zouga's design the centre of the field depicted a bull elephant, with long tusks raised and ears spread wide. The upper border bore the legend, 'Mzilikazi Nkosi Nkulu'. And the lower border carried the translation, 'Mzilikazi, King of the Matabele'.

He started experimenting with various materials, clay and wood, but the results did not please him – and the following day he asked Gandang for permission to send a party of his porters under Jan Cheroot back to the ancient workings at the Harkness mine to retrieve the ivory buried there.

It took two days of careful consideration for Gandang to agree, and when he did, he sent fifty of his men to accompany the caravan, with orders to kill them all at the first hint of flight or treachery. Jan Cheroot returned with the four huge tusks from the two bull elephants, and Zouga had not only the material from which to carve the King's great seal, but a gift fitting the King's importance.

Ivory was a treasure of which the Matabele had long ago realized the value in trade. It was, however, a scarce commodity, for even the bravest of men cannot kill a bull elephant with a stabbing spear. They had to rely on pick-ups from animals that had died of natural causes, or the very occasional victim of the pitfall.

Gandang's amazement when he saw the span and weight of the four tusks decided Zouga. The largest and most pleasingly shaped tusk would be Zouga's gift to Mzilikazi, if he were ever allowed to reach the great kraal at Thabas Indunas alive.

Not only was the threat of the King's wrath still hanging over him, but his supply of quinine powder was reduced to a few ounces. All around the camp steamed the swampy ground, fed each day by the interminable rains, and in the night he could even smell the evil fever-bearing vapours rising from the stagnant waters to waft over the camp. Yet he was forced to reduce his daily preventative dose of the bitter powder to perilously small quantities in an attempt to eke it out.

The inactivity and the dual threat of spear and disease wore on Zouga's nerves, until he found himself toying with suicidal plans to make an attempt to avoid the guarding impi and to escape on foot southwards. He thought of seizing Gandang and holding him as hostage, or using the remaining fifty pounds of black powder to manufacture a combustible with sufficient power to destroy the entire impi host at a blast. Reluctantly, one after another, he recognized these as plans of folly and abandoned them.

Gandang's impi came again in the dawn. Zouga was awakened by a stentorian voice calling him out of the thorn *scherm*. Zouga threw a fur kaross over his shoulders and went out into the grey and icy drizzle of rain, sloshing through the red mud to the gate, and he knew at a single glance that the King had at last sent his reply. The ranks of silent Matabele surrounded the camp, still as statues carved from the black wild ebony.

Zouga judged how swiftly he could reach the loaded gun beside his cot in the little thatched hut behind him, and guessed that he would probably be cut down before he could fire a single shot, yet he knew he would still make the attempt.

'I see you Bakela,' Gandang stepped from the dark and silent ranks.

'I see you Gandang.'

'The King's messenger has arrived—' Gandang paused a moment, solemn and stern, and then his perfectly square white teeth gleamed in the grey of dawn as he smiled. 'The King has given you the road, and bids you attend him at Thabas Indunas.'

The two men grinned at each other with relief, for both of them it meant life. The King had determined that Gandang had done his duty and correctly interpreted his commands, while Zouga had been accepted as an emissary and not as an enemy.

'We will march at once,' Gandang smiled still. 'Before

the sun!' The King's summons brooked of no hesitations or delays.

'Safari!' Zouga roused his camp. 'We march at once!'

A natural delicacy and tact had made Gandang keep little Juba out of sight and hearing of Zouga's camp, nor had he mentioned the girl's name to Zouga while sentence of death still hung over the white man. However, that first night of the journey to Thabas Indunas after they had camped, he brought Juba to Zouga's hut, and she knelt and greeted him as 'Baba-father' and then with Gandang seated between them and listening attentively, he allowed them to talk for a short while.

Zouga was avid for news of his sister, and he listened in silence to the account of Fuller Ballantyne's death and his burial. It was better this way, and Zouga was already preparing his own fulsome tributes to the memory of his father.

While relieved also to hear that Robyn was safe, Zouga was less happy with the swift progress she seemed to be making. It must have been nearly three months since Juba last saw her, almost within sight of the eastern mountain range, and by this time Robyn must have surely reached the coast, and could be on board a Portuguese trade ship, well on her way to Good Hope and the Atlantic.

He did not know how long he would be delayed by the Matabele King, and then how long the overland journey down half the length of Africa might take him. Robyn's manuscript could be in London a year or even more before his own.

Zouga had exchanged one worry for another, and on the following day's march he chivvied his porters, heavily laden though they were, to keep up with their captors. It was of little avail, and they struggled along behind the trotting

585

impi until Zouga demanded of Gandang that his own bearers be ordered to help with the load of ivory and the even greater weightier package of bark and plaited grass which contained the granite bird which Zouga had plundered from the tomb of the Kings.

W ith each day's travel towards the west, so the land became drier, and the forests thinned out and gave way to level pasturage with sparsely dotted acacias, graceful, mushroom-shaped trees from whose branches hung the big protein-rich, bean-shaped pods so dearly loved by game and domestic animals alike.

The relief from the driving and endless rain storms lifted their spirits, and the impi sang on the march, winding like a thick black serpent through the lovely park-like lands below the bald and rounded kopjes of granite.

Soon they came across the first of the King's herds. The small humpbacked cattle whose origins lay far back beyond the veils of history, perhaps it had taken them and their drovers four thousand years to travel down from the valley of the Nile or from the fertile plains enclosed by the twin rivers of the Euphrates and the Tigris.

The cattle were sleek, for the grass was dense and sweet, even here in the drier lands the rains had been good. The animals were of every colour and pattern, chocolate, red and black and white and yellow, piebald and skewbald, solid black or pure snowy white. They watched with a blank bovine stare as the column of men trotted past, and the small herd boys, naked except for the tiny apron of the *beshu*, came scampering to stare in silent wide-eyed awe at the fighting men in plume and tassel, for they were already pining for the day that they would be called into their own regiments and in turn follow the warriors' road.

They reached the first of the Matabele towns. It was

situated on the banks of the Inyati river. Gandang explained that it was the headquarters of his own impi, the Inyati impi, and that it was not the largest of the regimental towns. The settlement was laid out around the central cattle-pen, a vast enclosure large enough to hold ten thousand head of the King's herds. The dwellings were of identical thatched beehive construction in the tradition that the wandering tribe had preserved since leaving their native Zululand. The outer stockade was of cut mopani poles, set deep in the earth and forming a stout defensive wall. The villagers streamed out to welcome the returning impi, lining both sides of the route, a singing, clapping and laughing throng, mostly of women and children.

'Most of the men and the marriageable maidens have left already for Thabas Indunas. In the full of this moon, the *Chawala* dance begins, and all the nation will assemble at the King's kraal. We will rest here only one night and then take the road again to reach Thabas Indunas before the moon.'

The road from the Inyati westwards was now a populous highway, as the nation went in towards the King's capital city to celebrate the festival of the first fruits. The men marched in their regiments, their distinctive dress and ornaments, the colour of their war shields identifying each from afar, from the scarred and silver-headed veterans who had fought the Basuto, the Griquas and the Boer in the south, to the young bloods eager for their first kill, eager to learn in which direction the King would hurl his war-spear at the conclusion of the *Chawala* – for that was the direction in which they would find their reputation, their manhood, their glory and possibly their deaths.

The regiments of young unmarried women interspersed those of the warriors, and as they passed each other on the road, the girls preened and giggled, casting languorous sloe-eyes at the unmarried men, and the men pranced and leapt in the pantomine of battle, the *Giya*, showing how they would wash their spears in blood and earn the privilege to 'go in to the women' and take wives.

With each day's march towards Thabas Indunas the roads became more congested, and their pace was reduced by the throng. They might wait half a morning to take their turn across a ford of the river, for the regiments drove the cattle which were their food supply ahead of them and dragged their baggage train behind. Each warrior's finery, his tassels and plumes and feathers, were carefully packed and carried by the young apprentice who was his personal bearer.

At last in the sweltering noon of high summer, Zouga's little party, still borne along on the river of humanity, came over a crest of ground and saw laid out ahead of them the great kraal and capital city of the Matabele.

It was spread out over many square miles of open plain below the bald-headed granite hills that gave it its name 'The Hills of the Chieftains'. The furthest hill was the 'Place of Killing', Bulawayo, and from its sheer cliffs those condemned to die were cast down.

The stockades formed concentric circles, dividing the city into its separate parts. Always the huge open cattle-pens were the centre of Matabele life, their cattle the source and store of their wealth, and now that the outlying herds had been brought in for the festival, every pen was filled with the multi-coloured herds of fat beasts.

Standing at Zouga's shoulder, Gandang used his stabbing-spear to point out with pride the city's features. There were sections for the unmarried girls, and the unblooded regiments, and another huge area for the married quarters; the huts were uniform in size and laid out in orderly patterns, the thatched roofs shining golden yellow in the sunlight. The earth between them was swept clean and beaten hard by the passage of bare feet.

'There is the King's hut.' Gandang pointed out a single huge conical structure in its own separate enclosure. 'And that is the compound of the King's wives,' a hundred other huts, within a high guarding stockade, 'and it is death to any man who enters that gate.'

Gandang led Zouga down to a small grove of acacia trees outside the main stockade. There was a stream within a few yards of it, and for the first time in days they were free of the close press of humanity. Although the plain without the city walls was thick with the temporary dwellings of the visiting impis, the area around the grove was empty, as if it had been placed out of bounds to the common people.

'When will I see the King?' Zouga asked.

'Not until after the festival,' Gandang answered. 'There is ritual and cleansing that the King must undergo but he has sent you gifts, you are much honoured.' And he pointed with his blade at the line of young maidens that left a gate in the stockade. Each girl carried a large earthenware pot balanced easily upon her head, and she did not use a hand to maintain that balance.

The girls moved with that peculiar straight-backed grace, hips swinging in lazy rhythm, the hard unripe fruits of their breasts bouncing and jostling at each pace. They came in to Zouga's little camp in the grove and knelt to offer the gifts they bore.

Some pots contained thick millet beer, tart and effervescent, others the clotted and soured cow's milk, *imaas*, that was so much a staple of the Nguni diet, and others again, big chunks of fatty beef, roasted on the open coals.

'You are much honoured by these gifts,' Gandang repeated, apparently himself surprised by the King's generosity. 'Yet Tshedi, your grandfather, was always his good and trusted friend.'

Once the camp was set up, Zouga found himself again the victim of idleness, with long days of waiting to fill. Here, however, he was free to roam about the city and its surrounds, save only the forbidden areas of the royal enclosure and women's quarters. He sketched the fascinating bustle of preparation for the festival. During the heat of the day the banks of the rivers were lined with men and women, their velvety black skins shimmering with water as they bathed and cleansed themselves for the dances. Every tree for miles about was hung with the kilts and furs, the feathers and plaited ornaments that were airing, the creases and rumples of travel and packing were being allowed to drop out of them as they billowed and flapped in the light breeze.

He passed groups of young girls plaiting each other's hair, smearing and rubbing each other's bodies with oil and coloured clays, and they giggled and waved at Zouga as he passed.

At first the problem of hygiene that this huge assembly presented puzzled Zouga, until he realized that there was an area of thick undergrowth beyond the city walls where both men and women went at dawn and in the short twilight. This area had its own population of crows and kites, of jackal and hyena that served as the city's cleansing service.

Interested further in the running of the city, he found that all bathing and washing of clothing was allowed only below a certain spot on the riverbanks marked by a distinctively large tree or other feature, and that the women filled the water pots for drinking and cooking above this point.

Even the huge cattle-pen in its very heart helped to keep the city clean. It acted as a vast fly-trap. The insects laid their eggs on the fresh cattle dung, but before they could hatch most of them were deeply trampled by the hooves of the milling herds. The brilliant sunlight and the untainted wind completed the process of keeping the area relatively clean and the smells interesting but not unbearable.

Zouga should have been content to have reached this haven, instead of leaving his assegai-riven corpse for a hyena's feast in the wilderness, but he was not.

He set himself tasks to fill the waiting days. He drew sketches and maps of the city, noting weaknesses in the fortifications, and where an attack would have the best chance of penetrating these and reaching the King's private quarters. He sketched the uniforms of the various impis. He noted the colours of their shields and other means of identifying them in the field. Asking innocent-seeming questions of Gandang, he was able to estimate the number of warriors each regiment contained, the ages and battle experience of the warriors, the names and personal idiosyncrasies of their Indunas, and the location of their regimental towns.

He found that much had changed in Matabeleland since old Tom Harkness had drawn his map, and Zouga noted these changes and drew his conclusions from them.

As a further exercise to pass the waiting days he began drawing up a battle plan for a campaign against the old King Mzilikazi, the requirements in men and weapons, the logistics of supply and resupply, the lines of march, and the most expedient methods of bringing the impis to battle – for Zouga was a soldier, a soldier with a dream which might one day become reality only through military action.

Unsuspecting, Gandang was flattered by Zouga's interest and he answered every question with pride in his nation's might and its achievements. Despite this work Zouga had given himself, the days dragged past.

'The King will not give you audience until after the festival,' Gandang repeated, but he was wrong.

The evening before the festival began, two elderly Indunas, blue heron feathers nodding above the silvery woollen caps of close-cropped hair, entered the camp in the acacia grove and Gandang greeted them with deep respect, listened to them attentively, and then came to Zouga.

'They have come to take you to the King,' he said simply.

There were three small fires burning before the King's hut, and over the middle one crouched a wizened ape-like figure, who crooned a low incantation through toothless gums, rocking on his haunches and occasionally adding a pinch of powder, or a sprig of herb into one of the large earthenware pots that bubbled over the flames.

The witch-doctor was festooned with the grisly accoutrements of his profession, the dried skins of reptiles, the claws of eagle and leopard, the inflated bladder of a lion, the skull of ape and the teeth of crocodile, small stoppered gourds of potions and powders, the horn of duiker to be used as a blooding cup, and other unidentifiable charms and elixirs.

He was the orchestrator of the entire festival, the most important event in the Matabele calendar, the gathering of the first fruits of the harvest, the blessing of the nation's herds, and the setting of the warlike campaigns which would occupy the *amadoda* during the coming dry season. Thus the assembled Indunas watched his preparations with attention and awe.

There were thirty or so men in the squatting circle of elders, the senior Indunas of the nation, the King's privy council. The small courtyard was crowded. The tall thatched sloping side of the King's hut towered thirty feet or so above them all, the top of it lost in darkness.

The thatching was skilfully done, with intricate patterns worked into the grass, and before the low doorway stood an armchair of European design and construction. With a small prick of recognition Zouga realized that this must be the same chair given to the King by his own grandfather Moffat, Tshedi, nearly twenty years before.

'Bayete! Mzilikazi, the bull elephant of the Matabele.'

Gandang had coached Zouga in the correct etiquette, the formal greetings and the behaviour which the King would expect.

As Zouga crossed the narrow yard of bare earth, he intoned the King's praise names, not shouting them aloud, not crawling on his knees as a subject would have done, for he was an Englishman and an officer of the Queen.

Nevertheless, at a distance of ten feet from the King's chair Zouga squatted down, his own head below the level of the King's, and waited.

The figure in the chair was much smaller than he had expected for a warrior of such fearsome reputation, and as his eyes adjusted to the gloom, Zouga saw that the King's feet and hands were small and delicately shaped, almost feminine, but that his knees were grotesquely swollen and distorted by the gout and arthritis that was attacking them.

The King was an old man now, nobody knew how old, but he had been a fighting warrior at the turn of the century. His once fine muscles had sagged so that his belly bulged out on to his lap, the skin stretched and riven with stria like that of a pregnant woman.

His head seemed too big for the narrow shoulders, and the neck hardly strong enough to support it, but the eyes which watched Zouga intently from the seams of wrinkles and loose, bagged skin, were black and bright and lively.

'How is my old friend Tshedi?' the King asked in a piping high-pitched voice.

Zouga had last seen his maternal grandfather twenty years previously; the only memory that persisted was of a long flowing white beard.

'He is well and happy,' Zouga replied. 'He sends you his greeting and respect.'

The old man in the armchair nodded the big ungainly head contentedly.

'You may present your gifts,' he said, and there was a buzz of comment from the Indunas, and even the witch-doctor at the fire looked up as the ivory tusk was carried in by three of Gandang's warriors, staggering under its weight, and laid before the King's chair.

The witch-doctor clearly resented this interruption of the ritual cleansing and the diversion of interest and attention from himself, and now with two of his assistants to help him he made an officious show of carrying one of the steaming pots from the fire and placing it between the King's feet.

Then he and his assistants raised a large kaross of stitched leopard skins and spread it like a tent over the King and his chair so the steam from the pot was trapped beneath it. Within a minute, there was a paroxysm of gasps and coughing from under the fur blanket, and when at last the witch-doctor removed it, the King was streaming sweat and choking for breath, his eyes inflamed and pouring tears, but any demons had been expelled by the coughing, and impurities washed away by the sweat and tears.

The gathering waited in respectful silence, while the King recovered his breath, and the witch-doctor withdrew to prepare the next potion. With his breath still wheezing and whistling in his throat, Mzilikazi reached into the small chest beside his chair and brought out the sealed package which Zouga had sent him.

'Speak the words.' The King handed it to Zouga, demanding that he read the letter.

Although he was illiterate, the King understood clearly the uses of the written word. For twenty years he had corresponded with Zouga's grandfather, who always sent one of his mission students to deliver his letters, to read them to the King, and to record the King's reply.

Zouga stood erect and opened the package. He read aloud, translating from the English as he went along, and adding a few small embellishments to the original text.

When he had finished, there was a respectful silence from the tribal elders, and even the King studied the tall and magnificently attired figure before him with new attention. The firelight danced on the burnished brass buttons and badges of Zouga's dress uniform, the scarlet cloth of his coat seemed to glow like the very flames of the fire.

The witch-doctor would have intervened again, coming forward with a brew of steaming medicine for the King to swallow, but Mzilikazi waved him away irritably.

Knowing that this was the moment when the King's interest was at its zenith, Zouga asked smoothly, 'Does the King see these signs of my Queen? They are her special marks, and every ruler should have such a mark to prove his power and the rocklike nature of his words.'

Zouga turned and beckoned the bearer who knelt in the gateway behind him and the terrified man crawled to Zouga's feet, not daring to look up the King, and handed to Zouga the small tea caddy that contained the carved ivory seal and the sticks of wax.

'I have prepared one of these for the King, that his dignity and power may be known to all men.'

Mzilikazi was unable to contain his interest; he craned forward in his chair and called Zouga closer. Kneeling before him, Zouga prepared the wax, melting it on to the lid of the tea caddy with a taper lit from the fire. Then he made the impression of the seal upon it, and when it had hardened, handed it up to the King.

'It is an elephant.' The King recognized the beast with unconcealed amazement.

'The great black bull elephant of the Matabele,' Zouga agreed.

'Speak the words.' The King touched the lettering on the border, and commanded Zouga to translate it.

'Mzilikazi, Nkosi Nkulu!'

The King clapped his hands with delight, and passed the seal to his senior Induna. Soon they were all clucking and exclaiming over the imprint, the wax impression passing from hand to hand.

'Bakela,' the King told Zouga, 'you must come to me

again on the day following the *Chawala* ceremony. You and I have much to discuss.'

Then with a wave of his hand he dismissed the splendid young man from his presence, and patiently, resignedly, gave himself up to the ministrations of the hovering doctor.

The full moon rose well past midnight, and the fires were stacked with new logs to welcome it, and the singing and drumming began. No man nor woman had dared to pick a single grain of corn from the harvest before this moment, for the rise of the moon heralded the *Chawala*, the dance of the first fruits, and the entire nation gave itself up to rejoicing.

The ceremony began in the middle of the first morning. The massed regiments assembled before their King, filing in column into the vast arena of the cattle stockade, and the earth shook to the crash of bare feet lifted in unison, twenty-five thousand at a time, lifted to the level of the shoulder and then brought down with the full force of the muscular, hardened bodies of highly trained warriors.

'Bayete!' they greeted the King with the royal salute.

'Bayete!' the crash of feet once more.

'Bayete!' a third time, and then the dancing began.

One regiment at a time coming forward in swaying, singing ranks, to perform before Mzilikazi's armchair throne. The perfect timing and execution of the intricate steps made it seem as though they were a single living organism, the shields interlocking and revolving and twisting together like the scales of some gigantic reptile, the dust rising and swirling through their ranks like smoke so that they appeared as wraiths, and their cloaks of furs and their kilts of civets' tails and monkey-skins, of the pelts of foxes and cats, swirled about their legs so they appeared to be divorced from earth, suspended above it on the moving cloud of dust and the soft waves of fur.

From the ranks sprang the great champions and the heroes of each regiment, to *giya* in their pride. Leaping as high as their own heads, and stabbing furiously at the air, screaming challenge and triumph, the sweat greased their muscles and flew in explosive droplets in the sunlight.

Mzilikazi was caught up in the building sea of excitement, and he quaffed from the beer pots that the maidens brought him until his eyes rolled in his head and he could not contain himself further. He struggled from his chair, and hobbled out on his swollen and deformed legs, and the champions fell back to give him place.

'My father is the finest dancer in all of Matabeleland,' said Gandang, squatting beside Zouga.

The old King tried to leap, but his feet did not leave the ground. He shuffled back and forth, making little pawing gestures, hacking at the air with his toy war spear.

'Thus I struck down Barend the Griqua, and thus his sons died.'

The nation roared.

'The bull elephant dances, and the earth shakes.' And the slamming of ten thousand feet goaded the King to circle in a painful and pathetic parody of the young champions' wild gyrations.

'Thus I spurned the tyrant Chaka, and thus I cut the plumes from the headdress of his messengers and sent them back to him,' squealed Mzilikazi.

'Bayete!' thundered the nation. 'He is the father of the world.'

Exhausted within minutes, the ancient King sank into the dust, and Gandang and two other of the King's sons leapt to their feet from the half circle of Indunas and raced to his side.

Gently they bore him up and carried him back to his chair, and Lobengula, the King's senior son, held a beer pot for him to drink from. The beer dribbled from his chin and ran down the King's heaving chest.

'Let the nation dance,' gasped the King, and Gandang returned to Zouga's side and squatted beside him.

'After war, my father loves best the dance,' he explained.

The maidens came, rank upon lovely rank, their naked skin shimmering in the glaring sunlight of noon. The tiny beaded apron that barely covered their little triangular sex was all they wore, and their singing was sweet and clear.

Mzilikazi hoisted himself from his chair once more, and hobbled out to dance with them, passing along the foremost rank, directing the singing with his ritual war-spear pointed to the skies. The King danced until he dropped once more, and was again carried back to his chair by his sons.

By nightfall, Zouga was exhausted. His sweaty neck was chafed by the high stock of his dress coat, and sweat had soaked through the thick scarlet serge in dark patches. His eyes were bloodshot and inflamed from the dust and the glare, his head ached from the cacophony of drums and the roar of Matabele voices, his tongue felt thick and furry from the draughts of sorghum beer that had been pressed upon him and his back and legs ached from the unfamiliar squatting attitude he had been forced into all that day – but the King was still dancing, hobbling and prancing and squeaking on those twisted and deformed legs of his.

The following morning Mzilikazi was on his throne again, so undaunted by the previous day's exertions, that when the *Chawala* bull was loosed into the arena his sons had bodily to restrain him from rushing out to slay it with his bare hands.

A champion from each regiment had been chosen, and stripped down to a loin-cloth. They waited in squatting ranks on each side of Mzilikazi's chair.

The bull came into the ring at the charge, horned head held high, red dust spurting from under his hooves, and his wild eyes glaring. He was pure, untainted black, with a huge humped back and glossy burnished hide.

Carefully picked from all the King's herds, he was the finest animal in the whole of Matabeleland and he made an arrogant circuit of the arena, stepping high, snorting and dropping his head to hook with curved horns at anything in his path.

The King, held by his sons, but struggling against their grip, was almost incoherent with excitement, and now he lifted his spear and his arms shook violently as he screamed, '*Bulala inkunzi!* Kill the bull!'

The waiting men leapt to their feet, saluted the King, and then raced out, spreading into a half-moon shape line, instinctively adopting the *jikela*, the movement of encirclement.

The black bull swung to meet them, came up short on locked front legs, his head swinging as he measured his charge; then the great rounded quarters bunched under him and he surged forward, picking a man in the centre of the line and thundering down on him.

The man he had chosen stood his ground, spreading his arms in a welcoming gesture and the bull dropped his head and hit him. Clearly Zouga heard the brittle snap of bone as the warrior absorbed the shock of impact against his chest, and then locked his arms around the animal's neck and held on.

The bull tried to toss him, hooking and throwing his head high, but the man held on, and was cruelly thrown about, but his body blocked the bull's sight and brought him to a halt. The racing line closed about him instantly, and abruptly the bull's great humped body was smothered by the rush of naked black men.

For long seconds the bull struggled to remain upright, but they bore him down and tore his legs out from under him so he hit the dusty earth with a heavy thump and a groaning bellow. A dozen men seized the long horns, and, using their leverage, began to twist them against the massive inert weight of his pinned body. Slowly, sweating and straining, they forced his head around, and the bull kicked his hooves in the air, his bellows becoming more desperate and strangled.

The King leapt up and down in his chair, screaming with excitement, and the roar of voices was like the sound of surf on a gale-driven coast of rock.

Inch by inch, the huge horned head revolved, and then suddenly there was no longer resistance. Zouga heard the crack of the vertebrae, sharp as a musket shot even above the thunder of the assembled nation. The horned head flicked through another half turn, the legs stiffened skywards for a moment, and the bull's bowels voided in a liquid green stream.

The sweat-drenched warriors lifted the carcass shoulder high and bore it bodily across the arena to lay it at Mzilikazi's feet.

On the third and last day of the ceremony, Mzilikazi stalked out into the centre of the cattle-pen. He made a frail and bent figure in the vast open space, and the noon sun burned down from above so that there was almost no shadow under him. The nation was quiet, forty thousand human beings watching one old man and there was not a whisper, nor a sound of breathing.

In the centre of the arena Mzilikazi paused and raised his war-spear above his head. The watching ranks stiffened as he revolved slowly, and then stopped facing towards the south. He drew back his spear arm, poised for a moment while the tension in the watchers was a palpable emanation from forty thousand charged bodies.

Then the King gave a little hop, and began slowly to revolve, the crowd sighed and swayed and then grew silent as again the King poised with his blade pointed towards the east. Then another little hop as he teased them deliberately, drawing out the moment with the timing of a natural showman.

Then suddenly his spear arm shot forward and the tiny toylike weapon flew from his hand in a high sparkling parabola, and dropped to bury its point in the baked earth.

'To the north!' thundered the nation. 'Bayete! The great bull has chosen the north!'

Ｗe go northwards to raid the Makololo,' Gandang told Zouga. 'I will leave with my impi in the dawn.' He paused, and then smiled briefly. 'We will meet again, Bakela.'

'If the gods are kind,' Zouga agreed, and Gandang laid one hand on his shoulder, squeezed briefly, and then turned away.

Slowly, without looking back, Gandang walked away into the darkness which was clamorous with the singing and the sound of the drums.

Ｙour guns would be terrible weapons if they did not have to be reloaded,' Mzilikazi piped in his querulous old man's voice. 'But to fight with them a man must have fast horse, so that he may fire and then gallop away to reload.'

Zouga squatted by the King's chair in the royal enclosure as he had for almost thirty successive days. The King sent for him each day, and he must listen to Mzilikazi's wisdom and eat huge quantities of half-raw beef washed down by pot after pot of beer.

'Without a horse my warriors will overrun them before they can reload, even as we did to the Griquas, and afterwards we picked up over three hundred of their precious guns from the battlefield.'

Zouga nodded his agreement, smiling inwardly as he imagined the *amadoda* trying those tactics on a square of British infantry.

Mzilikazi broke off to lift the beer pot. and then as he lowered it the sparkle of one of Zouga's tunic buttons caught his eye, and he leaned across to pluck at it. Resignedly Zouga took the clasp knife from his pocket and carefully cut the threads that held the button. He handed it to the king, and Mzilikazi grinned with pleasure and held it to the sunlight.

'Only five to go,' Zouga thought ironically. He felt like the Christmas turkey being plucked a feather at a time. His lapel badges and field officer's pips had long since been taken by the King, as had his belt buckles and helmet badge.

'The paper,' Zouga started and the King waved airily, dismissing the reference to the concession.

He might be on the verge of senility and certainly he was an alcoholic, drinking seven gallons of beer each day, by Zouga's count, but still Mzilikazi possessed a cunning and devious intellect with a natural grasp of his own bargaining weaknesses and strengths. He had teased Zouga for thirty days, just as he had teased the watching nation on the third day of the *Chawala*, while they had waited for him to hurl the war-spear.

Now the King turned away from Zouga at the mention of the concession, and transferred his attention to the young couple who knelt before him. They had been accused, and come before the King for judgement.

That day, in between chatting with Zouga, Mzilikazi had received emissaries bearing tribute from two of his vassal chieftains, he had rewarded a young herdboy for saving his herd from a marauding lion, he had sentenced to death another who had been seen drinking milk directly from the udder of one of his charges, he had listened to the reports of a messenger from the impis campaigning in the north against the Makololo, and now his attention was on the accused couple before him.

The girl was a lovely creature, with long delicately formed limbs and a sweetly rounded face with full flaring lips over small very white teeth. She kept her eyes tightly closed, so as not to look upon the King's wrath, and her body was shaken with tremors of terror as she knelt before him. The man was a finely muscled young warrior, from one of the unmarried regiments, who had still to win the honour of being allowed to 'go in to the women'.

'Rise up, woman, that the King may see your shame,' the voice of the accuser rang out, and hesitantly, timidly, still with her eyes tightly closed, the girl lifted her forehead from the dusty earth, and sat back on her heels.

Her naked stomach, drum tight and round as a ripening fruit, bulged out above the tiny beaded apron.

The King sat hunched in his chair, brooding silently for many minutes, then he asked the warrior,

'Do you deny this thing?'

'I do not deny it, Nkosi Nkulu.'

'Do you love this wench?'

'As I love life itself, my King.' The man's voice was low and husky, but firm and without a quaver to it.

The King brooded again.

Zouga had sat by and watched the King give judgement on a hundred such occasions. Sometimes the decision had been worthy of a black Solomon, at others Zouga had been appalled by the barbaric savagery of the sentence.

Now the King stared at the man before him, and he fiddled with the toy spear in his right hand, frowning and shaking his head softly, then he reached a decision, and leaning forward he proffered the weapon to the man who knelt before him.

'With this blade, open the womb of the woman you love, take out from it that which offends against law and against custom – and place it in my hands.'

603

Zouga did not sleep that night, and once he threw off his blankets and hurried to the fringe of the acacia grove to retch and heave, vomiting up his horror at what he had witnessed.

In the morning the memory of the girl's screams had not faded but the King was jocular and garrulous, pressing pot after pot of sour beer on Zouga's rebellious stomach, recounting episodes from his long and eventful life, describing vividly the scenes from his childhood and youth in far-off Zululand with a certain old man's wistful nostalgia.

Then suddenly, without any prior hint that he would do so, he commanded Zouga, 'Speak the words of your paper.'

He listened silently as Zouga translated the terms of the concession he was seeking, and at the end he mused a moment.

'To hunt elephant and to dig a hole,' he mumbled at last. 'It is not so very much you wish for. Write that you will do these things only in the land below the Zambezi, east of the Inyati River and above the Limpopo.'

Not quite convinced that the King was this time serious, Zouga quickly added the proviso to the foot of his home-grown legal document.

Then he directed the King's trembling hand to form the big uncertain cross below it.

'Mzilikazi: his mark.'

The King's pleasure in affixing the wax seal beside his mark was childlike and unaffected. After it was done he passed it to his Indunas to admire, and leaned forward towards Zouga.

'Now that you have what you want from me, you will wish to leave.' There was regret in the rheumy eyes, and Zouga felt a pang of guilt, but he replied directly.

'I cannot hunt in the rains, and there is much work for me to accomplish in my own land across the sea. I must go, but I shall return, Nkosi Nkulu.'

'I give you the road to the south, Bakela the Fist. Go in peace and return to me soon, for your presence pleases me, and your words are wise for one still so young.'

'Stay in peace, Great Elephant.' Zouga rose and left the King's courtyard, and his step was as light as his spirits were gay. He had the concession buttoned into the breast-pocket of his tunic, fifty-six pounds of native gold in his chest, the stone bird of Zimbabwe, and three fine tusks of ivory to buy his way. The road to the south, to Good Hope and to England lay open before him.

The wind was offshore, a faint backlash of the monsoon, but the sky was low and grey and the twisting squalls of rain fell from it like pearl dust.

Ensign Ferris, *Black Joke*'s most junior officer, was taking his sights off the traverse board as the gunboat closed the land, calling them quietly to the signals yeoman who swiftly worked out the distance offshore, and wrote this down on the navigation slate, so that at any moment the Captain could glance down at it to confirm his own observations.

There was a man in the bows with a leadline, chanting the depth as he read it from the markings on the line, then twirling the weight and hurling it out ahead of *Black Joke*'s creeping bows, letting it sink and then reading the mark again as the ship passed over it.

'By the deep six.'

Clinton Codrington was conning his ship in by the leadsman's chant, by the angles that Ferris was shooting off the land features which they had identified, by the colour shadings of the water, by the break and swirl of the tide on shoal and bank, and by a seaman's instinct. The chart surveyed thirty years previously by Captain Owen, R.N., he trusted not at all.

'Bring her around a point,' he told the helmsman quietly, and as the ship swung towards the land, they smelt it on the wind.

'Slave stink!' exclaimed Denham fiercely, and as he said it the masthead lookout sang out sharply,

'Smoke! Smoke from the right bank of the river.'

'How far up stream?'

'Two miles or more, sir!'

For the first time since sailing from Zanzibar harbour, Clinton allowed himself to believe that he was in time, that he had reached the Rio Save in time to answer the heartrending appeal of the woman he loved.

'We will clear for action, if you please, Mr Denham, but don't run out the guns.' He kept his voice level and formal, but his Lieutenant grinned at him.

'A fair cop, by Jove. Congratulations, sir,' and the crew laughed and skylarked as they lined up at the arms chest to receive their pistol and cutlass.

As Black Joke breasted the breaking white water of the bar, touched sand bottom for a moment and then broke the grip of it, and surged forward into the dark green and still waters of the estuary, Clinton nodded at Denham.

'You may run out the guns now.'

He had delayed until this moment to avoid altering the ship's trim during the critical passage of the bar. With the carriages rumbling ominously, Black Joke bared her fangs and under fighting sail, her bronze propeller thrashing exultingly, her crew armed and eager, she went into the labyrinth of the Rio Save like a ferret into the warren.

Clinton took her around the first bend, following the deep green sweep of the channel between the paler sand-banks. It was two hours past low tide and the flood was pushing strongly, the leadsman calling good bottom and Clinton was trying to conceal his impatience and eagerness by the calm controlled tone in which he called his orders to the helm.

'Would you just look at that!' exclaimed Ferris, and pointed over the side. Beside them floated what looked like an ebony log. It was only as they passed and it bobbed and rolled in their wake that Clinton realized that it was a human corpse, stomach distended, round and shiny with gas and limbs twisted like the branches of a tree blasted by lightning. Clinton transferred his attention back to the ship, with a small grimace of distaste.

'Meet her,' he told the helmsman and then as they tore around the broad curve of water between the mangrove forests and the full stretch of the estuary opened ahead of them, he said, 'Midships!'

His voice was flat, unemotional, neither triumphant nor dejected. Smoke rose from the bank of the river, and through his glass he could see the smouldering remains of a series of long low buildings, roofs burned through and collapsed inwards. They seemed to have been deliberately fired.

In the smoke soaring and circling on spread wings rose a host of birds, buzzards and kites and vultures and the carrion-eating marabou storks. They seemed to reach up with the smoke to the belly of the lowering monsoon clouds, dimming the light of day with their wings. The oppressive towering silence was broken only by the faint high cries of the birds, and the river was empty.

Clinton and his officers stared silently at the wide deserted sweep of the Rio Save, green and smooth from bank to mangrove-clad bank. No one spoke as *Black Joke* bore on, close in to the ruined and blackened barracoons. They stared at the piles of corpses, their faces stiff and expressionless, concealing their horror at the foul plague that lurked there in the palm groves, concealing also their disappointment and chagrin at finding the anchorage deserted and *Huron* gone.

'Stop engine,' Clinton broke the silence. 'Let go port anchor.'

Denham and Ferris turned to look at their Captain, their carefully controlled expressions cracking with disbelieving dread; he was going to send a party ashore. They would come back aboard carrying the plague with them, they would all be doomed.

The bow anchor hit the mud of the estuary bottom, and *Black Joke* turned sharply, the flow of the tide swinging her across the narrow channel in her own length, the anchor holding her head until she was pointed directly back down the river towards the sea. Immediately Clinton gave the order, 'Slow ahead,' and, as the ship breasted the tide, 'Up anchor,' and the steam winch clattered merrily. The officers relaxed visibly, and Denham allowed himself to smile with relief. The Captain had used the anchor merely to turn the ship as swiftly as possible without going through the dangerous business of backing and filling in the narrow channel against the push of the tide.

As the anchor came back aboard with the gluey black mud clinging to its flukes, Clinton gave another series of orders.

'Half ahead.' That was as fast as he dared push for the open sea.

'Secure from action stations.'

There was no enemy to fight and as the heavy long-barrelled thirty-two pounder guns were run in, *Black Joke* handled more easily.

'Mr Ferris, we will fumigate the ship.'

The smoke from the pails of burning sulphur would redden all their eyes and for days would taint the food and water, but Clinton's dread of smallpox outweighed such small considerations as comfort, and, besides, he smiled wryly to himself, any change to the taste of *Black Joke*'s bullied beef and hard bread could only be an improvement.

The smile was fleeting. He was sickened by his brief glimpses of the barracoons, and his anger was a cold sharp thing like the blade of the cutlass on his belt.

'Mr Denham,' he said quietly, 'please plot a course for Good Hope. We'll come on to it the moment we clear the land.'

He moved to the rail, half his mind on the job of conning *Black Joke* out of this stinking green river into the open sea, the other half of his mind running swiftly over the problem of bringing his ship into action against *Huron*.

How long was the start that the tall clipper had over his gunboat? Robyn Ballantyne's letter had been dated 16th November, today was the 27th November. Eleven days. That was too much. He could only hope that Robyn had been able to delay the sailing, as she had promised.

Clinton glanced back over his shoulder at the blue smoke column that blurred the horizon. How long had those fires in the barracoons been burning? Not more than three or four days, he guessed, with more hope than conviction. Yet even that was still too long a start. He had seen *Huron* on a wind and she was swift as a swallow, and light as a witch. Even with eight hundred slaves in her holds, and her water casks filled, she would toy with *Black Joke* in any wind better than a light breeze. His only advantage would be in the offing that Huron must make to get on the trades and weather the bulge of the continent, while *Black Joke* could cut across the other arm of the triangle, hugging the land. It was small advantage, a few hundred leagues, and in the end it would all depend on the wind.

Clinton found he was pounding his clenched fist on the ship's rail, that he was glaring ahead with such ferocity that the idle hands of the lee rail were watching him curiously.

With an effort he forced his features to relax, and linked his hands at the small of his back, beneath the tails of his uniform coat, but his eyes still glittered pale sapphire blue and his lips were chalky with stress. It seemed too long for his patience before he could at last bring *Black Joke* on to her true course, and bend to the engine room voice-tube.

'I want all the speed you can squeeze out of her,' he told

his engineer. 'There is twenty thousand pounds of prize money just below the horizon, but it's running like a fox and I'll need every pound of steam you can give me.'

He straightened up and the wind whipped his pale golden hair against his sun-gilded cheeks and forehead. He glanced up at the sky and saw the ponderous roll of the monsoon clouds on the wind.

Huron would have it full and boisterous on her port beam, with that hull and rigging it was probably her best point of sailing.

He knew that there was not the remotest chance of finding the clipper on the open sea, there were millions of square miles of ocean to cover – not even a battle fleet with a full squadron of frigates ideally placed ahead of the fugitive would have much chance of finding a single ship in that infinity of water.

Clinton knew that his only chance was to reach the southern tip of the continent ahead of *Huron* and to take up station athwart the narrow shipping lane that was threaded around Cape Point. However, once *Huron* rounded the southern cape, the whole of the wide Atlantic would lie ahead of her and she would be gone again. Clinton's jawline clenched spasmodically at the thought of the American escaping into that limitless expanse of sea. If she had indeed eleven days' lead on him with this wind holding fair, she could even at this moment be raising the rocky cliffs of Cape Point. He put the thought from his mind and concentrated on coaxing every inch of speed from his ship over the long days and nights ahead.

R obyn had searched in vain for a subterfuge to delay *Huron* departure from the Rio Save, even though she knew that if she succeeded she would be endangering the lives of all aboard her. However, Mungo St John had swiftly regained his energy and determination as his wound healed and the effects of the inoculation passed. Affected by Robyn's warning about the danger of battlefield plagues, he forced the pace of his crew in loading the slaves. They worked through the night by the light of tar-dipped rope torches, as well as by day, the crew every bit as eager as their Captain to be away from this cursed river. Four days after the loading began, *Huron*'s slave-decks were in place and her cargo of slaves aboard, and that evening at the full of the tide, using the last of the light and the first gusts of the offshore night breeze, *Huron* slipped over the bar, shook out her reefs and settled down to make her offing during the night.

At dawn they caught the steady push of the trades and Mungo brought her round on to a more southerly heading, close-hauled to make good their eastings before running for the bulge of Agulhas with the wind on the beam.

The sweet clean air of the open sea that had travelled thousands upon thousands of miles since last having touched land, swept through the ship, cleansing it of the dreadful stench of the plague barracoons, and Mungo's strict hygienic discipline helped to keep the holds free of the slave stink and impressed even Robyn, though reluctantly, with his forethought and his precautions.

By sacrificing a single slave deck, raising the space between each from the traditional 20 inches to 32 inches, he allowed not only greater comfort but easier access. In these mild conditions of wind and sea, the slaves were exercised during all of the daylight hours, the higher headroom and wider ladderways enabled even those in the lowest decks to be brought up on deck in batches of fifty at a time. On the main deck they were forced to dance to the rhythm of a tribal drum beaten by a tattooed naked tribesman, the rattle and clink of their chains made a mournful background music to the drum and the sweet sound of the singing slaves.

'Funny business, that tattooing.' Nathaniel stopped to chat with Robyn as she stood watching the dismal show. 'They started tattooing their children to make them repulsive to us slavers, some of them file or knock out their teeth like that one there.' He pointed at a muscular black man in the circle of dancing slaves whose teeth had been filed to sharp points like those of a shark. 'Some of them put bones through the noses of their daughters, and others stretch their tits – begging your pardon at the plain speaking, ma'am – or they put rings of copper around their necks until they look like giraffes, all so the slavers will leave them alone. They do say now that these have become marks of beauty amongst the heathen. No accounting for taste, is there, ma'am?'

Robyn saw how the extra space and regular exercise would affect the well-being of the slaves, and while they were up on the maindeck, in the open air, their empty slave decks were flushed out with sea water from the ship's pumps and then scrubbed down with a strong lye mixture. Though even this was not enough to prevent the slave stink slowly impregnating the ship.

Each slave spent two hours on deck each second day, and while they were there, Robyn held clinic and examined each of them for any signs of disease or injury. Before going below once more, they were each forced to drink a decoction of molasses and lime juice to supplement the plain diet of boiled farina and water, and to ward off the dreaded scourge of the scurvy.

The slaves responded well to this treatment, and, incredibly, began to put on the weight that they had burned off during the fevers that were the result of the inoculation against the smallpox. The mood of the slaves was resigned and compliant, although there were isolated incidents. One morning, while a batch of slaves were being brought up, one of them, a fine-looking naked woman, managed to work free the shackle of her chain and the moment she reached the deck she rushed to *Huron*'s side and leapt over it, into the creaming blue wake of the racing clipper.

Despite the fact that she still wore the iron cuffs on her wrists, she managed to keep afloat for many long minutes; her struggles were pitiful to watch as she was slowly drawn down lower and lower in the water.

Robyn had run to the rail to watch the woman's efforts, expecting Mungo to heave-to and lower a boat to rescue her, but he remained detached and silent on his quarterdeck, barely glancing over the stern before occupying himself once more with the management of his ship, while *Huron* tore away and the woman's head dwindled to a speck on the blue water, then was drawn inevitably below the surface by the weight of her iron cuffs.

Although Robyn realized that it would have been impossible to stop the clipper and reach the woman before she drowned, yet she glared at Mungo across the length of the deck, wishing that there were words to express her fury and indignation.

That night she lay awake in her tiny cabin, hour after hour, racking her imagination for some ruse that could be used to delay the tall clipper's full flight towards the Southern Cape.

She thought of stealing one of the ship's boats and casting herself adrift during the night, forcing Mungo to turn back and to search for her. It took only a few minutes' reflection to realize that it would take a dozen strong men to free the whaler from its lashings and lower it on its davits over the ship's side – and even if she managed to accomplish that, it was far from certain that Mungo would delay even a minute. He was more likely to sail away and leave her, as he had left the slave woman.

She thought of setting fire to the ship, overturning a

lantern in the mainsail locker, and creating so much damage that *Huron* would be obliged to call at the nearest port, Lourenço Marques or Port Natal, to effect repairs, and to give *Black Joke* an opportunity to come up with her. Then she imagined eight hundred chained slaves burning to death in the hold when the fire got out of control, and she shuddered, thrust that idea from her and hopelessly composed herself to a sleep which would not come.

In the end her opportunity came from a most unexpected source, Tippoo. The huge mate had a weakness, a single weakness that Robyn could observe. He was a trencherman and, in his own taste, a gourmet. Half the lazaretto was filled with delicacies that Tippoo had hoarded and which he shared with no others. There were dried and smoked meats and sausages, cheeses the smell of which brought tears to the eyes, and wooden crates of cans and bottles of preserved foods, though as a strict Muslim, he took no alcohol. He made up what he lacked in the glass with his spoon.

His appetite was one of the ship's jokes, and Robyn had heard Mungo chaffing him across the wardroom table.

'Were it not for the tucker you brought aboard, Mr Mate, we'd have room for another hundred blackbirds in our hold.'

'I'll wager that belly of yours costs you more to sustain than a harem of extravagant wives.'

'Sweet merciful heavens, Mr Tippoo, but whatever you are eating should have been given a Christian burial a month ago.'

One of Tippoo's favourite appetizers was a particularly virulent bloater paste, packed in half-pound tins. The instruction printed on the tin read 'Spread thinly on biscuit or toast', and Tippoo would spoon it directly from the tin, consuming the entire half pound without once breaking the rhythmic dip and lift of his soup spoon, his eyes half closed and a cherubic smile buckling the wide line of his toad's mouth.

The fourth night out of the Rio Save he began his dinner

with a can of bloater paste, but as he pierced the lid with his clasp knife, there was a sharp hiss of escaping gas, and Mungo St John glanced up from his own plate of pea soup.

'It is blown, Mr Tippoo. I would not eat it if I were you.'

'No,' Tippoo agreed. 'But you not me.'

They called Robyn a little before midnight. Tippoo was in convulsions, doubled up with agony, his belly swollen and hard as a yellow agate boulder. He had vomited until now he was retching only a little blood-flecked bile.

'It's ptomaine poisoning,' Robyn told Mungo St John. It was the first time she had spoken directly to him since that morning in the Rio Save, and her voice was cool and formal. 'I do not have the medicines to treat it. You will have to put into a port where he can receive treatment. There is a military hospital at Port Natal.'

'Doctor Ballantyne,' Mungo answered her as formally, but his infuriating smile lurked behind the gold-flecked eyes. 'Mr Tippoo's mother was an ostrich, he can digest stones, nails and lumps of broken glass. Your concern, touching as it may be, is entirely misplaced. He will be ready to fight, flog, or devour an ox by noon tomorrow.'

'And I tell you that without proper treatment he will be dead in a week.'

However, Mungo's prognosis proved to be correct, for by morning the vomiting and retching had abated and Tippoo seemed to have purged his bowels of the poisoned fish. Robyn was forced to a decision which she made on her knees in her cabin.

'Forgive me, oh Lord, but there are eight hundred of your children chained below decks in this foul prison, and I will not kill him – at least, with your help, I will not kill him.'

Then, off her knees she went briskly to work. She used a solution of peppermint tincture to disguise the taste, and let fall fifteen drops of essence of ipecacuanha into the medicine glass, which was three times the recommended dosage for the most powerful emetic known to medical science.

'Drink it down,' she told Tippoo. 'It will soothe your stomach, and cure the diarrhoea.'

Late that afternoon she repeated the dose, but the wardroom steward had to help her lift Tippoo's head from the bolster and pour the draught down his throat. The effect was enough to alarm even Robyn.

An hour later she sent for Mungo, and the steward came back with the message, 'Captain says as how the ship's safety demands all his attention at the moment, begging your pardon, Doctor.'

When Robyn herself went on deck, Mungo was at the weather rail, sextant in hand, waiting for the sun to appear in a gap in the clouds.

'Tippoo is dying,' she told him.

'And this will be my first sight of the day,' Mungo replied without taking his eye from the eyepiece of the instrument.

'I at last believe that you are a monster with no human feelings,' she whispered fiercely, and at that moment blazing sunlight struck the deck, as the sun showed briefly through the ragged hole in the cloud.

'Stand by the chronometer,' Mungo called to the signals yeoman, and then 'Mark!' as he brought the sun's image down to bounce lightly as a green rubber ball on the dark line of the horizon.

'Excellent,' he murmured with satisfaction, as he lowered the sextant and read off the height of the sun, and called it to the yeoman to mark on his slate. Only then did he turn back to Robyn.

'I am sure you have misjudged the severity of Tippoo's ailment.'

'See for yourself,' she invited.

'That is my intention, Doctor.'

Mungo stooped into Tippoo's cabin, and paused. His expression changed, suddenly the light mocking smile was gone. It was evident that Tippoo was indeed dangerously ill.

'How are you, old friend?' Mungo asked quietly. It was

the first time Robyn had ever heard him use that form of address. He lifted his hand and laid it on the mate's sweat-beaded forehead.

Tippoo rolled the bald yellow cannon ball of his head towards Mungo, and he tried to smile. It was a brave effort. Robyn felt a terrible guilty pang, at the suffering she was inflicting and at being the witness to this private, and strangely intimate, moment between these two hard and dangerous men.

Tippoo tried to lift himself, but the effort brought a long dragging groan rattling up his throat, and he clutched at his stomach with both hands, drawing his knees up with agony, and then desperately twisting his head as a fresh bout of heaving and retching racked his body.

Mungo snatched the bucket off the deck and with one arm around Tippoo's shoulders held it for him – but all that Tippoo could bring up was a little splash of blood and brown bile, and when he fell back on the bunk he was gasping unevenly, bathed in fresh sweat and his eyes rolled back in his skull until only a little half moon of the iris still showed.

Mungo stood beside his bunk for a full five minutes, bowed and attentive, swaying slightly to the ship's movement, but otherwise still and silent. His brow was creased with thought and his gaze remote, and watching him, Robyn knew he was making the dire decision – the throw of the dice of life, friendship against the loss of his ship and perhaps even his own liberty, for to go into a British port with slaves in his holds was an awful risk.

Strange, now that he was showing this gentler side of his nature, that her affection came flooding back at full strength, she felt mean and cheap for playing on his deepest emotions, and for torturing the huge yellow Muslim on the narrow bunk.

Mungo swore quietly but decisively and, still stooped under the low deck, strode from the cabin.

Robyn's affection turned to disgust and disappointment.

Disgust that even the life of an old and loyal friend meant nothing to this cruel and merciless man whom she was doomed to love, and intense disappointment that her ruse had failed, that she had inflicted this dangerous suffering on Tippoo to no avail.

She sank down wearily and bitterly beside his bunk, took a cloth soaked in sea water and bathed the sweat-basted yellow head.

During their long voyage down the Atlantic, Robyn had grown sensitive to all *Huron*'s moods, to the feel of the deck underfoot at every point of sailing, and to the sounds that her hull made in different sea and wind conditions, and now abruptly she felt the deck cant sharply beneath her. She heard the stamp of bare feet on the deck above as her yards were trained around and *Huron*'s action became easier, the sounds of her hull and rigging muted as she took the wind in over her stern quarter and rode more easily.

'He's altered course towards the west,' she breathed as she lifted her head to listen. 'It worked. He is going in to Port Natal. Oh thank you Lord, it worked.'

Huron anchored well offshore, out on the thirty-fathom line of the shelving coast, so that she could not benefit from the shelter of the huge whale-backed bluff that protected Port Natal's natural harbour. Even with a powerful telescope, a watcher on the shore would be unable to make out any significant details of *Huron*'s cargo, nor of her true occupation. However, the ship paid for her offshore berth by taking the unfettered scend of the sea and the wind. She pitched and she rolled and she jerked at her anchor chain.

At her peak she flew the stars and stripes of her country – and below that the yellow 'Quebec', the plague flag which warned, 'Stay away from me! I have plague on board!'

Mungo St John placed an armed watch on both sides of the ship and others at her bow and stern, and, despite Robyn's strident protest, she was confined to her cabin for the duration of the ship's call, with another armed guard outside her door.

'You are very well aware of the reason, Doctor Ballantyne.' Mungo answered her protests calmly. 'I do not wish you to have any communication whatsoever with your countrymen ashore.'

The whaler, when it took Tippoo ashore, was rowed by men that Mungo selected personally, and they were instructed to inform the Harbour Master that there was smallpox aboard, and to request that no other vessel be allowed near Huron.

'I can only wait three days for you.' Mungo stooped over the litter on which Tippoo was carried on deck. 'That is all I can risk. If you are not sufficiently recovered by then, you will have to stay here until my return. That cannot be more than five months.' He tucked a leather draw-string purse under Tippoo's blanket. 'And that will pay your expenses in the meanwhile. Get well, Mr Tippoo, I need you.'

Robyn had administered another dose of the peppermint and ipecacuanha a few minutes previously, and Tippoo could reply only in an agonized whisper.

'I will wait for you, Captain Mungo, as long as she takes.'

Mungo's voice was husky as he straightened and spoke to the seamen carrying the litter.

'Handle him easy, you hear me.'

For three days Robyn sweated and fretted in the stuffy little cabin, trying to occupy her time with writing up her journals but distracted by any loud noise from the deck above, her heart pounding as she both hoped for and dreaded the hail from a British gun-boat, or the rush of a boarding-party coming in over Huron's side.

On the third morning Tippoo was rowed back to the ship, and he climbed up the side and in through the entry port unaided. Without further doses of ipecacuanha, his recovery had amazed the military surgeons ashore, but he was so thin that the skin hung in folds from his jowls like a bulldog, his belly had shrunk so that he had tied his breeches with a length of rope to keep them from sinking down past his flattened belly, but still they flapped around his shrunken buttocks.

His skin was the pale yellow of ancient ivory, and he was so weak he had to pause to rest when he reached the deck.

'Welcome aboard, Mr Tippoo,' Mungo called from the quarterdeck. 'And if you have finished your holiday ashore, I'll thank you to get this ship under way immediately.'

Twelve days later, having struggled with flukey and variable winds, Mungo St John played the field of his glass down the open gaping maw of False Bay. On his right hand rose the distinctive curved black peak of Hangklip, shaped from this angle like a shark's dorsal fin, and directly opposite it across the mouth of the bay the southernmost tip of the African continent, Cape Point, with its lighthouse perched high above the steep wet cliffs.

It was a magnificent Cape summer's day, a light and fickle breeze scratching dark patches on the surface of the rolling dark blue sea, leaving the rest of it with a satiny gloss. There were seabirds working, their wings twinkling like flurrying snow flakes in the sunlight, huge flocks of them that stretched low across the horizon.

Creeping along on the breeze, lying for minutes at a time completely becalmed, *Huron* took half a day to round the point and came on to west-north-west and a point north, the course that would carry her up the Atlantic, across the equator and finally into Charleston Roads.

Once they were on their new course, Mungo St John had leisure to inspect the other sails that were in sight. There were nine, no, ten other vessels in view now, for there was another far out to sea, just her topsails showing. They were small fishing craft out from Hout Bay and Table Bay, and the seabirds clouded the air about them, most of them were between *Huron* and the land, and all of them were bare-masted or under working sail as they plied their lines or their nets. Only the vessel furthest out was carrying topsails, and though she was hull down she gave to Mungo's seaman's eye the impression of being a bigger ship than the rest of the fishing fleet.

'There's a ship for you!' Tippoo exclaimed, touching Mungo's arm to draw his attention and when he swung his glass back towards the land Mungo murmured with pleasure as a square-rigged East Indiaman came into view around the headland that guarded the entrance to Table Bay itself.

She was as splendid a sight as *Huron* was herself, canvas piled to the sky and her paintwork gleaming in snowy white and Burgundy red, the two lovely ships on reciprocal courses passed each other by two cables' length, the officers eyeing each other through their telescopes with professional interest and appraisal as they paid passing honours.

Robyn was also at *Huron*'s rail, pining towards the land. The proximity of the beautiful ship interested her hardly at all, it was that flat-topped mountain from which she could barely tear her gaze. It was so very close, marking as it did her one hope of succour, her friends there, the British Governor and the Cape Squadron, if they only knew that she was a prisoner aboard this slave ship.

The thought was interrupted by an abrupt movement that she caught from the corner of her eye – strange how receptive she was to Mungo St John's smallest movement, to his slightest change of expression – and now she saw that he had turned his back on the East Indiaman as she dwindled away astern, and instead he was peering intently over *Huron*'s port side, his expression rapt, his whole body seemed charged with latent energy, and the hands that gripped the barrel of the telescope were ivory knuckled with tension.

Quickly she followed his gaze, and for the first time

noticed a tiny shred of white on the horizon that did not fade like the white caps of the waves, but stood constant and bright in the sunlight, though even as she watched it seemed to alter its shape slightly and – was it her imagination, or was it a thin dark wavering line that seemed to appear behind it and spread slowly away in the direction of the wind?

'Mr Tippoo, what do you make of that sail?' She heard the timbre of concern and alarm in Mungo St John's tone, and her heart leapt wildly, with hope and a Judas dread.

For Clinton Codrington it had been a desperate run down the eastern coastline of southern Africa, long days and sleepless nights of unceasing strain, when hope and despair pendulumed against each other. Each shift or change in the wind either alarmed or encouraged him, for it would either aid or hinder the tall clipper ship he was racing to head. The calms elated him, and the revival of the south-easterly prevailing winds sent his spirits plunging.

On the last days there was another worry to plague him. He had burned his coal like a spendthrift on the long thousand-mile run southwards, and his engineer came up on deck, a small red-headed Scot with the grease and coal-dust etched into his skin so that he seemed to be suffering from some debilitating and incurable disease.

'The stokers' shovels be hitting the bottom of the bunkers already,' he told Clinton with mournful relish. 'I warned ye, sir, that we'd not make it if you—'

'Burn the ship's furniture if you must,' Clinton snapped at him. 'You can start with my bunk, I'll not be needing it.'

And when the engineer would have argued further Clinton added, 'I don't care how you do it, Mr MacDonald, but I want a full head of steam on your boiler until I reach

Cape Point, and another full head of steam when I bring this ship into action.'

They raised the Cape Point lighthouse a few minutes after midnight the following night, and Clinton's voice was hoarse with fatigue and relief as he stooped over the voice pipe.

'Mr MacDonald, you can let your fires damp down, but keep your furnaces warmed and ready to stoke. When I ask for steam again, I'll need it in a hurry.'

'You'll be calling at Table Bay to take on fresh bunkers, of course, sir?'

'I'll let you know when,' Clinton promised him, and snapped the lid of the speaking tube closed and straightened up.

The Cape naval base, with all its amenities lay only a few hours steaming away. By dawn he could be filling *Black Joke* with coal and water and fresh vegetables.

However, Clinton knew that within minutes of dropping anchor in Table Bay, Admiral Kemp or one of his representatives would be on his way out to the ship, and Clinton's term of independent command would be over. He would revert to being a very junior commander, whose recent actions needed a great deal of explanation.

The closer that Clinton drew to Admiralty House, the louder the warnings of Sir John Bannerman rang in his ears, and the more soberly he was forced to review his own position. The excitement of storming Arab barracoons and of seizing slave-laden dhows on the high seas had long ago cooled, and Clinton realized that once he entered it he would not be able to escape again from Table Bay for weeks, or possibly months. It would not even suit his immediate plans to be seen and recognized from the land, for a boat would immediately be sent out by Admiral Kemp to order him in to face judgement and retribution.

Clinton felt not the least trepidation about the Navy's

ultimate judgement, he was so indifferent to the threat hanging over his career that he surprised even himself. There was only one desire, one object in his mind, that overshadowed all else. He must have his ship in position to intercept *Huron* as she rounded the Cape, if she had not already done so. Nobody and nothing must prevent him from doing so. After that he would face his accusers with complete equanimity. *Huron* and Robyn Ballantyne first, beside them all else was pale and insignificant.

'Mr Denham,' he called his Lieutenant across the dark deck. 'We will take up night patrol station ten miles off Cape Point, and I am to be called immediately the lights of any ship are sighted.'

As Clinton threw himself down, fully dressed and booted, upon his bunk, he experienced the first peace of mind since leaving Zanzibar harbour. He had done all that was in his power to reach Cape Point ahead of *Huron*, and now the rest was in the hands of God – and his trust in God was implicit.

He fell asleep almost instantly, and his steward woke him again an hour before dawn. He left the mug of coffee to grow cold beside his bunk and hurried on deck, reaching it a few seconds ahead of Lieutenant Denham.

'No ships during the night, sir.' Ferris, who had the watch, saluted him.

'Very well, Mr Ferris,' Clinton acknowledged. 'We will take up our daylight patrol station immediately.'

By the time the light was strong enough for a watcher on the shore to make out any details, *Black Joke* had retreated tactfully below the horizon and it would have taken a sharp eye to pick out the occasional flash of her topsails, let alone to identify the gunboat and to speed a report to Admiral Kemp that his prodigal had returned.

From *Black Joke*'s masthead the land was a low irregular distortion of the horizon, but a ship rounding the Cape would be many miles closer than the land. *Huron*'s mainmast was almost one hundred and fifty feet tall, her sails would shine like a flaming beacon and as long as the fog did not come down, which was unlikely at this season of the year, Clinton was confident that she could not slip by him.

The only disquiet that scratched him like a burr as he paced his deck, and the gunboat settled down into the regular four-sided box-patterned legs of her patrol, was that *Huron* had long ago flown northwards on this fine wind that at last bore steadily out of the south-east at almost gale force, and that she was already lost in the endless watery green wastes of the southern Atlantic Ocean, leaving *Black Joke* to guard the gate of an empty cage.

He was not left long to brood, the first sighting was called down to the deck from the look-outs in the crow's nest at the main peak, and Clinton's nerves jumped tight and his expectations flared.

'What do you make of her?' he called up through the voice trumpet.

'Small lugger—' and his expectations plunged. A fishing boat out of Table Bay, there would be many of them, but each time he could not control the wild surge of excitement when another sail was sighted, so that by nightfall his nerves were ragged, and his body ached with exhaustion as he gave the order for *Black Joke* to take up her inshore patrol station for the night.

Even then he could not rest, for three times during the night he was called from his bunk, and he stumbled owl-eyed on deck as *Black Joke* went down to investigate running lights that winked ruby red and emerald green out of the darkness.

Each time the same leap of expectation, the steeling of nerves for instant orders and swift action, and then the same let down as the lights proved to belong to small trading vessels, and the gunboat sheered away quickly, lest she be recognized and her presence off the Cape be reported in Table Bay.

In the dawn, Clinton was on deck again, as the gunboat moved further offshore to take up her daylight patrol station. He was distracted by the sighting reports as his masthead look-out picked up the first sails of the fishing fleet coming out for the day's business, and by the lugubrious report of his coal-stained Scottish engineer.

'Ye'll not last out the day, sir,' MacDonald told him. 'Even though I am burning just enough coal to keep the furnace warm – we've not more than a bucket or two left.'

'Mr MacDonald,' Clinton interrupted him, trying to keep his temper under control and to disguise his exhaustion. 'This ship will stay on station until I give the order, I don't care what you burn, but you are to give me steam when I ask for it – or kiss good-bye to the fattest bundle of prize money that will ever come your way.'

Despite this brave promise and threat, Clinton's hopes were sinking swiftly. They had been on station for more than a day and a night already – he could not bring himself to believe that he had beaten the swift clipper to the Cape by that margin, not unless she had been somehow miraculously delayed, and every hour increased the certainty that she had run clear away from him, taking her cargo and the woman he loved out of his life for ever.

He knew he should go below to rest, but his cabin was stifling in the rising summer heat, and in it he felt like a trapped animal. He stayed on deck, unable to keep still for more than a few moments at a time, poring over the chart table and fiddling with the navigational instruments before throwing them down and resuming his nervous pacing, shooting quick glances up at the masthead, and then roaming the ship so obviously intent on finding fault or criticizing the ship's running that his officers followed his lanky figure with troubled expressions, while the watch on deck was silent and subdued, not one of them daring to glance in his direction until Clinton's voice rose in a coldly furious cry that froze them all.

'Mr Denham,' the Lieutenant almost ran to the summons, 'this deck is a pig sty. What animal is responsible for this filth?'

On the white holystoned deck planking was a brown splash of tobacco juice, and Denham stared at it for an instant before wheeling away to bellow a series of orders that had a dozen men scampering. The activity was so intense, the atmosphere electric, as Captain and Lieutenant stood over four men on their knees scrubbing furiously at the offending stain while others carried buckets of sea water and still others rigged the deck pump, that the hail from the masthead was almost ignored,

It was left to Ferris to acknowledge it, and to enquire through the voice trumpet,

'What do you make of her?'

'She's hull down, but she's a four-masted ship, square rigged—'

The activity on the deck ceased instantly, every head lifting as the look-out went on elaborating on his sighting.

'She's on a course to weather the Cape, now she's coming round on to a heading of north-north-west or thereabouts.'

Clinton Codrington was the first to move. He snatched the telescope out of Lieutenant Denham's hand and ran to the ratlines. With the telescope tucked in his belt he began to climb, hand over hand.

He went up steadily, never pausing nor faltering, not even when he reached the futtock shrouds and for a few moments hung over backwards one hundred feet above the swaying deck. However, when he reached the crow's nest at the main peak and tumbled into it thankfully, his breath was sawing dryly in his throat and the blood sang in his ears. He had not climbed like that since he had been a midshipman.

The look-out tried to make himself as small as possible, for they were crowded together in the canvas bucket, and he pointed out the ship to his Captain.

'There she be, sir.'

Black Joke's roll was emphasized up here on the tall pendulum of the mast, and the horizon swung giddily through the field of Clinton's telescope as he tried to keep it focused. It was an art that he had never completely mastered, but that was of little consequence for the first time the little regular white pyramids popped up in the field of his glass the last doubts were dispelled, and Clinton felt his heart leap fiercely against his ribs.

His voice was strangled with triumphant emotion as he shouted down at the tiny foreshortened figures on the gleaming white quarterdeck far below.

'Bring her round to due east, Mr Denham. Order a full head of steam on the boiler—'

Though he had not yet fully recovered his breath, he threw himself out of the crow's nest and scrambled down faster than he had climbed. In his haste he slid the last fifty feet down the backstay and barely noticed the rough hempen rope scorching the palms of his hands.

By the time his feet hit the deck, *Black Joke* was coming around on to her new heading, and in anticipation of Clinton's next order, Denham had already called the watch below, and they came boiling out on to the deck.

'And we will clear for action, Mr Denham, if you please,' gasped Clinton, his face dark with blood under his deepwater tan and the sapphire eyes alight with battle lust.

All *Black Joke*'s officers carried swords on their belts, only Clinton had selected a cutlass, for he preferred the stouter and heavier weapon, and he fiddled with the hilt even now as he spoke to them quietly and seriously.

'Gentlemen, I have documented proof that the ship ahead of us has a cargo of slaves aboard her.'

Denham coughed nervously and Clinton forestalled him.

'I am also aware that she is an American vessel, and in ordinary circumstances we would be helpless to oppose her passage.' Denham nodded with relief, but Clinton went on remorselessly, 'However, I have received an appeal from one of Her Majesty's subjects, Dr Robyn Ballantyne whom you all know well, who is being abducted aboard the *Huron* against her will. I am certain what my duty is in these circumstances. I intend to board her, and if she resists me, I intend to fight her.' He paused and their faces were shocked, strained. 'Those of you who have objections to this course of action should immediately note them in the ship's log, and I will sign them.'

Their relief and gratitude was immediate, few other captains would be so lenient.

He signed his name neatly beneath the entry in the ship's log, and then returned the pen to its holder.

'Now that the formalities have been seen to, gentlemen, shall we get on with earning our hire?' And Clinton was smiling for the first time since they sailed from Zanzibar harbour as he indicated the pile of snowy white sails that was now clearly visible from the deck ahead of *Black Joke's* bows, and as he spoke there was an eruption of dark tarry-smelling smoke from the tall single stack above them, and the engine telegraph clanged sharply as the pointer moved to 'Engine Standing By' position on the repeater. *Black Joke* had steam in her boiler.

Clinton stepped to the telegraph, thrust the handle fully around the dial to the 'Full ahead' position, and the deck vibrated under his feet as the propeller shaft began to spin and *Black Joke* lunged forward eagerly, breaking the swells off her shoulders in explosions of white spray.

'By God, he's got us pinned against the land—' Mungo drawled the words out nonchalantly, even smiling slightly at Tippoo as he lowered the glass from his eye for a moment and polished the lens on his shirt sleeve. 'We'll have to run like the very devil to get round him, and make the open sea. Mr Tippoo, will you be good enough to shake out every last reef and crack on all the sail we can carry right up to the sky sails?' He lifted the glass to his eye again as Tippoo began bellowing the orders. 'It's a little bit too much luck for any one man,' Mungo murmured aloud. 'It's too much coincidence that the one man I would not wish to meet be lying in the one place on all the oceans where I would not wish to meet him.' Again he lowered the glass and strode to the poop rail to look down on to the maindeck.

Robyn Ballantyne was at the ship's side, staring out across the indigo blue waters at the sail and the smear of dark smoke, still far out to sea, but every moment converging on them, headed to a point far ahead of *Huron*'s elegant bows where the courses of the two ships would meet. She sensed Mungo's eyes upon her, and she lifted the shawl off her head so that her dark russet hair tumbled out and snapped and danced around her cheeks. Her skirts were flattened against her legs by the wind, so that she had to lean slightly forward to balance herself.

She lifted her chin and stared up at Mungo, her expression defiant and she watched him as he carefully bit the end off one of his long thin black cheroots and cupped his hands around the sulphurous flare of the Vesta match and lit the cheroot without once breaking the steady gaze that held her own.

Then he sauntered easily down the poop-ladder to her side.

'A friend of yours, Doctor Ballantyne?' The smile was on his lips alone. His eyes were frosty.

630

'I have prayed every night for him to come. Ever since I sent the letter that summoned him.'

'You do not deny the betrayal?'

'I am proud I was able to perform my Christian duty.'

'Who carried your letter?'

'No member of your crew, sir. I sent it by the master of the Omani buggaloo.'

'I see.' His voice was low, but stinging as dry ice. 'And what of Tippoo's illness, would a physician, per chance, stoop to poisoning a patient?'

She dropped her eyes, unable to meet that accusation.

'Will you be so kind, Doctor Ballantyne, as to return to your cabin immediately and stay there until I give you permission to leave it. There will be an armed guard at the door.'

'I am to be punished?'

'No man would blame me for dropping you over the side, and leaving you to be picked up by your countrymen. However, it is your safety I am thinking of. This deck could become an unhealthy spot in the very near future, and we will all be too much occupied to care for you.'

His attention left her, and he was staring ahead, and then glancing back at *Black Joke*'s smoke, judging speeds and angles with a seaman's eye. Then he smiled.

'Before you go, I would like you to know that all your efforts have really been a fine waste of time – look there!' He pointed ahead along the sheer and mountainous coastline, and following his arm she saw for the first time that ahead of them the sea was as black and broken as new coal cut from the face, glittering with wild jumping wavelets, each flecked with pretty white crests.

'There is the wind,' Mungo said. 'That's where it comes through the mountains, and we will be into it before you are safely tucked up in your cabin.' He chuckled now comfortably, confidently. 'Once we get on that wind, there are few ships on all the oceans, either steam or sail, that would match *Huron* for a moment, and God knows, there are none who could run her down.' He gave her a small mocking bow, parody of gracious southern manners. 'Take a good look at that ugly little steam packet, before you go, ma'am, you'll not be seeing her again. And now, if you'll be good enough to excuse me—'

He turned from her and ran lightly back up the poopladder.

Tears of anger dimming her eyes, Robyn clutched the rail and stared across the narrowing strip of sea at the bustling, puffing little gunboat, already she could catch glimpses of the hull, and its painted chequer-board gunports. She began to hope that Mungo St John's boast had been mere bravado, for *Black Joke* seemed to be keeping pace with the tall clipper, the wind was still a long, long way ahead.

There was a respectful touch on Robyn's shoulder, and old Nathaniel stood beside her.

'Captain's orders, ma'am, and I am to see you safe in your cabin.'

Clinton leaned forward, all his weight on the balls of his feet, almost as though he were trying to urge his ship on by the balance of his body, the way a rider leans forward into the jump. He also had seen the wind pattern on the surface of the sea, and knew what it heralded.

The tall clipper ship looked lazy now, indolent in her flounces and ruffles as a lady of high fashion, and *Black Joke* was snorting and snuffling busily down the short leg of the triangle. If they both maintained this speed and course they would meet about eleven nautical miles ahead. Clinton could visualize the exact spot clearly just beyond the spit of land marked on his chart as 'Bakoven Point'.

Huron was committed to her present course. She could not bear up for the land lay close under her weather rail, and the chart showed breaking shoals well offshore – there was one of them now, close on her starboard beam, showing its round black granite back and blowing like a whale. *Huron* was in a trap, and her only escape would be to find the extra turn of speed which would carry her away out of gunshot range – and there it was, the wind, less than three miles ahead of her.

There was a loud banging from the deck below Clinton's feet, and he glanced irritably at Ferris.

'See what that is,' he snapped, and turned his full attention back to the clipper. Three short miles she had to go, but even as Clinton stared at her through his glass he saw her huge square mainsail quiver and then shake gently as she luffed in the flukey airs below the mountains.

'Please God!' Clinton whispered, and *Huron*'s speed bled off perceptively, all her sails were losing their taut clean shape and she checked away, baulking like a weary animal.

'She's found a hole in the wind,' Lieutenant Denham called exultantly. 'We've got her now, by God!'

'I'll thank you not to blaspheme on this quarterdeck, Mr Denham,' Clinton told him sharply, and Denham's expression was instantly crestfallen.

'I beg your pardon, sir.' At that moment Ferris arrived breathlessly back on deck.

'The stokers, sir,' he panted. 'They are knocking out all the furniture from the officers' quarters. Your bunk's gone, sir, and your desk also.'

Clinton barely glanced at him, he was studying the clipper, evaluating each yard of difference in their speeds, judging as finely as he dared the angle of his interception.

Yes, he decided, with *Huron*'s loss of speed, he could edge in a touch more.

'Bring her up a point to starboard,' he told the helmsman,

and then he glanced up at his own sails that were helping the big bronze screws under the stern to hurl *Black Joke* forward. The alteration of course had affected them.

'Mr Ferris, see to the trim of your jib, if you please.'

Ferris bellowed an order to the foredeck watch, and watched critically as they hardened up the long triangular sail.

All *Huron*'s sails shivered, and then refilled, once more taking on their true shape, and she spurted away, a curl of white sparkling under her bows. The dark windswept water was much closer ahead of her.

It had been Denham's unnecessary blasphemy, Clinton was sure of it, and he glanced darkly at his Lieutenant, and then reluctantly gave the order.

'Let her fall off a point.' *Huron* was head-reaching again. If *Black Joke* held on she would be aiming to cut behind the clipper's stern. The alteration was an acknowledgement of the advantage changing hands once again.

The engine voice pipe squealed, and it was a relief to have even the small distraction.

'Engine room, Captain,' he snapped into the mouthpiece.

'Coal long ago gone, sir. Pressure down to 100 pounds, sir – and falling.'

'Burn everything you can find.'

'Wood goes up like paper, sir. No body to it – and it chokes the flue.' MacDonald seemed to relish the gloomy news, and Clinton felt his irritation turn to anger.

'Do your best, man, nobody can ask more than that,' and he snapped the voice pipe closed.

Was he close enough yet to try a shot with his bow-chaser, he wondered? The long sixteen-pounder had almost twice the range of the big thirty-two pounders that made up *Black Joke*'s main armament. A lucky shot might carry away one of the clipper's spars, might even bring down a yard – and at that moment he distinctly felt the change in the engine vibrations coming up through the deck, *Black Joke* was faltering, the steam in her boilers losing pressure.

'Mr Ferris, break out the colours, if you please.'

As the Ensign unfurled at the peak, spread gloriously against the pale blue sky, crimson and silky white, shouting a challenge on the wind, Clinton felt that tightness in his chest, that swell of pride which never failed him.

'*Huron*'s replying,' Denham muttered, and Clinton lifted his glass, and watched the American's colours bloom like a flower high above her shimmering piles of white canvas.

'And be damned,' Denham interpreted the show of colours. *Huron* was scorning their challenge.

'Mr Ferris, we'll give her a gun now,' Clinton decided grimly. 'Put one over her bows.' And Ferris scrambled away to the bows to supervise the loading and laying of the bow-chaser.

The shot when it came was a puny little pop of sound, muted by the wind and the long spurt of grey powder smoke was whipped away almost before it could form. Though they were all watching avidly through the telescopes, not one of them could spot the fall of shot, and Denham spoke aloud for all of them.

'She's not altering. She's ignoring us.'

'Very well.' Clinton kept his voice low. 'We'll try one into her rigging.'

The sixteen-pounder banged again, like an unlatched door in a high wind, and this time they both exclaimed in unison. A pinprick of light appeared in one of the *Huron*'s studding sails, pierced by shot; it held its shape for an instant and then burst like a paper packet and was blown to tatters.

Clinton saw the bustle of *Huron*'s seamen on the decks and yards, and before the bow-chaser could be reloaded, the ruined sail was hauled down and another clean new sail spread open in its place. The speed with which the sail change was made impressed even Clinton.

'The devil is a good sailor, I'll grant him that—' And then he broke off, for *Huron* was turning boldly, seeming to aim to cut the gunboat's bows, and Clinton realized what her Captain was doing. He was anticipating the rush of the wind, and as Clinton watched it struck her.

It came roaring aboard the clipper, howling through her rigging, like a pack of hunting wolves, and the tall ship heeled and seemed almost to crouch, gathering herself like a blood stallion feeling the cut of the lash, and then she hurled herself forward and was away.

The dark wind-scoured sea burst open before the long lithe knifing hull, and joyously *Huron* tossed the dashing white spray over her bows.

'She's making twenty knots,' Denham cried with disbelief, as *Black Joke* seemed to come up dead in the water, a wallowing log when compared to the swift and lovely ship that cut daringly across her bows, just out of cannon shot, and dashed away into the open Atlantic Ocean.

Through his glass Clinton saw that the seamen who lined *Huron*'s yards were gambolling and waving their caps, mouths wide open as they cheered and jeered, and then he focused the glass on *Huron*'s deck.

There was a tall figure at *Huron*'s near rail, clad in a plain dark blue jacket. Clinton could not make out the man's features at this distance, but he recognized the set of wide shoulders, the arrogant carriage of head, that he had last seen over the sights of a duelling pistol.

The acid bile of hatred rose to scald his throat as the figure lifted a hand in a laconic salute, a taunting gesture of farewell, and then turned away from the rail unhurriedly.

Clinton snapped his telescope closed.

'Stern chase!' he ordered. 'We'll keep after her!'

He did not dare look at his officers' faces, lest one of them wore an expression of pity.

L ying on her bunk, her arms held stiffly at her sides and hands clenched painfully, Robyn heard the creak and squeal from the deck below her that meant *Huron* was altering course, and that the eight-inch thick rudderlines were running through their blocks as the helmsman spun the wheel. It was a sound she had long become accustomed to, and she braced herself instinctively as the rudder lines attached to the pintles trained around the enormous wooden rudder under *Huron's* stern and the ship altered her action through the water.

Seconds later there was a thunderous commotion from the deck above her, the blustering roar of the gale socking into the rigging, the crash of tackle coming up taut, the slam of the great sails as their awesome power was transferred into the hull, and Robyn was almost hurled from her bunk as *Huron* heeled wildly.

Then the cabin was filled with the exultant thrumming of the hull through water, as though she were the body of a violin as the bow was drawn across the bass strings, and *Huron* trembled with life, lifting and dropping to the new urgency of her run.

Very faintly Robyn could hear above it all the sound of men cheering. She jumped from the bunk and clutching for handholds crossed the cabin and pounded her fist upon the door.

'Nathaniel,' she called. 'Answer me this instant.'

'Captain says as how I'm not to talk to you.' His voice was muffled.

'You cannot torment me so,' Robyn yelled back. 'What is happening?'

A long pause while Nathaniel considered his duty and then weighed it against his affection for this spirited young woman.

'We are on the wind, ma'am,' he told her at last. 'And going like all the devils of hell with a crackerjack tied to their tails.'

'What of *Black Joke?*' she pleaded. 'What of the British gunboat?'

'Ain't nothing will catch us now. Reckon the puffing Billy will be out of sight before nightfall. From here she looks like she's dropped her anchor.'

Slowly Robyn leaned forward until her forehead pressed against the planking of the door. She closed her eyes very tightly, and tried to fight down the black waves of despair that threatened to overwhelm her.

She stayed like that for a long time until Nathaniel's voice roused her. It was rough with concern.

'Are you all right then, missus?'

'Yes, thank you, Nathaniel. I'm just fine,' she replied tightly, without opening her eyes. 'I'm going to take a little nap now. Don't let anybody disturb me.'

'I'll be right here, missus. Ain't nobody going to get past me,' he assured her.

She opened her eyes and went back to the bunk, and knelt before it and began to pray – but for once she could not concentrate. Jumbled images kept intervening, and, when she closed her eyes, the face of Clinton Codrington was there, with those pale beautiful eyes in the darkly tanned mahogany of his face that accentuated the sun-bleached platinum of his hair. She longed for him as she had never done before – he had become a symbol for her of all that was good and clean and right.

Then her mind darted away and it was that distant mocking smile, the taunting gold-flecked eyes of Mungo St John. She trembled with humiliation, the man who had violated her and turned her own emotions traitor, who had dallied with her and allowed her to hope, nay, to pray that she could bear his children and become his wife. Her despair turned to hatred once again, and hatred armed her.

'Forgive me, Lord – I'll pray later – but now I *have* to do something!'

She started to her feet, and the cramped little cabin was

a cage, suffocating and unbearable. She hammered her fists on the door and Nathaniel replied immediately.

'Nathaniel, I cannot bear it in here a moment longer.' she cried. 'You must let me out.'

His voice was regretful but firm. 'Can't do that, missus. Tippoo would have a look at my backbone!'

She flung away from the door, angry, confused, her mind in a turmoil.

'I cannot let him carry me away to—' She did not go on, for she could not imagine what awaited her at the end of this voyage unless – and she had a vivid mental image of *Huron* coming into dock, while standing on the quay was a beautiful tall and aristocratic French woman in crinolines and velvets and pearls with three small sons standing at her side waving up at the tall arrogant figure on *Huron*'s quarterdeck.

She tried to close her mind to it, and she concentrated instead on the sound that *Huron* made as she bore away joyously on the wind, the drumming of her hull, and the pop and creak of her planking, the clatter of tackle and the stamp of bare feet on her deck as a party of seamen walked away with a fall, training one of the yards more finely to the wind. From beneath her feet came another muted squeal, like a rat in the cat's jaws, as the helmsman made a small adjustment to *Huron*'s heading, and the rudder tackle ran protestingly through the blocks.

The sound triggered a memory, and Robyn froze, trembling again, but this time with anticipation. She remembered Clinton Codrington describing to her how as a young Lieutenant he had been in command of a cutting-out party sent into a river estuary that was crammed with small slaving craft, buggaloos and dhows.

'I didn't have enough men to take them all as prize at once, so we jumped from one to the other, cut their rudder lines and left them drifting helplessly, until we could pick 'em up later – those that hadn't gone aground, that is.'

Robyn roused herself from the memory and rushed to the corner of her cabin. She had to wedge her back against the bulkhead and push with both her feet to move her wooden chest into the centre of the cabin. Then she dropped to her hands and knees.

There was a small trap-door in the deck, so neatly fitted that its joints were knife edges, but there was a small iron ring let flush into the woodwork. Once on the long voyage down the Atlantic, she had been disturbed by a very apologetic carpenter's mate and she had watched with interest while he had dragged her chest aside and opened the hatch, to descend through it with a grease pot.

She tried now to open it, but the hatch was so tight-fitting, that it resisted her efforts. She snatched a woollen shawl from her chest, and threaded it through the iron ring. Now she could get a fairer purchase. Once more she strained back, and the hatch moved inchingly and then abruptly flew open with a crash that she was sure must have alerted Nathaniel. She froze again, listening for a half minute, but there was no sound from beyond the cabin door.

On her hands and knees again she peered into the open hatchway. There was a faint breeze of air coming up out of the dark square hole, and she could smell the thick grease, the reek of the bilges and the awful slave stink that not all the lye and scrubbing had been able to cleanse – her gorge rose at the taint. As her eyes adjusted, she made out the low and narrow tunnel that housed *Huron*'s steering gear. It was just high enough for a man to crawl along, running fore and aft along the hull.

The rudder lines came down from the deck above, ran through heavy iron blocks bolted into one of *Huron*'s main frames, and then changed direction and ran directly astern down the narrow wooden tunnel. The pulley wheels of the blocks were caked with black grease, and the rudder lines were of new yellow hemp. They seemed as thick as a man's leg, and she could sense the enormous strain on them, for they were as rigid as steel bars.

She looked around for a means of damaging them, a knife, one of her scalpels, perhaps, and almost immediately realized the futility of anything so puny. Even a strong man with a double-headed axe would be hard pressed to hack his way through those cables, and there was no room in which to swing an axe in that narrow tunnel. Even if a man had succeeded in severing one of them, he would have been cut to bloody tatters as the cable whiplashed.

There was only one means, one sure means, and she quailed at the thought of what would happen if it got out of control, and if *Black Joke* was not very swiftly alongside to render assistance with her steam-driven pumps and hoses. She had once already rejected the idea of using fire, but now with help so close astern, with the last chance rapidly fading, she was ready to accept any risk.

She reached across and pulled off the wooden bunk one of her grey woollen blankets and wadded it into a bundle, then she stood up and lifted the oil lamp from its gimbals in the deck above her. Her fingers were clumsy with haste as she unscrewed the cap of the oil reservoir in the base of the lamp.

She soaked the blanket, and then looked round for anything else that was inflammable – her journals? No, not them, but she pulled her medical manuals out of the chest and ripped the pages out of them, crumpling them so they would burn more readily, and she made a sack of the oil-soaked blanket and wadded the paper into it.

She stuffed it down the hatch and it fell across the straining rudder lines, and entangled itself in the iron pulleys.

The mattress on the bunk was filled with coir, the dry coconut fibre would burn fiercely; she dragged it off the bunk and pushed it into the hatch. Then the wooden slats

641

off the bunk followed it, then the navigational books from the narrow bookshelf behind the door. She looked about her swiftly, but there was nothing else in the cabin that would burn.

The first Swan Vesta that she dropped burning down the hatchway flickered once and then went out. She tore the end sheet out of her journal and twisted it into a spill, when it was blazing strongly she let it fall into the dark square opening, and as it floated down it illuminated the gloomy recesses of Huron's bilges, and the rough planking of her underbelly.

The burning spill landed on the oil-soaked blanket, and pale blue flames fluttered over it as the evaporating gases flashed off, then a crumpled ball of paper caught and little orange flames peaked up and danced merrily over the blanket and the linen covering of the mattress. A rush of heat came up through the hatchway, scorching Robyn's cheeks, and the sound of the flames was higher than that of the rushing seas along the outside of the hull.

Using all her strength, Robyn swung the hatch cover over and let it drop back on to its seating with a thump that alarmed her anew, but immediately the sound of the flames was cut off.

Panting with the effort and a savage excitement, Robyn backed away and leaned against the bulkhead to rest. Her heart was pounding so fiercely that the blood in her ears nearly deafened her, and suddenly she was afraid.

What if Black Joke had abandoned the unequal contest, and there was nobody to rescue the eight hundred miserable souls chained below Huron's decks before the flames reached them?

That first wild assault of the wind, as it came boiling down off the mountains, had settled to a steady blast, not so furious, but constant and reliable.

'There will be no flukes or holes in this gale of wind,' Mungo thought with satisfaction, pausing in his pacing to look up at the small scudding wind-torn shreds of cloud that seemed to scrape the tops of his masts, and then turning to survey an indigo Atlantic that stretched to every corner of the horizon, dark with the wind rush and dappled with the prancing white horses that curled from every wave crest.

His leisurely survey ended over *Huron*'s stern rail. The land was already out of sight, so swiftly had *Huron* run the great flat-topped mountain below the horizon, and *Black Joke* was hull down. Only her topsails showed, not a trace of furnace smoke.

The absence of smoke puzzled Mungo a little and he frowned, considering it, and finding no plausible answer, he shrugged and resumed his pacing. *Black Joke* would be out of sight, even from *Huron*'s towering masthead, before sunset, and Mungo was planning the evolutions he would make during the night to confuse thoroughly any pursuit, before settling on to his final course to run through the doldrums and cross the equator.

'Deck, masthead.' A faint hail reached him, breaking his line of thought, and he stopped again, threw back his head and with both hands on his hips stared up at the masthead as it dipped and swung across the sky.

Tippoo answered the hail with a bull bellow, and the look-out's voice was strained, his anxiety evident even against the wind and at that remove.

'Smoke!'

'Where away?' Tippoo's voice was angry, the reply should have given both distance and bearing from *Huron* – already every man on *Huron*'s deck was twisting his head to sweep the horizon.

'Dead astern.'

'That will be the gunboat,' Mungo thought comfortably. 'She's got her boiler going again – and much good may it do her.' He dropped his fists from his hips and took one more pace before the look-out's voice rang out again.

'Smoke dead astern, we are trailing smoke!'

Mungo stopped dead in his tracks, his foot still an inch from the deck. He felt the icy spray of fear chill his guts.

'Fire!' bellowed Tippoo.

It was the one most dreaded word to men who lived their lives in the tinder hulls of wooden ships, whose seams were caulked with tar and pitch, and whose sails and rigging would burn like straw. Mungo completed that suspended pace, spinning on the ball of his foot as it struck the deck, and the next pace carried him to *Huron*'s rail. He leaned far out, peering back over the stern, and the smoke was a pale wisp, thin as sea fret, lying low against the dark blue sea, drifting away behind them, and dissipating even as he watched it.

Dry oak planks burn with a fine clean flame and little smoke, Mungo knew that, he knew also that the first thing he must do was starve the flames of air, heave the ship to, to reduce the wind of her passage while the extent of the flames could be explored and the ship's pumps—

He turned again, his mouth opening to begin shouting his commands. The quartermaster and his mate stood directly ahead of him, both of them balancing easily before the massive mahogany and brass wheel. Larger than the driving-wheel on a steam locomotive, it required the strength of two men to hold *Huron*'s head in this wind and on this point of sailing, for the huge spread of her canvas was opposed by the massive oak and copper rudder under her stern.

Down in the steering tunnel, the flames were being fed by the strong breeze that the canvas scoops on *Huron*'s foredeck were directing down into her own slave-decks in an attempt to keep them sweet.

The draught forced its way through the companionways and ladderways, through the ports and cracks in *Huron*'s bulkheads, and this steady breeze at last found its way into the long narrow tunnel that housed her steering-gear.

The bright rustling flames were almost smokeless, but intensely hot. They frizzled the loose fibres of hemp off the thick hairy rudder lines, and then swiftly blackened the golden brown cords, began to eat through them so that here a strand parted with a snap that was lost in the rising crackle of burning timbers and the strand unravelled, spinning upon itself and bursting in another tiny new explosion of light.

The two quartermasters were ten feet from where Mungo stood, poised to give his commands, composing the orders in his head, when suddenly the massive wheel no longer resisted the thrust of the brawny men who held it over.

Deep down in *Huron*'s hull, in the long wooden tunnel that had been turned into a raging blast furnace, the rudder lines had burned through, and as they snapped, they snaked and whipped viciously, smashing through the burning deck timbers, scattering flaming brands into the hold below, and letting in a fresh whistling gush of air that forced the flames higher.

Under the helmsman's hands the wheel dissolved in a spinning blur of glittering brass and the quartermaster was hurled the length of the deck, striking the bulwark with jarring force that dropped him to the planking wriggling feebly as a crushed insect. His mate was less fortunate, his right arm was caught in the polished mahogany spokes of the wheel and it twisted like a strip of rubber, the bone of his forearm breaking up into long sharp splinters whose points thrust out whitely through the tanned skin, and the head of the long humerus bone was plucked from its socket in the scapula and the whole upper arm screwed up in a twist of rubbery flesh.

With the press of the rudder under the stern no longer

controlling the rush of *Huron*'s hull through the water, the tremendous pressure of the wind in her sails took over unopposed, and *Huron* became a giant's weather-cock. She spun in almost her own length, her bows flying up into the wind and every man on her deck was hurled to the planking with stunning force.

The yards came crashing about, tackle snapping like cotton, one of the upper yards tearing itself loose, falling in a twisted web of its own canvas and rigging, and *Huron* was taken full aback, the neat geometrical pyramids of her sails disintegrating into flapping, fluttering chaos, wrapping around the stays and halyards, flogging against their own yards and masts.

With the gale of wind flying fully into the front of the sails, from the diametrically opposite direction to that for which they had been designed, the tall masts flexed and arched dangerously backwards, the backstays drooping slackly, adding to the confusion of sail and rigging, while all the forestays were humming with unbearable tension – and one of them parted with an ear-jarring crack and the foremast shifted a few degrees and then hung askew.

Mungo St John dragged himself to his feet and clung to the rail. The screams of the maimed helmsman dinning in his ears, he looked about him, and disbelief turned to bitter despair, as he found his beautiful ship transformed to an ungainly shambles. Wallowing drunkenly, *Huron* was beginning to make sternway, as the wind pushed her backwards and the waves came tumbling aboard her.

For long seconds Mungo stared about him numbly. There was so much damage, so much confusion, and so much mortal danger, that he did not know where to begin, what his first order must be. Then over *Huron*'s heaving bows, in the opposite direction to which he had last seen it, the distant but suddenly dreadfully threatening speck that was *Black Joke*'s topsails showed in a pale flash above the horizon, and it galvanized Mungo.

'Mr Tippoo,' Mungo called. 'We'll reef the mains and send down all her top hamper.'

The logical sequence of orders began to arrange themselves in his mind, and his voice was calm and clear, without the strained and panicky timbre they had expected.

'Mr O'Brien, go below and give me a fire report, quick as you like.'

'Bosun, rig port and starboard pumps, and stand by to hose down the fire.'

'Mr Tippoo, send a party to batten all her hatches and strike the air scoops.' They must try to prevent air reaching the flames, he was sealing the hull.

'Coxswain, have the whaler off her davits and launch her.' He would attempt to tow the heavy boat astern, to act as a drogue, a sea-anchor. He was not sure that it would provide a solution, but he intended to work *Huron*'s bows around with the delicate use of her forward sails – and with the drogue holding her tail in place of the rudder, he might be able to run directly before the wind. It was not his optimum course, and it would be fine and dangerous work with the deadly risk of gibing and broaching, but at the least it would give him respite while he rigged the emergency steering tackle to her useless rudder, and get *Huron* under control once more.

He paused for breath, but once more glanced forward. *Huron* was moving rapidly astern, dipping and staggering into the swells so they came flurrying aboard her in spray and solid green gouts of water, while over her bows the British gunboat was closer – so close that Mungo glimpsed a little sliver of her painted hull, and it seemed that her action through the water was more boisterous and cocky, like a game rooster erecting its coxcomb and ruffling its feathers as it bounces across the sandy floor of the cockpit.

Unable to endure the company of his junior officers a moment longer, stifled and filled with a sense of helplessness by his inability to prevent the tall American clipper from romping away from him, desperate for some activity to help his nerves from fraying further, Clinton Codrington had taken his telescope and gone forward into *Black Joke*'s bows.

Oblivious to the spray that splattered over him, soaking the thin linen shirt and chilling him so that his teeth chattered even in the brilliant sunlight, Clinton clutched for a hand-hold in the ratlines, balancing on the narrow bulwark and staring ahead through eyes that swam not only with the stinging spray and wind, but as much with humiliation and frustration.

Just perceptibly, *Huron*'s tower of canvas was sinking below the irregular watery horizon, by sunset she would be gone. She and Robyn Ballantyne. His chance had come and he had missed it. His spirits could sink no lower.

To add to his suffering, his streaming eyes were playing him false, and what he could still make out of the clipper became distorted, changing shape as he still stared after her. Then the hail from the look-out high above him broke the grip of his despair,

'Chase is altering!' Clinton could not yet believe the high-pitched shriek from the masthead. 'She's coming about!' The hail was almost incoherent with excitement and surprise.

Clinton whipped the telescope to his eye, and once again doubted his eyesight. *Huron*'s masts had been almost dead in line, but now they showed individually. She was coming about, already *Huron* was almost broadside, and Clinton stared. For a few moments more the orderly mass of sails retained their perfect snowy shape, and then the pattern began to break up. The ponderous belly of the mainsail wobbled and trembled, then began to flutter and shake like a pennant in the gale, it spilled its wind and collapsed like a bursting paper bag and lashed itself in a petulant fury around its own mainmast.

Huron was a shambles. Through the glass Clinton could see her beginning to tear herself to pieces, sails ripping, yards tumbling – her foremast sagging out of true, and he still could not believe it was happening.

'She's taken full aback.' He heard Denham's triumphant yell, and other voices took up the cry.

'She fast in irons!'

'We've got her, by God, we've got her now!'

Though his vision blurred and the wetness running down his cheeks was not all splattered spray, Clinton went on staring incredulously through the telescope.

'There is smoke, she's on fire!' Denham again, and Clinton picked up the fine pale mist of smoke spreading away from her; and at that moment a fresh burst of spray over the bows drenched the lens of his telescope, and he lowered it.

He took a silk bandanna from his hip pocket, and wiped his face and eyes of spray and the other wetness, then he blew his nose noisily, stuffed the handkerchief back into his pocket, jumped on to his deck and strode back to his quarterdeck.

'Mr Ferris,' he said crisply, 'please send up a flag hoist under *Huron*'s name and make the following "I am sending a boarding party to you."' The pale sapphire eyes shone with a zealot's intensity. '"If you resist I shall fight you."'

It was a long message, and while Ferris called for the pennants from the flag locker, Clinton turned to Denham. His voice shook with passion.

'Please clear the ship for action, Mr Denham – and we'll run out our guns now.'

Above the gale Clinton heard the clatter of the opening gunports, the rumble of the gun carriages, but all his attention was concentrated ahead upon the crippled slave ship.

He saw and understood the desperate attempts that her Captain was making to get her before the wind. He knew what a feat it had been to take down that tangled mast of canvas and rope in such a short time, yet he felt no admiration, only cold fighting fury.

Huron was showing only a storm jib.

St John was clearly trying to break the grip of the gale upon her for she was fast 'in irons' – her bows to the wind, and he was attempting to bring her round, but the tall ship that was usually so compliant and obedient was baulking, resisting him, and every minute *Black Joke* was swarming down upon her, closer and still closer.

'She's got serious structural damage,' Denham gloated aloud. 'I'd hazard a guess that she's lost her rudder.'

Clinton did not answer him, he strained ahead, half exultant, half fearful that St John's efforts would succeed and he would watch helplessly as *Huron* turned her stern to him once more, and went plunging away at the speed which *Black Joke* could never hope to match.

Then, as he watched, it happened. *Huron* swung her long, low length nearly broadside to him, beam on to the wind once again, and hung there for infinite seconds, then she shuddered and shook herself free of the gale's grip and went through the eye of the wind. Instantly the scraps of sail on her foremast snapped open, she came around presenting her stern to *Black Joke* and was sailing again.

Even in his bitter chagrin, Clinton could at last feel admiration for that barely credible feat of seamanship, but beside him his officers were struck dumb, paralysed with disappointment to see their prey slipping away from them once more.

More sails bloomed upon her tall bare masts, and the gap between the two ships was no longer narrowing; instead, it began to widen once more; slowly, infinitely slowly, *Huron* was forging away, and the night was coming.

'She's streaming a warp behind her,' Denham lamented quietly.

'It's a small ship's boat,' Ferris corrected him. They were already close enough to make out such details, *Huron* was only three or four nautical miles ahead of them, all her hull was in plain sight and they could even make out the tiny human figures on her decks with the naked eye. 'Damned clever, what!' Ferris went on with professional interest. 'Who would have believed it would work. Like as not the damned Yankee has the legs of us still.'

Clinton's chagrin turned to anger at his junior's unnecessary commentary.

'Mr Ferris, instead of chattering like a washerwoman, will you not read the signal *Huron* is flying?'

Huron's signal flags were blowing almost directly away from the watchers on *Black Joke*'s deck, making them difficult to spot and interpret, and Ferris, who had been fixing all his attention on the towing whaler, started guiltily, and then dived for his signal book and began busily scribbling on his slate.

'*Huron* sends under our name, "Stay clear of me, or I will fire upon you."'

'Good.' Clinton nodded and drew an inch of bare steel from the scabbard of his cutlass to make sure the weapon was free before thrusting it back to the hilt. 'Now we all know where we stand!'

But, slowly, inexorably, *Huron*, even partially crippled, and steering only by the sails on her foremast, was drawing away from them, and she was still far out of random cannon shot.

'The fire has taken hold in the steering-gear under the doctor's cabin.' The third mate came hurrying back on deck to make his damage report. 'I got her out of there.' He jerked a thumb as Robyn came up on deck clutching her black leather valise into which she had hastily crammed her journals and other small valuables.

'It's got through into the cable tier and the lazaretto, it will be into the sternquarters in a minute.' The mate's arms and face streamed with oily sweat, and the soot had blackened them like a chimney sweep.

'Put the hoses in through the poop companionway,' Mungo told him calmly. 'And flood the stern section abaft the main hold.'

The mate hurried away and within seconds there was the tolling clangour of the pumps as a dozen men threw their combined weight on the handles and the canvas hoses filled and stiffened, ejaculating solid jets of seawater down the stifling ladderways where already the air was trembling with heat like a desert mirage. Almost immediately hissing clouds of white steam began to boil from the ports and stern lights.

Satisfied, Mungo turned away, shot one glance over the stern to make sure that the gunboat was still falling away behind the limping clipper, then let his gaze linger a moment longer on the thick hawser that was secured to the port stern stanchions and ran through the fairlead to the bobbing whaler that *Huron* was dragging half a cable's length astern. The whole complex arrangement of the wind and sails and drogue was critical and unstable, the slightest change might upset it. He decided he could not risk hoisting another square inch of canvas, nor could he send a party below to rig a jury tackle on the useless rudder until the fire was brought under control.

He lit a cheroot, frowning with concentration over the simple and familiar task, and then he raised his eyes to look directly at Robyn for the first time since she had come up on deck.

For a second they stared at each other, and then Robyn looked astern at the ugly little gunboat that was still plugging along after them.

'I keep making the mistake of trusting you,' Mungo said beside her.

'I only made that mistake once – with you,' she replied, and he inclined his head slightly, accepting the riposte.

'How did you get into the steering gear,' he began to enquire, then snapped his fingers irritably at his own oversight. 'Of course, the inspection hatch. Yet, your ingenuity, Doctor, has been of no avail. Your friends still cannot hold us and as soon as it is dark, I will have the rudder cables repaired.'

For the last minute Mungo had been studying her face, oblivious to the sea and the ship and the gale. He did not see the fresh squall racing down upon *Huron*. When it struck, there was no helmsman to hold her. She saw the flash of alarm in his eyes, the realization of danger. His voice, as he yelled an order down the length of the deck, had for the first time the crack of fear in it.

'Get the sails off her, Mr Tippoo. Quick as you can!'

For the squall had upset the nice balance of *Huron*'s drogue and sail. The ship lunged forward sharply, the long bellied length of cable trailing astern lifted itself above the broken surface of the sea, straightening and coming under such strain that the seawater spurted from the hemp cords in tiny feathery jets.

The empty whaler, with her tarpaulin cover still lashed down over her in an attempt to keep her dry, was at that instant canted steeply over the crest of a breaking swell. The shattering impact transmitted by the taut cable to her bows pitched her forward and heaved her clear of the crest, so that for a moment she was airborne, like a leaping porpoise and then she struck bows first and was snatched below the surface.

For an instant *Huron* staggered to the enormous increase

in drag upon the trailing cable, and then the whaler disintegrated in a boiling flurry of white disturbed water. Her broken planking popped to the surface and the cable freed of its wearying weight flicked high in the air like the tail of an angry lioness. Without restraint, *Huron* gibed fiercely, spinning once more across the wind, and this time being blown flat, her tall bare masts swinging over almost parallel to the sea's surface.

The lee rail dug deeply into the sea, and the water came aboard her in a sweeping torrent, like a bursting dam wall.

It caught Robyn and hurled her against Mungo St John's chest; if it had not done so, she would have been carried overboard, but he caught her to him and held her as they were tumbled down the steeply canted deck – and then *Huron* was righting herself again, the water cascading off her in silver spouts.

She wallowed helplessly, taking the gale-driven seas on her beam, her desperate rolling accentuated by the pendulum of her high bare masts, but at least that drenching wall of sea water had poured into her hull through every opening and had extinguished her fires on the instant.

Mungo St John dragged Robyn by the wrist across the flooded deck, sloshing and slipping knee deep with loose tackle slithering and floating around them.

At the break of the poop he stopped, both of them panting for breath, their clothing and hair streaming sea water, the deck heaving and dropping crazily under them so he had to cling to the weather rail for support. He stared across at *Black Joke*.

The race was run. The gunboat was crowding down upon them exultantly, so close that he could see the cannon protruding from her open ports and the heads of the gunners above the bulwarks. Her challenging flag hoist still flew in her rigging, gaudy and gay as Christmas decorations. She would be up to the wallowing clipper in minutes, long before Mungo could ever hope to get his ship sailing again.

Mungo shook the water from his sodden dark locks like a spaniel coming ashore, and he filled his lungs.

'Mr O'Brien, a pair of slave cuffs here,' he bellowed, and Robyn, who had never heard him raise his voice, was stunned by the volume of sound that came up out of that muscular chest. She was still dazed and confused as she felt the cold kiss of iron on her wrists.

Mungo snapped the cuff on her left wrist, took two swift turns of chain around *Huron's* rail and then snapped the second cuff on her right wrist.

'I have no doubt your friends will be delighted to see you – in the cannon's mouth,' he told her, his face still set with anger, the rims of his nostrils white as bone china. He turned from her, running his fingers through his dark curls, throwing the hair back from his forehead and eyes.

'Mr O'Brien, muskets and pistols to every hand. Run out the guns and load with ball, we'll change to grape as the range closes.' The mate shouted his orders as he ran, and the crew scattered from the futile task of attempting to bring the clipper under control. They stumbled across the wave-swept deck, dodging fallen and broken tackle, hurrying to arm themselves and to man *Huron's* guns.

'Mr Tippoo!' Mungo's voice cut through the cacophony of gale and shouted orders.

'Captain Mungo!'

'Bring up the first deck of slaves.'

'We deep-sixing them?' Tippoo asked, for he had served before under slave captains, who, when capture by a naval vessel was imminent, would deep-six their cargo of blackbirds, drop them overboard, chains and all, and rid themselves of the most damning evidence against them.

'We'll chain 'em to the weather rail, Mr Tippoo, with the woman.' Mungo used neither Robyn's title nor her name. 'Make the limejuicer think a spell before he opens fire.' And Tippoo let an explosive chuckle of laughter come bouncing up his throat as he bounded away on those thick bowed legs, to get the gratings off the main hatch.

'Sir!' Denham's voice was incredulous, shocked. 'Sir!'

'Yes, Mr Denham,' Clinton answered him quietly, without lowering his telescope. 'I have seen it—'

'But, sir, that's Doctor Ballantyne—'

'And black slaves.' Ferris could hold his tongue no longer. 'They're chaining them to the rail.'

'What manner of man is that Yankee!' Denham burst out again.

'A damned clever one,' Clinton answered him quietly. He was watching the woman he had come to rescue through the glass. He could already recognize her features. Her eyes seemed too large for her deathly white face, her sodden and rumpled clothing stuck to her body. Through a rent in her blouse he could see the pale skin of her shoulder and upper arm gleaming with a pearl-like lustre in the sunlight.

'Mr Denham,' Clinton went on speaking. 'Warn the crew that we will be receiving fire in about five minutes, and we will be unable to return it.'

He watched the ranks of naked black slaves still coming up on to the clipper's main deck and taking their place along the rail, their gaolers fussing about them, chivvying them into place and securing their chains.

'We are fortunate in having a gale of wind, so we will be exposed to fire for a short period, but warn the men to lie flat upon the deck below the bulwark.'

Black Joke's fragile eggshell plating would give some protection at extreme range, but as they closed with the slave ship, he could expect even grape shot to penetrate their sides. One blessing, they would be spared the lethal flying splinters that were so much dreaded in wooden ships.

'I am going to lay her alongside the Yankee's stern,' Clinton went on. That way she would be exposed to the clipper's broadsides while the two ships were bound to each other. 'But she stands taller than we do. I want your best men with the grappling irons, Mr Denham.' *Huron*'s main deck was ten feet higher than the gunboat's. There would be nice work ahead when they leapt the gap, and scrambled up *Huron*'s stern with its pronounced tumble home.

'By God! She's running out her guns. She means to fight us after all,' Denham cut in, and then, penitently, 'I beg your pardon, sir.' He excused himself for the interruption and the blasphemy.

Clinton lowered the telescope. They were so close now that he no longer needed it.

The clipper had six light cannon on each side of her, mounted on the main deck. The barrels were almost twice as long as *Black Joke*'s own heavy carronades. However, the bore of the muzzles was much smaller in diameter, and as Clinton watched, they began to train around towards him, one at a time beginning at the stern.

Even without the glass, Clinton could make out the tall lean figure in the plain blue jacket moving at a deceptively leisurely pace from one gun to the next, laying each of them personally, gesturing at the gun crews to strain on the tackles and traverse the long cannon on to their target.

Clinton saw St John reach the bow gun and make a careful adjustment, working over it a few seconds longer than he had the others, and then he leapt to the clipper's bulwark and balanced there with the assurance of an acrobat against the rudderless hull's unpredictable movements.

The scene engraved itself upon Clinton's mind, it seemed so theatrical, like the cast of a stage production lined up at the end of the performance to receive the applause of the spectators. The file of naked black bodies, standing almost shoulder to shoulder with their arms extended in unison, like the trained chorus, their wrists locked to the teak rail by the slave cuffs. Then the principal, the figure of the woman, slim and somehow delicate and tiny in their midst. The bodice of her dress, a buttercup yellow, was a gay spot of colour that drew Clinton's eye irresistibly. It was a distraction that he could not afford at this moment.

The American seemed to be watching Clinton, seemed to have singled him out from the group of officers, and even across the wide stretch of water that still separated them, Clinton was aware of the mesmeric pull of those golden-flecked eyes, the eyes of a predator, a leopard perhaps, poised with a lithe and patient grace upon the bough above the waterhole, awaiting the moment when the prey moved beneath him.

At the level of Mungo St John's knees were the heads of his gun crews, little knots of pale tense faces, contrasting starkly with the quiescent rank of black slaves. They crouched over their weapons, and the long slim barrels were reduced to small dark circles as Clinton stared directly down the bores.

There were men also in *Huron*'s rigging, roosting in the cross-trees of the yards and masts, and the long barrels of their muskets were clear to see against the backdrop of the wind-driven sky. They would be picked marksmen, *Huron*'s best, and their preferred and special target would be the small group of officers on the gunboat's quarterdeck. Clinton hoped that the clipper's wild action in the gale would throw out their aim.

'Gentlemen, I advise you to take cover until we can bring the ship into action,' he told Denham and Ferris quietly, and felt a little prick of pride when neither of them moved. It was the tradition of Drake and Nelson not to flinch from the coming storm of fire, and Clinton himself went on standing at his ease, hands clasped at the small of his back, calling a small adjustment to the helm as *Black Joke* drove in eagerly, the terrier going for the hold on the bull's nose.

He saw the American move his head, a final judgement

of range, considered against the clipper's rolling and beside him Ferris murmured the age-old blasphemy which Clinton this time could not find it in him to resent, for it was also a part of the great tradition.

'For what we are about to receive—' said Ferris, and as if he had heard the words, the American drew the sword from the scabbard on his belt, and raised it above his head. Involuntarily all three naval officers drew breath together and held it. *Huron* was at the bottom of her roll, her cannon pointing down into the sea close alongside, then she was coming up, the barrels rising – levelled, and the sword arm fell.

The six cannon leaped together, in perfect concert, and the startling white gusts of smoke shot fifty feet from her sides, completely silent, for the sound had not reached them, and for a fleeting part of a second they could believe that *Huron* had not loosed her broadside.

Then the very air beat in upon them, shocking the eardrums, seeming for a moment to suck their eyeballs from the sockets with the vast disruption of air caused by passing shot, and close above Clinton's head a stay parted with a whiplash crack.

That was one ball high, but under Clinton's feet, the deck jumped with the multiple impact of ball into her, and the hull rang like the strokes of a gigantic brass gong.

A single ball came through at deck level. It struck a burst of sparks from the steel hull, like Brocks fireworks at Crystal Palace, brilliant orange even in the strong sunlight, and the hole it tore through *Black Joke*'s plating was fringed with bare jagged tongues of metal like the petals of a silver sunflower.

A seaman in striped vest and baggy canvas breeches, who was kneeling behind the bulwark, took the ball full in his chest.

His severed limbs were strewn untidily across the gunboat's spotless deck and the ball went on to strike the foot of *Black Joke*'s mast, shivering it like a tall tree struck by lightning, and tearing a long white splinter from the seasoned Norwegian pine. Then, with its force mainly spent the ball rolled the length of the deck, smoking and stinking of scorched metal until it thumped into the scuppers and rolled idly back and forth. Only then, seconds after the broadside struck, did the crash of the discharge reach their ears across the turbulent waters that separated the two vessels.

'Not bad shooting for a Yankee,' Ferris grudged them, raising his voice above the gun thunder, and Denham had his watch out and was timing how long it took for the clipper's gun crews to reload.

'Forty-five seconds,' Denham intoned, 'and not a single gun run out again. A bunch of fairground tinkers could do better.'

Clinton found himself wondering if it was merely bravado, or complete indifference to danger and violent death which allowed the two younger officers to chat so casually, while the seaman's severed arms still twitched on the deck, not twenty feet away.

Clinton was afraid, afraid of death and afraid of failing in his duty and afraid of being seen to be afraid, but then he was older than they, for despite their manly airs Ferris was a boy and Denham barely twenty – so perhaps it was not courage but ignorance and lack of imagination.

'Fifty-five seconds!' Denham grunted scornfully, as the next ragged broadside crashed into *Black Joke*'s iron hull, and somebody started to scream below decks, a high mindless keening like steam from a kettle.

'Send somebody to stop that fellow,' Ferris murmured to the seaman who crouched nearby, and doubled over still the man hurried away. Seconds later the screaming stopped abruptly.

'Good work,' Ferris told the seaman as he took his place at the bulwark again.

'Dead, sir, he is, poor devil.'

Ferris nodded without change of expression, and moved closer to listen to his Captain.

'Mr Denham, I am going to lead the boarding-party. You are to be ready to sheer off and leave us to it, should there be any danger to the ship—' There was a sharp fluting sound, like the flight of a giant insect past their heads, and Clinton glanced up irritably. The marksmen in *Huron's* rigging had opened fire, the pop of their muskets seemed muted and without menace. Studiedly Clinton ignored them and went on issuing his final orders, raising his voice to compete with the crash and roar of shot and the strike of it into the gunboat's hull.

As Clinton finished speaking, Denham blurted abruptly, 'It's hell not being able to reply.' He was staring across at the clipper whose silhouette was blurred with a bank of pale gunsmoke that even the gale could not disperse rapidly enough. 'It's bad for the men,' he corrected himself swiftly, and Clinton had his answer. Denham was afraid as he was, and the knowledge gave him no comfort at all. If only they could do something, anything, instead of having to stand here in the open and make studied conversation, while *Black Joke* tore across the last few hundred yards of white crested sea that still separated them.

The crash of cannon shot tearing into *Black Joke's* vitals was almost continuous now as the fastest gunners aboard *Huron* outstripped the others. The bow cannon that the American Captain was supervising and laying was firing three times to the other's twice, Clinton had been counting the plumes of muzzle smoke, this would be the sixth ball they sent into the little gunboat since the American had given the order to fire as many minutes ago.

He watched the gunsmoke bloom again from the cannon's maw, and this time the gunboat's deck was swept as if by hail stones, but leaden hail stones as big as ripe grapes that pierced the thin steel bulwarks with pricks of sunlight and clawed chunks of wood from the main deck, a deck that was now threaded with meandering scarlet snakes of blood and slick little puddles of it that spread from beneath the inert and crumpled figures that seemed to be scattered in purposeless profusion wherever Clinton looked.

Black Joke was taking merciless punishment, perhaps already more than she could afford, but they were close now, very close, seconds only left to go.

He could hear the cheering of the clipper's gun crews, the terrified wailing of the slaves who were huddled down in pathetic little heaps upon *Huron*'s decks, he could clearly hear the rumble of the sixteen-pounders run out against the straining tackles, and hear the shouted commands of the gun captains.

The girl at the rail still stood rigidly erect, staring white-faced across at him, and she had seen and recognized him now. She tried to raise a hand to wave a greeting, but the iron slave cuff on her wrist hampered the movement. As Clinton stepped forward the better to see her, something tugged sharply at the sleeve of his jacket and behind him Ferris gasped.

Clinton looked down at his arm, the sleeve was torn and the white lining showed in the tear – it was only then he realized that it was a musket ball fired from *Huron*'s cross-trees which might have struck him squarely had he not moved, and he turned quickly to Ferris.

The boy was pressing a wadded handkerchief to his chest, standing very upright.

'You are wounded, Mr Ferris,' said Clinton. 'You may go below.'

'Thank you, sir,' wheezed Ferris. 'But I'd just as soon not miss the kill.' As he spoke, a droplet of blood formed in the corner of his mouth, and with a chilling little jolt, Clinton realized that the boy was probably mortally struck – blood in the mouth would almost certainly mean a lung hit.

'Carry on then, Mr Ferris,' he said formally, and turned away. He must not let the doubts assail him now, he must

662

not question whether his decision to board *Huron* had been correct, or if his execution of the attack had been properly carried out – or if he was responsible for those dismembered corpses that littered *Black Joke*'s deck, or for the dying lad who still determinedly kept erect. He must not let his resolve weaken.

Instead he slitted his eyes against the lowering sun that outlined *Huron*'s bare masts with golden haloes, and stared across at her with true hatred. It was then he realized suddenly that at last her bow cannons could no longer bear, the terrible battering of close-range grape shot was abating as they sailed into *Huron*'s stern quadrant.

'Bring her up two points,' he snapped, and the *Black Joke* cut in sharply under *Huron*'s stern. She loomed suddenly high above them and there was no more cannon fire from her while the clipper's tall hull shielded them from the gale. The sudden silence was ghostly, unnerving, as though the cannon fire had damaged his ears and he was rendered deaf.

Clinton shook off the sense of dreamlike unreality, and strode down the length of his deck.

'Up the Jokers!' he shouted, and his crew rose from where they had been crouching under the fragile bulwark.

'You've shown you can take it, boys, now let's show those damned Yankees we can hand it out.'

'A tiger for Tongs!' yelled a voice, and suddenly they were all cheering, crowding the gunboat's side, so he had to cup his hands to his mouth to give the order.

'Helm a lee, let fly all!' and *Black Joke* spun up sharply under *Huron*'s counter, spilling her wind, while the seamen in her rigging stripped the canvas off her.

The two ships came together with a rending crackling crash, the gnashing of steel plate against timber and the shattering of glass in *Huron*'s sternlights.

A dozen of *Black Joke*'s sailors hurled the three-pronged grappling hooks high over the clipper's gunwale with the lines snaking up after them, and then heaved them up tight and made them fast to the portside cleats, and a swarm of seamen cheering wildly went up *Huron*'s stern, like a troop of vervet monkeys pursued into the trees by a hunting leopard.

'Take command, Mr Denham,' Clinton shouted above the hubbub.

'Aye, aye, sir,' Denham's lips moved and he saluted as Clinton thrust his cutlass back into his scabbard and headed the next rush of his men, those who had been freed by the heaving-to of her sails.

The two hulls were working against each other as viciously as millstones, grinding and bumping, the gap opening and closing as the wind and the seas tore at them.

At *Huron*'s rail a dozen of her crew were hacking at the grappling lines with axes, the clunking of the blades into her timber blended with the popping of pistols and muskets as their mates blazed down on the swarm of seamen climbing up from the gunboat's deck.

One of *Black Joke*'s sailors climbed swiftly hand over hand, pushing off with his feet from the clipper's raked stern like a mountaineer, and he had almost reached the rail when an American sailor appeared above him, an axe held high and then sent thudding down into the woodwork, severing the line at a single blow.

The sailor dropped like a windblown fruit from the bough into the gap between the hulls. He floundered for an instant in the surging water and then the two ships came together again, with the shriek of rending timbers, chewing the man like a pair of monstrous jaws.

'More lines,' Clinton howled, and another grappling hook flew over his head, thrown by a stout British arm, and the line whipped around Clinton's shoulders.

He seized it, heaved once upon it to set the hook and then swung across the gap and his boots thudded on *Huron*'s stern. He had seen his seaman drop on the severed line, so he climbed with the strength and agility of terror and desperation, and only as he swung one leg over *Huron*'s rail did the battle rage seize him, the world changed colour before him, seen through a reddish haze of hatred and fury, hatred for the slave stink that rose to offend his soul and fury for the death and punishment that his ship and his seamen had suffered.

His cutlass sprang from its scabbard with metallic rasp, and there was a man rushing down upon him, and the man naked to the waist with a bulging hairy belly and thick heavily muscled arms. He was brandishing a double bladed axe above his head, and Clinton uncoiled his lanky frame as though it were the 'S' in an adder's body, straightening as it strikes. He drove the point of the cutlass through the axe man's furry silvered beard and the axe flew from his raised hands and went sliding away across the deck.

Clinton stood over the man he had killed, placed his foot on his chest and yanked the cutlass blade out of his throat. A bright scarlet carotid fountain followed the blade out, splattering Clinton's boot.

Half a dozen of his seamen had reached *Huron*'s deck ahead of Clinton, and without a word of command spoken, they bunched to guard the grappling lines over the stern, holding off *Huron*'s axe men with cutlass and point-blank pistol fire. Behind them, *Black Joke*'s men came swarming aboard, unopposed, surging forward, their cheers rising into a triumphant chorus.

'At 'em the Jokers!'

'All together, boys,' howled Clinton, the madness had taken complete hold of him now. There was no fear, no doubts, not even conscious thought. The madness was infectious, and his men howled with him, hunting as a pack like the wolf or the wild dog they swept across *Huron*'s deck to meet the wave of her own seamen rushing back from *Huron*'s bows.

The two waves of running, screaming men met just below the break of the poop, were transformed into a struggling

665

mass of closely locked humanity, and their cries and curses mingled with the animal howl of terrified slaves. The pistols and muskets had all been discharged and there was no change to reload. It was steel against steel now.

Black Joke's crew were battle-hardened, they had fought together fifty times in the past year, they had stormed glacis and barracoon and withstood fire and steel. They were blooded fighting men and proud of it.

Huron's men were commercial seamen, not warriors, most of them had never swung a cutlass nor fired a pistol at another man before, and the difference was evident almost immediately.

For a minute or less the closely engaged mass of men swayed and churned like the meeting of two strong currents at the tidal line of the ocean, and then *Black Joke*'s sailors began to forge forward.

'Up the Jokers!' They sensed their advantage.

'Hammer and tongs, boys. Give 'em hell!'

At only one point was the tide of British sailors checked, at the foot of *Huron*'s mainmast two men stood almost shoulder to shoulder.

Tippoo seemed immovable on the solid foundation of those massive bared legs. Like a Buddha carved from solid rock, he spurned the press of men around him, and their ranks parted and drew back.

His loin-cloth was drawn up between his legs, and his smooth belly bulged over it, again as hard as mountain rock, with the deep cyclops eye of his navel in its centre.

The golden thread of his embroidered gilet sparkled in the sunlight, and he held his great round head low on his shoulders as he swung a double-bladed axe as easily as though it were a lady's parasol, and the axe hissed fiercely at every stroke and *Black Joke*'s seamen gave him ground.

A pistol ball had nicked the scarred flesh of his bald head, and blood poured copiously from the shallow wound, turning his face to a glaring gory mask.

His wide toad's slit of a mouth opened as he laughed and shouted his contempt of the men who swarmed about him like pygmies about an ogre.

Beside him fought Mungo St John. He had stripped off his blue jacket to free his sword arm, and his white linen shirt was open to the belt, the buttons torn from their threads by a clutching enemy hand. He had knotted a silk bandanna around his forehead to keep the sweat from running into his eyes, but sweat poured down his naked chest, and had soaked through his shirt in patches. He had a sword in his right hand, a weapon with a plain silver steel guard and pommel, and he cut and parried and thrust without a break in the rhythm of his movements.

He was unmarked, the droplets of thrown blood that stained the full sleeve of his white shirt and which had been diluted by his sweat to a dirty brown, were not his blood.

'St John!' Clinton called to him. They were both tall enough to look over the heads of the men between them, and they started at each other for a moment.

Clinton's eyes were pale fanatical blue, and his lips white with fury. Mungo's expression was grave, thoughtful almost, and his gaze troubled, almost grieving, as though he knew he had lost his ship and that his life and the lives of most of his crew were forfeit.

'Fight me!' Clinton challenged him, his voice strident, ringing with triumph.

'Again?' Mungo asked, and the smile touched his lips fleetingly, then was gone.

Clinton shouldered his way roughly through the press of his own men. The last time it had been pistols, Mungo St John's choice of weapon, but now Clinton had the familiar weight and balance of a naval cutlass in his right hand – his weapon, that he had first swung as a midshipman of fourteen years old, and the whipcord muscles in his long right arm were seasoned to the use of the blade, and every evolution in its use, every trick and subterfuge had been drilled until they were instinctive.

As they came together, Clinton feinted and cut backhanded and low, going for the hip to cripple and bring the man down. As the stroke was parried he felt the strength of Mungo's sword arm for an instant before he disengaged, and switched his attack fluidly, going on the thrust leading with his right foot, a full stroke, and again the parry was strong and neat, but only just strong enough to hold the heavier, broader steel of the cutlass.

Those two brief contacts were enough for Clinton to judge his adversary and find his weakness; the wrist. He had felt it through the steel the way a skilled angler feels the weakness of the fish through line and rod – it was the wrist. St John did not have the steely resilience that comes only from long and dedicated exercise and practice.

He saw the flare of alarm in St John's strangely flecked eyes. The American, too, had felt his own inadequacy, and he knew he did not dare to draw out the encounter. He must try to end it swiftly, before the Englishman's superiority could wear him down.

With the swordsman's instinct, Clinton translated the little flicker of the golden yellow eyes. He knew that Mungo St John was going on the attack, so that as the stroke came an instant later, he caught it on the broad curved blade of the cutlass; then he shifted his weight forward and, with a twist of his own iron wrist, prevented the disengagement, forcing St John to roll his own wrist, the two blades milling across each other, the steel screeching sharply on a harsh abrasive note that set their teeth on edge. Clinton forced two and then three turns, the classic prolonged engagement from which Mungo St John could not break without risking the thunderbolt of the riposte, and Clinton felt the other's wrist give under the strain. He lunged his weight against it, slid the guard of the cutlass high up the blade and used the rolling momentum of two blades and the leverage of his wrist and the curved guard of the cutlass to tear the hilt out of Mungo St John's fingers.

The American's sword clattered to the deck between them, and Mungo St John threw up both hands, sucked in his belly and flung himself back against the mainmast in an attempt to avoid the thrust of the heavy cutlass blade which he knew would follow. In the fury of hatred that possessed Clinton, there was no thought of giving quarter to the man whom he had disarmed.

The thrust was full-blooded, driven by all the strength of wrist and arm, of shoulder and of Clinton's entire body weight, the killing stroke.

Clinton's whole being had been concentrated on the man before him, but now there was movement in the periphery of his vision. Tippoo had seen his Captain disarmed in the same instant that he had just completed a swing with the axe. He was off balance, it would take only a shaded instant to recover that balance, to raise the axe again, but that instant of time would be too long, for he saw the cutlass stroke already launched, and Mungo St John trapped helplessly against the mainmast, his belly unprotected and his empty hands held high.

Tippoo opened his huge paws and let the axe go spinning away, like a cartwheel, and then he reached out and seized the gleaming cutlass blade in one bare hand.

He felt the blade run between his fingers and the terrible sting of the razor-sharp edge cutting down to the bone, still he heaved with all his weight, pulling the point away from the helpless man against the mainmast, deflecting it but unable to hold it for the tensed tendons in his lacerated fingers parted, and the blade ran on driven by the full weight of the tall platinum-headed naval officer.

Tippoo heard the point of the cutlass scrape over one of his ribs, and then a numbness filled his chest, and he felt the steel guard of the hilt strike his rib cage, a thud like that of a butcher's cleaver striking the chopping board, as the cutlass blade reached the limit of its travel.

Even the savage force of that blow was not enough to knock Tippoo off his feet, though it drove him back a pace. He stood solidly. His eyes screwed up into slits of skin, staring down at the blade that transfixed his chest, his bleeding hands still clutching the guard of the cutlass.

Only when Clinton leaned back and pulled the blade from his flesh, did Tippoo begin to sag slowly forward, his knees buckling, and he fell, his body slack and unresisting.

Clinton freed his cutlass blade, and it was thinly smeared along its full length, so that it blurred redly as he went on to the forehand cut, going once more for the man who was still pinned against the mainmast.

Clinton did not complete the stroke. He arrested it in mid-air, for Mungo St John had been borne to the deck beneath a wave of struggling British seamen.

Clinton stepped back and rested on his cutlass. The fight was over, all around him the crew of *Huron* were throwing down their weapons.

'Quarter, for the love of God, quarter!'

They were dragging Mungo St John to his feet, two seamen on each of his arms. He was unwounded, and Clinton's hatred was unabated. It took an enormous effort to prevent himself driving the point of the cutlass into Mungo's belly. Mungo was struggling to throw off the hands of the men who held him, straining to reach the massive body of the half-naked Moslem mate that lay at his feet.

'Let me free,' Mungo cried. 'I must see to my mate.' But they held him remorselessly, and Mungo looked up at Clinton.

'In the name of mercy,' he was pleading, and Clinton had never expected that. He took a deep ragged breath, the madness began to fade.

'I give you my word, sir,' Mungo was stricken, there was

no mistaking his consuming grief, and Clinton hesitated. 'I am your prisoner,' Mungo told him. 'But this man is a friend—'

Clinton let out his breath slowly, and then he nodded to the seamen who held Mungo St John.

'He has given his word.' And then to Mungo, 'You may have five minutes.' And the seamen released him. Mungo sank swiftly to his knees beside the inert figure.

'Old friend,' he whispered, as he tore the bandanna from off his own head and pressed it to the obscene little slit between Tippoo's ribs, 'old friend.'

Clinton turned away, slipping the cutlass back into its scabbard and he ran across the deck to the weather rail.

Robyn Ballantyne saw him coming and she strained towards him, unable to lift her arms for the slave cuffs that still bound her, but as he embraced her she put her face against his chest and her whole body trembled and shook as she sobbed.

'Oh, I give thanks to God—'

'Find the keys,' Clinton ordered brusquely, and as the cuffs fell from Robyn's wrists he snatched them up and handed them to one of his men. 'Use these on the slaver's Captain,' he ordered.

With that gesture, the last of his madness was gone.

'Forgive me, Doctor Ballantyne. We will speak later, but now there remains much to be done.' He bowed slightly and hurried away calling his orders.

'Carpenter's mate, go below immediately, I want the damage to this ship repaired at once. Bosun, disarm her crew, and have them sent below under lock and key with a guard on the companionway. Two men on her wheel, and a prize crew to work her. We'll sail her into Table Bay with the dawn, my boys, and there'll be prize money for your fancy.' His men were still drunk on excitement and battle lust, and they cheered him hoarsely as they rushed to obey the string of orders.

671

Rubbing her chafed wrists, Robyn picked her way across the littered deck and through the throngs of bustling British seamen as they hustled their captives and the still-chained files of slaves below.

Almost timorously she approached the ill-assorted pair at the foot of the ship's mainmast. Tippoo lay on his back, the mound of his naked belly pressing upwards like a woman in labour, the soiled bandanna hiding the wound. His eyes were wide, staring up at the mast that towered above him, and his lower jaw sagged.

Mungo St John held the huge bald cannon-ball head on his lap. He sat with his legs thrust out straight ahead of him, his back against the mast and as Robyn approached, he closed the lids over Tippoo's staring eyes, one at a time, with his thumb. His head was bowed, his hands gentle as those of a mother with her infant as he lifted the bandanna and used it to bind up the sagging jaw.

Robyn went down on one knee and reached out to Tippoo's chest, to feel for the heart beat, but Mungo St John raised his head and looked at her.

'Don't touch him,' he said softly.

'I am a doctor—'

'He no longer needs a doctor,' Mungo's voice was low and clear, 'especially if that doctor is you.'

'I am sorry.'

'Doctor Ballantyne,' he told her, 'you and I have no reason to apologize to each other, nor for that matter to speak to each other, ever again.'

She stared at him, and his face was cold and set, the eyes that stared back at her were devoid of all emotion, and it was in that moment she knew she had lost him, irrevocably and for ever. She had thought that was what she wanted, but now the total knowledge left her devastated, without the strength to break her gaze, without the power of speech, and he stared back at her remotely, hard and unforgiving.

'Mungo,' she whispered, finding at last the strength and

will to speak. 'I did not mean this to happen, as the Almighty is my witness, I did not mean it.'

Rough hands dragged Mungo St John to his feet, so that Tippoo's dead head slipped from his lap and the skull thumped against the wooden deck.

'Captain's orders, me old cock, and you are to 'ave a taste of your own chains.'

Mungo St John did not resist as the slave cuffs were fastened on his wrists and ankles. He stood quietly, balancing to *Huron*'s wild gale-driven lunges, looking about the fire-blackened ship with its decks covered with fallen and tangled rigging, stained with the blood of his crew, and though his expression did not change, there was a limitless grieving in his eyes.

'I am sorry,' whispered Robyn, still kneeling beside him. 'I am truly sorry.'

Mungo St John glanced down at her, his wrists fastened at the small of his back by the cold black links of chain.

'Yes,' he nodded. 'So am I.' And a seaman thrust the palm of a horny hand between his shoulder blades, shoving him away towards the *Huron*'s forecastle, and the slave chains clanked about his ankles, as he staggered.

Within a dozen paces he had recovered his balance, and shrugged off the hands of his gaolers. He walked away with his back straight and his shoulders thrown back, and he did not look back at Robyn kneeling on the blood-stained deck.

Mungo St John blinked at the brilliant sunlight as he followed the scarlet uniform coat and white cross-straps of his escort out into the courtyard of the Cape Town castle.

He had not seen the sun for five days; the cell in which he had been confined since he had been escorted ashore, had no external windows. Even in midsummer, the dark and chill of the past winter still lingered in the thick stone walls, and the air that entered through the single barred opening in the oaken door was stale and sullied by the gaol odours, the emanations from the dozen or so prisoners in the other cells.

Mungo filled his lungs now, and paused to look up at the ramparts of the castle. The British flag spread jauntily above the Katzenellenbogen redoubt, and beyond it the seagulls planed and volleyed on the fresh south-easterly wind. Force five and standing fair for a ship to clear the bay and make the open Atlantic, Mungo noted instinctively.

'This way please.' The young Subaltern who commanded the prison escort urged him on, but Mungo hesitated a moment longer. He could hear the murmurous song of the surf-break upon the beaches just beyond the castle walls, and from the ramparts he would have a clear view across Table Bay to Bloubergstrand on the far curve of the land.

Huron would be lying at anchor close inshore, still under her prize crew, and he longed for just a single glimpse of her, longed to know if the stern quarters were still smoke-blackened and gutted, or if O'Brien had been allowed to make the repairs to her hull and her steering-gear.

'If only Tippoo,' he began the thought, and then stopped himself, shivered briefly in the sunlight not only from the prison chill in his bones. He squared his shoulders and nodded to the Subaltern.

'Please lead the way,' he agreed, and the hobnailed boots of the escort gnashed the cobbles as they crossed the courtyard and then climbed the broad flight of steps to the Governor's suite of offices.

'Prisoner and escort, halt.'

Upon the portico a naval Lieutenant waited to receive them in his navy-blue and gold jacket, white breeches and cocked hat.

'Mr St John?' asked the Lieutenant. He was old for his

rank, grey and worn-looking, with a weary disinterested eye, and Mungo nodded disdainfully.

The Lieutenant turned to the officer of the escort.

'Thank you, sir, I will take over from here,' and then to Mungo. 'Kindly follow me, Mr St John.'

He went in through the magnificent teak doors, carved by the master craftsman Anreith, into the Governor's antechamber with its polished floors of butter-coloured Cape deal, the high hewn rafters of the same timber, and with the thick walls hung with the treasures of the Orient gathered so assiduously by that great plunderer, the Dutch East India Company, which had in turn succumbed to an even more powerful predator.

The Lieutenant turned right, avoiding the brass and mahogany double doors of the Governor's private office to which Mungo had expected to be led; instead they went to a less pretentious single door set in a corner of the ante-chamber. At the Lieutenant's knock, a voice bade them enter, and they went in to a small office, clearly belonging to the Governor's Aide-de-camp whom Mungo had met before.

The Aide-de-camp sat at the plain oak desk facing the door, and he did not rise nor did he smile as Mungo entered. There were two other men in the room, both seated in armchairs.

'You know Admiral Kemp,' said the Aide-de-camp.

'Good morning, Admiral.'

Slogger Kemp inclined his head, but made no other gesture of recognition.

'And this is Sir Alfred Murray, Chief Justice of the Supreme Court of the Cape Colony.'

'Your servant, sir.' Mungo neither bowed nor smiled, and the judge leaned forward slightly in his armchair, both hands on the gold and amber handle of his walking-stick, and stared at Mungo from under beetling white brows.

Mungo was pleased that an hour previously his gaoler had provided him with hot water and razor and that he had been allowed to contract with the ex-slave Malay washerwoman who laundered for the castle's officers. His breeches were clean, his boots polished and his shirt crisply ironed and snowy white.

The Aide-de-camp picked up an official document from the desk before him.

'You are the Captain and owner of the clipper *Huron*?'

'I am.'

'The ship has been seized as prize by the Royal Navy under Articles Five to Eleven of the Treaty of Brussels, and presently lies under prize crew in British territorial waters.'

That did not need reply, and Mungo stood silently.

'The case has been considered by the Courts of Mixed Commission for the colony, under the presidency of the Chief Justice, and after hearing evidence from the Officer Commanding the Cape Squadron and others, the Court has determined that as the *Huron* was taken on the high seas, the Cape Colony has no jurisdiction in this matter. The Chief Justice has recommended to His Excellency, the Governor of Cape Colony, that the – ahem—' the Aide-de-camp paused significantly, 'the cargo of the clipper ship *Huron* be impounded by Her Majesty's Government, but that the clipper be released under the command and cognaissance of its owner and that the owner be ordered to proceed with all despatch to place himself and his vessel under the jurisdiction of a properly constituted American Court and there to answer such charges as the President of the United States deems fit to bring against him.'

Mungo let out a long slow breath of relief. By God's breath, the Limeys were going to duck the issue! They were not about to chance the wrath of the new American President-elect. They had taken his slaves, eight hundred thousand dollars worth, but they were giving him back his ship and they were letting him go.

The Aide-de-camp went on reading without looking up.

'The Governor of Cape Colony has accepted the Court's recommendation and has so decreed. You are required to make your ship ready and safe for the voyage with all speed. In this respect, the officer commanding the Cape Squadron has agreed to place at your disposal the repair facilities of the Naval Station.'

'Thank you, Admiral.' Mungo turned to him, and Slogger Kemp's brows came together, his face mottled with passion, but his voice was very quiet and clear.

'Sixteen of my men dead, and as many maimed by your actions – sir – each day the smell of your filthy ship blows in to the windows of my office.' Admiral Kemp lifted himself with an effort from his armchair, and glared at Mungo St John. 'I say rot you, and your thanks, Mr St John, and if I had my way we wouldn't be playing coy and cute with Mr Lincoln, and I would have you swinging at the mainyard of a British man-of-war rather than sailing out of Table Bay in your stinking slaver.'

Slogger Kemp turned away and went to stare out of the single window, into the courtyard of the castle where his carriage waited.

The Aide-de-camp seemed not to have noticed the outburst. He went on smoothly,

'A representative of the Royal Navy will accompany you aboard your ship and remain there until he determines that your vessel is seaworthy.'

The Aide-de-camp reached back and tugged the bell-pull behind his shoulder, and almost immediately the door opened and the naval Lieutenant reappeared.

'Just one other thing, Mr St John, the Governor has declared you to be an undesirable alien and you will immediately be arrested if you are ever again so rash as to set foot in Cape Colony.'

The tall figure came striding up the yellow gravel pathway, under the avenue of tall date palms, and Aletta Cartwright called gaily across the rose garden,

'Here comes your beau, Robyn. He is early today.'

Robyn straightened, the basket full of rose blossoms hanging on her arm, the wide straw hat shading her face from the flat glare of the Cape noonday. She watched Clinton coming towards her with the warmth of affection. He looked so gangling and boyish and impetuous, much too young ever to have led that rush of fighting seamen over *Huron*'s stern.

She had grown accustomed to him over the past weeks while she had been a guest once more of the Cartwright family, and each afternoon Clinton had walked up the hill from his modest lodgings in Waterkant Street. She looked forward to his visits, to their serious conversations after the frivolity and inconsequences of the Cartwright daughters. She found his admiration and his adoration flattering and deeply comforting. She felt it was something that would never change, something constant, a pole-star in the confusion and uncertainty that had been her life to this time.

She had learned to value his good sense, and his judgement. She had even allowed him to read the manuscript which was occupying most of her days now, and his comments and criticisms were always well based.

Then she had found that he filled a part of her life that had been empty for much too long. She needed something or someone to cherish and protect and comfort, somebody who needed her, someone on whom to lavish the bounty of her compassion.

'I do not believe I could ever live without you, my dear Doctor Ballantyne,' he had told her. 'I do not believe I could have endured this terrible period of my life without your help.'

She knew it was probably true, not just the hyperbole of

678

the love-sick swain, and Robyn was entirely unable to resist the appeal of anybody in pain or in suffering.

It was many weeks since that heady day when *Black Joke* had sailed into Table Bay with her bulwarks and upperworks riddled with shot, her rigging in heroic and picturesque ruins, and her huge captive, blackened with smoke and limping under jury rigging and makeshift steering-gear, herded submissively under the menace of her carronades to an anchorage close inshore at Rogger Bay.

How the townsfolk had swarmed to the beachfront to gawk and exclaim, and how the other naval vessels in the bay had lined their rails and yards with seamen to cheer them in.

She had been standing at Clinton Codrington's side when the two contingents of naval officers from the Cape Squadron's headquarters had been rowed out to *Black Joke's* anchorage. The first had been headed by a naval Commander, some years junior to Clinton.

'Captain Codrington,' he saluted. 'I am under orders to take over command of this ship from you forthwith, sir.'

Clinton accepted this without change of expression.

'Very well, sir, I will have my gear removed, and in the meantime we should complete the formalities, and I will introduce you to the remaining officers.'

When Clinton had shaken hands with his officers and his sea chest was at the entry port, the second longboat which had been lying on its oars a few yards off, now came alongside and a senior Captain came aboard. Everyone on *Black Joke's* deck knew what was about to happen, and Denham stepped close to Clinton and said softly, 'Good luck, sir, you know you can count on me when the time comes.'

They both knew what he was referring to – the day they would meet again in the court-martial chamber.

'Thank you, Mr Denham,' Clinton replied, then he went forward to where the senior Captain waited.

'Captain Codrington, it is my duty to inform you that you have been called upon by the Officer Commanding the Squadron to answer certain charges concerning the conduct of your duties. Therefore, you are to consider yourself under open arrest and to hold yourself in readiness to answer those charges as soon as a court martial can be convened.'

'I understand, sir.'

Clinton saluted him, and then preceded him through the entry port and down the ladder into the waiting boat.

A single voice called out, 'Give 'em hell, Tongs.'

And suddenly they were all cheering. *Black Joke*'s crew lined her side and hung in her rigging and they cheered as though their throats would crack.

'Hammer and Tongs!' They tossed their caps on high.

'At 'em the Jokers!'

As the boat pulled away and rowed for the beach, Clinton Codrington stood in the stern and stared back at them without expression, and his bare head shone like a beacon fire in the sunlight.

That had been so many weeks ago. Still the opportunity of assembling enough senior officers in a small station like the Cape Colony to act as his judges might not occur for weeks still or even months.

Clinton had spent his nights in the cheap lodgings on Waterkant Street. Ostracized by his brother officers, he had spent most of his days alone upon the waterfront staring out at the little gunboat that was making her repairs at the anchorage, and at the bare-masted clipper.

He had watched while the slaves were brought ashore from *Huron*'s holds, and their chains were struck off by a blacksmith from the castle. He had seen the bewildered blacks put their marks upon the indenture contracts, and then be led away by the Dutch and Huguenot farmers to learn their new duties, and he had wondered at this other fate to which he had delivered them.

Then in the afternoons he had climbed the hill to the Cartwright mansion set in its green and pleasant garden to pay his court to Robyn Ballantyne.

This day he was early, the noonday gun banged from the top of Signal Hill as Clinton came striding up the pathway, almost breaking into a run when he saw Robyn in the rose garden. He left the pathway and cut across the velvety green carpet of the lawn.

'Robyn! Doctor Ballantyne!' His voice was strange, and his pale eyes wild.

'What is it?' Robyn handed the basket to Aletta and hurried across the lawn to meet him.

'What is it?' she repeated with concern, and he seized both her hands in his.

'The slaver!' he was stuttering with the force of his emotion, 'The American – *Huron*!'

'Yes?' she demanded. 'Yes?'

'She is sailing – they are letting her go!'

It was a cry of outrage and despair, and Robyn froze, her face suddenly pale.

'I do not believe it.'

'Come!' said Clinton. 'I have a carriage at the gate.'

The coachman whipped the horses at the slope, with Clinton shouting to him to hurry still, and they came out on the crest of Signal Hill in a lather, with froth splattered on their chests and forelegs.

The moment the coach braked, Clinton jumped down and led Robyn to the side of the roadway facing down the steep hillside out over the bay. The tall American clipper slid silently and gracefully over a green sea that was speckled by the dancing white caps of the south-easterly wind.

As she cleared the low dark shape of Robben Island, she altered her heading a fraction and more sail bloomed upon her yards, white as the first flowers of spring. Silently, the man and the woman stared after the beautiful ship, and neither of them spoke as she merged with the milky sea fret, became a ghostly silhouette, and then quite suddenly was gone.

Still in silence the couple turned back and climbed into the waiting carriage, and neither of them spoke until it drew up before the gates of the Cartwright estate. Clinton looked at her face. It was completely bloodless, even her lips were ivory white and quivering with suppressed emotion.

'I know how you feel. After all we endured, to see that monster sail away. I share your distress,' he said quietly, but she shook her head once vehemently and then was still again.

'I have other news,' he told her when at last he judged she had recovered, and a little colour had returned to her cheeks.

'There is a Rear Admiral on the passenger list of the East Indiaman that anchored in the bay yesterday. Slogger Kemp has asked him to make up the numbers at the court martial. It begins tomorrow.'

Immediately she turned to face him, her expression softening with concern and alarm.

'Oh, I will pray for you every moment.' She reached out her hand impulsively, and he seized it with both of his, and clung to it.

It was as though the contact had loosened something in her that she had locked away tightly and at last the tears welled up in her hot dry eyes.

'Oh my dear Doctor Ballantyne,' Clinton whispered. 'Please do not fret for me.' But through the tears Robyn was still seeing the ghostly image of a tall and beautiful ship fading away into the pearly curtain of sea fret, and the first sob shook her body.

T he floor of the ballroom of Admiralty House was laid out in chessboard squares of black and white marble, and the human characters like chess pieces were ranged upon it haphazard, as though by the vagaries of a hard-fought end game.

Robyn Ballantyne in skirt and blouse of sober green stood by the head of the board, a solitary queen, while arranged opposite her were the rooks of the legal council: two naval officers in full uniform and sword who were playing the roles of prosecutor and defender. They had been chosen arbitrarily, and neither of them relished the unfamiliar task.

They had isolated themselves from the rest of the company, and each of them busied himself with the sheaf of documents he carried, not looking at the man whom they were destined to save or condemn, depending on the deliberations of the senior officers who were even now closeted behind the tall double doors at the far end of the ballroom.

The other witnesses, Denham of *Black Joke* bearing the ship's log under his arm, MacDonald the engineer hiding his grey coal-stained hands behind his back, the colony's agent and Honorary Consul for the Sheikh of Omani, a prosperous Asian trader, were like the scattered pawns of the game around the edge of the board.

Only the officer accused and on trial for his life was not at rest. Captain Clinton Codrington paced at random about the ballroom floor, his heels clicking on the marble slabs, his cocked hat clasped under one arm, his pale blue eyes staring dead ahead. He paced without pattern, like the roving knight of the chess board.

The tension seemed to charge even this huge room, increasing rather than lessening with every minute. Only the two red-coated marines on each side of the double doors seemed totally unaffected.

They stood stolidly, their musket butts grounded beside

the polished toe caps of their right boots, their expressions blank and their eyes fixed directly ahead.

Once Clinton stopped in front of Robyn and drew his watch.

'Fifty minutes,' he said.

'It could be hours yet,' she answered quietly.

'I can never thank you for the evidence you gave.'

'It was nothing but the truth.'

'Yes,' he agreed. 'But without it—' He broke off, and resumed his restless pacing.

The prosecuting officer, who had attempted for the two previous days to damn him and send him to the gallows, glanced up at Clinton, and then hurriedly, almost guiltily, returned his eyes to the documents he held in his right hand. Robyn was the only one who watched him openly, and her eyes were dark with worry and concern, yet when he caught her eye again a few minutes later, she smiled at him bravely, trying to hide her doubts.

The four senior officers, before whom she had given her evidence, had listened attentively, but she had seen no warmth nor compassion in their faces.

'Madam,' Admiral Kemp had asked her at the end, 'is it true that you obtained a medical degree by impersonating a man, and if your answer is 'Yes', would you not then believe us justified in doubting your allegiance to the truth?'

Robyn had seen the faces of the senior officers flanking Kemp harden, their eyes become remote. The Sultan's agent had been blatantly hostile, as the prosecuting officer had led him dutifully through a long list of aggressions and warlike acts against his master's sovereign territory and against his subjects.

Denham and MacDonald could only recite the facts, and their own repudiation of their Captain's orders was recorded in the ship's logbook.

The only thing that surprised Robyn was that the court had deliberated so long, and then she started involuntarily as, with a crash that echoed around the walls of the empty ballroom, the double doors were thrown open, and the two marine guards stamped to attention.

Through the doors she could see the naval officers seated down the length of the long dining room table facing the ballroom. Their frogging and epaulettes gleamed with gold lace and Robyn was too far to be certain of their expressions. Though she took a step forward and craned to see the polished top of the table in front of the grim line of judges, she could not be certain of the hilt and point of the single weapon that lay upon it, and then her view was blocked by the backs of the three men who lined up facing the doors.

Clinton was in the centre with the prosecution and the defending officers flanking him. At a muttered command, they marched briskly through the open doors. The doors closed behind the trio, and still Robyn could not know which way the naval dirk on the table was pointing, whether it was in its sheath or if the blade was naked.

Clinton had explained to her the significance of that weapon. It was only placed upon the table when the judges had reached their decision. If the blade was sheathed, and if the hilt was pointed towards the prisoner when he entered, then the judgement was 'not guilty'. When the bare blade was pointed towards him, then he knew that the wrath of the service was about to descend upon him – and he might be called to pay his penalty upon the flogging grating, or upon the gallows itself.

Clinton kept his gaze fixed upon a point above Admiral Kemp's head, while the doors were banged closed behind him, and he and the officers flanking him came to attention five paces from the long polished table behind which sat his judges.

Only then did he allow himself to glance down at the

dagger upon the table top. The bare blade glinted a bluish-silver in the late sunlight that slanted in from the tall french windows, and the bright point was aimed at Clinton's stomach.

He felt the cold drive of despair in his guts, as though the dagger had been plunged through them. The shock of the injustice of the verdict, the disbelief that his whole life had been brought down at a single stroke, the shame and disgrace of a career shattered and a reputation indelibly besmirched, left him numbed and blind to all but the wicked blade before him, and deaf to all but the voice of Admiral Kemp.

'Guilty of flagrant disregard of the orders of his superior officer.'

'Guilty of acts of piracy upon the high seas.'

'Guilty of destroying the property of the subjects of a friendly power.'

'Guilty of flouting the terms of a treaty between her Britannic Majesty's Government and the Sultan of the Omani Arabs.'

It must be death, Clinton realized, the verdict was too detailed, the list of his transgressions too long and his guilt too serious. It must be death by the noose.

He lifted his eyes from the accusing weapon, and he stared out of the french windows beyond his judges. The high stock of his uniform collar felt as tight as the hangman's noose as he tried to swallow.

'I have never feared death, Lord,' he prayed silently. 'There is only one thing I will regret – that I must leave the woman that I love.'

To be deprived of honour, and of life was sufficient punishment, but to lose his love as well was the final injustice.

'The Court has deliberated at length on the sentence,' Admiral Kemp paused and shot a sideways glance at a lean, tanned and silver-haired Rear-Admiral beside him, the passenger from the visiting East Indiaman, 'and has heard and been swayed by the eloquent arguments of Admiral Reginald Curry.'

He paused again and puffed out his lips, indicating clearly that he did not agree with those eloquent arguments, before going on.

'The sentence of this Court is that the prisoner be stripped of all rank, privileges and pay and that the Queen's Commission which he holds shall be withdrawn, and that he be *dishonourably* discharged from the naval service.'

Clinton steeled himself, the stripping of rank and discharge would precede the main body of the sentence.

'Furthermore,' Kemp paused and cleared his throat. 'Furthermore it is the sentence of the Court that the prisoner be taken from here to the castle and that he be there—'

The castle was the place of execution, the gallows would be erected on the parade ground before the main gates.

'That he be there imprisoned for a period of one year.'

The judges were standing up, were filing out of the room. As the lean silver-haired Admiral came level with Clinton, a small conspiratorial smile touched his lips, and for the first time Clinton realized that it was not death.

'A year,' said the Lieutenant who had prosecuted, as the door closed, 'not a flogging, nor a hanging – damned generous, I'd say.'

'Congratulations.' Clinton's defending officer was grinning incredulously. 'It was Curry, of course, he commanded the west coast anti-slavery squadron himself. What a stroke of luck to have him on the Board.'

Pale, voiceless, swaying slightly on his feet, Clinton was still staring blindly through the open windows.

'Come on, my dear fellow, a year will soon be past,' the defending officer touched his arm, 'and after that, no more bully beef and hard bread – do pull yourself together.'

Twenty miles a day since leaving grandfather Moffat's mission-station at Kuruman, Zouga had pushed the mules and his servants hard all the way, and now at the crest of the pass he reined in the tall sway-backed mule and stared out across the sweeping panorama of the Cape peninsula.

Directly below him was that strange pale hill of smooth rock, Die Paarl as the Dutch burghers had named it, 'the Pearl', and it shone with an almost translucent lustre in the Cape sunlight of high summer.

Beyond that the wheatlands and vineyards dotted the flat land that stretched away to the Paarde Berg, the Horse Mountains, where once the wild mountain zebra had roamed, and the Tyger Berg. The leopard to the Dutch burghers was a tiger and the zebra was a horse.

'Nearly home now, Sergeant,' Zouga called to Jan Cheroot.

'Just look at that—' The little Hottentot pointed to the smoky blue flat-topped mountain that stood up tall and massive against the southern horizon.

'We will be there before dark tomorrow night.'

Jan Cheroot puckered his lips and blew a kiss towards it. 'Pull the cork and tell the Cape Town ladies that my mama didn't call me big cheroot for nothing.'

His mule flicked its long hairy ears to the sound of his voice and gave a little half-hearted buck. 'You feeling it too, you old thunder!' Jan Cheroot chuckled. 'Let's go then!' and he whipped the animal up and went clattering away down the steep and rocky roadway.

Zouga stayed to watch the battered little two-wheeled Cape cart follow him at a more sedate pace, carrying its precious burden of ivory and sculptured green soapstone, as it had for a thousand miles and more.

It was a month before Robyn was allowed to pay her first visit to the castle. After the guard at the gates inspected her pass, she was led to a small whitewashed guard room, devoid of all furniture except three high-backed uncushioned chairs.

She remained standing for ten minutes before the low door opposite her was opened and Clinton stooped through it. He stopped, facing her, and she was struck instantly by the prison pallor of his skin. His deep-water tan had faded to a tobacco stain of yellow, and the roots of his hair, no longer bleached by salt and strong sunlight, had darkened.

He looked older, tired and dejected.

'You at least have not deserted me in my disgrace,' he said simply.

The Subaltern of the guard took the third chair and tried to look as though he was not listening to their conversation. Robyn and Clinton sat facing each other stiffly, on the uncomfortable chairs, and their conversation was at first as stilted, a polite series of enquiries after each other's health.

Then Robyn asked, 'Have you received the newspapers?'

'Yes. The warder has been good to me.'

'Then you have read what the new American President has promised at his inauguration.'

'Lincoln was always a staunch enemy of the trade,' Clinton nodded.

'He has granted the ships of the Royal Navy the right of search at last.'

'And six of the Southern States have seceded already,' Clinton told her grimly. 'There will be fighting, if he tries to force it.'

'It's so unfair,' Robyn cried. 'Just a few short weeks and you would have been a hero instead of a—' she broke off with her hand to her mouth, 'I am sorry, Captain Codrington.'

'Captain no longer,' he said.

'I feel so much to blame – had I not sent that letter—'

'You are so kind, so good,' then he blurted abruptly, 'and so beautiful that I can scarcely bear to look at you.'

Robyn found herself blushing hotly, and she glanced at the listening guard officer. He was studying the rough plaster ceiling of the cell.

'Do you know what I thought when I entered the chamber and saw the dirk pointed at me?' Clinton went on, and she shook her head. 'I thought I was going to lose you. That they would hang me and I would never see you again.' His voice was shaking with such emotion, that the listening officer rose to his feet.

'Doctor Ballantyne, I will leave the room for five minutes,' he said. 'Do I have your word that you will not attempt to pass a weapon or a tool to the prisoner in my absence?'

Robyn nodded jerkily and whispered, 'Thank you.'

The moment the door closed, Clinton launched himself across the gap between them and dropped to his knees before Robyn. He encircled her waist with both arms and pressed his cheek to her bosom.

'But now I have nothing to offer you, I have nothing to share with you but my disgrace.'

Robyn found herself stroking his hair as though he were a child.

'Soon I will go back to that beautiful land below the Zambezi river. I know now that is where my destiny lies,' she said quietly. 'To minister to the souls and the bodies of those who live there.'

She paused a moment and looked down fondly on the dense pale locks of his hair.

'You say you have nothing to offer, nothing to share, but I have something to offer you, and to share with you.'

He raised his head and looked up at her questioningly, hope starting to dawn in his pale sapphire eyes.

'Will you not offer yourself to be ordained in God's service as a missionary, and come with me into the wilderness, to the land of Zambezia?'

'To share my life with you, and with God.' His voice was hush and choked. 'I never dreamed I was worthy of such an honour.'

'T'he fellow is a prig,' said Zouga firmly. 'And, damn me, but now he is a gaol-bird to boot. Neither of you will be able to hold up your heads in society.'

'He has a true and noble spirit, and now he has found his true calling in God's service,' Robyn replied hotly. 'Neither of us intend spending much of our time in society, you may be certain of that.'

Zouga shrugged and smiled. 'Of course, that is your affair. At least he has made a pretty packet of prize money which they can't take away from him.'

'I assure you that money had nothing to do with my decision.'

'I will believe that.' Zouga's smile infuriated her, but before she could find a scathing enough retort, he turned away and sauntered the length of the long veranda under the trellised vines and stood with his hands thrust into his pockets, staring out across the Cartwrights' gardens to the far glimpses of blue bay seen through the oaks and the rustling palms.

Robyn's anger subsided and gave way to regret. It seemed now that the two of them must always be squabbling, their desires and their motives always directly opposed.

At first her relief at his safety had been almost as strong as her sisterly delight at seeing him again. She had barely recognized him as he rode the bony sway-backed mule up the path to the Cartwright mansion. It was only when he dismounted and lifted the stained old hat from his head that she screamed with joy, leapt up from the luncheon table and ran down off the terrace to hug him.

He was so lean and hard and bronzed, and somehow

endowed with new authority, charged with purpose and presence, that she glowed with pride as he recounted his experiences and all the company hung avidly on each word.

'He is like a Greek god!' Aletta Cartwright had whispered to Robyn, which was not an original description – but then Aletta did not run much to original thought, and Robyn had to agree that in this case it was accurate.

She had followed his description of the land of the Matabele, and of the long trek southward with all her attention, asking such acute questions that Zouga had asked sharply, 'I hope that you will not be using any of this in your own account, Sissy?'

'Of course not,' she assured him, but still that had been the first sour note, and he had not spoken further of his adventures, except to give her the greetings and news of their grandfather, Robert Moffat at Kuruman.

'You would never believe that he was seventy-five years old this past December. He is so bright and alive that he has just finished translating the Bible into the Sechuan language. He gave me every courtesy and help, and it was he who arranged for mules and for the cart which made the last portion of the journey so much easier. He remembered you as a little girl of three years old, and he has received your letters and gave me this in reply.' It was a thick packet. 'He tells me that you have asked him about leading a missionary expedition to Zambezia or Matabeleland.'

'That is correct.'

'Sissy, I do not think that a woman on her own,' he had begun, but she had forestalled him.

'I shall not be alone. Captain Clinton Codrington has decided to seek ordination as a missionary, and I have consented to become his wife.'

That had led to the explosion which had once more marred their relationship. As her anger faded, she made another determined effort to avert the new clash of temperament.

'Zouga,' she went down the length of the terrace and took his arm. 'I would be grateful if you would consent to give me away at the wedding.'

Some of the hardness went out of his arm as his muscles relaxed.

'When will that be, Sissy?'

'Not for seven months. Clinton has that much longer of his sentence.'

Zouga shook his head. 'I will not be here. I have booked passage on the P. & O. steamer that sails for home at the beginning of next month.' They were both silent, and then Zouga went on, 'But I wish you joy and happiness – and I apologize for the remark I made about your future husband.'

'I understand.' She squeezed his arm. 'He is a different kind of man from you.'

Zouga almost exclaimed, 'Thank God for that,' but caught the blasphemy before it reached his lips, and again they were silent.

Zouga was considering the problem that had concerned him so intimately since his return to Cape Town – how to find out from Robyn what she had written in her manuscript, and if possible to influence her into amending those portions of it which might offend the family reputation.

Now that he had learned that she would not be returning to England, the natural opportunity had presented itself.

'Sissy, if your manuscript is prepared, I will be happy to take it with me and to make certain that it is delivered safely to Oliver Wicks.'

The voyage to England would give Zouga ample opportunity to study Robyn's work, and if the delivery was delayed for a month or so after his arrival, then Zouga's own published account of the expedition would skim the cream off the pool of interest and critical literary attention.

'Oh, did I not mention it to you?' Robyn lifted her chin, and her smile was spiced with a little spiteful relish. 'I sent my manuscript on the mail steamer a month before your arrival here. It will be in London already, and I should not be surprised if Mr Wicks has not published it already. I expect he will have sent the reviews, and we will have them on the next mailship.'

Zouga jerked his arm out of her grip, and his eyes were steely as he glared down at her.

'I really should have mentioned it,' she added sweetly. His reaction confirmed her suspicions. and she knew that what last small chance they had was finished. From now on they would be enemies, and somehow she knew that the centre of their enmity would always be the land and peoples of that faraway country between two great rivers which Zouga had named Zambezia.

At the end of the Woodstock road, on the bank of the Liesbeeck river, not far from the domed roof of the Royal Astronomical Observatory, stands the Cartwright warehouse. It is a rambling whitewashed building of a burnt Kimberley brick with a corrugated iron roof.

Against the rear wall of the main storeroom stood three articles left there in storage and for later collection by Major Morris Zouga Ballantyne, presently on passage aboard the Peninsular and Orient Steamship S.S. *Bombay* from India to the Pool of London. The three bulky articles were almost completely screened by the bays and hillocks of bales and crates, and of barrels and bags, which reached almost to the high ceiling.

The two huge elephant tusks formed a perfect frame with their curved yellow ivories for the third package. The carved soapstone figure was still contained in its protective covering of plaited elephant grass and twisted bark rope. It stood upright on its wide heavy base, and it was merely chance that it faced towards the north.

The grass covering had been torn away from the head by

careless handling and long months of travel on the shoulders of porters and on the buckboards of an unsprung Cape cart.

The cruel proud head of the bird of prey protruded from its wrappings. The stony sightless eyes stared across forest and mountain and desert, one thousand five hundred miles, to a walled and ruined city and the words of the Umlimo's prophecy seemed to hover in the air above the graven head of the bird like living things.

'The white eagle has stooped on the stone falcons and cast them to earth. Now the eagle shall lift them up again and they will fly afar. There shall be no peace in the kingdoms of the Mambos or the Monomatapas until they return. For the white eagle will war with the black bull until the stone falcons return to roost.'